20th International Conference on Soil Mechanics and Geotechnical Engineering (ICSMGE 2022)

A Geotechnical Discovery Down Under

Post-Conference Volume

Sydney, Australia
1 – 5 May 2022

Editors:

Md Mizanur Rahman
Mark Jaksa

ISBN: 978-1-7138-7962-6

Printed from e-media with permission by:

Curran Associates, Inc.
57 Morehouse Lane
Red Hook, NY 12571

Some format issues inherent in the e-media version may also appear in this print version.

Printed with permission by Curran Associates, Inc. (2023)

For permission requests, please contact Australian Geomechanics Society
at the address below.

Australian Geomechanics Society
PO Box 955
St. Ives, New South Wales
2075 Australia

Phone: +61 2 9144 7519

https://australiangeomechanics.org

Additional copies of this publication are available from:

Curran Associates, Inc.
57 Morehouse Lane
Red Hook, NY 12571 USA
Phone: 845-758-0400
Fax: 845-758-2633
Email: curran@proceedings.com
Web: www.proceedings.com

PROCEEDINGS OF TWENTIETH INTERNATIONAL CONFERENCE ON SOIL MECHANICS AND GEOTECHNICAL ENGINEERING (ICSMGE 2022) – POST-CONFERENCE VOLUME, A GEOTECHNICAL DISCOVERY DOWN UNDER, SYDNEY, NEW SOUTH WALES, AUSTRALIA, 1-5 MAY 2022

Proceedings of 20th International Conference on Soil Mechanics and Geotechnical Engineering – Post-Conference Volume

Editors

Professor Md Mizanur Rahman
UniSA STEM (Science, Technology, Engineering and Mathematics)
University of South Australia, Adelaide, Australia

Professor Mark Jaksa
School of Architecture and Civil Engineering
University of Adelaide, Adelaide, Australia

AUSTRALIAN GEOMECHANICS SOCIETY

SIMSG ISSMGE

Published by: Australian Geomechanics Society
PO Box 7, The Gap, QLD 4061, Australia
e-mail: secretary@australiangeomechanics.org
www.australiangeomechanics.org

Proceedings of the 20th ICSMGE-State of the Art and Invited Lectures– Rahman and Jaksa (Eds)
© 2022 Australian Geomechanics Society, Sydney, Australia, ISBN 978-0-9946261-7-2

Preface

The Conference Organising Committee are delighted to have been granted the great honour of organising and hosting of the 20th International Conference on Soil Mechanics and Geotechnical Engineering (20th ICSMGE), which was held in Sydney, Australia between May 1 and 5, 2022. This, the third of three volumes, contains one technical paper which was inadvertently omitted from Volume 2, five Technical Committee general reports, and the papers from the ISSMGE's two Council Meetings, during the 2017–2022 presidential term, which were held in Capetown on October 6, 2019 and in Sydney, in hybrid mode, on May 1, 2022.

We trust that this three-volume set of proceedings will be of great value to the geotechnical engineering community, now and for many years to come.

Professor Md Mizanur Rahman
University of South Australia, Australia

Professor Mark Jaksa
University of Adelaide, Australia

Table of contents

Technical Paper

Effects of bio-mediation on deformation and yielding responses of compacted residual soil as studied by acoustic emission

Effets de la bio-médiation sur les déformations et les réactions d'élasticité du sol résiduel compacté, étudiés par émission acoustique.

Yasuo Tanaka, Jun Xian Lim & Siaw Yah Chong
Department of Civil Engineering, Lee Kong Chian Faculty of Engineering and Science, Universiti Tunku Abdul Rahman, Malaysia, [yasuo@utar.edu.my, ytgeotec@tiger.kobe-u.ac.jp]

Min Lee Lee
Department of Civil Engineering, Faculty of Science and Engineering, University of Nottingham Malaysia, Malaysia

Ying Hui Ong
Department of Chemical Engineering, Faculty of Science and Engineering, Universiti Tunku Abdul Rahman, Malaysia

ABSTRACT: Bio-mediation has recently become a popular method to improve the mechanical properties of coarse grained materials. The application of the ground improvement technique has largely been limited to uniform sands, while its applicability for residual soils that are commonly encountered in the tropical countries such as Malaysia is still rarely reported. The main objective of this paper is to examine the deformation and yielding characteristics of a bio-mediated residual soil (weathered granite) in Malaysia. Acoustic emission (AE) measurement was employed to monitor the microscopic structural changes of the soil under loading. A series of undrained triaxial tests were carried out on compacted specimens under isotropic-consolidation with the AE measurement. From the test results, it was found that the bio-mediation treatment with calcite contents of < 5% resulted in a soil structure that has a higher lateral resistance and elastic rebounding during isotropic compression than its untreated counterpart. This indicated that the treatment resulted in a behaviour of denser soil state. The increase of undrained strength in the bio-mediated soil became less significant when the soil was consolidated to a higher level, i.e. greater than 100kPa. From the AE measurement, it was also concluded that the bio-mediated soil fabric still affected the yielding behavior and shearing resistance even when the soil was pre-consolidated to a high stress level.

RÉSUMÉ : La bio-médiation est récemment devenue une méthode populaire pour améliorer les propriétés mécaniques des matériaux à gros grains. L'application de la technique d'amélioration du sol a été largement limitée aux sables uniformes, tandis que son applicabilité pour les sols résiduels qui sont couramment rencontrés dans les pays tropicaux tels que la Malaisie est encore rarement rapportée. L'objectif principal de cet article est d'examiner les caractéristiques de déformation et d'élasticité d'un sol résiduel bio-médiatisé (granit altéré) en Malaisie. La mesure d'émission acoustique (AE) a été employée pour surveiller les changements structurels microscopiques du sol sous chargement. Une série d'essais triaxiaux non drainés a été réalisée sur des spécimens compactés sous consolidation isotrope avec la mesure AE. A partir des résultats des tests, il a été constaté que le traitement de bio-médiation avec des concentrations de calcite inférieures à 5% a pour résultat une structure de sol qui a une résistance latérale plus élevée et un rebondissement élastique pendant la compression isotrope. Cela indique que le traitement entraîne un comportement de sol plus dense. En outre, l'augmentation de la résistance non drainée observée pour le sol bio-traité à un faible niveau de consolidation est devenue moins significative lorsque le sol a été consolidé à un niveau plus élevé, c'est-à-dire plus de 100kPa. De la mesure AE, il a été également conclu que le tissu de sol bio-médié affecte toujours le comportement de yielding et la résistance de cisaillement même lorsque le sol a été préconsolidé à un niveau de contrainte élevé.

KEYWORDS: MICP treatment, residual soil, acoustic emission (AE), yielding, undrained triaxial test

1 INTRODUCTION

Over the past decade, there was a rapid development in the use of natural biological processes in geotechnical engineering applications. New terms such as "Biogeotechnics" or "Bio Inspired Geotechnics" are used to promote more applications of this technology in engineering practices where a global view on the sustainability and the safety of future society is becoming upmost concerns and significances, (DeJong et al. 2013; CBBG 2021). Among the techniques available in the biogeotechnics, ground improvements or strengthening of soil properties using microbially induced calcite precipitation (MICP) is the most popular subject that have recently received considerable research works.

Geotechnical properties of soil can be improved by the MICP treatment in terms of mechanical and hydraulic responses of soil against external forces. Recently several state of the art papers have been published highlighting the most recent research progresses made on the MICP treatments, such as Terzis & Laloui (2019), and Rahman et al. (2020). For example, Rahman et al. (2020) have reviewed soil modelling and analytical studies, such as constitutive laws and DEM models, to simulate the mechanical behaviour of MICP treated soil that can closely resemble naturally cemented soils. In the development of soil mechanics, parallel progresses between experimental and analytical works are the key engines for establishing a sound engineering tool and knowledge base for geotechnical engineering practices. To enhance such progresses, we need experimental studies to understand the unique behaviours of bio-mediated soil as a particulate material and its fabric that are affected by the action of MICP treatment.

Although there have been a rapid development in the research works on the MICP treated soil, the studies have been mostly on clean sands and not much on the tropical residual soils that are abundant in the tropical regions such as the Southeast Asia where notable economic growth over the decades has occurred. In this work, the effects of MICP treatment on the compacted residual soil that represents the state of soil as a construction fill as frequently used are studied by focusing on the basic soil particle responses during consolidation and undrained shear with the measurement of acoustic emission (AE). The use of AE measurement is to enrich the understanding of soil fabric changes as brought by the MICP treatment. Our early work on the effects of consolidation pressure on the mechanical responses of the MICP and untreated compacted residual soils has been presented by Lim et. al (2020). Although there is a significant difference between the in-situ and the artificially compacted states of residual soil with respect to their engineering behaviours, we aim to identify the effect of MICP on the residual soil by first focusing on the behaviour changes at the compacted state because of the scarcity of such study. The term "residual soil" will be retained hereinafter to represent the soil at compacted state that has an altered fabric resulted from re-compaction while the mineralogy of soil likely be retained.

2 SOIL PREPARATION AND TEST PROCEDURES

2.1 *Soil properties*

The residual soil used in this study was sampled from Bandar Sungai Long area of Selangor, Malaysia. The residual soil was a clayey sand that can be geologically classified as weathered granite soil. The basic physical properties and gradation curve of the soil are shown in Fig. 1. Triaxial specimens were produced by compacting moist soil in a specimen mold with and without bio-mediation. The target density was set at 90% of the maximum dry density (i.e. 1.795 g/cm^3). Details of bio-mediation treatment are described in the following section.

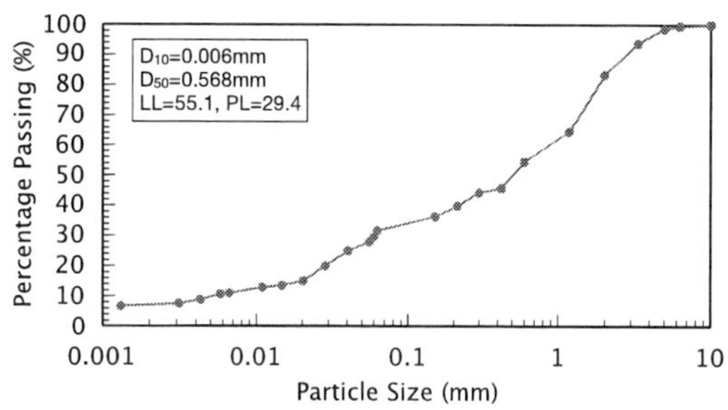

Figure 1. Gradation curve of residual soil.

2.2 *MICP treatment*

In this study, the bacterium used in the experiment is *Sporosarcina Pasteurii* (DSM-33) which was formerly known as *Bacillus Pasteurii*. The details of the MICP treatment for the residual soil have been described by Lim et. al (2020). The outlines of MICP treatment procedures are listed below:

i) Prior to mixing with soil, the bacteria solution was cultivated and prepared at a microbial density of 6.4×10^7 colonies forming unit (cfu) per ml.
ii) The bacteria solution was mixed with the air-dried soil to attain 15.6% moisture content prior to compacting it in a specimen mold to the target dry density.
iii) The bacteria-mixed and compacted specimen was attached a flow control device to facilitate permeation of both nutrient and cementation reagents through the specimen under a low hydraulic head of 2 m. The concentration of cementation reagent was prepared at 0.5 mol/L while the concentration of nutrient broth solution was 8g/L.
iv) The prepared specimen was then placed in a triaxial test apparatus for subsequent mechanical testing.
v) After the triaxial testing, the sheared soil specimen was used to evaluate its calcite content by washing the soil with hydrochloric acid (at a concentration of 1 Mol/L) to dissolve the calcium carbonate precipitated inside the MICP-treated specimen.

It should be noted that the third procedure above took a long period of time (about 3 days) to treat the low permeability ($\approx 9 \times 10^{-5}$ m/s) residual soil. Based on these MICP treatment procedures that are found to be practical for the studied residual soil, it was suggested that only a low degree of calcite content (less than 5%) was attainable for the MICP-treated residual soil in Malaysia.

2.3 *Test apparatus and undrained triaxial tests*

In this study, AE measurement was used to examine how the MICP treatment affected the deformation and the yielding responses of residual soils by performing the undrained triaxial shear tests on consolidated samples with and without bio-mediation. The AE measurement was made by mounting an accelerometer (Fuji Ceramics PA 12C) at the base of triaxial cell, and an assemble of AE signal amplification and digital recording/counting system was built by integrating mostly readily available units in house. The triaxial shearing was carried out by increasing the axial load monotonically through an air cylinder at the top of loading frame. The details of test apparatus and data processing procedures have been described by Lim (2021).

Lim (2021) has conducted a series of undrained triaxial tests on the MICP and the untreated residual soil as well as on clean sand. In this paper, a number of his triaxial test results on the residual soil will be used to study the effects of bio-mediation on the deformation and strength properties. Table 1 lists the triaxial tests that were analysed in this paper.

Table 1. List of undrained triaxial tests analysed.

Test Name	Maximum Consolidation Pressure (kPa)	Initial Void Ratio	Final Void Ratio	Calcite content (%)
UR C40	40	0.635	0.635	Nil
UR C80	80	0.635	0.634	Nil
UR C120	120	0.635	0.596	Nil
UR C220	220	0.635	0.560	Nil
UR OC5	220	0.635	0.571	Nil
BR C40	40	0.580	0.580	3.5
BR C80	80	0.585	0.583	3.2
BR C120	120	0.595	0.571	2.5
BR C220	220	0.592	0.566	2.7
BR OC5	220	0.616	0.583	1.2

3 TEST RESULTS

A series of consolidated undrained triaxial shear tests were performed both on the residual soil with MICP treatment and without treatment. The maximum consolidation pressure applied was 220 kPa, and in one set of tests, the soil was over-consolidated (OCR = 5) and sheared undrained as shown in Table 1. The symbols **BR** and **UR** denote the residual soils with

MICP (Bio) treatment and the untreated residual soil respectively.

3.1 *Soil behaviors during isotropic consolidation*

As noted earlier, the effects of consolidation pressure on the mechanical responses and their differences between BR and UR soils have been previously presented by Lim et. al (2020). A different anisotropic structural response of two soils during the isotropic consolidation has been identified, namely on the weaker lateral response of UR soil under an isotropic loading while BR soil exhibited a higher lateral resistance due to the calcite precipitation (Lim et. al, 2020). In the present study, this aspect of different deformation response between BR and UR soils will be further identified and examined by the analysis presented herein.

Figs. 2(a) and (b) compare the different changes of the volumetric, axial and radial strains between the BR and UR soils during the isotropic consolidation and unloading to OCR = 5 (BR OC5 and UR OC5 Tests in Table 1). Also shown in Figs. 2(c) and (d) are the compressive responses of dense and loose states, respectively of a clean sand, Ham River Sand (HRS), as studied by Kuwano (1999). It may be noted that the plotted values of HRS are manually measured from the graphs presented in her Ph.D. thesis, and therefore they were handicapped by limited accuracy. Her results will be used to examine the MICP effects on soil behaviors and relate them to soil densification effect.

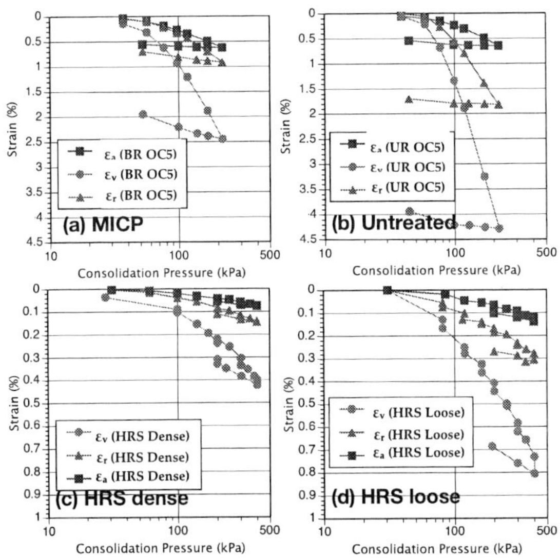

Figure 2. Volumetric, axial, and radial strains during isotropic consolidation for MICP/untreated residual and dense/loose Ham River Sand (HRS): (a) MICP, (b) Untreated, (c) Dense HRS, (d) Loose HRS.

Fig. 3 compares the relationship between the normalized consolidation pressure and the normalized volumetric, axial, and radial strains between the BR and UR soils of OC5 tests in the present study, and also those studied by Kuwano (1999) using the Ham River Sand (HRS), a clean river sand, of dense and loose states. The solid red lines corresponded to those of residual soil, while the dotted blue lines indicated the HRS. The normalized pressure was computed as the measured pressures divided by the maximum consolidation pressure, that was 220kPa for the present studied residual soil and 400kPa for Kuwano (1999)'s HRS. Numerical values of the maximum volumetric strain (e_{vmax}) for the BR and UR soils were 2.45 and 4.28, respectively, and the UR soil was about 1.7 times compressible.

Two observations were notable from the Fig. 3. Firstly, the increase of the radial strains against the axial strain was smaller for the MICP soil and also for the dense HRS soil as compared with those of the UR and the loose HRS soils. This indicated that both BR and dense HRS have higher lateral stiffnesses than the UR and loose HRS, respectively. Secondary, during unloading of the MICP soil, the radial strain rebounded significantly greater relative to the axial strain, while those strains were almost identical in UR soil. These results implied that the MICP treated soil has a greater elastic rebound in the radial direction than in the vertical direction, and this behavior could be caused by calcite crystallization within the soil structure. Further examination on this observation will be made later in this paper.

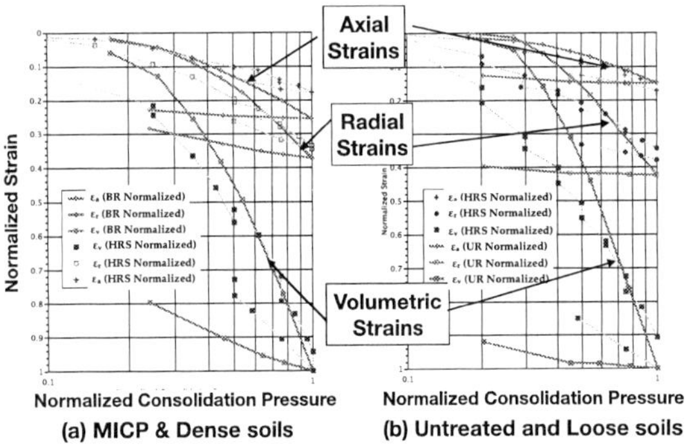

Figure 3. Comparison of normalized volumetric, axial, and radial strains during isotropic consolidation: (a) MICP residual soil vs dense HRS, (b) Untreated residual soil vs loose HRS.

Regarding the first observation, Kuwano (1999) has discussed the anisotropic responses of HRS sands and noted that a loose deposition of sand particle would result in a less stiff horizontal response during an isotropic consolidation as follows; "The ratios of radial to axial strain increment, $d\varepsilon_r/d\varepsilon_a$, at p_c of 100 kPa were, for example, 3~4 for loose HRS and 1.5~2.5 for dense HRS, respectively. The value of $d\varepsilon_r/d\varepsilon_a$ decreased with pressure increase, falling to around 1.5~2.5 at p_c=400 kPa."

To compare the ratios between the radial to axial strain increment between the MICP and untreated residual soils during the consolidation, Fig. 4 is produced using the OC5 test. Because of the low permeability of the BR and UR soils, each step of consolidation takes time to reach a steady value of the ratio.

Also it should be noted that at a low consolidation pressure below 60 kPa, the ratio for the UR soil was lower than unity. As can be seen from Fig. 4, the $d\varepsilon_r/d\varepsilon_a$ ratio of the untreated residual soil was much higher than that of the MICP treated soil and the ratio ranged between 2.5 and 4.5, while the BR soil ranged between 1 and 2.

Because of page limit, the time changes of the $d\varepsilon_r/d\varepsilon_a$ ratio for the other two tests (UR C220 and BR C220 tests) will not be presented here but a similar trend of lower ratio for the MICP residual soils was observed. It was also observed that the higher the calcite content, the lower the value of the ratio.

An important finding that can be obtained from the above observations is that bio-mediation tends to make the studied residual soil to behave as a dense soil. The similarity in the deformation behavior between the BR soil and the dense HRS was clearly seen, while the untreated residual soil behaved similarly to the loose HRS soil.

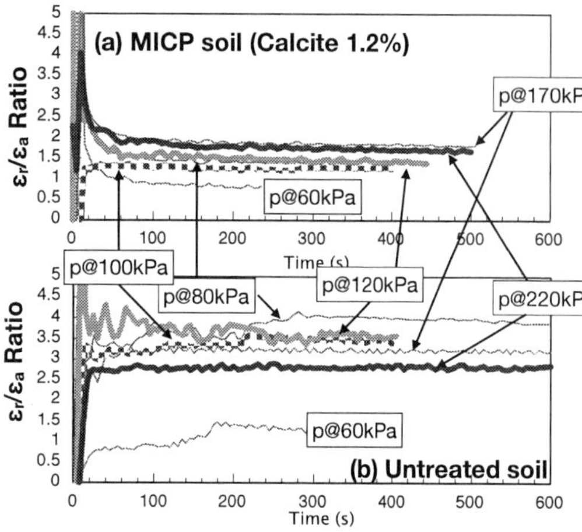

Figure 4. Changes of radial(ε_r)/axial(ε_a) strain ratios with time during isotropic consolidation: (a) MICP Soil, (b) untreated soil.

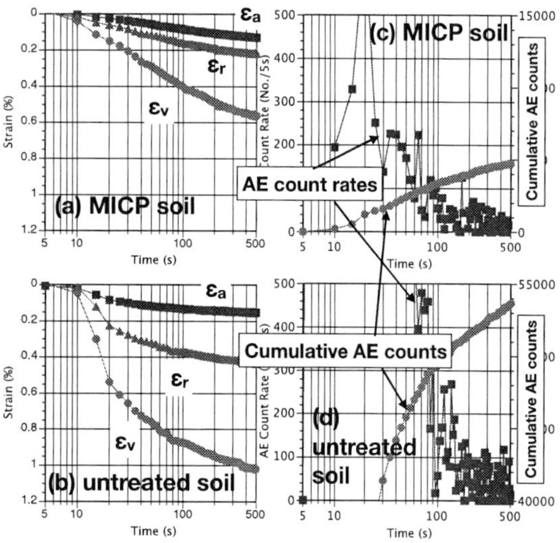

Figure 5. Changes of strains and AE with time at a constant loading step during isotropic consolidation: (a) ε_v, ε_r, ε_a of MICP soil, (b) ε_v, ε_r, ε_a of untreated soil, (c) AE count rate & cumulative AE counts of MICP soil, (d) AE count rate & cumulative AE counts of untreated soil.

Regarding the second observation on the elastic rebound of radial strain during unloading of the MICP soil, AE measurements taken during the consolidation was used to further examine this finding. Fig. 5 (a) & (b) show the changes of volumetric, axial, and radial strains with time during the last increment of consolidation (i.e., 170 to 220 kPa) for the BR & UR OC5 tests, while Fig. 5 (c) & (d) for the time changes of AE count rate & cumulative AE counts for the BR & UR OC5 tests, respectively. Logarithmic time scale was used to highlight the secondary compression behaviors of the soils.

It can be seen from Figs. 5 (a) & (b) that the amounts of the three strains during the secondary compression for both the BR & UR soils were nearly identical and this is numerically tabulated in the Table 2 by comparing the increases of the three strains over the time of 100 and 500 seconds. However, the AE measurements, both the count rate and the cumulative count, showed a large difference (nearly 3 times) between the BR and UR soils for the same time period.

Table 2 Amounts of volumetric, radial, and axial strain changes from 100 to 500 sec.

@100-500 s	$\partial\varepsilon_v$ (%)	$\partial\varepsilon_r$ (%)	$\partial\varepsilon_a$ (%)
MICP	0.157	0.056	0.046
Untreated	0.142	0.060	0.021

The AE measurement should indicate the amount of particle slippage and this is a plastic and irrecoverable movement. A lesser amount of the AE measured for the BR soil at nearly the same amount of strain as the UR soil indicated that a portion of the BR soil movements were of an elastic nature and recoverable. This interpretation would be argued favorably by the second observation on the large elastic rebound of radial strain of the BR soil as discussed above. The calcite formation in the residual soil tested herein would therefore not only result in the densification effects but also the nature of particle bonding being enhanced with the elastic stiffness increases.

3.2 *Increase/decrease in undrained shear strength of MICP / untreated residual soil*

The results above showed the effect of bio-mediation on the compressibility behavior of residual soil. After the isotropic consolidation and/or unloading, the BR and UR samples were subjected to undrained shearing with AE measurements. The followings describe the observations made during the undrained shear. Figures 6 and 7 present the increases of deviator stress and excess pore water pressure with axial strain for the tests of isotropically consolidated to pressures ranging from 40 kPa to 220 kPa, and the tests unloaded to 50 kPa after the maximum consolidation of 220 kPa, respectively. In those figures, the blue square and the red circle symbols or lines represent the results of BR and UR soils, respectively.

It can be seen from Figure 6 that the undrained strengths of isotropically consolidated MICP specimens were higher than those of untreated residual soil for all the ranges of consolidation pressure. The strength gains, however, decreased with the increase in consolidation pressure, as reported by Lim et al. (2020). The gain of undrained strength due to the MICP

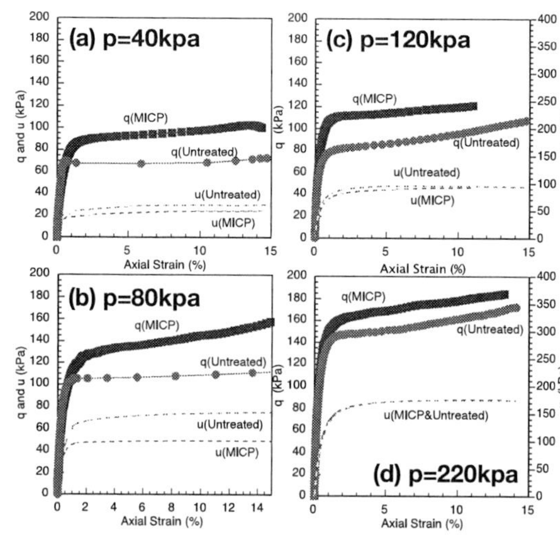

Figure 6. Comparison of stress (q)-axial strain and excess pore water pressure (u)-axial strain relationships between isotropically consolidated untreated (UR) and MICP (BR) residual soils: (a) p = 40 kPa, (b) p = 80 kPa, (c) p = 120 kPa, (d) p = 220 kPa.

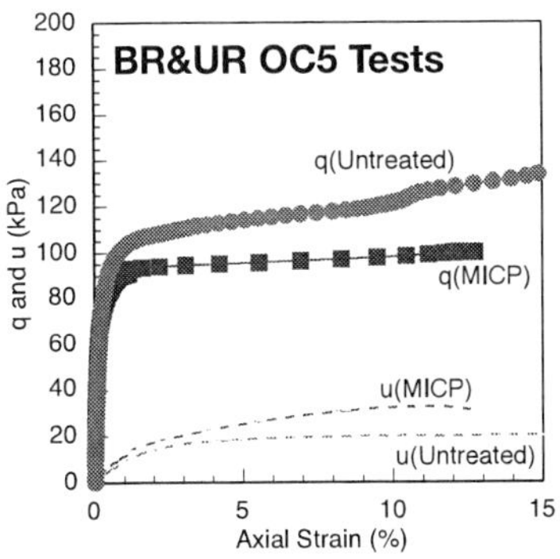

Figure 7. Comparison of stress (q)-axial strain and excess pore water pressure (u)-axial strain relationships between over-consolidated untreated (UR) and MICP (BR) residual soils.

effect is about 30% under low consolidation pressures, while the gain decreased to 10-20% under high consolidation pressures.

The magnitude of the excess pore water pressure generated under shearing seemed to have controlled the difference in the undrained shear strength. Under low consolidation pressures, the MICP soil showed relatively low excess pore pressure because of the densification effect of MICP. However, under high consolidation pressures, the difference of pore water pressures generated became negligible.

The undrained strength of over-consolidated MICP treated specimen was lower than that of the untreated residual soil as can be seen from Figure 7. The difference again seemed to be controlled by the amount of the excess pore water pressure generated. For the MICP soil with OCR = 5, a high excess pore water pressure was observed. In order to examine this reverse trend of the undrained shear strength and the excess pore water pressure, the AE measurement result was used to explain how the particle movements under shearing differed between the BR and UR soils after being subjected to a stress history of over-consolidation.

Figs. 8 (a) and (b) show the changes of the AE count rate with the increase of q/p' stress ratio over the entire and the initial loading ranges, respectively. The increments of the axial strain and the excess pore water pressure against the q/p' ratio were depicted in Fig. 8 (c). The vertical axes of the figures were arranged such a way to conveniently distinguish the two or multiple data points in the figures.

Tanimoto & Tanaka (1986) used the AE measurement to identify the yielding stress of sandy soil under drained triaxial test by subjecting the specimen to various degrees of over-consolidation. According to their works, the onset of AE as indicated in Fig. 8 (b) should indicate the start of irrecoverable particle slippages, and therefore the deformation responses above the corresponding stress level should be of plastic nature.

The yield stress level as judged from the figure was nearly the same for both the MICP and untreated residual soils, but the AE and the pore pressure / strain occurrences were slightly different between the two soils. The increase of AE for the MICP soil was more rapid and sharper, while for the untreated soil the AE was sporadic when the emission started and gradually increased after the initial yield. Their pore pressure and axial strain responses were also lagging behind those of the MICP soil. In any case, the generation of higher excess pore pressure of the over-consolidated MICP soil under undrained shearing would result in a lower level of undrained strength than that of the untreated soil. It was also noted that the densities (void ratios) of the two soil specimens prior to the undrained shearing were not identical as the MICP soil rebounded more and elastically upon the unloading to OCR = 5. This difference would have contributed to the difference in the undrained strength as well.

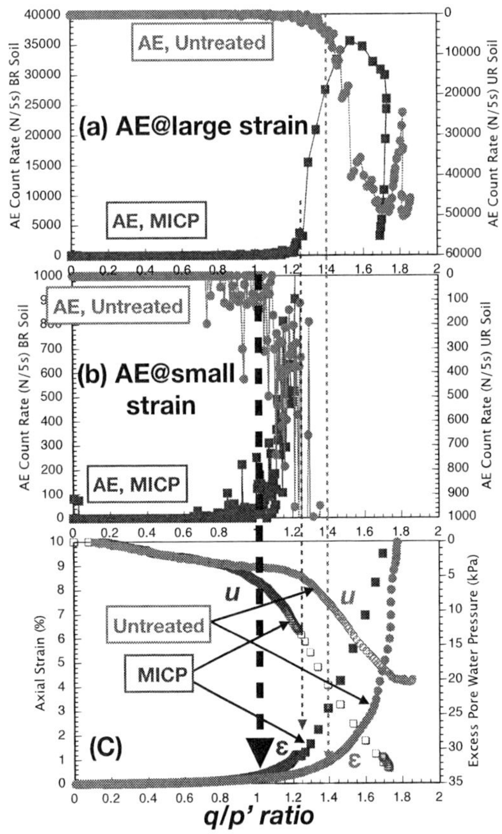

Figure 8. Relationships among AE count rate, excess pore water pressure (u), axial strain (ε) against increase of q/p' stress ratio between over-consolidated untreated (UR) and MICP (BR) residual soils: (a) AE over the entire q/p' range, (b) AE over the initial increase of q/p' range, (c) u & ε changes with q/p' ratio.

Although it was difficult to reach a clear interpretation of how the particle contacts for the MICP and untreated residual soils were responding to the shearing based on a single set of OCR = 5 test, the particle contacts for the MICP would possess a property rich with an elastic rebound as discussed earlier in the consolidation analysis. On the contrary, the particle contacts for the untreated soil would possess a property as resulted from large plastic fabric alterations. Further studies on the comparative studies between MICP and untreated residual soils are warranted by using the AE measurement and by subjecting the soils to various stress histories in order to enrich the understandings above.

4 CONCLUSIONS

In this paper, the effect of bio-mediation on the deformation and yielding behaviors of residual soil, which is commonly found in the tropical countries, such as Malaysia, were investigated. Particular attention was paid on the behaviors arising from the particle movements that are affected by the formation of calcite bonding within the residual soil fabric by using the AE measurement. The following conclusions can be derived from the study presented herein;

1) The bio-mediation treatment resulted in a soil structure that has a higher lateral resistance and elastic rebounding during isotropic compression than its untreated counterpart. There was a close similarity in the compression changes and the anisotropic behaviors between the sand from the dense to loose states and the residual soil from the MICP to untreated states. This indicated that the treatment resulted in a behaviour of denser soil state.
2) The AE measurement taken during the isotropic consolidation indicated that a non-negligible portion of the MICP soil compressions were of an elastic nature and recoverable. The calcite formation in the residual soil tested herein would therefore not only result in the densification effects but also the nature of particle bonding being enhanced with the elastic stiffness increases.
3) The undrained shear strength of the MICP residual soil under an isotropically consolidated state is higher than that of the untreated, but this difference decreases as the soil being consolidated to a higher stress level. However, this trend is reversed when the soil is in an over-consolidated state. The former trend can be understood by the MICP effect through densification while the increase in confining pressure tends to suppress the dilation effect. However, more works are needed to completely understand the latter trend because how the MICP effects on particle behavior remain after re-compression is not known well. Further studies on the comparative studies between MICP and untreated residual soils are warranted by using the AE measurement and by subjecting the soils to various stress histories in order to enrich the understandings above.

5 REFERENCES

CBBG. 2021. [Online]. Learning from nature to develop more sustainable, safer, and more resilient civil infrastructure systems, Available at: https://cbbg.engineering.asu.edu/about/cbbg-fact-sheet/ (Accessed: 26 May 2021).

DeJong, T.J.; Soga, K.; Ian, K.; Burns, S.; Paassen, I.A.V.; Qabany, A.A.; Aydilek, A.; Bang, S.S.; Burbank, M.; Caslake, I.F.; et al. 2013, Biogeochemical processes and geotechnical applications: Progress, opportunities and challenges. *Géotechnique*, 63, 287–301.

Kuwano R. 1999, The Stiffness and Yielding Anisotropy of Sand, Ph.D. Thesis, the University of London (Imperial College of Science, Technology and Medicine).

Lim, J.X., Chong, S.Y., Tanaka, Y., Ong, Y.H. and Lee, M.L. 2020. Influence of Consolidation Pressure on Stress-Deformation Responses of Biomediated Residual Soil in Malaysia. International Journal of Geotechnical Engineering, DOI: 10.1080/19386362.2020.1836113.

Lim, J.X. 2021, Mechanical and Acoustic Behaviors of Bio-Mediated Residual Soil in Comparison with Sand, Ph.D. Thesis (Submitted), Universiti Tunku Abdul Rahman, Malaysia.

Rahman, M.M., Hora, R.N., Ahenkorah, I., Beecham, S., Karim, M.R. and Iqbal, A. 2020. State-of-the-Art Review of Microbial-Induced Calcite Precipitation and Its Sustainability in Engineering Applications, Sustainability 2020, 12, 6281; doi:10.3390/su12156281.

Terzis, D., Laloui, L. 2019. A decade of progress and turning points in the understanding of bio-improved soils: A review, *Geomechanics for Energy and the Environment* 19 (2019) 100116.

General Reports from Technical Committees

Proceedings of the 20th ICSMGE – Post-conference Volume – Rahman and Jaksa (Eds)
© 2023 Australian Geomechanics Society, Sydney, Australia, ISBN 978-0-9946261-7-2

TC103 – Numerical methods in geomechanics – Organized sessions in the 20th International Conference on Soil Mechanics and Geotechnical Engineering

TC103 – Méthodes numériques en Géomécanique – Sessions organisées à la 20e Conférence internationale sur la mécanique des sols et l'ingénierie géotechnique

Akira Murakami & Ryosuke Uzuoka
Kyoto University, Japan

Francesca Ceccato
University of Padua, Italy

Helmut Schweiger
Graz University of Technology

ABSTRACT: Technical Committee 103, who is devoted to numerical methods in geomechanics, organized 5 sessions at the 20th ICSMGE conference. One session was in-person and 4 sessions were on-line. The present paper offers a report of the organized sessions. We received a significant number of papers, but only 20 could be selected for oral presentation and discussion.

KEYWORDS: please provide a relevant keyword list (up to 5 keywords, single line only). TC103, Numerical methods

1 INTRODUCTION

TC103 (Numerical Methods in Geomechanics) organized 4 technical sessions in the 20th International Conference on Soil Mechanics and Geotechnical Engineering (ICSMGE 2022) (https://icsmge2022.org/), which was held in "hybrid mode" with in-person at International Convention Centre Sydney, Australia and with virtual online during 3 to 5 May 2022. TC103 received 83 papers, of which 20 papers with in-person registration, 43 papers with virtual registration, and 20 papers with no registered author. After careful review of the submitted papers, TC103 selected 5 papers with in-person registration and 15 papers with virtual registration for presentation. Finally, one in-person session and three virtual sessions were scheduled and the summary of the sessions is described below.

2 SESSION 1

For the in-person session, chaired by Prof. Helmut Schweiger, the title of the contributions and the presenters are listed below:

- Experimental and DEM study of monotonic and cyclic simple shear testing of an angular sand (Dr Matteo Ciantia)
- Fabric anisotropy of granular soils: a DEM study (Mr Farid Khayyer)
- Development and application of a GPGPU-parallelized hybrid finite-discrete element method for modelling geo-structure collapse and resultant debris flow (Dr Hongyuan Liu)
- A Bounding Surface Constitutive Model with Moving Projection Centre for Highly Overconsolidated Clays (Dr Mohammad Rezania)
- Influence of constitutive model on undrained passive earth pressure (Prof Franz Tschuchnigg)

We had 5 presentations in the session (no recordings, all in person). Time allowed for each presentation was 10 minutes + 2 minutes for discussion. Authors kept to time limit, so a short discussion was possible for each presentation. DEM-modelling was a main topic coming up in 3 presentations, showing the importance of this methodology to better understand soil behaviour. All in all an interesting session.

The brief descriptions of each presentation are the following. Ciantia presented experimental and micromechanical numerical analysis (DEM) of monotonic and cyclic simple shear of an angular sand. Khayyer used three-dimensional DEM to explore the properties of the critical state by taking the critical fabric anisotropy into consideration to provide further evidence to support the concepts of the Anisotropic Critical State Theory. Liu developed GPGPU-parallelized HFDEM to investigate the collapse process of 3D irregular-shaped and non-cohesive soils heaps under gravity. Rezania enhanced a simple bounding surface constitutive model with more advanced mapping rule and projection centre for highly overconsolidated clay. Tschuchnigg showed that the constitutive model adopted plays a significant role under undrained conditions by solving a simplified earth pressure problem.

3 SESSION 2

This is a Virtual Session dedicated to micromechanics and coupled approaches, held on 3 May chaired by Prof. Ryosuke Uzuoka. The title of the contributions and the presenters are listed below:

- Pressure grouting of microfine cements in soils: micromechanical processes (Miss. Katia Boschi)
- Hierarchical Multiscale SPH (HM-SPH): A New Continuum Approach to Model Desiccation Cracking in Clay Soils (Prof Ha H. Bui)
- A coupled particle-fluid numerical analysis of failure initiation of non-cohesive granular soils induced by upward seepage flow (Dr Yutaka Fukumoto)
- Simulation of Pile Penetration in Sand Employing Lagrangian and CEL Based FE Approach: A Comparative Study (Dr Mousumi Mukherjee)
- Three-dimensional discrete element analysis of triaxial tests on crushable pumice sand (Prof. Rolando Orense)

TC103 virtual session on coupled approaches had 5 selected presentations related to micromechanics, large or discontinuous deformation behavior with coupled approach of soil-water and soil-structure interactions. All authors provided their recording videos, which were shown during the virtual session by the conference staff. During the video presentations, questions and answers were shown in Discussion Forum on the Virtual Platform, and the participants discussed on the platform in 13 minutes after the video presentations by the end of session.

Orense investigated the monotonic drained and undrained response during triaxial tests of crushable pumice using 3D DEM. The influence of number of particles, mass conservation and particle crushing effect during the DEM simulations were discussed. Mukherjee presented effectiveness of updated Lagrangian (UL) and Coupled Euler-Lagrangian (CEL) formulations for replicating the pile penetration process. The contact model and interaction between pile and soil during large deformation were discussed. Fukumoto focused on sand boiling phenomenon using coupled Lattice Boltzmann and Discrete Element (LB-DE) methods. The limitation of number of particles in LB-DE method was discussed. Boschi analyzed the microfine cement injection considering soil grains-grout local interaction by DEM modeling. The solidification process in microfine cements was discussed. Bui and his co-author presented their latest development of an advanced continuum computational approach (HM-SPH) to predict complex desiccation cracking in clay soils. The numerical instability was discussed after the presentation. All numerical methods in the presentations were advanced and useful to study geomechanical coupled behaviour.

4 SESSION 3

This is a Virtual Session dedicated to seismic and slope stability, held on 4 May, chaired by Prof. Kazunori Fujisawa. The title of the contributions and the presenters are listed below:

- Structure-soil-interactions during maintenance works of the flood barrier "Eidersperrwerk" in Germany (Denis Maier)
- Numerical study on aseismic reinforcement method for leaning-type retaining wall (Dr Keita Abe)
- Numerical approach for predicting large deformation of soil structures using the Convected Particle Domain Interpolation (CPDI) method (Prof Christian Moormann)
- Two types of water absorption failure mechanism of unsaturated silt triaxial specimen (Dr Takahiro Yoshikawa)
- Simulation of reliquefaction of sandy soil using combined loading elasto-plastic constitutive model (Dr Shotaro Yamada).

The session was originally scheduled during 16:30 to 17:30, but additional 10 minutes were allowed due to relatively long video streaming (57 minutes) of all the 5 presentations. Total 70 minutes of 57 minutes' video streaming and 13 minutes' discussion were assigned.

The first speaker was Dr. Denis Maier, who presented the numerical analysis of structure-soil interaction around the flood barrier. His presentation was focused on the effect of gas inclusion into a saturated bed on the dynamic response of pore water pressure subjected to tidal waves. The second presentation was delivered by Dr. Kaita Abe, who showed the numerical analysis of the ground deformation behind a leaning-type retaining wall with/without a reinforcement for seismic loading. He employed MPM (Material Point Method) for his numerical analysis, which enabled the large deformation of the ground behind the tilted retaining wall. Prof. Toshihiro Noda delivered the third presentation instead of Dr. Takahiro Yoshikawa, and their numerical study considered the two types of water absorption leading to failure of unsaturated soils. They showed that shrinkage of the yield surface involved with volumetric contraction due to suction increase as well as dilation due to plastic deformation can cause the failure. The fourth speaker is a PhD student supervised by Prof. Christian Moormann. He introduced Convected Particle Domain Interpolation (CPDI) method which can improve the accuracy and efficiency of MPM, and presented the numerical analysis of pile penetration into the ground by the sophisticated numerical scheme. The final speaker was Dr. Shotaro Yamada, who proposed a combined loading elasto-plastic constitutive model. The prominent ability of describing complex stress-strain relationships realized the numerical simulation of re-liquefaction of sandy soils.

After the five video-streaming presentations, the 13 minutes' discussion was moderated by the chair. The questions given to the speakers in the discussion were as follows.

To Dr. Maier: How much gas inclusion can be significant to the delayed response of pore water pressure?

To Dr. Abe: How did he deal with the interface between soil and wall? (In the presentation of Prof. Moormann, the interface between the pile and the soil was modeled via a numerical scheme of the contact mechanics.)

To Dr. Noda: How can we distinguish or predict the two types of water absorption in actual cases?

Since the session was managed punctually on the virtual platform and was closed exactly in 70 minutes.

5 SESSION 4

This is a Virtual Session dedicated to foundations and soil improvement, held on 5 May chaired by Prof. Ha H. Bui. The title of the contributions and the presenters are listed below:

- Numerical simulation of driven displacement piles with respect to installation effects on the basis of in-situ test results (Mrs Ann Kristin Hundertmark)
- Large deformation finite element analysis to predict penetration resistance of offshore pipelines (Prof Samanthinka Liyanapathirana)
- Hierarchical Multiscale SPH (HM-SPH): A New Continuum Approach to Model Desiccation Cracking in Clay Soils (Prof Ha H. Bui)
- Numerical simulation of heat and mass transfer during the in-situ thermal remediation process of NAPLs-impacted soil (Mr Zhixin Chen)
- Thermo-hydro-mechanical hypoplastic modelling of bentonite buffers for nuclear waste disposal: model calibration and performance (Dr Gianvito Scaringi)

The session consists of five selected virtual talks, each of which focuses on a specific geotechnical aspect related to the topic of foundations and soil improvements. Two papers by Hundertmark and Liyanapathirana presented their recent work on the application of large deformation finite element analysis to model soil-structure interactions (i.e. pile-driven and offshore pipelines), showing the importance of large deformation analysis in geotechnical designs. Bui and his co-author presented their latest development of an advanced continuum computational approach (HM-SPH) to predict complex desiccation cracking in clay soils. On the other hand, papers by Chen and Scaringi focused on the multiphysics interactions and showcased their proposed models to describe the influence of heat on the mass transfer and coupling hydro-mechanical response of nuclear waste disposal.

All authors provided their recorded videos, which were shown during the virtual session, and strictly followed the guideline to allow a short discussion at the end of the session. All in all, this is an interesting session, which shows the importance of advanced mathematical models and computer simulations to study complex geotechnical systems.

6 CONCLUSIONS

The session organized by TC103 (Numerical methods in Geomechanics) has been very well participated and we received a large number of contributions of good quality. We regret that we could not have the opportunity to assign a presentation to more authors (only 20 papers over 63 could have an oral presentation). The discussions were interesting and we believe that the event was successful.

Proceedings of the 20th ICSMGE – Post-conference Volume – Rahman and Jaksa (Eds)
© 2023 Australian Geomechanics Society, Sydney, Australia, ISBN 978-0-9946261-7-2

TC201 report (2017–2023)

Rapport TC201 (2017–2023)

Cor Zwanenburg
Deltares, The Netherlands, Cor.Zwanenburg@deltares.nl

Norma Patricia López Acosta
Instituto de Ingeniería, Universidad Nacional Autónoma de México, Mexico

TERMS OF REFERENCE:

1. Discussion state of the art and developing the science and engineering in order to increase fundamental understanding and modelling of strength, deformation and erosion behaviour of soils and failure processes of dikes, dunes, (underwater) dams, reclaimed land, man-made islands and ports in river and coastal zone.
2. Contributions to the international congresses of the society to increase the quality of design by including experience, such as large scale tests, case histories, failures, measurements, simulations and artificial knowledge.
3. Encouraging international collaboration of experts in multidisciplinary developments of new techniques for constructing and maintaining safe and reliable protection works.

Producing technical reports and encouraging international training and meeting in this research field and exchange experience and ideas on improving communication to other researchers, public and politics for more alertness and awareness.

1 COMPOSITION OF TC201

Members:

1) M.K. Sharp (USA), 2) Y. Watabe (Japan), 3) M. Van (Netherlands), 4) C. Zwanenburg (Netherlands), 5) T. Abdoun (USA), 6) E. Asanza (Spain), 7) L. De Vos (Belgium), 8) R. Hutchison (UK), 9) A. Kort (Norway), 10) N.P. López Acosta (Mexico), 11) K. Maeda (Japan), 12) L. Nagy (Hungary), 13) D. Peluso (USA), 14) P. Smith (UK), 15) A. Tarantino (Italy), 16) J.A. Teeuw (Netherlands), 17) R. Tourment (France), 18) A. Pinto (Portugal), 19) J. Vlaanderen (Portugal), 20) E. Imre (Hungary), 21) H. Mori (Japan), 22) M. Pohl (Germany), 23) P. Mengé (Belgium), 24) B. Trak (Turkey), 25) D. Zhang (Australia), 26) L. Tonni (Italy), 27) G. Gottardi (Italy), 28) Y. Higo (Japan), 29) D. L. Shah (India), 30) S. Das (India), 31) X. Li (China), 32) J. Ji (China), 33) M. Achmus (Germany), 34) F. Mathijssen (Netherlands), 35) H. Jitno (Indonesia), 36) M. Sottile (Argentina), 37) P. Simonini (Italy), 38) E. Koch (Hungary), 39) J. Wang (Hong Kong), 40) I. Muhammad Alatas (Indonesia), 41) D.S. Park (South Korea), 42) S. Cola (Italy), 43) A. Dabska (Poland), 44) T. Le (UK).

2 ACTIVITIES PRIORITIZED BY TC201

2017

- TC201 meeting in Seoul, South Korea during the 19th International Conference on Soil Mechanics and Geotechnical Engineering.
- Workshop on seepage at Imperial College London, United Kingdom.

2018

- Paper "Lessons learned from dike failures in the past" was published in the International Journal of Geoengineering Case Histories.
- European working group on levees under ICOLD (EUCOLD).

2019

- TC201 Workshop during the 16th Pan American Conference on Soil Mechanics and Geotechnical Engineering, Cancun November 2019.
- TC201 Session in the 17th ECSMGE 2019
- TC201 Workshop during the 17th ECSMGE 2019.
- TC201 Committee meeting during the 17th European Conference on Soil Mechanics and Geotechnical Engineering (17th ECSMGE) Reykjavik, September 1st - 6th, 2019.
- European working group on levees under ICOLD/EUCOLD.

2020

Due to the COVID-19 pandemic, planned public events were canceled. In the background TC201 started to work on the joint report on failure paths for levees, which was finalized in 2022.

2021

- on line TC201 meeting via Microsoft Teams on Tuesday, June 15th, 2021, including the presentation of the ISSMGE Time Capsule Project 2021.
- Online meeting EWG-IE 2020 "Internal erosion at the field scale" on Thursday, February 4, 2021.
- working on report *Failure Paths for Levees*.

2022

- Finalizing joint report *Failure Paths for Levees*, DOI: 10.53243/R0006
- Joining the Time Capsule initiative
- On line TC201 meeting via Microsoft Teams on Tuesday, June 14th, 2022
- Session in the ISSMGE conference, Sydney
- Start new joint report on upgrading and retrofitting needs for levees.

3 REPORT ON ACTIVITIES

3.1 *2017*

3.1.1 *TC201 meeting in Seoul, South Korea during the 19th International Conference on Soil Mechanics and Geotechnical Engineering*

The 19th ISSMGE conference was held in Seoul during September 17th to 22nd 2017. One session was dedicated to dikes and levees. In total, 8 papers were selected for a presentation out of 13 submitted papers. In total an audience of 40 people visited the session. The presented topics showed a wide range of topics of interest for our working field. In total 11 papers were presented for which a TC-201 member is author or co-author.

3.1.2 *Report on findings of ISSMGE survey on State of Practice (SoP) versus State of the Art (SoA) in respect to TC201*

The aim of the survey is to bring academics and industry together. The results gave an interesting view on the differences between state of the art and state of practice. Regarding the working field of TC-201, the differences between SOA and SOP are felt very differently for the different respondents. For example, on the question on the cooperation between universities and industry, respondents answered in the range of non-existent to very intense.

3.1.3 *Workshop on seepage at Imperial College London, United Kingdom*

On August 31 and Sept 1 a workshop on seepage induced instabilities was held at Imperial College London.

- Seepage Workshop
- Seepage Induced Geotechnical Instability
- Workshop exploring recent developments considering seepage and instability in cohesionless soil.

This workshop was run in conjunction with the ISSMGE Technical Committees: TC105 Geo-Mechanics from Micro to Macro, TC201 Dykes and Levees, TC213 Scour and Erosion. The workshop launched the EPSRC funded research project. Delegates included academics, civil engineering consultants and representatives from some UK dam owners.

3.1.4 *European Levee Community of Practice Initiative*

The European Levee Initiative was launched. An important item will be the library/database of case histories, which is also presented in the workshop as the SAFElevee initiative-wg4. This initiative was supported by the European working group on levees under ICOLD (EUCOLD). TC201 actively supported this initiative and was encouraged TC members to come forward with relevant cases.

3.2 *2018*

3.2.1 *"Lessons learned from dike failures in the past" paper published in the Int. Journal of Geoengineering Case Histories.*

One of the TC201 ambitions is to encourage writing (joint) publications. As announced in previous newsletters a selection of TC201 members has produced a joint paper entitled: "Lessons learned from dike failures in the past" for the *ISSMGE International Journal of Geoengineering Case Histories.*

Fig. 1. Failed canal dike at Wilnis, August 2003.

3.2.2 *European working group on levees under ICOLD (EUCOLD)*

TC201 continuously participated actively in this initiative and was encouraged TC members to come forward with relevant cases.

3.3 *2019*

3.3.1 *TC201 Session in the 17th ECSMGE 2019*

During the 17th European Conference on Soil Mechanics and Geotechnical Engineering (ECSMGE-2019) in Reykjavik, Iceland, TC201 organized a session on dikes and levee. The 17th ECSMGE was held on September 1st – 6th, 2019 (http://www.ecsmge-2019.com/). The organizing committee received about 37 papers that by the authors themselves were indicated to fall in the TC201 interest. The Technical Discussion Session took place on Wednesday 4th September 2019, from 16:10 to 18:00 h.

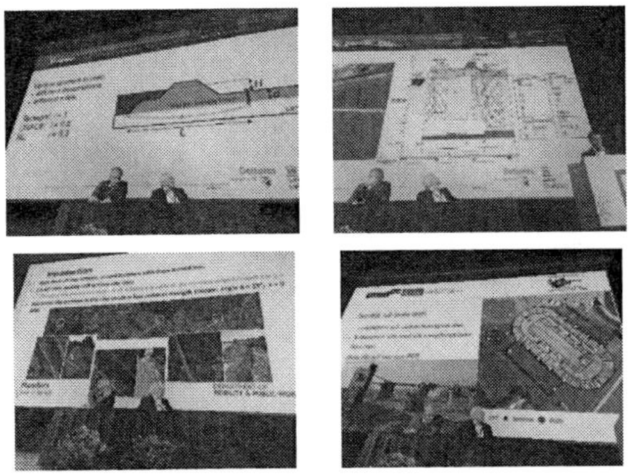

3.3.2 *TC201 Workshop during the 17th ECSMGE 2019*

TC201 had also given the opportunity to organize a workshop in advance of the Reykjavik conference. The titles of the papers sent in for the TC201 session in the 17th ECSMGE-2019 conference suggested that some of the contributions with regard topics that are open for discussion. The workshop had the following program:

- Ulrich Föster:
  Large scale experiments with a coarse sand barrier as part of a multi-scale investigation of a novel remediation technique against backward erosion piping.
- Rémy Tourment:
  I. Levees breaching scenarios and the different deterioration/damage mechanisms; link with failure modes analysis and risk analysis.
  II. Levee and dike survey using drones: the DIDRO project.
- Joost Breedeveld:
  The impact of the Eemdijk full-scale field test programme.
- Jonathan Black:
  Internal erosion of earth flood embankments.

3.3.3 *TC-201 Workshop during the 16th Pan American Conference on Soil Mechanics and Geotechnical Engineering, Cancun November 2019*

TC201 organized a workshop during the XVIth Panamerican Conference on Soil Mechanics and Geotechnical Engineering. The workshop was held on Wednesday, November 20th. The workshop was concentrated around five topics:

a) The international levee handbook
b) Codes and Standards
c) Historical case studies
d) Evaluation and of instability
e) Experimental test

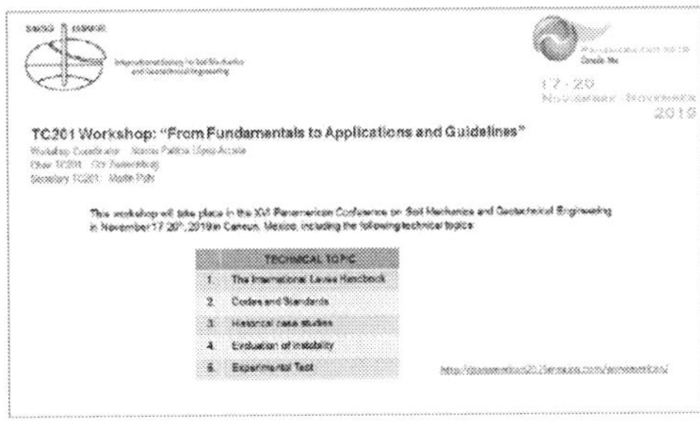

The workshop took place on Wednesday November 20th, 2019 in room Coral, from 10:15 to 16:15 h, with the following program:

Lectures	Tittle	Lecturer
10:15-11:45	"Small dams: some case histories"	Ivan Vanicek
	"The International Levee Handbook"	Maureen Corcoran
	"Experimental testing on levees"	Anne Lemnitzer
14:45-16:15	"Geomembrane sealing systems for the construction of new embankment dams"	Pascual Perazzo
	"lessons learned from full scale tests"	Cor Zwanenburg
	"Risk assessment on levees subjected to flooding through fragility curves"	Norma Patricia López-Acosta

3.3.4 *European working group on levees under ICOLD/EUCOLD*

The ICOLD/EUCOLD levees group participated on a dams – levees comparison with a workshop in Paris, 12-13 February. This report is, with the levees situation report, a major basis for future activities on levees in ICOLD.

The ICOLD/EUCOLD levees group organized two events in this year:

- 2-4 Oct 2019, European Club Symposium Crete, preliminary symposium themes:
  Topic A. Dam Safety and Risk Management. Social impact and awareness.
  Topic B. Dams to address Climate change issues.
  Topic C. European directives on dams and the water sector.
  Topic D. Managing ageing infrastructure.
  Topic E. Dam monitoring /surveillance and rehabilitation.
  Topic F. Advances in dam engineering. https://www.eurcold2019.com
- 9-14 June 2019, ICOLD Annual Meeting 2019 Ottawa,

abstract deadline is 15/9/2018.
Conference Theme is "Safety and Environment for Global Benefit". http://icoldcigb2019.ca/

All TC members were invited to join the above mentioned events.

3.4 *2020*

Due to the COVID-19 pandemic, planned events were canceled.

3.5 *2021*

3.5.1 *Virtual TC201 meeting via Microsoft Teams on Tuesday June 15th, 2021Virtual TC201*

Last June 15, 2021 (7:00 h Mexico/ 14:00 h Delft/ 22:00 h Seoul) the members of the TC201 Technical Committee had an online meeting, with the following agenda:

- I. Opening
- II. Minutes of last TC201 meeting
- III. Presentation joined document on failure paths
- IV. TC201 activities
- a) 20th ICSMGE-2022 (May 1 to 5, 2022, Sydney, Australia)
  - o A pre-conference workshop
  - o A conference session
  - o Our annual meeting
- b) Field trip to water retaining structures in Australia (additional activity during the 20th ICSMGE-2022)
- V. The time capsule project
- VI. Any other business
  - o New members
  - o 6th International Conference on Geotechnical and Geophysical Site Characterization 26-29 September 2021, Congress Center, Budapest, Hungary.
- VII. Closing

The attendants were:

1. Cor Zwanenburg (Chairman TC201), The Netherlands
2. Norma Patricia López-Acosta (Secretary TC201), Mexico
3. Meindert Van, The Netherlands
4. Michael K. Sharp, USA
5. Philip Smith, United Kingdom
6. Rémy Tourment, France
7. Alexandre Pinto, Portugal
8. Joana Vlaanderen, Portugal
9. Alessandro Tarantino, United Kingdom
10. Emőke Imre, Hungary
11. Hirotoshi Mori, Japan
12. Leen Vincke, Belgium
13. Patrick Mengé, Belgium
14. Adda Athanasopoulos-Zekkos, USA
15. Thibaut Mallet, France
16. Esther Rosenbrand, The Netherlands
17. Jean-Jacques Fry, France

The TC201 committee members very well received the virtual meeting. It seems a good idea to hold virtual meetings in the meantime for the next face-to-face conferences.

During this virtual TC201 meeting the chairman Cor Zwanenburg showed a presentation about the ISSMGE Time Capsule Project (TCP) 2021, which intend connecting and interacting with individual members of the ISSMGE all around the world about the work of the ISSMGE (past, present and future). The ISSMGE Time Capsule 2021 has been conceived to help achieve this aim. Many paths to communicating with individual members already exist: Web site/App, Library (open source), Bulletin, Geoworld, Virtual University, Technical Committees, Member Societies, Young members (YMPG), Corporate Associates (CAPG), Time Capsule blog. And additionally, ISSMGE affiliated Conferences (4 year international, 4 year regional, individual technical committee, Local/sub regional).

Rémy Tourment says that TCP is a very interesting project but is too ambitious in terms of time, because dates are very close.

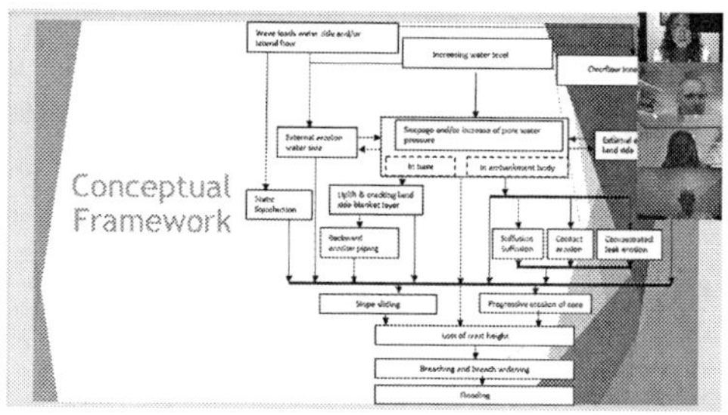

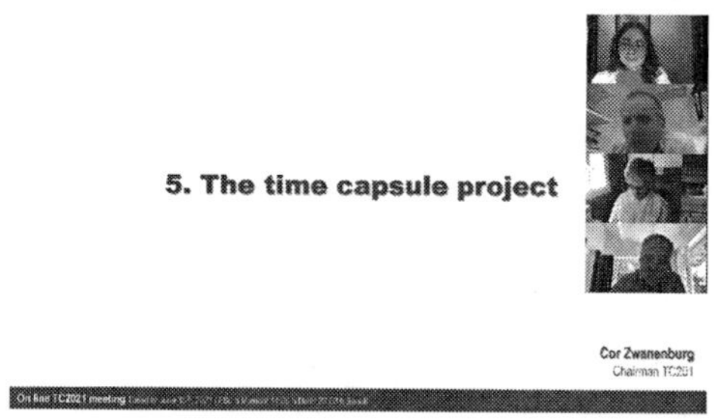

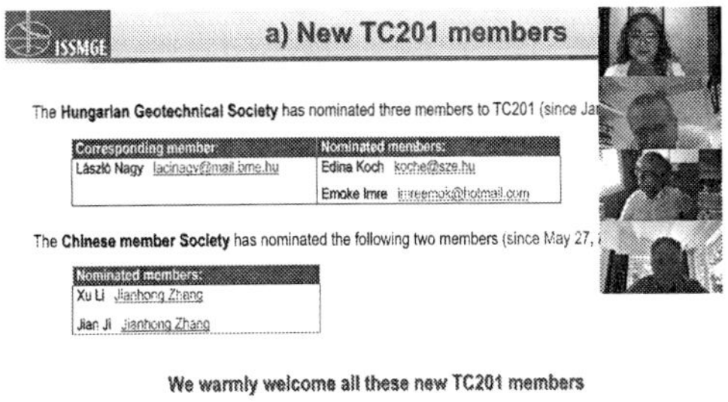

3.5.2 *Online meeting EWG-IE 2020 "Internal erosion at the field scale" on Thursday February 4, 2021*

Due to the COVID-19 pandemic, the annual meeting of the European Working Group on Internal Erosion (EWG-IE) was replaced by a series of online meetings. This working group, which was set up in 1993 and is linked to the ICOLD Technical Committee on Embankment Dams, consists of internal erosion experts and scientists from over 30 countries in Europe and all other countries. The group facilitates knowledge exchange knowledge between practitioners and scientists, and between experts, young engineers and researchers.

3.5.3 *Final draft of the "Failure paths for levees" report*

The final draft of the "Failure paths for levees" report was shared with TC201 members. This report can be downloaded in the TC201 sub-website.

This TC201 report was made possible by the TC201 member contributions and the fruitful discussion at the TC201 meeting in June. With our group, Rémy Tourment, Philip Smith, Meindert Van, Cor Zwanenburg and Esther Rosenbrand spent a lot of time thinking about terminology and the overall framework and we hope that this report helps to facilitate the use of failure paths.

3.6 *2022*

3.6.1 *Finalizing joint report*

In the first quarter of 2022 the joint report "Failure paths for levees" was finalised. The report is made traceble on the internet by: https://doi.org/10.53243/R0006, hosted by the ISSMGE website.

The report was also presented in the conference Geohazards 8, Quebec, Canada, June 2022. (Van M.A., Rosenbrand E., Wopereis L., Zwanenburg C. (2022) Failure paths for levees, an international framework. In Proc. GeoHazards 8 Quebec Canada).

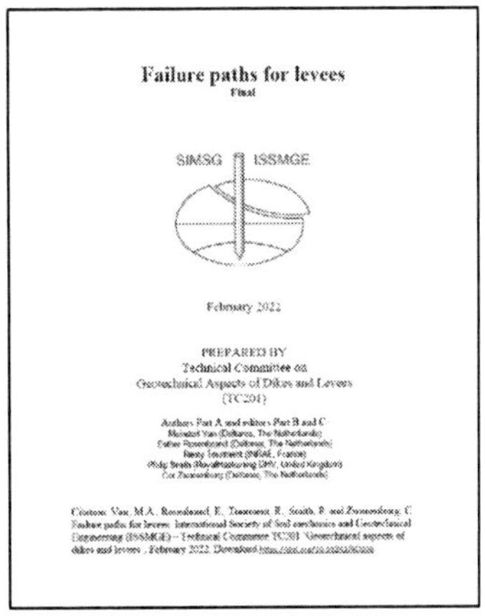

Failure paths for levees
Final

SIMSG ISSMGE

February 2022

PREPARED BY
Technical Committee on
Geotechnical Aspects of Dikes and Levees
(TC201)

3.6.2 *Time Capsule Project*

TC201 joined the Time Capsule Project by sending the report "Failure paths for levees".

3.6.3 *New joint report*

With the finalization of the report on failure paths a new initiative for a joint report was started. The new joint report will focus on upgrading and retrofitting techniques. The idea was launched in 2022. The activities started in 2023 with an inventory.

3.6.4 *On-line TC201 meeting*

Last June 14th, 2022 (7:30 h Mexico/ 14:30 h Delft/ 22:30 h Seoul) the members of the TC201 Technical Committee had an online meeting, with the following agenda:

I. Opening
II. Minutes of last TC201 meeting
III. Brief concluding comments on the final joined document on failure paths
IV. TC201 next activities
  a) XXXI Mexican National Conference on Geotechnical Engineering (XXXI MNCGE) to be held in Guadalajara, Jalisco, Mexico in Friday November 18th, 2022 (15:30 - 17:30 h).
  b) XVII Panamerican Conference on Soil Mechanics and Geotechnical Engineering (XVII PCSMGE).
V. Any other business
  - Members suggestions
  - New members
VI. Closing

The attendants were:

1. Cor Zwanenburg (Chairman TC201), The Netherlands
2. Norma Patricia López-Acosta (Secretary TC201), Mexico
3. Meindert Van, The Netherlands
4. Michael K. Sharp, United States
5. Martin Pohl, Germany
6. Rémy Tourment, France
7. Francois Mathijssen, The Netherlands
8. Joana Vlaanderen, Portugal
9. Alessandro Tarantino, United Kingdom
10. Imre Emőke Mária, Hungary
11. Hirotoshi Mori, Japan
12. Truong K. D. Le, United Kingdom
13. Edina Koch, Hungary
14. Simonini Paolo, Italy
15. Guido Gottardi, Italy
16. Yosuke Higo, Japan

The TC201 committee members very well received the

virtual meeting. It seems a good idea to hold virtual meetings in the meantime for the next face-to-face conferences.

3.6.5 *Participation in Sydney ICSMGE 2022 (1-5 May, 2022)*

Last May 5th, 2022 (9:45 – 10:45 h Australia) the TC201 Technical Committee had a virtual technical session in the ICSMGE 2022:

- The virtual session was named "Latest developments in dyke technology".
- Cor Zwanenburg chaired this session.
- A large portion of the presentations were held virtually.
- The virtual session included 5 presentations of 10 min each and they were available, on-demand, throughout the conference, and for 6-months afterwards for participants who registered for the on-line conference.
- This virtual session was successful.

3.7 *2023*

3.7.1 *TC201 management meeting 2023*

Due to the great success of the previous on-line meetings in 2021 and 2022, it is intended to have the next meeting as an on-line meeting. This meeting will be held on June 27th, 2023 (7:30 h Mexico/ 14:30 h Delft/ 22:30 h Seoul).

Details of this meeting as well as an invitation to the on-line meeting will follow later.

3.7.2 *Announcement of TC201 activities during the XVII Panamerican Conference on Soil Mechanics and Geotechnical Engineering (La Serena, Chile)*

We notify you that during the XVII PCSMGE-2024, the TC201 will organize activities:

- A technical session, if this is possible, of our TC201 technical committee on "Dykes and levees". With this objective, we invited the TC201 members to send abstracts for this session.
- A pre-conference workshop "Dykes and levees". We invite you to send us suggestions for this workshop.
- Our annual meeting (both face-to-face and online).

Important information about this conference is provided below:

- The XVII Panamerican Conference on Soil Mechanics and Geotechnical Engineering (XVII PCSMGE) will be held from November 12 to 17, 2024, in La Serena, Chile.
- Website: https://panamgeochile2024.cl/
- Preliminary general program:
  Tuesday 12: Workshops/Short courses/Ice breaker
  Wednesday 13 to Friday 15: XVII PCSMGE
  Saturday/Sunday 16 to 17: Sightseeing/technical visits
- Papers:
  Presentation of abstracts: June 2023
  Acceptance of abstracts: July 2023
  Presentation of papers: November 2023
  Acceptance/comments papers: January 2024
  Deadline for final version of papers: April 2024

3.7.3 *New members*

- Prof. Dąbska Agnieszka (Poland, since April 14, 2022)
- Prof. Le Truong K. D. (United Kingdom, London; since April 21, 2022)
- Msc Ing. Mauro Sottile (SAIG, Argentina)

We warmly welcome all these new TC201 members.

3.7.4 *Newsletter from the EUCOLD WG on Levees and flood defences, issue number 10 (ICOLD TC LE)*

The ICOLD TC LE last issue of the newsletter can be downloaded from the following link https://lfd-eucold.inrae.fr/index.php/wg-newsletter/

3.7.5 *Some upcoming events*

- ASCE Geo-Congress 2023. Los Angeles, California, USA, 26th – 29th March 2023. https://2023.geocongress.org/
- 8th International Conference on Unsaturated Soils, Milos island, Greece between the 2nd and 5th of May 2023. Contact person: Dr Michael Bardanis (mbardanis@edafos.gr)
- 9th International Congress on Environmental Geotechnics, Chania, Crete island, Greece, between the 25th and 28th of June 2023, https://iceg2023.org
- XX Technical Dam Control International Conference, Safety of hydraulic structures, the jubilee edition of the conference.

  The conference will be held from 12 to 15 September 2023 in Hotel Diament Arsenal Palace in Chorzów, in the Silesia region of southern Poland, near Katowice. The conference is organized in a hybrid formula, with the possibility of online or in-person participation. Participants will have the opportunity to take part in a technical tour to the Goczałkowice reservoir located in the Silesia region, as well. Registration for participation in the conference has just been opened. You can register via the conference website: https://tkz.is.pw.edu.pl/registration/.

  The deadline for registration and submission of abstracts expires on February 28, 2023. The template of the abstract you can find on the website: https://tkz.is.pw.edu.pl/templates/ You can submit an abstract via the conference website: https://tkz.is.pw.edu.pl/manuscripts-form/. Submitted papers will be published in journals: "Energetyka Wodna", "Gospodarka Wodna", "Inżynier Budownictwa", "IMGW Monograph" series, and for an extra charge in "Archives of Civil Engineering" and "Journal of Water and Land Development".

  Detailed information about the conference is available at: www.tkz.is.pw.edu.pl
- 5th International Conference on Geotechnics for Sustainable Infrastructure Development, Vietnam, Hanoi, between the 14th and 15th of December 2023, https://geotechn.vn/

3.7.6 *Next TC201 Newsletter*

The next newsletter will be sent in June 2023. Please provide all available information, like TC related publications, academic papers, news, research, among others, to the Secretary TC201 Norma Patricia López-Acosta, nlopeza@iingen.unam.mx

4 FINAL COMMENTS

Currently, the TC201 is made up of 44 members who actively participated in several conferences and symposiums around the world during the period 2017-2022.

Despite the global contingency due to COVID-19 pandemic, the TC201 continued working and disseminating research with the publication of articles, technical reports and presentations in conferences through digital platforms and virtual events.

Proceedings of the 20th ICSMGE – Post-conference Volume – Rahman and Jaksa (Eds)

Resilient evaluation and control in geotechnical and underground engineering

Gang Zheng, Xuesong Cheng, Haizuo Zhou, Tianqi Zhang, Yu Diao, Wei Guo
MOE Key Laboratory of Coast Civil Structure Safety, Tianjin University, Tianjin 300072, China and Department of Civil Engineering, Tianjin University, Tianjin 300072, China

Ruozhan Wang & Fan Yi
Department of Civil Engineering, Tianjin University, Tianjin 300072, China

ABSTRACT: Resilient cities and societies are significantly required for modern cities, social security, and sustainable development. However, lots of structures in geotechnical and underground engineering are of low resilience due to the uncertainties of soil properties and the surcharges, and the complex interaction between soil-water-structures. The damage accidents or even progressive collapse accidents happened from time to time in deep excavation, shield tunnel, pile foundation, and embankment projects all over the world. Therefore, it is important to reveal the development and termination mechanisms of the progressive failure to resist natural or human-induced serious disasters, maintain its functions to an extreme extent, and restore its functions as soon as possible after disasters. It is of great importance to build the evaluation and controlling design theories of resilience against progressive failure, improve the resilience of structural units, systems, and the associated large-scale urban in geotechnical and underground engineering, improve their disaster resistance to disasters or accidents, adaptation and recovery capacity, ensure the safety of construction projects for the development of resilient cities. This paper summarized the connotation and development of resilience and pointed out that its evaluation and control methods can be divided into the following four levels of design: permissible stress design method, reliability-based design method, robustness method, and recoverable performance method based on the failure probability, ability to prevent progressive collapse under unforeseen events and its recoverability from disasters. The recent research studies on each of these levels in geotechnical and underground engineering structures are reviewed, summarized, and analyzed. The progressive collapse in the deep excavation engineering, underground engineering, shield tunnels, and column-supported embankment engineering are taken as examples to discuss the existing research studies on the resilience against the progressive collapse to provide references for establishing design theories and frameworks in geotechnical and underground engineering.

KEYWORDS: Excavation engineering; Shield tunnel engineering; Resilience; Progressive collapse; Reliability; Robustness; Recoverability

1 INTRODUCTION

The utilization of underground space is one of the most effective ways to solve traffic congestion, shortage of land resources, expansion of urban spaces, and alleviation of environmental deterioration. It is also an important way for the economy and human society development to build a sustainable, resource-saving, and environment-friendly society. For this purpose, underground spaces have been intensively developed in large and medium-sized cities, such as underground shopping malls, transportation hubs, complexes, parking lots, subway tunnels and stations, and comprehensive pipeline corridors. The transportation infrastructures are also being optimized and developed to provide strategic support for stable economic growth, coordinate developments of regional urban and rural areas, developments of land and space, improvements of productivity distribution, and optimizations of industrial structures.

Geotechnical engineering is the science that explains the mechanics of soil and rock and their applications to the development of humankind. It includes, without being limited to, the analysis, design, and construction of slopes, excavations, soil and foundations, tunnels, and other systems that are made of or are supported by soil or rock. Underground structures are often related to the structures built beneath the earth's surface, such as tunnels, subway stations, underground shopping malls and car parks, and so on. The development of underground space and the construction of transportation infrastructure involves many geotechnical and underground engineering problems which are almost ubiquitous. The large-scale engineering structures are supported or embedded in soil with their system safety largely depending on the interactions between soil and structures. The safety of geotechnical structures has a significant influence on large-scale engineering structures. Geotechnical engineering is being recognized as one of the high-risk construction fields as it often faces high risks of complex engineering geological and hydrogeological conditions and harsh surrounding environments. These risks are roughly reflected in the following two aspects:

(1) Low integrity and high risk of progressive collapse

Compared with other engineering structures, geotechnical engineering is a complex interaction system composed of structure, soil, and groundwater. Soil is one of the most complex engineering materials because of its uncertainties in geotechnical parameters, the physical analyzing models, and the environmental conditions (excess heat, cold, snow, earthquakes, fire, and flooding) during the construction and service procedures. The integrities of the connections between each component of the structure are much lower than that of the architectural structures, resulting in lower integrity and robustness of the system.

Take the deep excavation project, for example, the whole retaining system is composed of the weakly connected or not connected remaining piles (walls) and scattered struts systems. However, the current design methods are still based on the component design, without considering the integrity and robustness of the entire retaining system. Huang and Gu (2008) also pointed out that the safety stock of the retaining system is relatively low because they are mostly designed as temporary structures. Another example is the shield tunnel which is made up of concrete segments and spliced together using lots of bolts which are the weak and sensitive parts of the shield tunnel (Huang et al., 2012, 2016). Therefore, the entire shield tunnel is relatively low in integrity and high in fragility.

It can be seen from the above discussions that some geotechnical structures, such as the deep excavation and the shield tunnels structures, can be considered low-resilience structures. Once local failure occurs, there is a high risk for the surrounding structures to be damaged in progressive failure. The robustness of the intervention ability before the disaster is poor.

(2) hard to rescue, repair, and rebuild after damage, but easy to cause cascading damage to adjacent structures under or above the ground

As the underground structures are often deeply embedded underground, the local failure induced accident develops rapidly with a large range of damage. It is extremely hard to intervene, rescue, and control the disaster during the damage developing process. The damaged structures generate a large number of solid rubbish which is deeply buried underground. After the damage occurs, most or all the engineering functions are lost and difficult to repair. With the continuous development of urban underground space, many adjacent unground structures system have been constructed. Once an underground structure is damaged, it will lead to different degrees of change in the constraints of the surrounding soil. A severe change may result in excessive deformation or even collapse of the adjacent structures forming a chain of damage to the adjacent projects.

These risks make the geotechnical and underground structures one of the well-known engineering fields with the highest risks of damage. A large number of engineering accidents in the world show that once damaged, the accidents for geotechnical and underground engineering structures rapidly develop with large scopes of destruction. The current design theory cannot predict the damage location and damage range, nor effectively evaluate and control the damage. As a result, underground engineering accidents occur frequently, resulting in serious social impact and huge economic and property losses.

In 2019, the 13th Chinese National Conference on Soil Mechanics and Geotechnical Engineering (CNCSMGE) with the main theme of "Quality Improvement and Sustainability Development of Geotechnical Engineering" held in Tianjin, gave the consensuses: the prospects of Geotechnical Engineering should have the quality of Resilience, Green, Intelligence and Humanism. The "Resilience of geotechnical engineering" is defined as: "Geotechnical engineering is closely related to people's livelihood, and geotechnical engineering is almost ubiquitous in engineering construction. Breakthrough traditional safety concepts, establish a new safety concept for geotechnical engineering and improve rock and soil engineering. The resilience performance of geotechnical engineering to resist serious disasters caused by natural disasters or man-made factors, maintain its functions to the greatest extent, and restore its functions as soon as possible after a disaster is an important quality of geotechnical engineering to support the construction of a resilient city and a resilient society."

Improving the resilience of geotechnical and underground engineering structures and preventing them from causing large-scale collapse due to local small-scale damage in unexpected situations is of great significance to the safety of urban and major infrastructure constructions. How to improve the safety and resilience of structures in geotechnical and underground engineering, improve their ability to resist local failure caused by unexpected factors and its induced progressive failure and collapse, restore their functions as soon as possible and realize the resilient design for underground structures are the major problems that geotechnical and underground engineers must solve.

This paper summarized the connotation and development of resilience and pointed out that its evaluation and control methods can be divided into the following four levels of design: permissible stress design method, reliability-based design method, robustness method, and recoverable performance method based on the failure probability, ability to prevent progressive collapse under unforeseen events and its recoverability from disasters. The recent research studies on each of these levels in geotechnical and underground engineering structures are reviewed, summarized, and analyzed. The progressive collapse in the deep excavation engineering, underground engineering, shield tunnels, and column-supported embankment engineering embankment are taken as examples to discuss the existing research studies on the resilience against the progressive collapse to provide references for establishing design theories and frameworks in geotechnical and underground engineering.

2 CONNOTATION AND DEVELOPMENT OF RESILIENCE

2.1 *Concept and development of resilience*

The term resilience originated from the Latin "resiliere". Holling (1973) firstly proposed the concept of ecosystem resilience stands for an ecosystem with the ability to resist changes from the outside world and maintain the number of organisms and their relationships. The concept of resilience has since been extended to various academic fields such as ecology, materials science, psychology, economics, engineering, etc.

The concept of resilience was extended to the field of urban planning in the 1980s after that basic concepts such as "Resilient Cities" and "Urban Resilience" emerged (Hao and Xu, 2015). In the field of engineering, resilience is also understood to be related to its recoverability, that is, the ability of a system to functionally recovered from external disturbances.

The exploration of the eternal question of "how safe is enough to be safe" drives the continuous development of engineering design concepts. Shadabfar et al. (2022) believed that the engineering design concepts were started from allowable stress design (ASD), gradually developed to load and resistance factor design (LRFD), then performance-based design (PBD), and eventually resilience-based design (RBD). Among them, the ASD and LRFD concepts make the safety levels to be acceptable in practice using the structural reliability analysis method, without directly considering the performance of the entire system, nor the accidental disasters of the system and the balance between the possible losses and the current construction cost. As a result, the PBD concept was started to appear, apply and develop in the field of earthquake engineering, and gradually expanded from the displacement-based design under the action of the earthquake to the risk-based design which incorporates the repair costs, downtime losses, casualties, and other loss probabilities into system performance metrics. However, the PBD concept often does not consider the post-disaster recovery of the infrastructures, from which the design concept is further developed into the RBD concept.

Bruneau et al. (2003) from the Multidisciplinary Research Center for Earthquake Engineering (MCEER) at the University of Buffalo introduced the recoverability theory into the field of civil engineering and proposed a conceptual framework for earthquake resilience. MCEER believes that resilience includes at least eleven aspects (see Fig. 1) to achieve three aims of high reliability, low disaster consequences, and rapid recovery; four dimensions of technical, organizational, social, and economic dimensions; four evaluation indexes of robustness, rapidity, resources, redundancy. The robustness and rapidity are to evaluate the resilience effect. The resources and redundancy are used to evaluate recoverability (Bruneau and Reinhorn, 2006). Therefore, resilience is a more comprehensive evaluation system for system security.

In the field of earthquake engineering, the concept of resilience has received extensive attention (NEES/E-defense, 2010; Lv et al., 2017). The 16th World Conference on Earthquake Engineering was held in Santiago, Chile, with the theme of "Resilience - A New Challenge for Civil Engineering" and pointed out that the design and construction of Earthquake Resilient Structures is the basic requirement for realizing resilient cities (Lv et al., 2017).

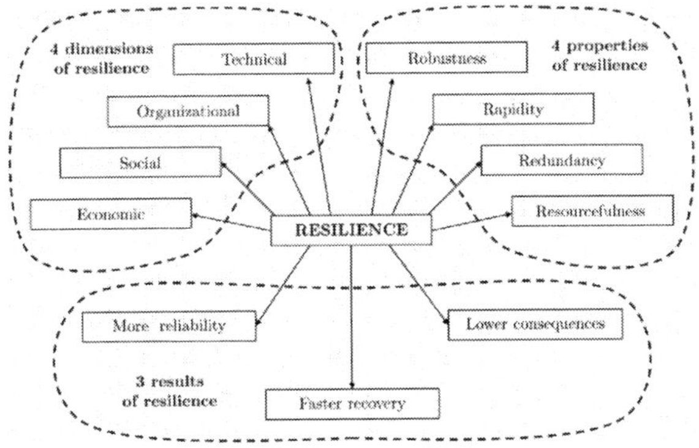

Figure 1. The targets, dimensions, and properties of resilience.

The concept of resilience has gradually extended from earthquake-resistant structures to resilient cities and resilient societies, with its connotations including robustness before a disaster, sustainability during the disaster, and recoverability after the disaster (Zhou, 2016). The robustness before a disaster mainly refers to the ability of the structure to withstand potential disaster and maintain its stability. Sustainability during disasters mainly refers to the ability of the system to self-adjust and adapt during disasters or to control and mitigate disasters under human prevention and intervention. The recoverability after a disaster refers to the ability to artificially restore the original state or achieve a new balance state to prevent future disasters. The problems to be solved for resilient cities and societies include not only earthquakes but also all unexpected events and sudden disasters, such as climate change, natural disasters (storms and floods, etc.), wars, and even epidemics of infectious diseases.

2.2 *Resilience assessment indicators and methods*

The connotation of resilience is larger than that of risk, and resilience assessments include not only risk assessment but also recovery assessment. A resilient system can deal with unexpected disasters in two ways: absorption and recovery. The studies believe that resilience includes four performance indicators, namely robustness, rapidity, resource, and redundancy, see Fig. 1 (Shadabfar et al., 2022). The robustness, resource, and rapidity play their roles in different stages after accidental disasters. The robustness is mainly reflected in the absorption stage. The higher the robustness, the less likely the damage will develop. The higher the resources and rapidity, the less the recovery time to restore the system performance, as shown in Fig. 2. Redundancy is also considered an important feature of resilient systems. When the redundancy is high, the system has spare components or force transmission paths for damaged components in unexpected disasters, thereby reducing the consequences of disasters and reducing the impact of disasters on the overall performance of the system.

In most projects, the system resilience when an unexpected disaster occurs should be evaluated in the dimensions of space and time. For example, Bruneau et al. (2003) pointed out that the system resilience index R can be used as the envelope of the performance curve after the disaster occurs, such as the shaded area under the integral of the performance curve Q(t) versus time (t) curve as shown in Fig. 2. In different fields, many scholars have proposed different definitions of the performance curve Q(t), including deterministic performance indicators and probability-based performance indicators. The used calculation methods of the toughness indicator R are similar to that shown in Fig. 2.

In the field of geotechnical and underground engineering, Huang and Zhang (2016) proposed a toughness evaluation method for shield tunnels under extreme loads (such as ground overload), using the lateral convergence of the tunnel as the tunnel performance indicator. The resilience index Re is defined as the ratio of the integrated time-history curve of the performance index under extreme overloading to that under normal overloading. They pointed out that the toughness index Re of the tunnel can be heavily improved if the tunnel performance recovery period is shorter.

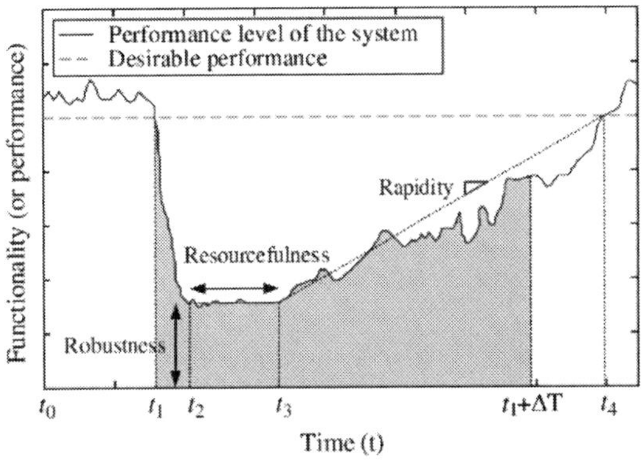

Figure 2. Schematic to calculate the system resilience index (Shadabfar et al., 2022).

2.3 *Framework of resilience design*

Two limit states of the structure in structural and geotechnical engineering, such as the serviceability limit state and ultimate limit state, are widely used for designing. But in fact, engineering structures are often facing the relatively serious large-scale progressive failure problem. The damage has the characteristics of "progress" and "disproportionality", more like the "domino" phenomenon. The subsequent damage follows the initial damage, and the scope of the final damage is much larger than the initial one (Ellingwood and Leyendecker, 1978). Breen (1975) pointed out that the design to resist progressive collapse is "an advanced limit state" for building structure, that is, consider the overall safety limit state of the structure (the limit state of preventing progressive failure) based on the serviceability and ultimate limit states.

According to the above-mentioned limit states, resilience can comprehensively evaluate the safety of the system. However, the present design methods in various engineering structures can only guarantee the resilience of the structure to a certain extent or from a certain aspect. From the perspectives of whether to consider the probability of failure, the ability to prevent continuous damage in unexpected situations, and the ability to recover after accidental disasters, the existing design methods to ensure the resilience of engineering systems can be generally divided into the following four levels: (1) the permissible stress design method using the factor of safety to prevent single or sub-item of the components and structures from reaching their ultimate limit states; (2) reliability-based design method using the theory of probability and statistics to evaluate the ability of the structure to complete the predetermined functions within the specified time and under the specified conditions, and predict the failure probability under the predetermined failure mode; (3) robustness method refers to the design method of ensuring the engineering structures withstand disproportionate collapse induced by unexpected events, or have the ability to resist progressive collapse which can also be called the resilience of the structure against progressive failure; (4) performance recoverable method refers to the abilities of the engineering structures to withstand extreme conditions, and to quickly restore their original service performance through reinforcement and engineering measures. The author believes that the above four levels of safety performance design methods constitute the system resilience of engineering structures in geotechnical and underground engineering.

3 RELIABILITY DESIGN IN GEOTECHNICAL AND UNDERGROUND ENGINEERING

3.1 *Uncertainties in geotechnical engineering*

The uncertainty of geotechnical engineering mainly includes the following aspects: (1) uncertainties of soil properties and their profile distributions; (2) uncertainties of analytical models, such as constitutive and analytical models; (3) uncertainties of surcharges and resistances; (4) uncertainties in construction procedure; and (5) uncertainties of boundary conditions, etc. Among them, the most basic feature of geotechnical engineering is the uncertainty of soil parameters, which originated from natural variabilities of soil properties and random test errors, insufficient test numbers, test methods, and field differences. Soil parameters have not only their variabilities but also spatial variabilities which can be described by the random variable model (Lumb, 1966) and the random field model (Vanmarke, 1977), respectively.

The reliability analysis methods used in geotechnical engineering are similar to those used in structural engineering. However, structure in geotechnical engineering is a complex system formed by soil, groundwater, and structure, and the complex interactions between them. The safety problems in geotechnical engineering have many unique characteristics in structural engineering, such as the large uncertainties of soil parameters, the complexity of the interactions between soil and structure, the soil arching effect and other unique stress distributed characteristics, various water-soil coupling effects such as large strain deformation, seepage, erosion, inrush, and instantaneous destruction. It is often accompanied by large deformation problems, soil erosion, soil collapse, and dynamic problems of instantaneous damage. Therefore, the safety evaluation and design of geotechnical structures are more complicated. In addition, the analyses of the interactions between soil and structure cannot separate the actions and resistances of the soil stratum. This is because the actions of the soil stratum are sometimes determined by its resistance, such as the effective earth pressure. The resistance of the stratum is sometimes determined by or related to the action of the stratum. For example, the bearing capacity of a shallow foundation depends on its applied surcharges.

3.2 *Permissible stress design method (Factor of safety method)*

The permissible stress design method (factor of safety method) is an important method for design and stability evaluation in geotechnical engineering. The parameters mostly used in geotechnical engineering are random variables with small changes that may lead to a different level of changes in soil properties. In the design of geotechnical infrastructures, it is often necessary to select a larger safety reserve to deal with possible deviations, and then use a factor of safety in the permissible stress calculation to express the degree of safety, which is also called the factor of the safety method. However, this method does not consider the uncertainty of soil mass which makes it hard to evaluate the reliability of the structures.

Take the stability analyses of slopes and foundations, for example, although the calculated factors of safety indicate that the infrastructures are stable, instability accidents often occur, such as the slope collapse of Liuhuangkou of the Ninghua Highway in Panzhihua City, Sichuan, China, in July 1997, the landslide accident at the entrance of the Dongronghe No. 1 tunnel of the Chengdu-Kunming Railway, in 1994 (Zheng 2007). An underwater slope with a height of 30 m, length of 75 m, and a slope ratio of 1:1 with a factor of safety of 1.25 collapsed during the LASH construction in the San Francisco Port in August 1970. The following reliability analysis using the Taylor series expansion method showed the failure probability of the slope was 18 % which indicated that the slope has a higher risk of instability failure. It also shows the limitations of using factors of safety method to evaluate stability problems in geotechnical engineering. The reliability analysis can reveal engineering problems from another aspect.

3.3 *Reliability-based design method*

In the late 1960s, researchers started to conduct reliability studies in the field of geotechnical engineering based on those methods used in structural engineering. Casagrand (1966) analyzed the risk problems in the design of foundation and geotechnical structures which starts the reliability analyses in geotechnical engineering. Lumb (1970) studied the probability distribution forms of the shear strength parameters and concluded that most of the physical and mechanical parameters of soil can be approximately described by the normal distribution function. Asaoka and Grivas (1981) analyzed the variability of undrained shear strength of cohesive soils.

Wu and Kraft (1970) applied the reliability theory to the typical geotechnical problems and the slope stability analyses. Alonso (1977) studied the uncertainty of the slope through a single probability-based slice method. Sivandran (1979) used the fixed value and the reliability analysis methods to study the slope stability of the embankment. The results showed that the reliability analysis method was more realistic.

Babu and Murthy (2005) analyzed the reliability of typical unsaturated soil slopes and concluded that the reliability index is better than the traditional factor of safety method for capturing the damaged area. Duncan (2000) compared the results from the safety factor method and the reliability analysis method and concluded the validity and superiority of the reliability analysis method in geotechnical engineering. He also pointed out that the reliability analysis is more suitable as a supplement to the safety factor method, rather than replacing it. Many scholars have discussed this finding. Cherubini et al. (2001) pointed out that Duncan's point is limited by current design methods. The reliability analysis is more urgently required when facing greater uncertainty problems in geotechnical engineering.

At present, the reliability analysis methods used in geotechnical engineering are various and constantly being developed. The classical reliability analysis methods include the first-order second-moment method, Monte Carlo (MC) method, stochastic finite element method, response surface methodology, Duncan method, and so on. Due to the limitations of classical methods, many scholars have conducted deep research on the reliability analysis for typical geotechnical engineering structures. For the tedious derivation problem in the first-order second-moment method, Xu et al. (2000) used a genetic algorithm to solve the equations which avoided the complicated derivation problem in varying degrees.

Aiming at solving the drawbacks of the MC method that the derived functions are hard to explicitly express and require a large amount of calculation time, Deng and Zhu (2002) proposed and demonstrated a Finite Element Monte- Carlo method using neural networks. This method can be directly used in the deterministic structural finite element analysis program which can greatly reduce calculation steps and calculation time to improve the efficiency of reliability analysis. Zheng et al. (2003) integrated the immune algorithm into the reliability analysis in geotechnical engineering and validate that using case studies. The method avoids the complicated derivation process for the nonlinear and complex functions. Ge (2007) compared the characteristics of the immune and genetic algorithms. He pointed out that the search target of the immune algorithm has the characteristics of dispersion and independence, while those of the genetic algorithm has the characteristics of singleness and exclusiveness. However, the optimization efficiency of the immune algorithm is lower than that of the genetic algorithm. Olsson and Sandberg (2002) introduced the Latin Hypercube Sampling (LHS) method to stochastic finite element analysis. They pointed out that LHS is more computationally efficient than the standard Monte Carlo sampling method. Zuo et al. (2013)

proposed a fourth-order moment method based on an artificial neural network and verified its effectiveness and accuracy.

3.4 *Robustness method*

In 2013, at the 18th International Congress of Soil Mechanics and Geotechnical Engineering (ICSMGE) held in Paris, France, the ideal geotechnical design standards for 2050 were discussed in the joint seminar of TC205. The robust design for geotechnical and tunnel engineering is mentioned. At the 19th International Conference on Soil Mechanics and Geotechnical Engineering held in Seoul, the joint seminar of TC205 and TC 304 have been identified two principal types of robustness by the group: (1) the ability of the final design to accommodate events and actions that were not foreseen or consciously included in the design; and (2) the sensitivity of the final design to variations of the known parameters within their anticipated range of uncertainty (ISSMGE, 2017). For the first type of robustness, there are many research results in the fields of structure and bridge engineering but small in the field of geotechnical engineering. This section mainly focuses the research development on the second type of robustness.

The reliability-based design method needs to determine firstly the random distribution of soil parameters, and then calculate the probability of failure. The random distributions of soil parameters cannot be accurately determined due to sampling and statistical errors because the variabilities of geotechnical parameters are often overestimated or underestimated (Juang et al. 2013). In addition, the model error and the uncertainty of the construction level are also difficult to be calibrated. Based on the inaccurate statistical law of input parameters, the results of the reliability design will be affected (Huang et al., 2014). This can further lead to over- or under-reliable reliability. The robustness analysis method can consider the variability of the standard deviation of soil parameters, and effectively evaluate their response to the sensitivity of the system.

The robustness design was first proposed by Taguchi (1986) for quality controlling design in the field of industry. The idea is to reduce the sensitivity of the system to the difficult controlling design parameters by adjusting the easily controlling parameters, see Fig. 3. Juang et al. (2012) introduced this concept to the infrastructure design in geotechnical engineering. Compared to the reliability design and factor of safety design methods, the robustness design method can more reasonably reflect the safety level and performance of the infrastructures in geotechnical engineering.

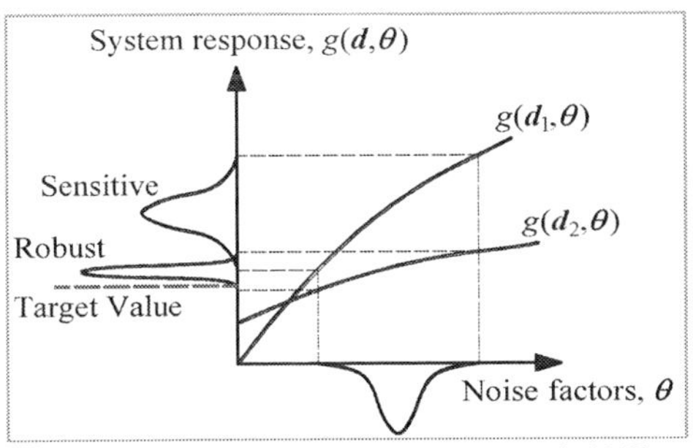

Figure 3. The core idea of robustness design (Juang et al., 2012).

The standard deviation of failure probability was used as the measurement criteria of robustness in the earliest robustness design method in geotechnical engineering (Juang et al., 2012; Huang et al, 2014). This method can also be called a feasibility level-based robustness index, which means the probability (or feasibility) that meets the design requirements (the target failure probability in reliability design) is taken as the robustness index, among which the higher the feasibility level of robustness, the more robust the design. Although this method is mathematically accurate, it requires too much calculation time. This is because the method used the first-order reliability method to calculate the failure probability, the point estimation method to calculate the standard deviation of the failure probability, and the multiple objective optimization algorithms to find the optimal design with the premise of robustness and cost control. Therefore, it is necessary to conduct more studies to further improve this method.

Gong et al. (2014) proposed a more efficient robustness design framework by adopting a gradient-based robustness measurement method. The system responses to the uncertainty of the inputs (i.e. the normalized gradient) are mainly used as the robustness level. The higher the sensitivity the less robust the design. Khoshnevisan et al. (2015) proposed a robustness design method that can consider simultaneously the performance requirements, design robustness, and cost-efficiency. The method reflects the robustness of the design through the variabilities of the function with its form of transforming and the performance requirements of the system calculated by the reliability. The robustness design can be obtained from the same calculation step or the so-called coupling calculation. The robustness design can be carried out efficiently using this coupling method. Peng et al. (2017) used the Monte-Carlo-based weight technology to calculate the failure probability of the geotechnical structures and to replace the multi-objective optimization with a series of single-objective optimizations. The method is more efficient and convenient for the robustness design in geotechnical engineering.

Huang et al. (2014) used the multi-objective optimization method and the confidence level to calculate the variability of the structural failure probability of the gravity retaining wall. They optimized the wall body through parameter variability and failure probability analyses. In the field of shield tunneling, Zhang et al. (2019) found the optimal control knee point in the design of steel plate reinforcement in large deformation tunnels through robustness design to optimize its robustness and cost. Wang (2019) applied the robustness design concept to the cross-section design of double-lined water-conveying shield tunnels and a large-diameter shield tunnel to optimize the structural design.

It should be pointed out that the concepts of robustness design currently used in geotechnical engineering are slightly different from those used in the field of structural engineering. The latter emphasizes the ability of the structure to resist progressive collapse in the event of local failure (Ye et al., 2008). The former mainly comes from the inability to accurately obtain the variability of geotechnical parameters. The uncertainty of parameter variability is usually considered to enable a reliability assessment of the failure probability in geotechnical engineering. However, local failure of soil and structures occurs frequently in geotechnical engineering which even leads to large-scale progressive failure. The research and evaluation of the robustness in geotechnical engineering should expand the connotation and extension to avoid progressive failure caused by local failure. The evaluation and control methods of the anti-progressive collapse ability of the geotechnical and underground engineering structures must be established to provide reasonable abilities to prevent progressive failure of the geotechnical structures. Reasonable control of the degree of damage provides a basis and guarantee of the functional maintenance, recoverability, and ease of repair after the progressive failure, even if the geotechnical engineering has a reasonable resilience.

3.5 *Performance recoverable method*

Recoverability is another type of resistance in structural engineering. A very important aspect of structural resistance is its recoverability after earthquakes which refers to the structures that could be restored to their designed functions without repair or a little repair after an earthquake (Lv et al., 2011). Compared to the traditional earthquake-resistant structures, it has better performance in protecting people's lives and properties during earthquakes and in rapid recovery after earthquakes. The function of restorable structures includes rocking structures, self-centering structures, repairable structures with replaceable members, and so on. In addition to these methods of using structural members, Anastasopoulos et al. (2010) proposed the concept of rocking isolation to allow the foundation to be lifted off, transfer the plastic hinge originally formed in the superstructure to the foundation soil, reduce or avoid plastic deformation or even collapse of the superstructure to ensure the safety of the superstructure (Wang et al., 2021).

Because the structures in geotechnical and underground engineering are deeply buried in the ground, it is difficult to repair once damaged. For example, a completed subway tunnel constructed by the shield method may be misaligned between the segments due to earthquakes, ground stacking, excavation of adjacent excavations, or tunnel leakage. If the excavation opening is large enough, it may also cause groundwater leakage or even lead to a large amount of soil gushing into the tunnel, as shown in Fig. 4(a), and even cause severe deformation, subsidence, misplacement, damage, or even collapse of the tunnel, as shown in Fig. 4(b). Figs. 4(c) and 4(d) show the leakage at the starting shaft of the tunnel resulting in a large amount of groundwater and sediment gushing into the tunnel. The tunnel segment was severely damaged because the ground outside the starting shaft sunk and cracked. After the accident, tens of meters of the tunnel could not be repaired but had to be excavated and rebuilt.

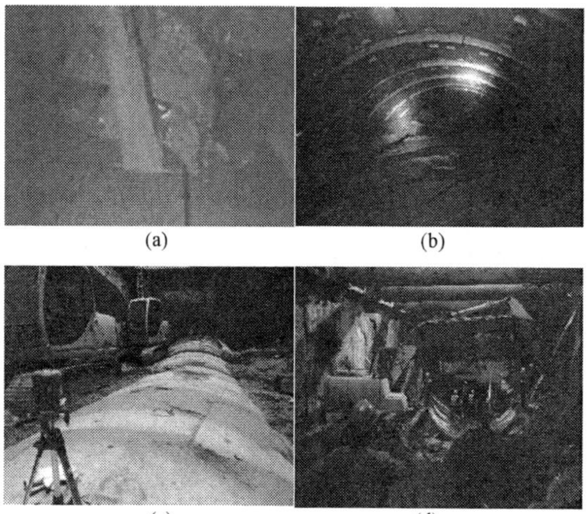

(a) (b) (c) (d)

Figure 4. Damage and repair process of shield tunnel accident (a) groundwater and soil gushing into the tunnel, (b) segments settling and misplacement, (c) tunnel settlement, (d) excavation of the damaged tunnel.

It is necessary to conduct the performance recoverable design methods for the shield tunnels, especially in (1) high construction risk sections such as the originating and receiving sections, and connecting passage sections, (2) the high geological risk sections, such as boundaries between soft and hard soil layers, ground fissures, (3) sections passing through extremely important infrastructures, such as the operating tunnels, high-speed railways, rivers and lakes, and other high environmental risk sections with extremely serious damage consequences in the event of an accident and so on. The performance recoverable design is to make the degree of large deformation or tunnel damage controllable, self-recovered, or easily repaired after the disaster caused by unexpected events such as water leakage, sand leakage, earthquake, and so on.

For the recoverability of geotechnical and underground engineering, the rapid repair technologies for damaged geotechnical and underground engineering structures are an important way to improve engineering toughness in addition to improving the recoverability of the engineering structure itself during the recoverable design stage. In 2016, Huang and Zhang (2016) reported a case of recovering the deformation of the tunnel by grouting on both sides of a shield tunnel that was damaged under extreme ground overload four years ago. They pointed out that the grouting greatly improved the tunnel performance. If the tunnel deformation can be recovered earlier and faster (shorten the tunnel performance recovery period), the toughness index can be greatly improved. In terms of active grouting and quick repair of tunnel deformation, Zheng (2021) pointed out that the grouting efficiency and recovery effects of the traditional grouting methods, such as sleeve valve pipe low, are poor for sandy soil with higher permeability or soft clay with strong structure. The active control technology of balloon expansion and deformation grouting method can achieve a better recovery effect.

Wu and Ou (2014) concluded through model tests that the shield tunnel segments connected by shape memory alloy (SMA) bolts can reduce the amount of joint opening. The Shantou Bay Tunnel, the first large-diameter sea-crossing tunnel in the eight-degree earthquake-resistant area in China with a total length of 6.68 km and a section of 3.05 km under the sea, has adapted the memory alloy bolts to connect segments to improve its self-recoverable abilities after earthquakes.

(a) Bending failure (b) Pull out to destroy (c) Bending shear failure (d) Shear failure

Figure 5. Photos of typical pile damage after the earthquake (Washi et al. 2003; Qingdao et al., 2004; Zhang et al., 2018).

In previous major earthquakes, it was found that many piles were damaged in varying degrees, especially at the pile caps and connections between piles and pile caps which are weakest (Huang et al., 2002, Tang, 2010). Take the 1995 Hanshin Earthquake, for example, the piles of the Hanshin Viaduct were bent and damaged which made some parts of the pile foundation separated from the bearing platform as shown in Fig. 5(a) to 5(b). In Kobe City, the action of inertial force made the top of the pile foundation of the high-rise apartment buildings suffer bending shear failure as shown in Fig. 5(c). In 1989, Loma Prieta earthquake made the pile foundation suffer shear failure near the pile top due to lateral displacement of soil as shown in Fig. 5(d). The historical earthquake damage indicated that the pile-cap joint can be damaged by the earthquake-induced soil deformation, or the bending, shearing, and uplifting caused by the inertial action of the superstructures (Zheng et al., 2013). As the pile foundation is designed to support the superstructures and is difficult to be repaired after damage, its seismic performance has received extensive attention.

The seismic capacity of the pile-cap connection joint has received considerable attention. The traditional connection nodes between piles and caps are often treated as "fixed connections" where large internal forces were generated under the action of earthquakes. As shown in Fig. 6, the bending moment of the pile top under the embedded condition is significantly larger than that under the free pile top condition. The joints are prone to damage when the pile top is fixed and the post-earthquake repair is difficult. With the development of seismic design, the rapid recovery of functions has become an important target for toughness design. However, it is difficult to ensure the rapid recovery of structural functions after earthquakes only by improving the seismic capacity of joints. To improve the recovery of the pile foundation, some scholars have proposed a rocking self-resetting structure (Guan et al., 2018; Antonellis, 2015; Ha and Kim, 2014; Ko et al., 2019). Since the pile-cap connection determines whether the system is allowed to sway and reset, the design of the pile-cap connection becomes an important part of determining the stress and deformation mode of the pile-superstructure system during earthquakes. Therefore, the seismic design of the pile foundation-cap connection is one of the key issues for the toughness improvement of the pile foundation.

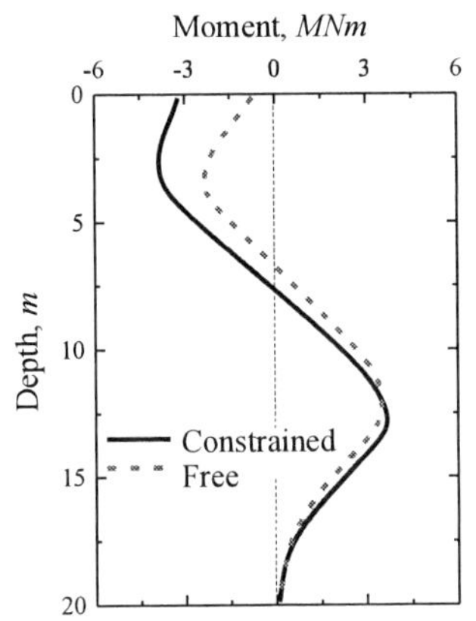

Figure 6. Influence of constraint conditions on bending moment distribution along with pile (Finn, 2015).

Studies in literature focused on the influence of anchoring types, steel bars, and filling materials on the seismic performance of the pile and pile-cap connection. Washi et al. (2003) proposed a simple connection between the pile and pile head to investigate the effects of anchoring types of reinforcement. A series of laboratory model tests are conducted to compare the effects of the proposed simple connection nodes without anchor reinforcement. They concluded that the proposed connection could reduce the rotational constraint by 70% more than those of traditional connections. Qingdao et al. (2004) proposed a method of using the conical gaps on the pile head to improve the rotational ability of the connection between piles and pile heads which can effectively reduce the bending moment at the pile head, see Fig. 7. Some scholars believed that strengthening the nodes is beneficial to improving the bearing capacity and energy dissipation capacity of the nodes. Zhang et al. (2018) and Larosche et al. (2014) studied the effect of embedded depth on the seismic performance of the pile and concluded that deeply embedded joints have a good seismic performance to effectively improve the stiffness, bearing capacity, and energy dissipation capacity of the nodes. However, because strengthening the nodes also similarly increases the surcharge onto the nodes, some scholars believed that weakening the connection of the nodes to a certain level is beneficial to earthquake resistance. The stress state of the nodes can be improved by installing filling materials at the connection between the pile head and the cap (Lehman et al. 2013).

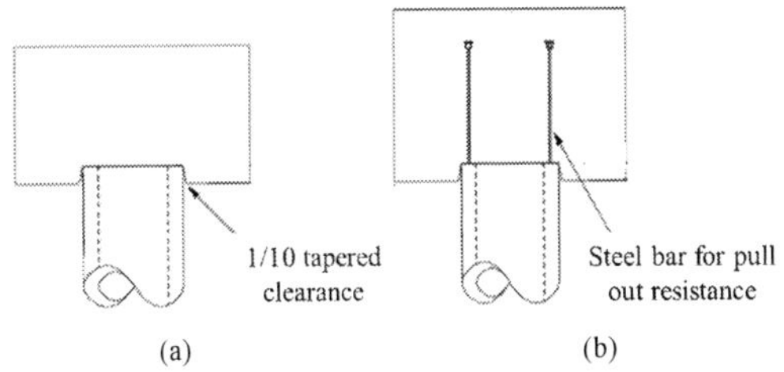

Figure 7. Improved simple connections (a) bearing pile, and (b) tension pile.

To improve the recovery toughness of the pile foundation, Guan et al. (2017, 2018) proposed a liftable pile group joint for bridges supported by pile-group foundations and verified using quasi-static tests. Compared with the traditional fixed pile group joint, the connection point of the liftable pile group joint can effectively reduce the residual displacement of the bridge structure, see Fig. 8, and the damage to the pile foundation. Ha et al. (2014) and Ko et al. (2019) studied the impact of pile foundation-cap connection on the superstructure and pile foundation through centrifuge tests. Disconnecting the pile and cap can effectively reduce the internal force of the pile body, reduce the structural responsiveness of the superstructure, prevent damages to the pile foundation and superstructure, and improves their recoverability. Antonellis (2015) studied the seismic response of swaying pile foundation bridges through numerical simulation. He found that, compared with the fixed foundation, the inelastic response of the upper structure under the dynamic action will decrease significantly when the cap on the upper part of the pile foundation is allowed to sway. All of this could help to improve the overall recoverability of the structure.

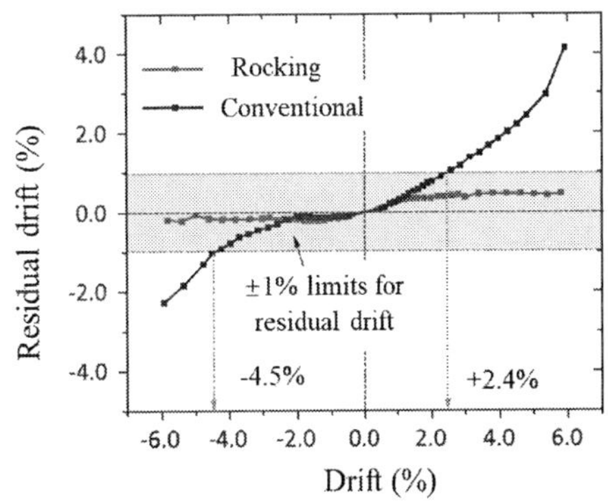

Figure 8. Residual deformation (Guan et al., 2017)

In general, studies in the literature improved the resilience of the connections between the pipe and pile cap to prevent pile head damage and improve the overall recoverability. For example, laminated rubber, spring, or other materials with better deformation properties are commonly installed within the region of the embedded pile in the pile cap to improve the rotational restraint of the pile cap and energy dissipation capacity of the joint, avoid damage to the pile and the bearing platform concrete and recover the position between the pile and the bearing platform after the earthquake. As shown in Figs. 9 and 10, the authors conducted a series of quasi-static laboratory tests to investigate the influences of the pile-cap connection on the bearing capacity of the pile foundation. The pile is made of an AB-type PHC pipe pile with a cracking bending moment of 138 kN·m and a bending capacity of 238 kN·m. The conventional connection and that installed with a rubber O-ring are used in test No. CT1 and CT2, respectively. The pile in test No. CT3 changed the angles of the anchored steel bar to reduce the resistance of the bending moment and the rotational constraint of the node. The hysteresis curves and secant modulus curves of the piles with the pile-cap connection using the three types of connections are shown in Figs. 11 and 12, respectively. Compared to the conventional connection method used in test No. CT1, the horizontal bearing capacity of the pile with Test No. CT2 and CT3 were reduced by about 23 % and 21 %, respectively.

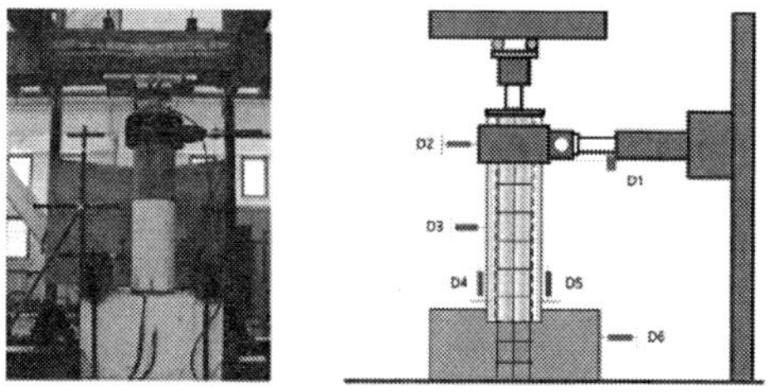

(a) (b)

Figure 9. Pile-cap connection in the quasi-static test (a) photo and (b) sketch of the model test setup.

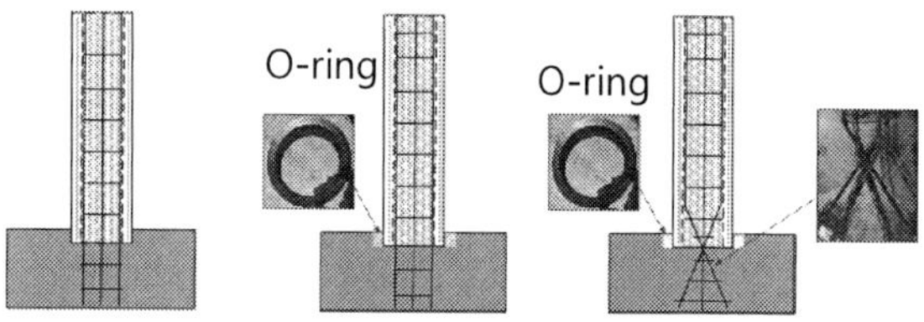

(a) (b) (c)

Figure 10. Three types of pile-cap joints (a) CT1, (b) CT2, and (c) CT3.

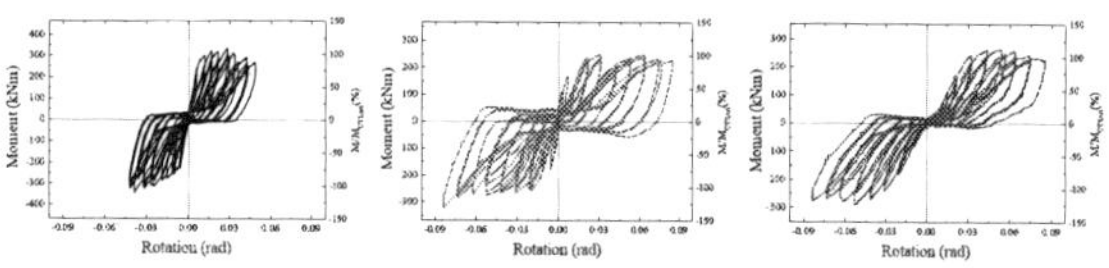

(a) (b) (c)

Figure 11. Hysteresis curve of the three kinds of pile-cap connections (a) CT1, (b) CT2, and (c) CT3.

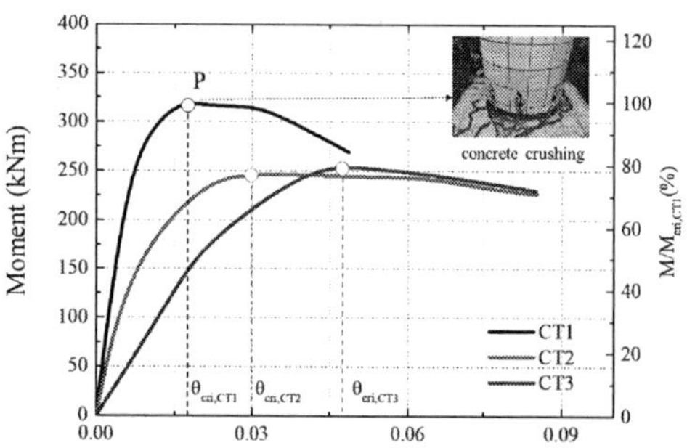

Figure 12. Secant modulus of the piles with three kinds of pile-cap connections.

The relationships between the bending moments and anchoring angles of the three-node connection forms in Fig. 12 show the peak bending moment of the pile with Test No. CT1 was 318kN·m when the angle of 0.018 rad. The concrete of the pile was significantly crushed at this time, accompanied by slight cracks on the pile surface. For the same angle, the bending moments of the pile with Test No. CT2 and CT3 are 216kN·m and 149kN·m which are smaller than those of Test No. CT1 of 32% and 53%, and are of 85% and 55% of their corresponding peak bending moments, respectively. This is indicating that the piles with Test No. CT2 and CT3 have larger strength recoverability than that with Test No. CT1.

The failure modes and ductility of the three types of node connections are analyzed through the bending moments versus the anchoring angles curves. When the peak bending moment is reached, the bending moment of the conventional connection is used in Test No. CT1 significantly decreased with its slope of 0.439kN·m/rad. The pile is brittlely damaged with the slopes of the bending moment of the components in Test No. CT2 and CT3 reduced to 0.109 and 0.170kN·m/rad, respectively. The components with the rubber O-ring can effectively maintain their bearing capacity and significantly increase their ductility. Compared to the conventional connection used in Test No. CT1 with its limit angle of 0.048 rad, the components used in Test No. CT2 and CT3 have not reached their limit magnitudes even with their limit angle of 0.086 rad.

In the later loading stage of model test No. CT1, the bending moment of the pile is not the controlling failure criteria because the concrete of the pile was significantly crushed accompanied by slight cracks on the pile surface. The model test results from No. CT2 and CT3 showed that the installation of the rubber O-ring could effectively increase the limit rotation angle of the pile head and improve the connection resilience of the pile cap by avoiding damage to the concrete at the pile head when the angle is large. In engineering practice, for the B-type PHC pipe pile with a higher flexural bearing capacity, i.e. cracking moment of 164 kN m, flexural bearing capacity of 311 kN m, the connection using those of test No. CT2 and CT3 could effectively improve the connection resilience of the pile cap by avoiding damage to the concrete at the pile head when the angle is large. For the A-type PHC pipe pile with low flexural bearing capacity, i.e. the cracking moment of 118kN m, the flexural bearing capacity of 176 kN m, the bending failure of the pile using the connection of test No. CT1 was occur. For the pile with node connections as those of test No. CT2 and CT3, the resilience of the pile is improved by avoiding the bending damage to the pile body at a large turning angle. For the A-type PHC pile filled with the core material, the ultimate bending resistance is increased.

The equivalent stiffness and equivalent damping ratio versus loading series curves of the pile with the three types of pile-cap connections are shown in Fig. 13. The loading series is the ratio of the actual loading displacement to the yield displacement. The equivalent stiffness is the ratio of the applied horizontal force to the horizontal displacement. It shows that with the increase of loading series, the structure is damaged with the equivalent stiffness reducing slowly. The pile with connections as test No. CT2 and CT3 reduced the equivalent stiffness of approximately 40%-50% under different loading series compared to the pile with conventional connection as test No. CT1. Jiang et al. (2014) concluded that the decrease of the restraint stiffness of the pile cap is beneficial to reduce the bending moment and the damage to the pile cap. The variation of the equivalent damping ratio versus the loading series curves for the pile with three types of connections is shown in Fig. 9(b). It shows that the equivalent damping ratio is related to the energy dissipated in the plastic deformation. For the pile with the conventional connection as that of test No. CT1 has relatively large damping due to the large plastic damage to the concrete. The damping of the pile with connection as that of model test No. CT2 is relatively small for the first two loading stages and reduces approximately 20% more than that of the pile with conventional connection as that of test No. CT1.

Since the damage to the pile-cap connection is difficult to be directly discovered, evaluated, and repaired during the earthquake, the pile-cap connections as shown in Fig. 10(b) and 10(c) are simple, versatile, and durable with higher resilience. Compared with the pile with conventional connection, the pile with new connections has advantages of higher redundancy, ductility, and reduction of pile top bending moment which is helpful to promote the seismic resilience of the pile.

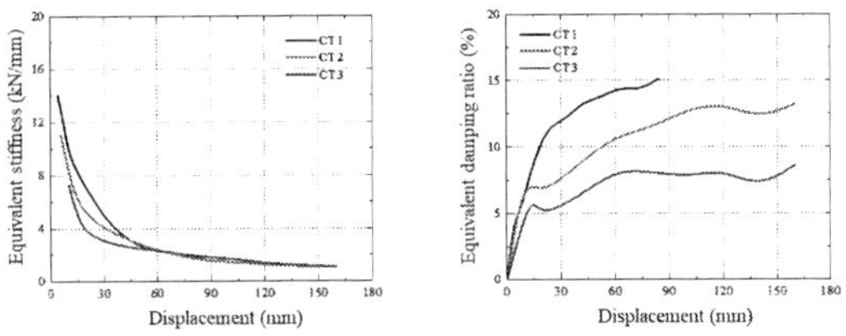

(a) Equivalent stiffness (b) Equivalent damping ratio

Figure 13. Equivalent stiffness and damping ratio versus loading series curves of the pile with the three types of pile-cap connections (Yang et al., 2014).

4 RESILIENCE AGAINST PROGRESSIVE FAILURE IN EXCAVATION ENGINEERING

4.1 *Limitations of existing design methods*

The retaining system of the deep excavation is composed of vertical piles (walls) and horizontal struts (anchors). The integrity of the retaining system is relatively weak due to the following reasons: (1) weak connections between vertical structures which are composed of rows of retaining piles and connected only through the capping beam on the pile head. Even if the diaphragm walls are installed, the connections between wall sections are still weak. (2) little mutual connections between the horizontal struts which are often made of steel pipes with no or weak connections. It is also true when using anchors as there is no connection between the anchors even when waler beams are installed. (3) weak connection between the vertical structure and the horizontal struts (anchors) because they are designed by assuming only bearing compression forces. In addition to the uncertainties of geotechnical engineering, the retaining system of the deep excavation has lower integrity than the superstructures of structural engineering.

In recent years, several collapse accidents of the deep excavation have occurred in Singapore (Xiao et al. 2009, 2010; Artola 2005; Whittle and Davies 2006), Hangzhou (Zhang and Li, 2010; Li and Li 2010), Cologne (Haak 2009) and other places. The analyses of these accidents showed that the instabilities of the deep excavation are complex failure processes, especially in the case of deep excavation, irregular plane shape, and complex supporting structures. The damage may be induced by local failure from retaining structures or stability failure of soil mass which in turn causes the retaining structures to collapse on a large scale. That is, the collapse of a deep excavation is a progressive failure process (Zheng et al. 2014, 2011, 2014; Cheng et al. 2016, 2017).

The retaining system of a metro excavation collapsed in Singapore, causing huge economic losses and serious damage to Nicoll Road (Xiao et al. 2009). The ninth level of struts first yielded, and the load transferred to the adjacent eighth level of struts, which failed in a "domino-like" progressive failure in the section. This led to the partial collapse and eventually resulted in substantial damage to an approximately 100 m section of the Nicoll Highway carriageway adjacent to the abutment of Merdeka Bridge (Xiao et al. 2009, 2010; Artola 2005; Whittle and Davies 2006). The excavation accident that occurred at Xianghu Station of Hangzhou Metro Line 1 was also due to the large-scale collapse induced by local failure (Zhang et al. 2010; Li and Li 2010). The collapsed length in this accident was up to 70 m, as shown in Fig. 14.

(a) (b)

Figure 14. Retaining system of deep excavation and its failure at Xianghu Station of Hangzhou Metro Line 1 (a) Excavation retaining system (b) Failure scene of the retaining system.

The current stability design theories simplified the excavation into a two-dimensional plane strain problem to calculate the stabilities of several cross-sections without considering the progressive failure induced by the failure of the excavation in the latitudinal direction (Liu et al. 1997; Chang 2000; Ukritchon et al. 2003; Zhang et al. 2006; Hsien et al. 2008; Huang et al. 2008; Liu and Wang 2009; Cheng et al. 2015). However, the excavation is a complex interactional system between supports, soil, groundwater, and structures (including piles) inside and outside the excavation. The damage transmission cannot reflect the development mechanism and process of progressive failure caused by local failure. In terms of the overall safety performance of the excavation, evaluation and control methods have not been systematically established. The existing reliability and robustness analyses in geotechnical and underground engineering cannot reflect the resilience performance of excavations against progressive failure developed under extreme accidental conditions.

4.2 *Progressive failure mechanism*

Local failure, progressive failure, and overall safety performance in excavation engineering have attracted more and more attention from scholars and engineers. Goh and Wong (2009) found through numerical analyses that the accidental failure of one to two levels of struts would not induce failure to the entire retaining system under the condition that the struts could provide enough compressive bearing capacity. Zheng et al. (2011, 2014) conducted numerical simulations and case studies and concluded that local failure can easily lead to progressive failure in excavation with a lower redundancy of the retaining system. Redundancy design theory is suggested for the design of excavations. The redundancy problems of the retaining system in excavation engineering are classified with corresponding design methods introduced to prevent progressive failure, such as increasing the force transmission path and reinforcing at certain intervals, and so on. Pong et al. (2012) pointed out that the design of retaining walls should have sufficient structural safety, robustness, and redundancy, which can avoid the progressive failure of structures induced by the overloaded failure of a single component. Three-dimensional and two-dimensional plane strain analyses were also carried out for the failure of the single strut. Itoh et al. (2016) simulated the failure of the anchor head of the anchored pile-supported excavation through a centrifuge test and concluded that both earth pressures in the active and passive zones increased during the excavation process. They also found that the diaphragm wall collapsed instantly once the anchor head exceeded its tensile strength. Goh (2018) conducted numerical analyses to study the failure problems of the multi-channel struts and pointed out the transferring path and percentage of the failure load. Zhao et al. (2018) summarized the influencing factors of the anchor failure in the excavation, analyzed the influence range of single anchor failure, and pointed out that the most dangerous position is at the pile head or near the bottom of the excavation. Han et al. (2018) took an anchored pile retaining deep excavation in a sandy soil layer as the engineering background to simulate the failure process of the single and two-level anchors. Xia et al. (2018) simulated the multi-level braced excavation, adopted the concept of stiffness (deformation) redundancy, verified their influence through the removal of struts at different layers, and back-calculated the redundancies of the structures.

Lu and Tan (2019) summarized typical excavation failure cases in China in the past 30 years and divided these failure cases into 15 failure modes. Choosrithong and Schweiger (2020) analyzed the strut failure problem using Plaxis software and found that the surcharge of a strut before failure was transferred to its adjacent struts. Ser and Sayin (2020) analyzed the reason and mechanism of the collapse of a certain tied-back excavation project and proposed control and improvement methods. Zheng et al. (2021) studied the load transfer mechanism of an anchored pile retaining system caused by local failure through numerical simulations.

The case studies of the deep excavation showed that the overall safety limit state (the limit state of preventing progressive failure) is equally important to the traditional ultimate limit state and the serviceability limit state. In recent years, Zheng et al. (2016, 2017, 2020), Han et al. (2018), Xia et al. (2018), and Cheng et al. (2016) have conducted a series of numerical and large-scale model tests to investigate the development of progressive failure along with the depth, width, and length of the excavation caused by local failure. They have preliminarily revealed the development and termination mechanisms of the progressive failure of cantilever contiguous piles, internal struts, and anchored pile retaining excavations. The studies laid the foundation for further research on resilience design theory in the field of excavation engineering.

4.2.1 *Cantilever pile retaining excavation*

For the most typical cantilever pile retaining system, Cheng et al. (2015, 2016) showed that the local failure of the retaining piles forms a significant soil arching effect in the active zone. The earth pressures are transferred to the adjacent intact retaining piles, resulting in a rapid increase in the soil pressure and internal force. The peak increase ratio of the bending moment in an adjacent pile induced by partial collapse is defined as the load transfer coefficient to evaluate the increase in the internal force and check the safety after the local failure occurrence (Cheng et al. 2016, 2017).

Zheng et al. (2016) and Cheng et al. (2016) proposed basic criteria to predict the occurrence of resilience failure only if the load transfer coefficient is larger than the factor of safety of the bearing capacity of the adjacent piles. The large-scale model test was conducted to simulate the progressive failure process of the adjacent piles induced by the local failure of contiguous piles, as shown in Fig. 15. The model tests showed that the local collapse of the excavation induced the soil pressure and internal force of the adjacent pile to increase instantaneously, and then the soil outside the excavation slipped into the excavation, inducing the unloading of the active zone of the adjacent pile. The unloading process is relatively time-lagged. With the increasing range of local collapse induced resilience to damage, the collapse terminated naturally instead of endless collapse when the unloading of the soil insufficiently induced the soil arching effect to generate damage to adjacent retaining piles. For cohesive soil, it can be judged that the cohesive soil outside the excavation does not necessarily slip into the excavation when the retaining pile partially collapses. The subsequent unloading effect induced by the collapse of the soil will be significantly reduced and time-lagged with possible lower safety of the retaining system.

Zheng et al. (2016) and Cheng et al. (2017) found through numerical studies that local failure often occurs in the middle edge of the square excavation, and the maximum load transfer coefficient first increased and then decreased with the number of failed piles. The progressive collapse terminated when it reached a certain range. The spatial effect may be another reason for the natural termination of progressive collapse.

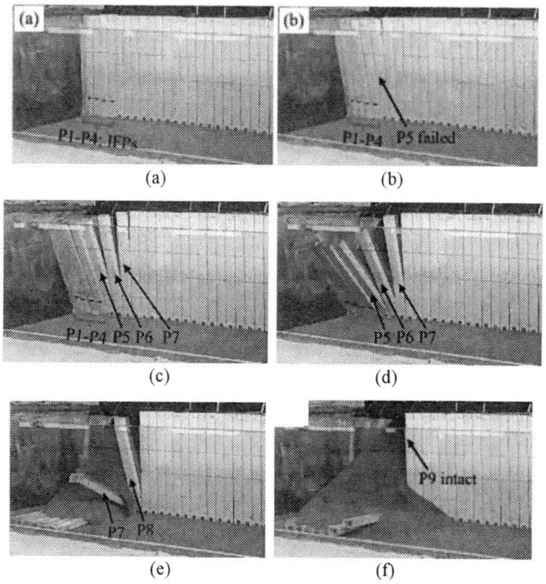

Figure 15. Collapse process after failure of four cantilever contiguous piles (Cheng et al. 2016) (a) before collapse (b) 2.0 s after the partial collapse, (c) 2.8 s after the partial collapse, (d) 3.2 s after the partial collapse, (e) 4.28 s after the partial collapse and (f) final stable status.

Even for the simplest cantilever pile retaining system, the mechanism of local failure-induced progressive failure is relatively complex. The initial failure range, soil strength and compressibility, capping beam, embedded depth of the retaining piles, and other factors will affect the occurrence and development of progressive failure (Cheng et al. 2016; Zheng et al. 2020). Assuming that the initial bending damage of the four piles occurred in the cantilevered pile, the embedded depth of the retaining pile heavily influenced the load transfer coefficients, as shown in Fig. 16. For the smaller embedded depth of the retaining pile with lower stiffness, the soil pressure transferred to the adjacent pile due to the redistribution of earth pressure, making the adjacent pile have greater additional displacement. This additional displacement further led to the redistribution of earth pressure to the other adjacent piles until stable deformation and stress redistribution were finally achieved. Therefore, the smaller the pile embedment depth is, the larger the influence scale of the internal force redistribution in the pile, the lower the load transfer coefficient of the adjacent piles when the local failure occurs, and the lower the possibility of progressive collapse (Cheng et al., 2016; Zheng et al., 2020). Conversely, the greater the embedded depth of the retaining pile with higher stiffness, the higher the possibility of progressive collapse.

The better the soil conditions are, the higher the lateral displacement stiffness provided by the soil in the passive area. After a local failure occurs, the load transfer coefficient of adjacent piles is higher, which indicates that progressive failure is more likely to occur under the same factor of safety or lower robustness and resilience.

Zheng et al. (2020) used the finite difference method to conduct a case study of an overturning and collapse accident of an excavation retained by double-row piles. The influence of local over-excavation on the internal force, deformation, and stability of the double-row piles is investigated. The ratio of the resisting moment to the overturning moment suffered by the pile during construction time was used to evaluate the stability of the piles. The mechanism of occurrence and development of progressive failure due to overturning failure was also studied. After local over-excavation, the overturning failure is mainly transferred outside the over-excavation area through the soil arching effect in the active zone and the loading transfer effect through the capping beam. The anti-overturning status value decreases instantaneously, with its minimum magnitude determining the occurrence and failure scale of overturning progressive failure. If the anti-overturning status value does not consider the load transfer effect of the capping beam, the range of overturning failure outside the over-excavation area will be seriously underestimated. For overturning progressive failure, the capping beam cannot reduce the risk of progressive failure but increases the degree of overturning outside the over-excavation area and expands the scale of overturning failure, which is different from that of bending progressive failure.

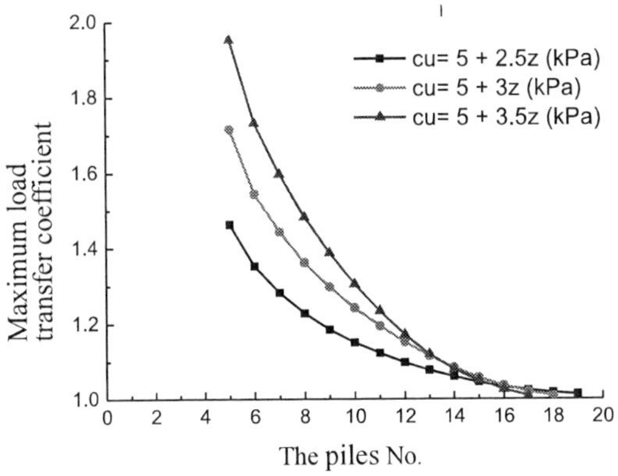

Figure 16. Load transfer coefficients after the failure of 4 piles (Zheng et al. 2016).

4.2.2 *Braced excavation engineering*

The progressive failure is often generated from a point of the retaining system and transferred along with the depth, latitude, and longitude directions of the excavation according to the case studies of the collapses of the deep, large, and long excavation.

(1) Progressive failure along the longitudinal direction
To further reveal the influence of local failure on the overall safety performance and the mechanism of progressive failure under complex retaining structures, Zheng et al. (2020a) studied the effect of local failure on the retaining system through model tests, as shown in Fig. 17. For the same excavation depth, the lateral displacements of the braced excavation were much lower than those of the cantilevered pile retaining excavation, with its lateral movement stiffness only provided by the soil in the passive zone. Therefore, the local failure induced greater additional earth pressure and internal forces in the retaining piles adjacent to the initial failure area, as shown in Fig. 18. After local failure, the redistribution of earth pressure behind the cantilever piles ranged significantly larger than those of the braced piles. When the released load due to strut failure cannot be transferred to the intact struts outside the failure area through the redistribution of internal forces but is concentrated on the adjacent struts as shown in Fig. 19, it can lead to strut failure due to these additional loads and then trigger widespread progressive failure.

(2) Progressive failure along latitude direction
Zheng et al. (2014) conducted a case study of excavation using the discrete element model to investigate the development and evolution process of progressive failure. The diaphragm wall on the left side of the excavation was sheared and broken due to over-excavation. A sliding surface was generated in the soil outside the excavation. The wall above the breakpoint of the diaphragm wall is rotationally deformed. This is because the

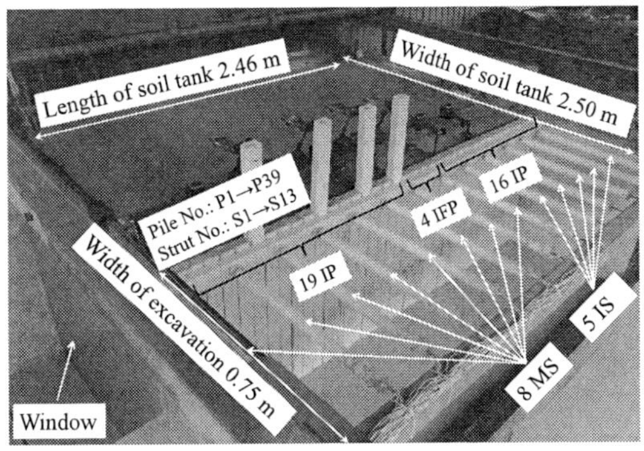

Figure 17. Platform of large-scale model tests and excavation model (Zheng et al. 2020a).

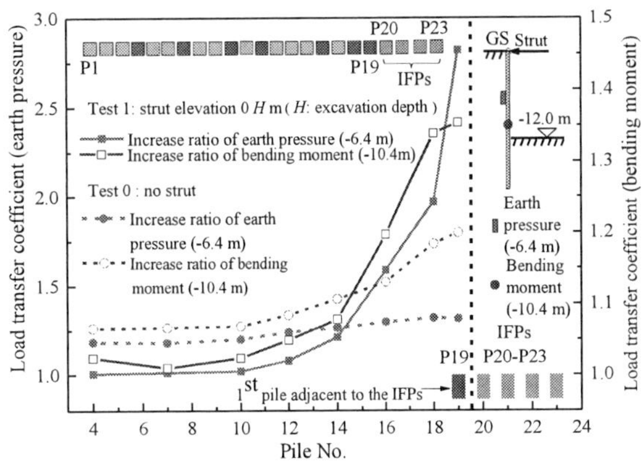

Figure 18. Comparison of earth pressures and load transfer coefficients of braced or cantilever contiguous piles (Zheng et al. 2020a).

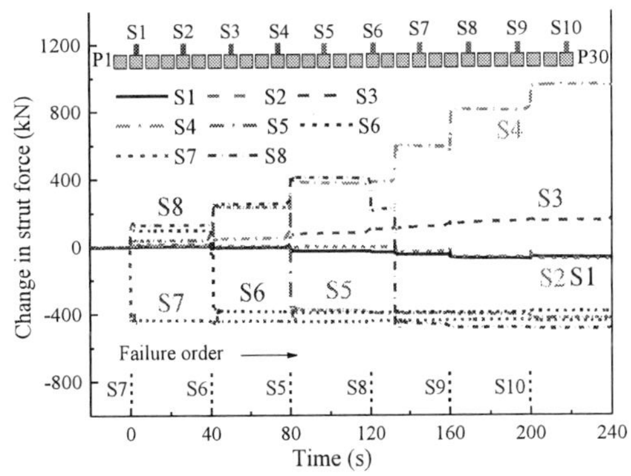

Figure 19. Time curves of strut forces with partial failure (Zheng et al. 2020a).

connection of steel struts and the diaphragm wall were designed as a compression node in traditional methods but did not effectively consider the displacement away from the struts, which made the struts quickly lose their efficiency. The diaphragm wall below the breaking point was transformed into a cantilevered pile and overturned into the excavation under the action of the earth pressure outside the excavation. As the diaphragm wall above the breaking point lost its efficiency, the diaphragm wall together with the soil mass slid into the excavation following the sliding surface through the breaking point. At the same time, the right wall turned into a cantilever structure with overturning failure. After the above-mentioned failure occurs in a section of the excavation, the progressive failure will further expand in the longitudinal direction of the excavation, resulting in large-scale collapse.

Zheng et al. (2021) conducted a case study on the metro excavation collapse of the Nicoll Highway carriageway in Singapore to investigate the progressive failure process of the retaining system induced by an initial failure under different ultimate bearing capacities and the load transferring process in the vertical direction under different initial failure degrees. The research results showed that the load transfer after the initial failure of the strut mainly acted on the adjacent levels, resulting in a rapid and large increase in its axial forces, while the other levels of struts were slightly affected. The initial failure is more likely to trigger the subsequent damage to the immediate level of the strut and develop upwards along the vertical direction. When a certain level of struts fails, the bending moment of the diaphragm wall exceeds its ultimate bending capacity. The diaphragm wall broke down which in turn induced the overall collapse of the excavation.

Therefore, the progressive failure of the excavation often transmits along with its depth and latitude directions with its process of local failure developing into overall collapse. The structure is first damaged somewhere. Then the damaged element releases its sustained load to its adjacent elements. After that, the following element is damaged due to the increased load. The transferring and overloading processes continue to happen until finally leading to the overall collapse of the excavation.

4.2.3 *Anchored pile-supported excavation*

Zheng et al. (2020b) studied the collapse caused by local anchor failure using finite difference software and large-scale physical model tests to reveal the earth pressure redistribution and load transfer along with the retaining system after local anchor failure. For the support using a single level of anchor, the anchor failure induces a significant increase in the axial load to the adjacent three to four levels of anchors and of the shear force and bending moment to the capping beam. With the increase in the failed anchors, the axial load transfer coefficient increases gradually until reaching the final constant magnitude. The pile within the damage range gradually transferred to the cantilevered form with the maximum bending moment first decreasing and then increasing to a constant magnitude. The load (bending moment) transfer coefficients of the pile are generally larger than those of the anchors. That is, the damage is only transmitted within the anchor for a small number of failed anchors. Otherwise, the failure progressively developed into the retaining piles. In addition, the greater the excavation depth is, the lower the soil strength and the higher the load transfer coefficient of the retaining piles and anchors. In the multi-level anchored pile excavation, the local failure of the first level of the anchor has a great influence on the adjacent intact anchors due to the decrease in the bending moment of the retaining pile. The opposite is true for the localized failure of the lowest level of the anchor. However, localized failure of one level of anchor among the multi-levels of anchors has a relatively smaller influence than that of a single level at the same depth. However, the effects of simultaneous local failure of a whole column of multiple levels of anchors are significantly greater than that of a single level of the anchor and are more likely to induce progressive failure. The design or construction measures, such as reinforcing the anchor at certain interval levels, can be used to control the failure of local anchors within a certain level. When the localized failure of several levels of anchors induces adjacent anchors to fail in sequence, the initial anchor failure position does not affect the development and transmission path of progressive failure. After the failure of a whole column of multi-level anchors, the progressive failure range expanded horizontally in an inverted trapezoid shape. The capping and waler beams are easily damaged which in turn induces the bending failure of the retaining piles and accelerates the progressive collapse procedure. Therefore, the design of the capping and waler beams must consider the effects of the localized failure of anchors to improve the overall safety performance of the excavation against progressive failure.

4.3 *Resilience design criteria*

For different types of retaining structures used in deep excavations, such as cantilever piles, braced piles, and anchored piles retaining systems, three-level resilience design criteria could be considered for resilience design, named (1) prevent local failure, (2) prevent the generation of progressive failure and (3) control range of progressive failure. The details of each design criterion are discussed in this chapter.

1st level: prevent local failure

This level of resilience can be achieved by improving the reliabilities and stabilities of the local components. Zheng et al. (2011, 2014) proposed the following methods to prevent local failure:

(a) reinforce the connecting nodes of the retaining system,
(b) ensure sufficient ductility of the components,
(c) carefully design the key components.

The key components are the components that make the excavation retaining system have low redundancy after their removal or destruction. However, there are no well-accepted methods to evaluate the redundancy and robustness of the excavation. It is necessary to conduct further studies for the identification and design of key components. Since the cost of eliminating the possibility of local failure is not cost-effective, it is necessary to consider the following second level of resilience design against progressive failure.

2nd level: Prevent the generation of progressive failure

This level of resilience ensures that progressive failure along the depth, latitude, and longitude directions cannot be generated after the occurrence of the local failure. To achieve this target, it is necessary to establish design theories to improve the redundancy and robustness of the entire retaining system. Zheng et al. (2014) proposed a redundancy design theory through the investigation of the mechanism and control measures of progressive failure in the early stage. The force transmission paths are optimized through a reasonable arrangement of the components of the retaining system to prevent the generation and reduce the degree of progressive failure. The purpose is to increase the force transmission path of the retaining system with no or little lost increase of the retaining system through the reasonable arrangements of the enclosure and retaining system and necessary connection measures to prevent the weakening and damaging of the local components and the large deformation, instability or even progressive failure of the entire retaining system.

Zheng et al. (2014a) used the discrete element method to simulate the progressive failure of the ring beam and radical retaining system. It is found that the probability and degree of failure of the system are significantly lower than those of the unbraced system due to more force transmission paths in the angled retaining system after a component is damaged under the applied surcharge, as shown in Fig. 20. A redundancy evaluation index was proposed, as shown in Eq. (1). After comparing the redundancy evaluation index of different ring beam supports, it is found that reasonably increasing the force transmission path can effectively improve the redundancy of the retaining system.

$$R_5 = \frac{L_{\text{intact}} - L_{\text{design}}}{L_{\text{intact}} - L_{\text{damage}}} \quad (1)$$

where L_{intact} is the ultimate bearing capacity of the complete structure, L_{damage} is the ultimate bearing capacity of the structure after the component is damaged, and L_{design} is the design bearing capacity of the original structure.

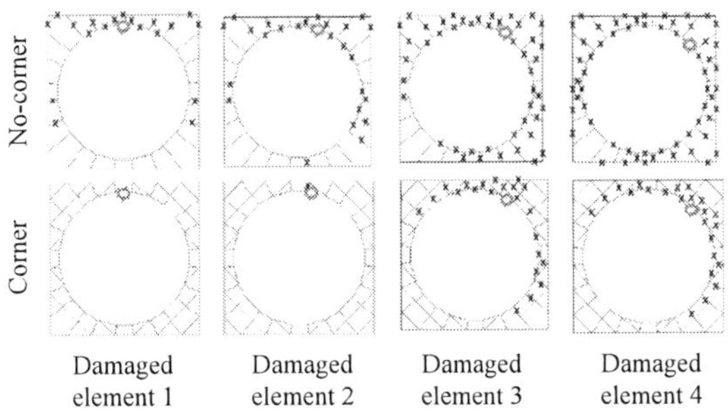

Figure 20. Progressive collapse of internal struts and ring beam.

For the multilayers braced excavation, Zheng et al. (2014b) conducted a case study on the Xianghu Station of Hangzhou Metro using the discrete element method to investigate the stability of using weak and strong connections between the braced piles and the diaphragm wall (see Fig. 21). In the case of a weak connection between the struts and the diaphragm wall, the diaphragm wall is rotated counterclockwise due to the instability of the left edge of the excavation, which makes the three struts fall off. In the case of a strong connection between the struts and the diaphragm wall, the joints can sustain large tensile and shear forces. Even if the left wall fails, the struts are still connected, which prevents the excavation from seriously collapsing. It can be concluded that improving the resilience of the retaining system greatly reduces the collapse degree of the excavation. For the multilayers of braced excavation, as shown in Figs. 10 and 17, the stability of the excavation can be fundamentally improved as long as the first level of strut and the head of the retaining pile (wall) are reliably connected.

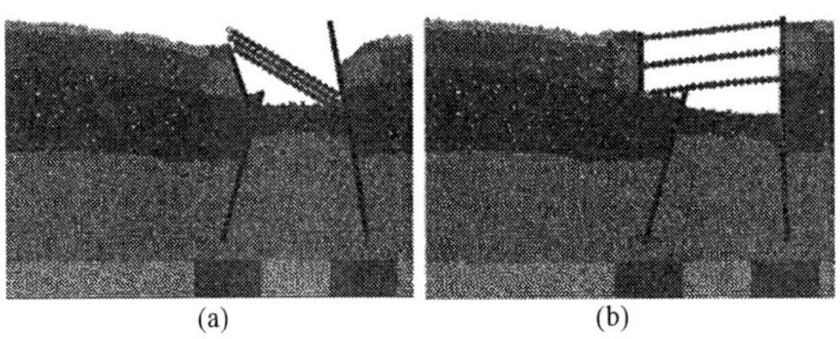

Figure 21. Effects of connections between inner struts and diaphragm wall on stability excavation (a) weak connection (b) strong connection.

The target of the second level of resilience design criteria is to prevent local failure from triggering progressive failure. From the points of balancing the cost of controlling and the loss of damage, the third level of resilience design criteria should also be considered. That is when local failure triggers progressive failure, can the range of development be controlled by appropriate structural measures?

3rd level: control range of progressive failure

If progressive failure has been generated in the excavation, efforts have to be made to control it within a limited range and degree according to the surrounding environmental conditions. Zheng et al. (2021) proposed a blocking element method to terminate the transfer of progressive failure. Taking the cantilever pile retaining excavation, as an example, a method of reinforcing the pile at certain intervals has been proposed to control the development of progressive bending failure. As shown in Fig. 22, several reinforced piles are installed at certain distances with greater embedded depth and pile strength to resist overturning and damage to the retaining piles. It is found that properly designed reinforced intervals can effectively control the development of progressive failure. When the number of reinforced piles is insufficient, the progressive failure crossovers the reinforced piles and continues to develop and in turn makes the reinforced piles fail. Taking the retaining structure shown in Fig. 22 as an example, the method of using two reinforced piles in each interval is insufficient to control the range of progressive failure as that induced by the local failure extends to the piles beyond the 5th~6th piles. If three reinforced piles are adapted in each interval, the progressive failure terminates at the 4th pile.

For the long strip excavation with a single level of the strut, it was pointed out earlier that the load released by the failed struts will concentrate onto their adjacent struts to induce progressive failure. To terminate the development of progressive failure, a certain number of struts can be reinforced at intervals.

However, the cantilever pile and single level of braced excavation are relatively simple. Further studies on the control of the range of progressive failure in complex retaining structures, such as multilayers of strut and anchor supported excavation, are still needed. According to the transmission path of progressive failure, rapid rescue after a local failure accident to terminate progressive failure is another research target in this level of resilience design.

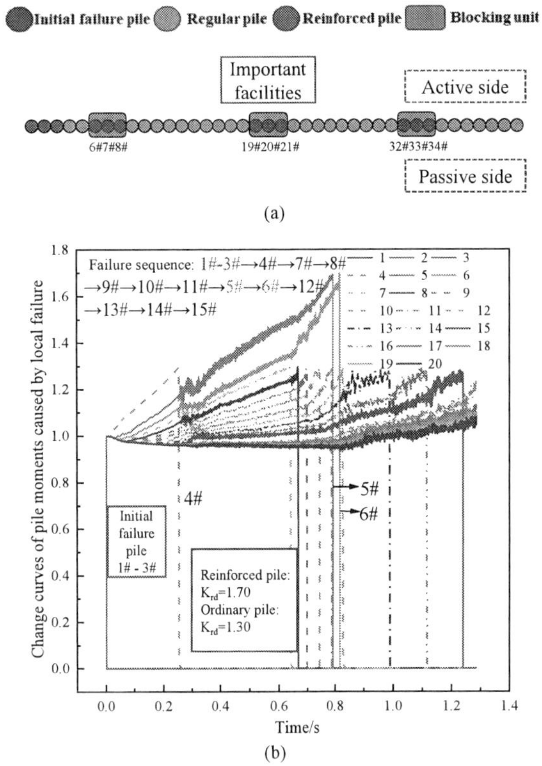

Figure 22. Effects of reinforced piles at certain intervals to terminate progressive failure in excavation retained by cantilever contiguous pile (a) schematic diagram, (b) increase of bending moment versus time curves.

5 RESILIENCE AGAINST PROGRESSIVE FAILURE IN SHIELD TUNNEL ENGINEERING

5.1 *Seismic resilience of shield tunnel*

Many scholars have conducted meaningful explorations and research on the seismic resilience of shield tunnels constructed with different materials, structural forms, and shock absorption measures. The traditional ways of improving the seismic resiliences of shield tunnels are generally achieved by increasing the thickness of the segments, setting up secondary linings, and strengthening the foundation which has limited effects of improvement. Xi (2017) and Liao (2019) reduced the seismic activity at the tunnel structure by setting SMA flexible damping nodes at the weak position of the pipe ring, see Fig. 23(a). They concluded that the SMA damping structure has a good damping effect to make the structure have a certain self-recoverability. Hu (2015) analyzed the seismic resilience of shield tunnels constructed using the steel fiber reinforced concrete lining, see Fig. 21(b). He concluded that the seismic performance of the steel fiber reinforced concrete lining was better than that of the ordinary concrete lining, and within a certain range, the greater the content of steel fiber to reinforce the concrete, the better the resilience of the lining.

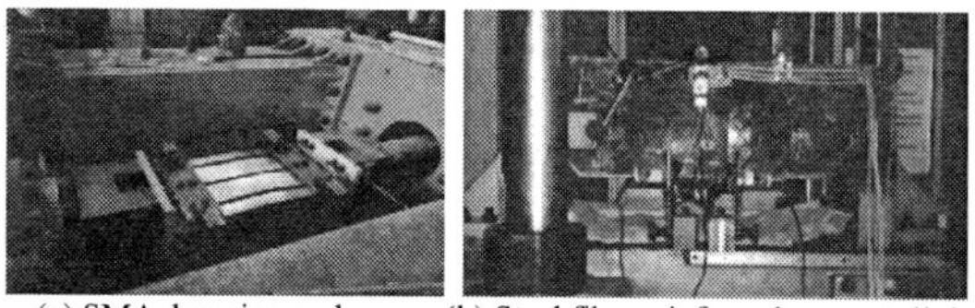

(a) SMA damping node (b) Steel fiber reinforced concrete lining

Figure 23. Methods to improve seismic resilience of shield tunnel.

Studies on the seismic performance of tunnels embedded in different soil conditions were conducted to optimize subway lines and measures of reinforcement. Wang et al. (2018) studied the influence of soil stiffness on the seismic response of the shield tunnel. They found that the dynamic response of the shallow tunnel in the weaker stratum was larger under the action of an earthquake. Li et al. (2021) used a three-dimensional finite element model to consider the interactions between shield tunnels and soil stratum to explore the seismic safety of shield tunnels passing through strata with varying stiffnesses. They concluded that sudden changes in stratum stiffnesses lead to the earthquakes response being significantly amplified which should be paid attention to in seismic design. Huang et al. (2009) used the finite difference program FLAC2D to analyze the shock absorption effect of tunnels after ground improvement. They found that the ground improvement can effectively reduce the internal force and deformation of the tunnel structures. The smaller the internal force and deformation of the structure, the more significant the shock absorption effect. Tian et al. (2019) used the finite element software MIDAS/NX to analyze the seismic response of the double-line shield tunnel in the form of spatial position. They found that the horizontal parallel tunnel had the largest dynamic response which is the most unfavorable combination of line seismic design. They improved once again that strengthening the soil around the tunnel can achieve the purpose of improving the seismic capacity of the tunnel.

5.2 *Progressive failure of shield tunnels*

By the end of 2020, urban rail transit lines with a total length of 7,978.19 km have been opened in 45 cities in China. In some mega-cities, the subway tunnel network sharply increased in the last decade. The cities are still expanding their subway networks now with the new shield tunnels often located in an environment with many buildings, dense population, roads, and pipelines. The effects of an accident on the surrounding environment may be catastrophic.

Shield tunneling is a highly mechanized tunnel construction method. Compared with traditional construction methods, the shield tunneling method is relatively safe and fast. As an engineering structure with a length much longer than its cross-sectional, the shield tunnel design is often based on the typical cross-sections embedded in the representative stratum to ensure the safety of the whole tunnel structure. This design criterion is based on the assumption that the working conditions of each segment among the typical cross-section are consistent. However, various unexpected events may result in localized damage to the lining segments or joints between them during the tunnel construction or operation stages. Local failure of segments of the shield tunnel embedded in the confined aquifer can even lead to a groundwater leakage disaster with the water and earth pressures significantly changed around the tunnel. The redistribution of loads and internal forces may lead to large-scale tunnel deformation, segment dislocation, joint opening, segment damage, and even collapse.

Cases of progressive damage or even collapse of shield tunnels induced by local failure have been reported many times. This kind of accident was being attracted great attention from scholars and engineers. On July 1, 2003, a large amount of groundwater and sand flowed into the tunnel during the excavation of the connecting passage between Pudong South Road Station and Nanpu Bridge Station at Shanghai Metro Line 4. The accident induced the collapse of the constructed tunnel with a total length of approximately 264 m (Zhang et al. 2017). Up to now, there are many examples of the shield tunnels reconstructed using the cut and cover method to remove difficult-to-repair tunnel structures and rebuilt using the cast-in-place method.

Furthermore, the parallel, overlapping, and overlapping of tunnels lines heavily increase the risk of tunnel construction. Because the tunnel is deeply buried in a complex surrounding environment, it will consume huge financial and material resources to carry out the repair work once the tunnel is damaged. The occurred cases of tunnel accidents due to accidental or human factors show that the progressive failure or collapse of the tunnel may be due to the localized damage induced soil loss and force concentration on the segments. The existing segment design theories consider the soil only as a design load but ignore the complex interactions between the soil and the tunnel segments during the dynamic developing process of the above-mentioned disasters.

Although the tunnel has sufficient safety reserve to resist the design load in serviceability and ultimate limit states, the stress distributions along the tunnel cross-sections heavily change once localized damage occurred. The performance of the segments in resisting the development of local damage or even collapse still needs further studies. Therefore, it is necessary to carry out in-depth studies on the mechanism of progressive failure and the overall safety performance of the shield tunnels.

5.3 *Progressive failure mechanism*

The research works in the literature related to the progressive failure during the construction of shield tunnels mainly focused on the reason analyses and developing process of the accidents, and the repairing measures after the accidents. However, there are relatively few studies on the developing and terminating mechanism of progressive failure caused by a local failure during the construction of shield tunnels in confined aquifers in soft soil, and their impact on tunnels and strata. Several studies have laid a foundation for the theoretical resilience design against progressive failure in shield tunnel engineering.

Referring to the research idea used in the progressive failure of the superstructure, Liu and Huang (2015) carried out a series of model tests to investigate the resistance of through-joint assembled concrete linings against the progressive collapse. A hydraulic loading system was used to apply multi-points concentrated forces to simulate the soil resistance, hydraulic and earth pressure, and ground surcharge onto the shield tunnel. They concluded that fully utilizing the plastic deformation of the structure was helpful to improve the overall safety of shield tunnels from the perspective review of the structural mechanics. Liu and Sun (2020) defined the progressive failure of underground shield tunnels by comparing them to the studies on building structures. By sorting out the case studies on the collapse accidents of the shield tunnels that occurred during the construction process, they summarized the process of progressive failure includes the following four stages: initial failure stage, ring joint failure, segment joint failure, and tunnel collapse.

Zhang et al. (2017) studied the hydraulic gradient distribution of seepages in shield tunnels using a FEM model and investigated the effect of localized leakage and erosion in different positions of the shield tunnel using a DEM model on the displacement and earth pressure on the tunnel cross-section, and the change of the ground settlement.

Ye et al. (2021) investigated the relationship between erosion-induced structural damage and lining displacement parameters for shield tunneling using a finite element method, which properly simulated the interaction between the eroded soil and tunnel structure as well as the details of the segments. The structural damage mechanism and its relationship with the lining displacement parameters were then clarified by conducting a series of parametric studies.

Zhao et al. (2017) conducted a case study of a leakage accident at Wuhan Metro Line 7 to investigate the effects of water leakage on the excavation surface, deformation of the segments, the stress on the segments and bolts, the opening and deformation of the transverse and longitudinal joints. It is found that the seepage of pore water can greatly decrease the hydraulic pressure onto the tunnel, increase the segment opening and misalignment, and even lead to damage to the tunnel segments and bolts. Zheng et al. (2014, 2016, 2017) analyzed the mechanism of occurrence and development of accidents induced by water leakage, soil erosion, water and sand leakages using model tests, numerical simulations, and a simplified theoretical model.

After the water and soil leakage accident happened in the shield tunnel, soil erosion can induce a localized cavity which can increase the earth pressure on the tunnel and even damage the tunnel structure. For example, the AB section of Metro Line 1 is a double-lane shield tunnel, soil erosion of 20 m^3 occurred at the waler of the 796th steel segmental lining on Jan 17, 2016. The accident made the concrete segments in the range of 798th to 802nd rings of the right-line tunnel compressed and spalled, as shown in Fig. 24 (Zheng et al. 2017). It can be seen that localized water leakage and soil erosion influenced the safety of the shield tunnel not only in its longitudinal direction but also in the adjacent tunnels. Zheng et al. (2017) investigate this case using the FEM modeling and found that the water leakage at the waler of the tunnel lining loosened the side segments. Parametric studies were also conducted to investigate the lateral deformation and force concentration on the tunnel with different loosening angles and degrees. The loosening areas firstly induced the concrete spalling under compression inside of the segment joints, and then made the top bolts yield in tension or even broken. Zheng et al. (2018) used the DEM method to simulate the influence of the cavity formed by soil erosion at different positions of the tunnel cross-section. The results showed that the soil erosion at the waler part of the tunnel loosened the surrounding soil and formed a soil arch outside of the loose area. The tunnel cross-section becomes more oblate with its internal force sharply increased. The soil erosion at the top and bottom of the tunnel cross-section may lead to the reversion of the directions of their bending moment. If soil erosion occurred at the tunnel shoulder, the position of the maximum bending moment shifted to the soil erosion position which endangered the safety of the adjacent segments.

Noticing the fact that a local failure may induce a large-scale progressive failure of the shield tunnel, Zheng et al. (2015) simulated the failure process using the PFC3D software. It was found that, after the failure of a middle ring segment, the internal force of the adjacent segments and bolts increased rapidly, and then progressively failed. In turn, the shield tunnel undergoes a domino-like progressive failure as shown in Fig. 25.

A series of model tests using the scaled shield tunnel embedded in sand embedment were conducted by Zheng et al. (2017) to study the effect of the local failure on the progressive failure along the longitudinal direction of the shield tunnel as shown in Fig. 26. The transmission mechanism of the internal forces on the adjacent segments and the development of the earth

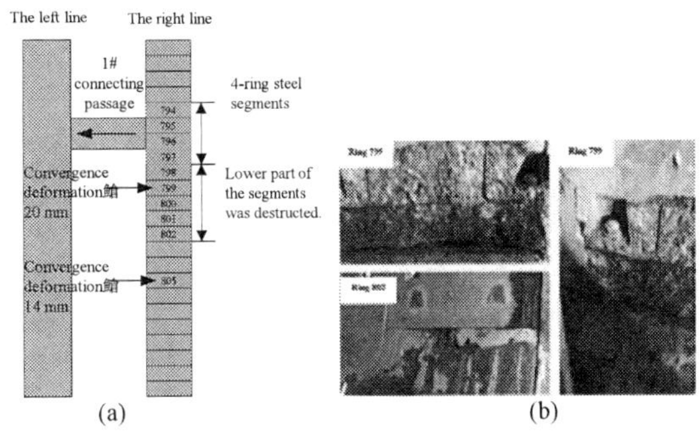

Figure 24. Damage to the shield tunnel due to soil erosion (a) sketch of the lining ring, (b) photos of the concrete spalling.

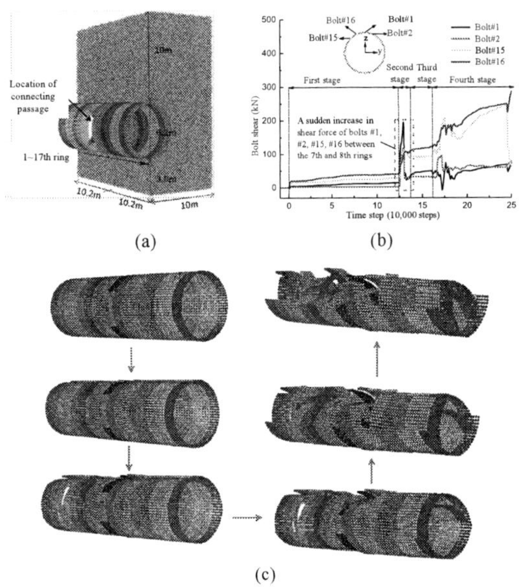

Figure 25. Numerical modeling of progressive failure in shield tunnel (a) PFC 3D Model and localized failed position (b) internal force development of bolts, (c) progressive failure process of the tunnel.

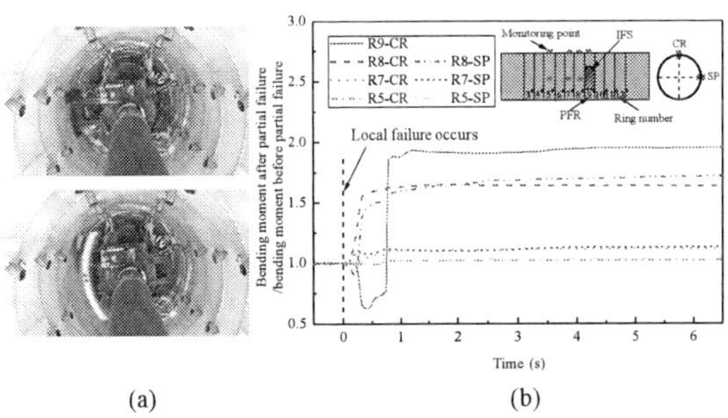

Figure 26. Progressive failure process of shield tunnel (a) photos of the model test (b) monitored bending moment on segments.

pressure after progressive collapse along the longitudinal direction were investigated. The results showed that the time duration of the soil arching effect induced by the local failure of the segment was very short with little influence on the adjacent rings. With the surrounding soil continuously pouring into the tunnel, the soil began to loosen, and the bending moment of the segment increased significantly. It is also found that the deeper the tunnel was buried the greater the increase in the bending moment of the adjacent segments and the higher the risk of progressive failure.

5.4 *Resilience design criteria*

The shield tunnel design is often based on the typical cross-sections embedded in the representative stratum without considering the resilience against progressive failure in the length direction of the entire shield tunnel. When local failure may induce large-scale progressive failure or even collapse of the whole tunnel, and the tunnel is constructed in some special conditions, such as passing through important infrastructures and facilities, large-scale damage or collapse to the tunnel is particularly serious, the resilience of the tunnel is suggested to be evaluated to ensure the toughness of the tunnel.

Similar to the resilience design against progressive failure in the excavation engineering, the same three-level of resilience design criteria are used in tunneling engineering, named as (1) prevent local failure, (2) prevent the generation of progressive failure, and (3) control the range of progressive failure to enable the tunnel to maintain the required safety and service performance after local disasters, such as local water leakage and soil erosion and structural damage, and so on.

1st level: prevent local failure
This level of resilience requirements could be achieved by improving the reliability of the segment and joints and preventing the known various risks. Mo and Chen (2008) and Chen and Mo (2009) summarized many engineering cases to sort the risk sources that may induce local failure during tunnel construction procedures. Engineers can take attention according to the forms of structural damage caused by these risk sources to avoid local failure.

2nd level: prevent the generation of progressive failure
The cross-sectional design method for the shield tunnels cannot reflect the change in the internal force of the adjacent structures after the localized damage. It is impossible to reinforce the tunnel structure with an unknown target. A design framework is proposed based on the beam-spring model to optimize the tunnel structure and improve its overall safety performance. As shown in Fig. 27, the design framework includes the following five steps: determine the forms of the local failure, (2) determine the key parameters, (3) build a numerical model, (4) determine the optimization direction, and (5) formulate an optimization plan.

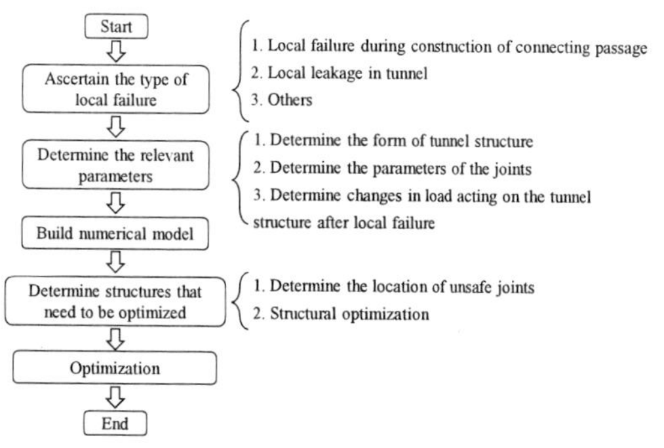

Figure 27. Design framework to optimize shield tunnel.

A Tianjin tunnel is used to illustrate the steps of using the proposed design framework. According to construction experience in Tianjin, the local failure usually happens at the location of the communication channel, tunnel entry, and exit holes. It is assumed that local failure occurs at the location of the connecting passage in step (1) as shown in Fig. 28(a). The pore water pressure and earth pressures are applied onto the segment lining as stress as shown in Fig. 28(b). It is assumed that after the local failure, an unloading area is generated near the shoulder of the first ring. All the loads acting on the tunnel structure disappear due to the generation of the unloading area as shown in Fig. 28(b). Taking the local failure of the inner joints adjacent to the first and second ring and their joints as an example, it is found that the shear forces of the inner ring and their joints significantly increased. A large tensile force is generated and exceeds its bearing capacity. It has to optimize and improve the shear and tensile bearing capacities of these joints. To formulate an optimization plan, the concrete below the bolt has to be reinforced. Trolleys and longitudinally connecting steel frames between segments can also be installed to improve the safety resilience of the shield tunnel. The proposed design framework can also be used for stress analyses after the soil and water leakage induced by different risk sources on the entrance and exit doors of the tunnel to optimize its overall safety.

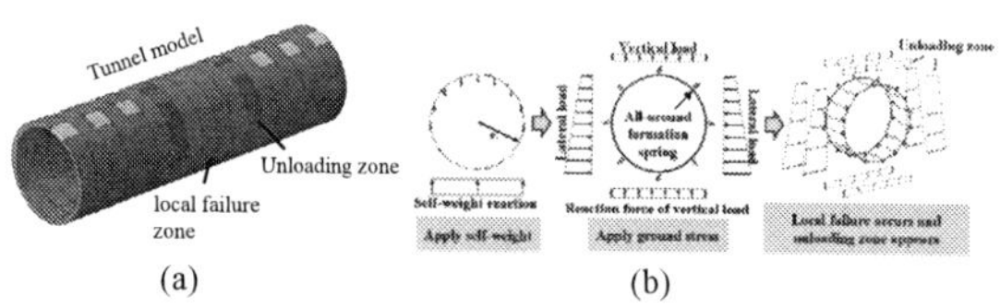

Figure 28. Load-structure model and simulation procedure (a) Schematic representation of the numerical model (b) Simulation process.

3rd level: control range of progressive failure
One method to control the range of progressive failure induced by the localized failure of the weak ring is to strengthen the weak ring and the rings around them. According to the burial depth of the tunnel, there are three types of reinforcement design for the concrete segments, such as shallow buried, medium buried, and deeply buried. As shown in Fig. 29, the 9th ring is found to be weak through numerical analyses. Two cases are studied to investigate the influence of the range of reinforced rings on the performance of the tunnel. One is only to reinforce the 8th to 10th rings, and the other is to reinforce the 6~8th to 10~12th ring. As shown in Fig. 30(a), the tunnel was still damaged with only one ring reinforced on both sides of the weak ring. The failure of the 9th ring still induces damage to the 8th and 10th rings, and then further extends to the 7th and 11th rings. This short-distance reinforcement could not effectively control the occurrence of progressive collapse. As shown in Fig. 30(b), the tunnel is not damaged with the second method of reinforcement. The maximum bending moment of the 9th ring trends to decrease which means a larger range of ring reinforcement could effectively reduce the occurrence of the progressive failure.

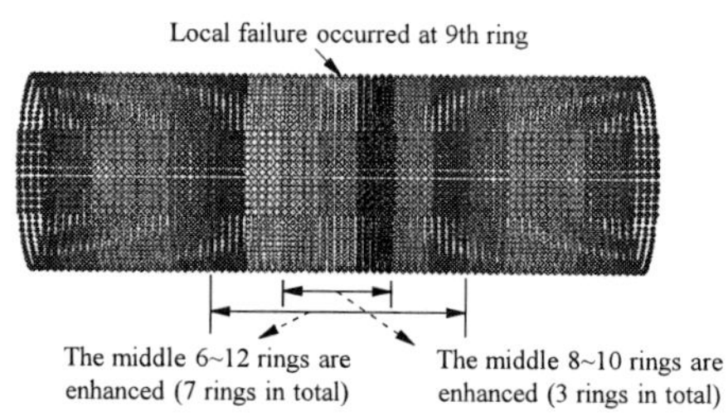

Figure 29. Schematic illustration of the range of reinforcement.

The second method to control the range of progressive failure induced by the localized failure of the weak ring is to install a temporary supporting frame. Take the same case as an example, lateral steel frames are installed onto the 8th and 10th rings as shown in Fig. 31. The time history diagram of the newly added vertical displacement value of each ring is shown in Fig. 32. It can be seen that the displacements of the 8th and 10th rings are controlled after localized failure, but those far from the weak rings increase slowly. If the lateral steel frame is not installed, the local failure induces a rapid increase in the lateral deformation of

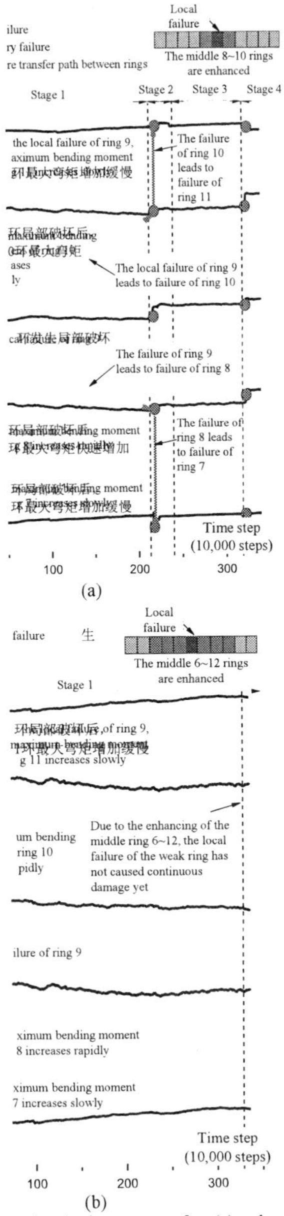

t developing curves after (a) only reinforce reinforce the 6~8th to 10~12th rings.

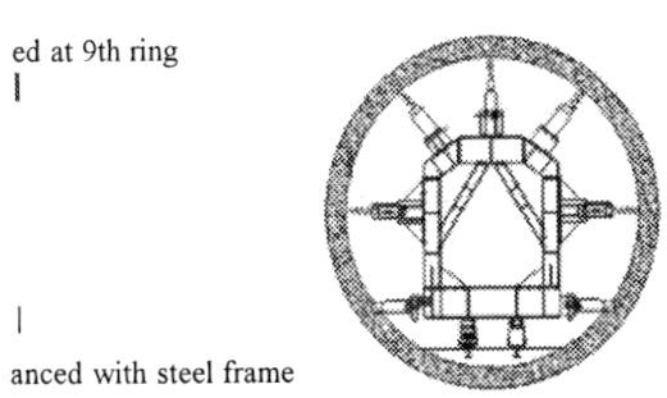

rcement using the lateral steel frame.

ally the progressive failure of the tunnel. ral steel frame at the weak rings could eformations, but its length of influence against the progressive failure along the the tunnel, the inner steel frames have d at certain intervals. The method of

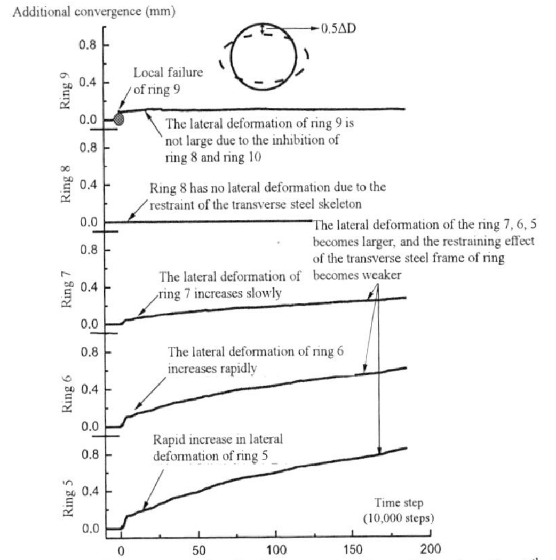

Figure 32. Displacement developing curves after reinforcing the 8th to 10th rings.

optimizing and improving the robustness of the tunnel structure mainly starts from the segments to strengthen them in advance the poor geological conditions or may be damaged in the later stage of construction, to against the occurrence of the progressive failure. The effect of adding the lateral steel frames is limited because they are mainly used as the temporary support structure in the parts of the tunnel where local damage or collapse has occurred or the dangerous sections during construction.

The third method to control the range of progressive failure induced by the localized failure of the weak ring is to reinforce the soil outside of the tunnel. After the local failure of segments, the soil outside the tunnel usually flows into it with soil arches generating around the tunnel. The unloading effect of the loose area and the loading effect of the arch foot area constitute the force transmission path along the longitudinal direction of the tunnel. To cut off the force transmission path against the development of progressive failure, the following measures can be taken: (1) reinforce the soil outside of the weak ring using freezing or grouting methods to let them meet the strength requirements, (2) install the diaphragm wall around the weak section to prevent a large range of soil erosion, and (3) reduce the overburden pressure at the top of the tunnel through excavation or replace them with lightweight materials. For example, soil excavation and backfill with EPS sheet were successfully used in the Ninghai West Road tunnel on Shanghai Metro Line 1.

6 RESILIENCE AGAINST PROGRESSIVE FAILURE IN EMBANKMENT ENGINEERING

6.1 *Composite foundation*

The composite foundation is one of the main soft soil treatment methods using the reinforced bodies constructed during the ground treatment together with the natural soil body around them to support surcharges, see Fig. 33. By utilizing the bearing capacity of the soil between the reinforced material, the composite foundation can effectively increase the bearing capacity and stability of the ground. It also has the advantages of reducing the differential settlement of the embankment, being cost-effective and construction time-saving. As there is no strong connection between the reinforced material, even if the horizontal geosynthetic layers are adapted to interlock them, the strength of their connections is still much lower than that of the superstructure. Therefore, the integrity of the composite foundation under the embankment is relatively weak.

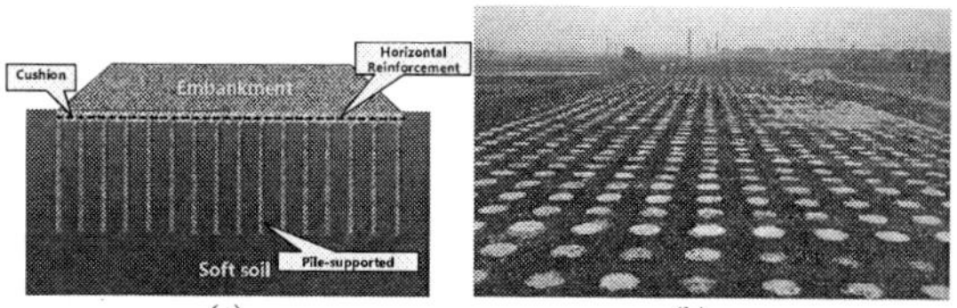

Figure 33. Rigid pile-supported embankment (a) sketch of the method (b) photo of the method.

Composite foundations are widely used in large-scale soft soil reinforcement projects to construct highways, railways, dockyards, and so on. It has been successfully applied in Beijing-Tianjin Intercity High-speed Railway, Beijing-Shanghai High-speed Railway, Mujia Passenger Dedicated Line, Japan Shinkansen, Malaysia High-speed Railway, LGV South Europe Atlantic high-speed railway, etc. The application of composite foundation can effectively improve the bearing capacity and disaster resistance of urban transportation facilities. The material used for the vertical reinforcement or their strength and stiffness after construction can be classified into four categories as shown in Table 2 (Liu and Sun, 2020; Zheng et al. 2012; 2020) with some photos shown in Fig. 34.

Table 1. Categories of composite foundation (Zheng et al. 2020).

Category / Strength characteristics	First species (granular columns)	Second species (Semi-rigid columns)	Third species (Rigid piles)	Fourth species (Combined columns)
	Sand column Stone column Sand-gravel column	Cement mixing column Lime column Rammed cement-soil column Lime-fly ash column Grouted gravel column Chemical churning column Sand bag well Geosynthetic-encased stone columns Bag-grouting columns	CFG pile Prestressed pipe piles Prestressed square pile Concrete plie Bored pile Spiral bored pile X-shaped concrete piles Y-shaped concrete piles Large-diameter tubular pile	Precast concrete stiffened cement mixing columns Profile steel stiffened cement mixing columns Cast-in-situ concrete stiffened cement mixing columns Concrete core sand columns Soil-cement core sand columns
Bond strength	None	Low-Medium	High	
Shear strength	Low-Medium	Low-Medium	High	Medium-High
Compressive strength	Low	Low-Medium	High	Medium-High
Tensile strength	None	None-Low	Low-High	Low-High
Bending strength	None	None-Low	Low-High	Low-High

Figure 34. Types of composite foundation (White and Pham, 2007)
(a) Granular column (b) semi-rigid column (c) rigid column
(d) reinforced rigid column.

The rigid column is an effective and efficient method of composite foundation technology to be widely used in engineering practice because of its ability in improving the bearing capacity and reducing the settlement of soft soils. In addition to the traditional rigid column with circular and square column cross-sections, the X-type and Y-type cross-sectional concrete columns have also been developed (Liu et al. 2008; Wang et al. 2011; Liu and Zhao, 2016). The composite column such as concrete-cored and shaped steel-cored deep cement mixing columns have been successfully proposed and applied in engineering practice due to their higher bearing capacity (Xiang et al. 2009; Fu et al. 2013; Chen et al. 2013; Kong and Zhou 2014; Liu et al. 2010).

6.2 *Development of embankment stability analyses*

Investigating the failure modes of embankments on different types of columns is conducive to evaluating the failure process of the embankment and its overall resilience assessment. Studies in literature have shown that the failure modes and post-destruction properties of different types of columns under embankments are quite different. The sand or gravel column undergoes shear failure under embankment loads because these types of reinforcements have little flexural stiffness to resist bending moments (Liu and Sun, 2020). Zheng et al. (2020), Zhang et al. (2014), and Abusharar and Han (2011) conducted a series of numerical studies to investigate the effects of column spacing, column diameter, properties of the column, and soil on the stability of gravel column. They concluded that the stability evaluation method of equivalent shear resistance may overestimate the stability of the embankment as the column and soil do not reach their limit states at the same time.

The semi-rigid column, such as jet grouting columns and cement mixing columns, may undergo several types of failure modes such as shear failure, bending failure, compression failure, inclination, lateral movement, and soil flow between columns (Kitazume and Maruyama, 2007; Han et al. 2004; Navin and Filz 2006; Zhang et al. 2014; Yapage et al. 2015a; 2015b). Kitazuma and Maruyama (2017) conducted a series of centrifuge tests and concluded that the cement mixing columns below the embankment suffered bending failure rather than shear failure. The columns at different positions failed progressively rather than simultaneously destroyed. The bending failure of the columns will not directly lead to the instability of the embankment which will lose its stability only after a certain range of progressive failure. Han et al. (2004), Navin and Filz (2006), and Zhang et al. (2014) found through numerical simulations that the failure modes of cement mixing columns, i.e., bending failure or overturning failure, heavily depends on the column strength. The Bishop's arc sliding method will heavily overestimate the stability of the embankment. Yapage et al. (2015a, 2015b) found that the column has an obvious softening behavior after failure. After bending failure of the columns, their cohesion and dilatation angles are reduced to a certain extent. The columns failed progressively at different positions.

The rigid columns, such as plain concrete columns, reinforced concrete columns, and prestressed pipe columns, are mainly suffered from bending failure with obvious brittle and progressive characteristics (Zheng et al. 2010, 2012, 2017). Zheng et al. (2017) proposed a model for considering the post-failure behavior of rigid columns. They observed the phenomenon of progressive failure and analyzed its influence on the stability of the embankment. The influence of one-column failure on the adjacent columns had been analyzed, and the key factors that control the overall stability of the embankments had been clarified. The results from the centrifuge and numerical studies showed that the columns below the embankment shoulder suffered a bending failure. This process made the tensile stresses and bending moments of the columns at the failure position. In the embankment failure cases, the bending failure phenomenon was also confirmed by the excavated columns as shown in Fig. 35(a) (Zheng et al. 2018; Yu et al. 2021). The analysis shows that, after the localized bending failure, the column and soil stresses redistribute within a certain range which will increase

Figure 35. Progressive failure of rigid column (a) and (b) photos of the excavated columns, (c) failure points of the column after centrifuge test (d) failure points of the column after numerical simulations.

the tensile stress and the bending moment of the adjacent columns and finally may lead to the progressive failure, see Fig. 35(b) to 35(d). With an increase in the embankment height, the columns below the embankment shoulder undergo secondary bending failure at a shallow position. Due to the large deformation of the soil in between the columns after damage, the low confining pressure induced by the soil was hard to confine the column. The secondary bending failure at these positions is more likely to induce the progressive failure of the adjacent columns and finally the instability of the embankment. At this time, a plastic sliding surface will be formed along with the bending failure position of the columns.

In the above failure modes of rigid columns, the stability of the embankment was determined by the strength of the columns after the shear failure and bending failure. However, in some other conditions, the instability of the embankment is not related to the strength of the column. Kitazume (2007, 2006) observed through centrifuge tests that the column bodies undergo bending failure when their tensile strength is small but undergo overturning failure when it is large. The centrifuge and numerical studies conducted by Toshinari et al. (2017) showed that the stability of the deep mixing column didn't depend on the column strength no matter whether it failed in shear failure or soil flow between columns. The deep mixing column was even under overturning failure when the underlying soil layer is inclined. Zheng et al. (2021) and Zhou et al. (2019) found that rigid columns easily undergo overturned failure when they are inserted into the inclined hard soil layer at a conventional embedded depth. Terashi et al. (1983) found through the centrifuge tests that the soft soil was easily flowing around the columns if the column spacing was large and the soil between them was very weak.

Based on whether the stability of the embankment is determined by the column strength, the failure modes of the embankment on the columns can be divided into the following two categories: (1) internal failure due to the loss of the strength of the column materials with their failure modes of shear failure, compression bulging failure, and bending failure; (2) external failure due to the loss of some contributions to the stability of embankment in a certain form with their failure modes of the soil flowing failure, overturning failure and horizontal slipping failure. The stability of the embankment mainly depends on the strength of the soil and the column-soil frictions.

6.3 *Resilience design methods*

6.3.1 *Case of embankment failure*

Even if various types of rigid columns are used to support the embankment on soft soil, landslide accidents of the embankment were still happened from time to time. On April 4, 2004, a large-scale landslide accident occurred in a section of a Luochang Expressway in Fujian Province, China, although the cemented fly ash gravel (CFG) columns are used to form the reinforcement in the deep silty clay. More than 70 m of the expressway were destroyed with the lateral influence distance of approximately 100 m, and the collapse depth of approximately 10 m, as shown in Fig. 36(a). The Tai-Hua Expressway used dry vibrating gravel columns for soft soil treatment. The supported embankment slipped during the filling process (Cao and Chen 2007). On a highway in the coastal area of Zhejiang province, the subgrade slipped in a road section during the filling process although the foundation was treated using granular columns because the soft soil foundation accounts for one-third of its total length. (Zhang et al. 2007). Although the soft soil foundation of a railway line was reinforced using cement mixing columns, the embankment suddenly collapsed when it was filled to a height of 5.2 m during the construction process (Qin and Wang 2007). A soft soil under the highway in the Pearl River Delta was treated using pipe columns and two layers of geogrids in the cushion on the surface of the pipe columns. When the subgrade was filled to 7 m, the subgrade collapsed with the pipe columns overturned and slid off the subgrade as shown in Fig. 36(b). Although the designs of the above cases satisfied the standards with their factor of safety being greater than 2.0, serious accidents still occurred because the current design standards cannot satisfy the resilience requirements.

(a) (b)

Figure 36. Photo of the embankment failure in (a) Luochang highway and (b) Railway subgrade.

6.3.2 *Stability analyses of column-supported embankments considering resilience*

The most widely used stability design method for embankments is based on the limit equilibrium theory. For a homogeneous soil slope, it is assumed that the sliding surface is cylindrical and the soil above that is rigid. The stability factor of safety is defined as the ratio of the average shear strength to the average shear stress acting on the entire sliding surface.

For the stability analyses of column-supported embankments, the traditional design method adapts the global safety factor to provide the safety reserve. The factor of safety of the subgrade slope is usually calculated using the Swedish method, the Janbu method, and the Bishop method. However, these design methods cannot quantitatively consider the random variations of load effects, geotechnical parameters, and soil resistance. The design parameters heavily depend on engineering statistics, laboratory tests, or engineering experience.

For the stability analysis of the embankment on the soft soil foundation treated by rigid column, it is usually calculated using the method for the granular column, i.e. circular arc sliding method, see Fig. 37. The shear strength on the sliding surface S_{sp} is calculated as:

$$S_{sp} = (1 - m)c_u + mS_p \cos\alpha \tag{2}$$

where S_p is the shear strength of the rigid column, α is the angle between the tangent direction on the sliding arc and the horizontal directions, c_u is the undrained shear strength of the soil, m is the area replacement ratio of the rigid column.

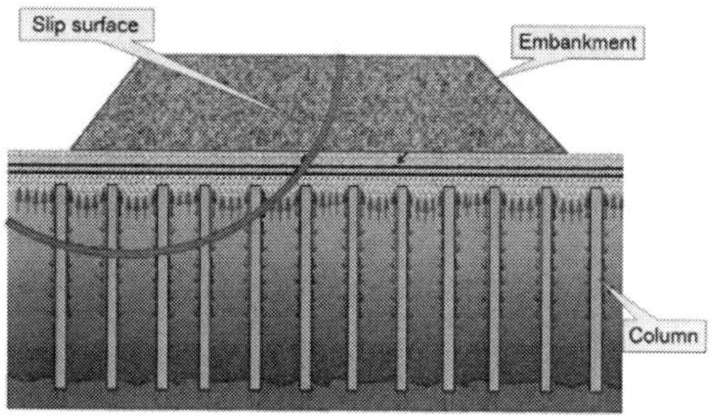

Figure 37. Slip surface of the column-supported embankments.

There are following three basic assumptions in this design method: (1) only shear failure happens in the column without considering the possibility of other failure modes; (2) the columns are simultaneously failed without considering the possibility of progressive failure; (3) the column is uniform and play a role without considering the possibility of external damage. Therefore, the circular arc sliding method for stability analysis may be seriously inconsistent with the failure characteristics of the actual project. The failure cases investigated by Cao et al. (2007), Zhang et al. (2007), and Qin and Wang (2007) showed that the design methods may make the engineering project unsafe or even induce failure accidents. It is necessary to analyze the real failure modes of different reinforcement methods, especially for rigid columns and high embankments constructed on the flexible columns in soft soil. The numerical and experimental studies have to be conducted to simulate the real failure modes and estimate the overall stabilities of the subgrade and embankment.

The three basic assumptions of the most commonly used arc sliding methods ignore the possibilities of the internal shear failure, progressive failure, and external failure of the column bodies. The calculation results overestimate the safety of the embankment to a varying degree. For the internal failure mode, the ultimate filling height of the embankment calculated using the potential shear failure mode is very close to that calculated using the arc sliding method but significantly larger than that calculated using the simultaneous bending failure. The smallest limit filling height of the embankment is obtained based on the progressive bending failure, which is almost consistent with the actual buckling height, as shown in Fig. 38(a). For the external failure mode, the embankment filling height based on the progressive bending failure mode is significantly larger than that based on the overturning failure mode, which is almost identical to the actual buckling height, as shown in Fig. 38(b).

The above studies show that the two cases are the progressive bending failure and overall overturning failure modes induced by the localized failure of a column. The traditional stability

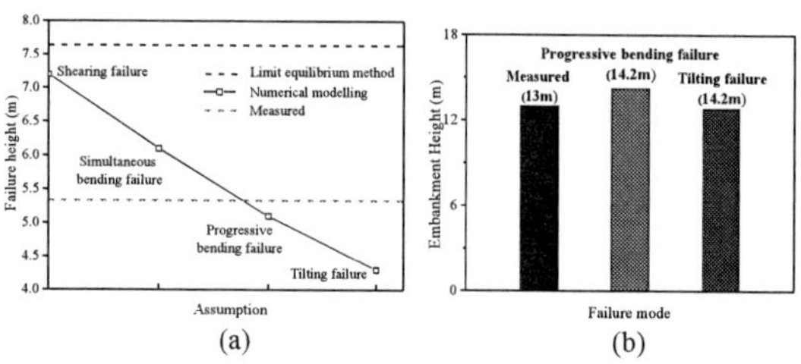

Figure 38. The failure modes of the embankment (a) Internal failure, (b) external failure.

analyses cannot guarantee resilience against the failure risks of the embankments. It is suggested to evaluate the stability of the embankment using different calculation methods based on its potential failure modes, establish the characterization of the failure evolution, identify the key factors and their thresholds, and finally propose analytical methods and technologies to improve the resilience of the embankment engineering. However, there are limited studies on the robustness and recoverability design methods of the embankment under extreme conditions.

The above studies show that the failure modes of the two engineering projects are progressive bending and overturning failures induced by the local failure of the first column. The traditional stability analytical methods could not guarantee the disaster resilience of the embankment which has potential instability risk.

It is therefore required to focus on the resilience of the embankment, realize the characteristic of its failure evolution process, identify the key influencing parameters and thresholds, and finally propose the evaluation methods according to the real failure modes of the embankment. Based on the progressive failure mechanism of the embankment, Zheng et al. (2020) defined the evaluation of the resilience against progressive failure, that is, the ability of the system to resist external disturbances during the progressive failure procedure. The sensitivity coefficient S_{ij} of each stage of progressive failure could be used to evaluate the resistance of the system, see Eq. (3). It can be found from Fig. 39 that the larger sensitivity coefficient S_{ij} represents the more serious column damage, i.e. $S_{ij} \leq 0.2$ for slight damage, $S_{ij} \leq 0.4$ for moderate damage, and $S_{ij} \leq 0.6$ for severe damage.

$$S_{ij} = \frac{\sigma'_{ij} - \sigma_{ij}}{\sigma_t - \sigma_{ij}} \tag{3}$$

where σ_t is the ultimate tensile strength, and σ_{ij}, σ'_{ij} are the tensile stresses of adjacent columns before and after local column damage, respectively.

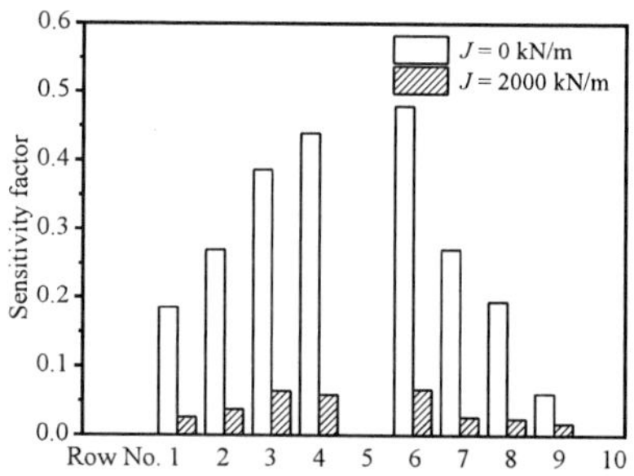

Figure 39. Sensitivity factor of resilient evaluation.

6.4 *Measures with resilience design methods*

For the granular column with a relatively lower strength, bearing capacity, and stability, Chen et al. (2015) proposed a design method for this type of composition foundation reinforced by geosynthetics. The stability of the column body is greatly improved because the geosynthetics reinforcement can coordinate the deformation of the system and prevent the column body from bulging. Zhao et al. (2014) conducted the laboratory model tests to analyze the bearing capacity and stability of the gravel column enclosed by geosynthetic within a certain depth from the ground surface and concluded that the geosynthetic enclosed gravel column has a better engineering performance. Zhao et al. (2016) proposed a two-way reinforced composite foundation using geocell and gravel columns to improve the deformation capacity of gravel columns. They found that geocells can effectively increase the bulge range and volume of the gravel columns with their maximum bulging position moving up to give full play of the bearing capacity. Wen et al. (2010) poured cement slurry into gravel columns to form slurry-solid gravel columns which have higher strength and rigidity to heavily improve the bearing and deformation performance of the embankment. The column body still keeps its characteristics of construction and environmental friendliness performance.

For the semi-rigid column, Broms (1991) installed the continuous column wall in the position where the column body is prone to be bending failed, such as under the embankment shoulder, to improve the bending resistance of columns. Han and Gabr (2002) analyzed the effect of horizontal reinforcement on the settlement characteristics of columns and concluded that horizontal reinforcements can effectively reduce their total and uneven settlements. The numerical study conducted by Navin (2005) showed that the installation of installing continuous column wall under the embankment shoulder could greatly improve the stability of the columns with the same replacement ratio, and maintain high-level reliability considering the spatial variability of materials, and optimize the safety performance of the embankment. Jamswang et al. (2015) found through engineering measurement and numerical simulations that the column body at the toe of the embankment is prone to bending damage which may lead to excessive lateral deformation. Installing grouting columns and continuous column walls on both sides of the embankment can significantly improve its deformation and stability performances.

For the rigid composite foundation, the column-net composite foundation, column-slab composite foundation, and column-raft composite foundation are widely used in engineering practices (Liu and Zhao 2016). Increasing the column-soil stress ratio could make the forces between the column groups more coordinated and thus improve the stability of the embankment. After the embankment is internally damaged, the rigid columns may suffer from local brittle bending failure and then a significant progressive failure. The rigid column may also experience external damage, such as overturning damage. When analyzing the stability of rigid columns under the embankment, the traditional overall stability analysis method assumes that the columns are sheared at the same time (Jamsawang et al. 2015). According to the loading characteristics of the rigid columns at different positions under the embankment, Zheng et al. (2010) divided the columns beneath embankments into four different areas, such as tension-bending area, and bending-shearing area, compression-bending area, and pressure-bearing area. There are significant differences in the anti-slipping mechanism of rigid columns in different areas. The performance requirements should be considered to optimize and improve the resilience of the embankment. The resilience design method could make the embankment on columns have a higher structural economy and safety with a reasonable spatial layout.

Zheng et al. (2018) proposed a model to simulate the post-failure behavior of rigid columns considering the characteristics. Their studies reflected the generation of tensile cracks and the weakening of the shear bearing capacity during the bending failure process of the rigid column. According to the column positions and their influences on the stability of the embankment, an unequal strengthen method was proposed to prevent the internal failure of the embankment on rigid columns. Reinforcing the key columns in the bending and shearing areas can effectively prevent the progressive failure of the rigid columns and improve the stability of the embankment.

Zhou et al. (2021) proposed a design method to consider the bending performance and post-failure of the rigid column. The numbers of strengthening columns versus ultimate bearing capacity curves are shown in Fig. 40. It can be seen that strengthening the three key columns can greatly improve the stability of the embankment with an effect similar to strengthening all columns. Since the steel reinforcements of the column determine the cost of the column, the number of columns used under the embankment is hugely reduced especially then the method was adapted for a long highway. Therefore, using the performance-based design method with key columns strengthened can significantly improve the embankment stability and save engineering costs.

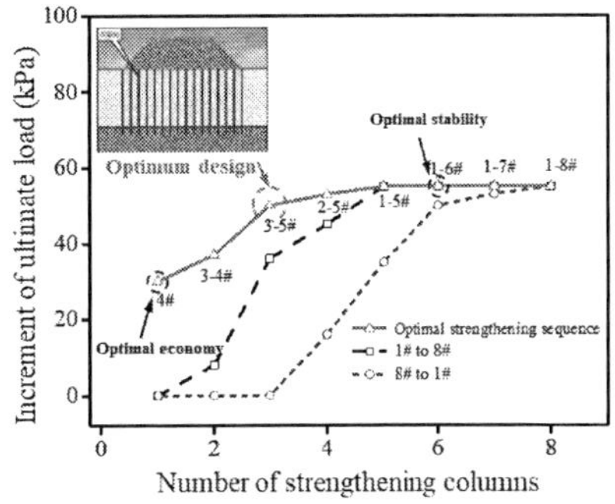

Figure 40. The numbers of strengthening columns versus ultimate bearing capacity curves (Zheng et al. 2018).

In the current stability studies of the embankment on the rigid column, the horizontal embedded layer is mostly considered. The rigid column is often designed to be embedded 0-2 times the column diameter into the hard soil layer. Zhou et al. (2019) conducted centrifuge tests and numerical studies and concluded that the rigid column body may suffer from overturning damage when a substratum is inclined. With the increase of the embedded depth of rigid columns, the failure modes change from the overturning failure to the bending failure. The mechanical characteristics of the columns below different positions of the embankment have different ways to influence the stability of the columns. A non-isometric performance-based design method was proposed based on the external failure of rigid columns. The surcharge loading versus the number of strengthened column curves is shown in Fig. 41. By increasing the spatial layout strategy of the embedding depth of the columns at key positions to fully use the characteristics of columns at different positions, the stability and resilience of the embankment can be economically and effectively improved.

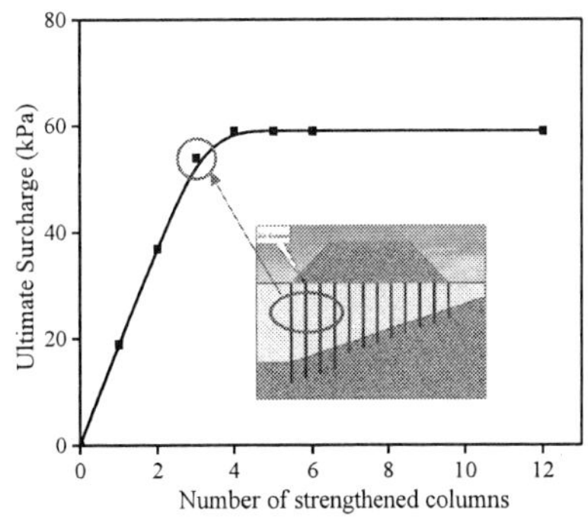

Figure 41. The surcharge loading versus the number of strengthened column curves.

7 CONCLUSIONS

The safeties of urban and major infrastructures in geotechnical and underground engineering are of great importance. To realize the design and construction of resilient underground infractions and cities, some major problems in geotechnical and underground engineering have to be solved to resist natural disasters or human-induced serious disasters, maintain their functions to the greatest extent, and restore their functions as soon as possible after disasters. This paper summarized the connotation and development of resilience and pointed out that the evaluation and control of resilience can be divided into the following four levels: permissible stress design method, reliability-based design method, robustness method, and recoverable performance method. Through this study, the following conclusions can be drawn:

(1) The factor of safety and reliability-based design theories have been fully developed and played important roles in ensuring the safety of infrastructures in geotechnical and underground engineering. However, the two methods are often adapted together in design which is required to be further developed.

(2) Based on traditional reliability-based design theories, the robustness design method has been proposed to consider the uncertainty of the geotechnical parameters, evaluate the sensitivity of system performance to the parameter variabilities and assess the failure probability of geotechnical engineering, which has deepened the traditional reliability design theories. More efficient and convenient methods for applying robustness design method to engineering and incorporating it into design specifications have to be developed.

(3) In geotechnical and underground engineering, large-scale progressive failure accidents often happened due to the local failures of soil and structures. Studies in the field of excavation engineering, shield tunnel engineering, and column-supported embankments have been reviewed to illustrate the mechanism, evaluation, and design theories of resilience against progressive failure. The evaluation and design methods of resilience against progressive failure in geotechnical and underground engineering still need further studies in various fields.

(4) The self-healing and rapid recoverability of engineering structures are very important for the improvement of their resilience. Because the infrastructures in geotechnical and underground engineering are deeply buried in the ground, it is difficult to repair once damaged. There are relatively few studies on the recoverability analyses of infrastructures in geotechnical and underground engineering. The cost of improving their recoverability is relatively high to rapidly recover the damaged underground structures. How to rapidly, economically, and effectively improve the recoverability of the geotechnical infrastructures after earthquakes and other accidents are required further in-depth studies.

(5) Resilience evaluation methods and indicators are the basis to establish systematic resilience design theories in geotechnical and underground engineering. Due to the uncertainties in geotechnical and underground engineering, the complex interactions between soil-water-structures, and the difficulties in repairing once damaged, the reasonable and convenient resilience evaluation indicators that can comprehensively consider the robustness (resilience against progressive failure), resources, rapidity, and redundancy of structures in geotechnical and underground engineering are required.

8 REFERENCES

Abusharar S.W. and Han J. Two-dimensional deep-seated slope stability analysis of embankments over stone column-improved soft clay, *Engineering Geology*, 2011, 120(1):103-110.

Alonso E.E. Risk analysis of slopes and its application to slopes in Canadian sensitive clays. *Géotechnique*, 1977, 26(26): 453-472.

Anastasopoulos I, Gazetas G, Loli M, et al. Soil failure can be used for seismic protection of structures. *Bulletin of Earthquake Engineering*, 2010, 8(2): 309–326.

Artola J. A solution to the braced excavation collapse in Singapore (Master thesis). Boston: Massachusetts Institute of Technology, 2005.

Asaoka A, Agrivas D. Spatial variability of the undrained strength of clays. *ASTM Geotechnical Testing Journal*, 1981, 108.

Babu G L S, Murthy D S. Reliability Analysis of Unsaturated Soil Slopes. Journal of Geotechnical and Geoenvironmental Engineering, 2005, 131(11): 1423-1428.

Breen J.E. Research Workshop on Progressive Collapse: Regulatory Approached to the Problem. National Bureau of Standards, Washington D.C., 1975.

Broms B.B. Stabilization of Soil with Lime Columns. *Ground Engineering*, 1991, 12(4): 833-855.

Bruneau M, Chang S E, Eguchi R T, et al. 2003, A framework to quantitatively assess and enhance the seismic resilience of communities. *Earthquake Spectra*, 19(4): 733 - 752.

Bruneau M., Reinhorn A., 2006, Overview of the resilience concept, Proceedings of the 8th US National Conference on Earthquake Engineering, 2006, pp. 18-22.

Cao Weiping, Chen Yunmin. Analysis of sliding failure mechanism and treatment for Taihua Highway embankment slope over soft soils, *Chinese Journal of Rock Mechanics and Engineering*, 2007, 26(7) : 1504-1510. (In Chinese)

Casagrande A. Role of calculated risk in earthwork and foundation engineering. *Journal of the Soil Mechanics & Foundations Division*, 1966, 91: 1-40.

Chang M. Basal stability analysis of braced cuts in clay. *Journal of Geotechnical and Geoenvironmental Engineering*, 2000, 126 (3) : 276-279.

Chen Fei, Wu Kaixing, He Shu. Field tests on load transfer performances of squeezed branch piles. *Journal of Geotechnical Engineering*, 2013, 35(s2): 990-993(In Chinese)

Chen J F, Li L Y, Xue J F, et al. Failure mechanism of geosynthetic-encased stone columns in soft soils under embankment. *Geotextiles and Geomembranes*, 2015, 43(5):424-431.

Chen, J. S., & Mo, H. H. Numerical study on crack problems in segments of shield tunnel using finite element method. *Tunnelling and Underground Space Technology*, 2009, 24, 91–102.

Cheng X.S., Zheng G., Diao Y. et al. Experimental study of the progressive collapse mechanism of excavations retained by cantilever piles. *Canadian Geotechnical Journal*, 54: 574–587 (2017).

Cheng Xue-song, Zheng Gang, Deng Chu-han et al. Study of the mechanism of progressive collapse induced by partial failure of cantilever contiguous retaining piles. *Chinese Journal of Geotechnical Engineering*, 2015, 37(7): 1249–1263. (In Chinese)

Cheng Xue-song, Zheng Gang, Huang Tiam-ming et al. Experimental study on mechanism of progressive collapse along length of excavation retained by cantilever contiguous piles. *Chinese Journal of Geotechnical Engineering*, 2016, 38(9): 1640–1649. (In Chinese)

Cheng, X. S., Zheng, G., Diao, Y., Huang, T. M., Deng, C. H., Lei, Y. W., and Zhou, H. Z. (2016). Study of the progressive collapse mechanism of excavations retained by cantilever contiguous piles. *Engineering Failure Analysis*, 71.

Cherubini C, Christian J T, Baecher G B, et al. Factor of Safety and Reliability in Geotechnical Engineering. *Journal of Geotechnical and Geoenvironmental Engineering*, 2001, 127(8): 700-721.

Choosrithong K , Schweiger H F . Numerical Investigation of Sequential Strut Failure on Performance of Deep Excavations in Soft Soil. *International Journal of Geomechanics*, 2020, 20(6):1-12.

Deng Jian, Zhu Hehua. Finite Element Monte- Carlo Method Using Neural Networks for Geotechnical Reliability Analysis. *Journal of Tongji University (Natural Science)*, 2002, (03): 269-272. (In Chinese)

Duncan J M. Factors of safety and reliability in geotechnical engineering. *Journal of Geotechnical and Geoenvironmental Engineering*, 2000, 126(4): 307-316.

Ellingwood, B. and Leyendecker, E.V. (1978), Approaches for Design Against Progressive Collapse, *Journal of the Structural Division*, Vol. 104, No. 3, pp. 413–423.

Fang JH, Zhang ZH, Zhang JY, 2009. Application of artificial freezing to recovering a collapsed tunnel in Shanghai Metro No. 4 Line. *China Civil Engineering Journal*, 42(8):124-128 (In Chinese)).

Finn W. D. L (2015). 1st Ishihara Lecture: An overview of the behavior of pile foundations in liquefiable and non-liquefiable soils during earthquake excitation. *Soil Dynamics and Earthquake Engineering*, 68(2015) 69–77.

Fu Qiang, Ding Xuanming, Liu Hanlong, et al. Dynamic analysis of PCC pile composite foundation under train vibration load. *Rock and Soil Mechanics*, 2013, 34(s2): 413-420 (In Chinese)

Goh A. and Wong K. S. Three-dimensional analysis of strut failure for braced excavation in clay. *Southeast Asian Geotech, Soc.* 2009, 40(2), 137–143.

Guan Z., Yang B., Dong K., Li J., Experiment on Hysteretic Properties of Bridges with Pile-group Foundations Allowing Rocking and Uplift of Pile Cap, *China Journal of Highway and Transport*, 2017, 30(10), 62-68.

Guan Z, Chen X, Li J. Experimental investigation of the seismic

performance of bridge models with conventional and rocking pile group foundations. *Engineering Structures*, 2018, 168:889-902.
Ge Hong. Comparison of immune algorithm with genetic algorithm. *Journal of Jinan University*, 2003, (01): 22-25. (In Chinese)
Goh A. T. C., Fan Z., Liu H., Zhang W., Dong Z. Numerical Analysis on Strut Responses Due to One-Strut Failure for Braced Excavation in Clays. 2018
Gong W P, Khoshnevisan S, Juang C H. Gradient-based design robustness measure for robust geotechnical design. *Canadian Geotechnical Journal*, 2014, 51(11): 1331-1342.
Gong Xiao-nan. Generalized composite foundation theory and engineering application. *Chinese Journal of Geotechnical Engineering*, 2007, 29(1): 1–13. (In Chinese)
Haack I A. Construction of the North-South-Metro Line in Cologne and the accident on March 3rd, 2009. *International Symposium on Social Management Systems (SSMS)*, 2009
Han J, Chai J C, Leshchinsky D, Shen S L. Evaluation of deep-seated slope stability of embankments over deep mixed foundations. *ASCE Geo-Support*, 2004.
Han J, Gabr M A. Numerical Analysis of Geosynthetic-Reinforced and Pile-Supported Earth Platforms Over Soft Soil. *Journal of Geotechnical & Geoenvironmental Engineering*, 2002, 128(1):44-53.
Han J. Recent research and development of ground column technologies, *Proceedings of the Institution of Civil Engineers Ground Improvement*, 2015, 168(4):1-19.
Han Jianyong, Zhao Wen, Jia Pengjiao et al. Mechanical Response of Anchored Pile Supporting Deep Excavation Influenced by Local Anchor Failure, *Journal of Northeastern University (Natural Science)*, 2018, 39(03): 426-430.
Ha J G , Kim D S . Evaluation of seismic behavior of soils under nuclear containment structures via dynamic centrifuge test. *Nuclear Engineering and Design*, 2014, 277(1):64–75.
Hao Yiwen, Xu Jiang, 2015. Understanding Urban Resilience: A Conceptual Analysis Based on Integrated International Literature Review. *Urban Planning International*, 30(02): 48-54. (In Chinese)
Holling C.S. Resilience and stability of ecological systems. Annual *Review of Ecology and Systematics*, 1973 (4):1-23.
Hsien P G, Ou C Y, Liu H T. Basal heave analysis of excavations with consideration of anisotropic undrained strength of clay, *Can. Geotech. J.*, 2008, 45:788-799.
Hu Lei. Seismic dynamic response research on steel fiber reinforced concrete in tunnel lining structure, Master thesis, 2015, Southwest Jiaotong University. (in Chinese)
Huang Hongwei, Gong Wenping, Zhuang Chuanxian, Zhang Jie. Robust Geotechnical Design of Gravity Retaining Wall, *Journal of Tongji University (Natural Science)*, 2014, 42(3): 377-385.
Huang Hongwei, Gu Leiyu, State-of-the-art of researches on risk management of foundation pit engineering, *Chinese Journal of Geotechnical Engineering*, 2008, (S1), 651-656 (In Chinese)
Huang Hongwei, Huang Xi, Schweiger F. H. 2012, Numerical analysis of the influence of deep excavation on underneath existing road tunnel. *China Civil Engineering Journal*, 45(03):182-189. (In Chinese)
Huang Hongwei, Shen Xianda, Wang Fei, Zhang Dongming, Vulnerability Analysis and Assessment of Segment Joints in Shield Tunnel 2016,44(02):198-206+219. (In Chinese)
Huang H., Gong W., Zhuang C., et al Robust design of gravity retaining wall. *Journal of Tongji University*, 2014, 42 (3):377- 385 (in Chinese)
Huang Juan, Peng limin, Li Xinglong, Transverse seismic response of Shiziyang shield tunnel in liquefiable soil and its shock absorption measures, *Chinese Journal of Geotechnical Engineering*, 2009, (10):1539-1546. (in Chinese)
Huang Mao-song, Song Xiao-yu, Qin Hui-lai. Basal stability of braced excavations in K0-consolidated soft clay by upper bound method. *Chinese Journal of Geotechnical Engineering*, 2008, 30(2):250-255. (In Chinese)
Huang Yu, Shu Xiang, Ye Weimin et al. Recent studies on seismic resistance of pile foundation, *Industrial Construction*, 2002, 32(7). 50-53,61 (In Chinese)
Huang H W, Zhang D M. Resilience analysis of shield tunnel lining under extreme surcharge: Characterization and field application.
ISSMGE. Joint TC205/TC304 Working Group on "Discussion of statistical/reliability methods for Eurocodes"– Final Report, 2017.
Itoh, K., Kikkawa, N., Toyosawa, Y., Suemasa, N., and Katada, T. Failure Mechanism of Anchored Retaining Wall due to the Anchor Head Itself Being Broken. *Proc., TC302 Symposium Osaka: International Symposium on Backwards Problem in Geotechnical Engineering & Monitoring of Geo-construction*. 2016.
Jamsawang P, Voottipruex P, Boathong P, et al. Three-dimensional numerical investigation on lateral movement and factor of safety of slopes stabilized with deep cement mixing column rows. *Engineering Geology*, 2015, 188:159-167.
Jiang Li-hong, Kong Ling-gang, Lin Gang, Chen Yun-min, Simplified method of laterally loaded single piles with partial pile head fixity, *Engineering Mechanics*, 2014, 31(11), 63-69
Juang C H, Wang L, Atamturktur S, et al. Reliability-based robust and optimal design of shallow foundations in cohesionless soil in the face of uncertainty. *Journal of Geoengineering*, 2012, 7(3): 75-87.
Juang C H, Wang L, Liu Z F, et al. Robust Geotechnical Design of Drilled Shafts in Sand: New Design Perspective. *Journal of Geotechnical and Geoenvironmental Engineering*, 2013, 139(12): 2007-2019.
Khoshnevisan S, Gong W P, Juang C H, et al. Efficient Robust Geotechnical Design of Drilled Shafts in Clay Using a Spreadsheet. *Journal of Geotechnical and Geoenvironmental Engineering*, 2015, 141(2).
Kitazume M, Kenji M. External stability of group column type deep mixing improved ground under embankment loading. *Soils and Foundations* 2006, 46(3): 323-340.
Kitazume M, Maruyama K. Internal stability of group column type deep mixing improved ground under embankment loading, *Soils and Foundations*, 2007, 47(3): 437–455.
Ko K W, Ha J G, Park H J, et al. Centrifuge modeling of improved design for rocking foundation using short piles. *Journal of Geotechnical and Geoenvironmental Engineering*, 2019, 145(8): 04019031.
Kong Gangqiang, Zhou Hang. Theoretical analysis of soil compaction effect of belled wedge pile China, *Journal of Highway and Transport*, 2014, 27(2): 9-16 (In Chinese)
Larosche A, Ziehl P, ElBatanouny M K, Caicedo J, Plain pile embedment for exterior bent cap connections in seismic regions. *Journal of Bridge Engineering*, 2014, 19(4), 04013016
Lehman D E, Roeder C, Stringer S J and Jellin A, Seismic performance of improved pile-to-wharf deck connections, PCI Journal, 2013: 62-80
Li Guang-xin, Li Xue-mei. The shear strength in stability analysis of subway pit in soft clay. *Geotechnical Investigation and Surveying*, 2010, 1: 1-4. (In Chinese)
Li Siming, Yu Haitao,Xue Guangqiao, Xu Lei, Analysis on seismic response of shield tunnels passing through soil-rock strata, *Modern Tunnelling Technology*, 2021, 58(5):65-72. (in Chinese)
Li X. Study on the load transfer mechanism induced by the failure of anchors in excavations retained by pile-anchor structure. Master thesis, Tianjin University, 2018. (In Chinese)
Liao D. Experiments and finite element analysis of underwater shield tunnel damping joints based on SMA, Master thesis, 2019. Guangzhou University. (in Chinese)
Liu G., Wang W. *Excavations Engineering Manual* (The second edition). Beijing: China Architecture and Building Press, 2009: 121-146. (In Chinese)
Liu Hanlong, Ren Lianwei, Zheng Hao, et al. Full-scale model test on load transfer mechanism for jet grouting soil-cement-pile strengthened pile. *Rock and Soil Mechanics*, 2010, 31(5): 1395-1401(In Chinese)
Liu Hanlong, Zhao Minghua. Review of ground improvement technical and its application in China, *China Civil Engineering Journal*, 2016, 49(1):96-115. (In Chinese)
Liu J., Hou X. *Excavations Engineering Manual*. Beijing: China Architecture and Building Press, 1997. (In Chinese))
Liu S., Yi Y., Zhu Z. Comparison tests on field bidirectional mixing column for soft ground improvement in expressway. *Chinese Journal of Rock Mechanics and Engineering*, 2008, 27(11): 2272-2280 (In Chinese)
Liu X., Huang X. Experimental study of collapse capacity of straight joint segmental tunnel lining structure. *Chinese Journal of Rock Mechanics and Engineering*, 2015(S2):3703-3714. (In Chinese)
Liu Xian, Sun Qihao. Case analysis on progressive collapse of shield tunnel linings. *Hazard Control in Tunnelling and Underground Engineering*, 2020, 2(2): 21-30. (In Chinese)
Lu Y, Tan Y. Overview of Typical Excavation Failures in China. *Eighth International Conference on Case Histories in Geotechnical Engineering*. 2019
Lu X., Chen Y., Mao Y., New Concept of Structural Seismic Design : *Earthquake Resilient Structures*, 2011, 39(7):941-948. (In Chinese)
Lumb P. Safety factors and the probability distribution of soil strength. *Canadian Geotechnical Journal*, 1970, 7(3): 225-242.
Lumb P. The Variability of Natural Soils. *Canadian Geotechnical Journal*, 1966, 3(2): 74-97.
Lv Xilin, Quan Liumeng, Jiang Huanjun. Research trend of earthquake resilient structures seen from 16WCEE, Earthquake Engineering and

Engineering Vibration, 2017, 01(3):1-9. (In Chinese)
Mo, H. H., Chen, J. S. Study on inner force and dislocation of segments caused by shield machine attitude. *Tunnelling and Underground Space Technology*, 2008, 23, 281–291.
Navin M P, Filz G M. Numerical stability of embankments supported on deep mixed columns. ASCE GSP 152, *Ground Modification and Seismic Mitigation*, 2006.
Navin M P. *Stability of Embankments Founded on Soft Soil Improved with Deep-Mixing-Method Columns*. 2005.
Nees / E-defense. Report of the seventh joint planning meeting of NEES / E-defense collaborative research on earthquake engineering. PEER 2010 /109. Berkeley: University of California, Berkeley, 2010.
Olsson A M J, Sandberg G E. Latin hypercube sampling for stochastic finite element analysis. *Journal of Engineering Mechanics-ASCE*, 2002, 128(1): 121-125.
Peng X, Li D Q, Cao Z J, et al. Reliability-based robust geotechnical design using Monte Carlo simulation. *Bulletin of Engineering Geology and the Environment*, 2017, 76(3): 1217-1227.
Pong, K. F., Foo, S. L., Chinnaswamy, C. G., Ng, C. C. D., and Chow, W. L. (2012). Design considerations for one-strut failure according to TR26 – A practical approach for practising engineers. *The IES Journal Part A: Civil & Structural Engineering*, 5(3), 166-180
Qin L., Wang Z. Fall into Unstable Analysis of a Certain Embankment on Soft Soil Ground in Railway, *Subgrade Engineering*, 2007(2) : 42-44.
Qingdao Kazuki, Yasuda Satoshi, Sadayuki Ishizaki, Hiroshi Shimada, Tsutomu Komuro, Ichizo Kawabata, Development of pile head semi-rigid junction construction method (F.T.Pile construction method), Taisei Construction Technology Center report, 2004, 37, 1~8
Ser, C., Sayin, B. Geotechnical assessment and rehabilitation of retaining structures collapsed partially due to environmental effects. *Eng Fail Anal* 119. 2020.
Shadabfar S, Mahsuli. M, Zhang Y., et al. Resilience-Based Design of Infrastructure: Review of Models, Methodologies, and Computational Tools. *ASCE-ASME Journal of Risk and Uncertainty in Engineering Systems, Part A: Civil Engineering*, 2022, 8(1): 03121004.
Sivandran C. Probabilistic analysis of stability and settlement of structure on soft Bankok clay. Ph.D. Dissertafion of Asian Institute of Technology.1979: 56-58P
Taguchi G. Introduction to quality engineering: Designing quality into products and processes. Asian productivity organization, Tokyo, 1986.
Tang Liang, P-Y model of dynamic pile-soil interaction in liquefying ground, PhD thesis, Harbin institute of technology, 2010. (In Chinese)
Terashi M, Tanaka H, Kitazume M. Extrusion failure of ground improved by the deep mixing method, *Proc. 7th Asian Regional Conference on Soil Mechanics and Foundation Engineering*, Haifa, Israel. 1983, 1: 313-318.
Tian Z., Xu Y., Study on the Law of Transverse Seismic Response of Two Paralleled Shield Tunnels, *Modern Tunnelling Technology*, 2019, (2):57-64. (in Chinese)
Tunnelling and Underground Space Technology, 2016, 51:301-312.
Toshinari Y, Matsuda S, Kitazume M, Nguyen B. Centrifuge Model Studies on External Stability of Column Type Deep Mixing Ground on Inclined Foundation. Grouting, 2017,227-23
Ukritchon B, Whittle A J, Sloan S W. Undrained stability of braced excavations in clay. ASCE, 2003, 8: 738-755.
Vanmarke E H. Probabilistic Modeling of Soil Profiles. *Journal of the Geotechnical Engineering Division ASCE*, 1977, 103(11): 1227-1246.
Wachi K., Asano S., Maeda N., Aoshima K., Nagai K., Mechanical characteristics of simplified connection of pile head for precast concrete pile, *Journal of Structure Concrete Engineering*, AIJ, 2003, 570, 85-91
Wang Qi, Yu Haitao, Dai Chunxiang, Wu Yongxin, Duan Keping, Seismic response analysis of ultra-shallow buried shield tunnel in soft soils, *Modern Tunnelling Technology*, 2018, 55(s2):493-500. (in Chinese)
Wang G., Wang Y., Yu H., Zheng N., Sun F. Shaking table tests on seismic response of rocking frame structure considering foundation uplift, *Chinese Journal of Geotechnical Engineering*, 2021,11(43):2064-2074. (In Chinese)
Wang Ruichuan. Prediction and analysis of shield boring parameters in a mixed ground based on BP neural network. Master thesis, Tongji University, 2019. (In Chinese)
Wang Xinquan, Chen Yonghui, Chen Long, et al. Comparative study on bearing characteristics of Y –section pile and common piles under embankment load. *Journal of Railway Engineering Society*, 2011(11): 46-52 (In Chinese)
Wen Shiqiang, Chen Yumin, Ding Xuanming, et al. Application of grouted gravel pile in soft subgrade improvement of expressway, *Rock and Soil Mechanics*, 2010, 31(5): 1559-1563. (In Chinese)
White D J, Pham H T V, Hoevelkamp K K. Support Mechanisms of Rammed Aggregate Piers. I: Experimental Results, *Journal of Geotechnical & Geoenvironmental Engineering*, 2007, 133(12):1503-1511.
Whittle A J, Davies R V. Nicoll Highway collapse: evaluation of geotechnical factors affecting design of excavation support system. *International Conference on Deep Excavations*, Singapore, 2006.
Wu B., Ou Y. Experimental study on tunnel lining joints temporarily strengthened by SMA bolts. *Smart Materials and Structures*, 2014, 23:1-10.
Wu T H, Kraft L M. Safety analysis of slopes. *Journal of the Soil Mechanics & Foundations Division*, 1970, 96(1):609-630. (In Chinese)
Xi Z. Research on longitudinal seismic performance of underwater shield tunnel and SMA flexible dissipation joint, Master thesis, 2017, Guangzhou University. (in Chinese)
Xia J., Gu J., Xu Y., You Y. Research on Foundation Pit Multiple Horizontal Bracing System Based on the Redundancy, *Bulletin of Science and Technology*. 2018, 34(08): 200-205. (In Chinese)
Xiang W., Liu S., JingFei, et al. Bearing capacity of composite foundation of soilcement deep mixing columns with different cross-section parts. *Journal of Southeast University: Natural Science Edition,* 2009, 39(2): 328-333(In Chinese)
Xiao X., Yuan J., Zhu Y\. Causation analysis of the collapse on Singapore MRT circle cine lotC824 (Part I)- Project background and process of collapse. *Modern Tunnelling Technology*, 2009, 46(05): 66-72. (In Chinese)
Xiao X., Yuan J., Zhu Y. Causation analysis of the collapse on Singapore MRT circle cine lot C824 (Part I)- Technically breaching back analysis and insufficient instrumentation. *Modern Tunnelling Technology*, 2010, 47(01): 22-28. (In Chinese)
Xu J., Shao J., Zheng Y. Application of genetic algorithm to reliability analysis of geotechnical engineering. *Chinese Journal of Geotechnical Engineering*, 2000, (05): 586-589. (In Chinese)
Yang Z. Research on seismic performance of prestressed high strength concrete pile, PhD thesis, Tianjin University, 2014.
Yapage N N S, Liyanapathirana D S, Kelly R B, et al. Numerical Modeling of an Embankment over Soft Ground Improved with Deep Cement Mixed Columns: Case History. *Journal of Geotechnical & Geoenvironmental Engineering*, 2015a, 140(11):04014062.
Yapage N N S, Liyanapathirana D S, Poulos H G, et al. Numerical Modeling of Geotextile-Reinforced Embankments over Deep Cement Mixed Columns Incorporating Strain-Softening Behavior of Columns. *International Journal of Geomechanics*, 2015b, 15(2):04014047.
Ye Lieping, Qu Zhe, Lu Xinzheng, Feng Peng. Collapse prevention of building structures: a lesson from the Wenchuan earthquake, *Journal of Building Structures*, 2008, (04): 42-50.
Ye Z, Liu H. Investigating the relationship between erosion-induced structural damage and lining displacement parameters in shield tunnelling. *Computers and Geotechnics*, 2021, 133, 104041.
Zhang D M, Zhai W Z, Huang H W, et al. Robust retrofitting design for rehabilitation of segmental tunnel linings: Using the example of steel plates. *Tunnelling and Underground Space Technology*, 2019, 83:231-242.
Zhang D., Gao C., Yin Z., Wang R., Yang T. Particle flow simulation of seepage erosion around shield tunnel, *Rock and Soil Mechanics*, 2017(S1): 435-444.
Zhang J., Wang W., Jiang D., Guan Z. Tests for aseismic performance of pile-to-pile cap joint of cast-in-place piles with steel casings, *Journal of Vibration and Shock*, 2018, 37(003):79-84. (In Chinese)
Zhang K., Li J. Accident analysis for '08.11.15' foundation pit collapse of Xianghu Station of Hangzhou metro. *Chinese Journal of Geotechnical Engineering*, 2010, 32(S1): 338–342. (In Chinese)
Zhang W., Ling S., Chen Y., Zhou M. Stability analysis of filled embankment laid on soft subsoil compacted with sand-stone pile, *Railway Engineering*, 2007(11): 46-49. (In Chinese)
Zhang Y., Chen X. Finite element analysis for slope stability with multi-braced retaining wall through reduction of strength parameters. *Chinese Journal of Geotechnical Engineering*, 2006, 28(11): 1952-1957.
Zhang Z, Han J, Ye G B. Numerical analysis of failure modes of deep mixed column-supported embankments on soft soils. *Ground Improvement and Geosynthetics*, 2014, 168: 78–87.
Zhang Z, Han J, Ye G. Numerical investigation on factors for deep-seated

slope stability of stone column-supported embankments over soft clay, *Engineering Geology*, 2014, 168(2):104-113.
Zhao M., Gu M., Zhang L., et al. Model tests on influence of vertical geosynthetic-encasement on performance of stone columns. *Journal of Geotechnical Engineering*, 2014, 36(9):1587-1593 (In Chinese)
Zhao M., Liu M., Ma B., Long J. Calculation for Stress Concentration Ratio and Settlement of Bidirectional Reinforced Composite Foundation Consisting of Geocell Mattress and Stone Column Under Embankment, *China Journal of Highway and Transport*, 2016, 29(05): 1-10. (In Chinese)
Zhao T, Liu W, Ye Z, et al. Effects of Water Inrush from Tunnel Excavation Face on the Deformation and Mechanical Performance of Shield Tunnel Segment Joints. *Advances in Civil Engineering*, 2017: 1-18.
Zhao W, Han J Y, Chen Y, et al. A numerical study on the influence of anchorage failure for a deep excavation retained by anchored pile walls. *Advances in Mechanical Engineering*. 2018,10(2):1-17.
Zheng G, Cheng X S, Diao Y, et al. Concept and design methodology of redundancy in braced excavation and case histories. *Geotechnical Engineering Journal of the SEAGS & AGSSEA*, 2011, 42(3): 13-21.
Zheng G, Cui T, Cheng X, et al. Study of the collapse mechanism of shield tunnels due to the failure of segments in sandy ground. *Engineering Failure Analysis*, 2017, 464-490.
Zheng G, Dai X, Diao Y, et al. Experimental and simplified model study of the development of ground settlement under hazards induced by loss of groundwater and sand. *Natural Hazards*, 2016, 82(3): 1869-1893.
Zheng G, Dai X, Diao Y, et al. Parameter analysis of water flow during EPBS tunnelling and an evaluation method of spewing failure based on a simplified model. *Engineering Failure Analysis*, 2015, 96-112.
Zheng G, Yang X., Zhou H, et al. Numerical modeling of progressive failure of rigid piles under an embankment, *Canadian Geotechnical Journal*, 2018(4).
Zheng G. Method and application of deformation control of excavations in soft ground. Chinese J. Geot. Eng., 2022, 44(1): 1-36. (in Chinese)
Zheng G., Cheng X., Diao Y. DEM simulation and redundancy analysis of excavation collapse. *Rock and Soil Mechanics*, 2014b, 35(2): 573-583. (In Chinese)
Zheng G., Cheng X., Zhang Y. Progressive collapse simulation and redundancy study of ring-beam supporting structures of excavations. *Chinese Journal of Geotechnical Engineering*, 2014a, 36(1): 105-117. (In Chinese)
Zheng G., Cui T., Cheng X., Diao Y. Transverse Deformation of Shield Tunnel Under Lateral Loosening Conditions, *Journal of Tianjin University*, 2017, 50(12): 1304-1313. (In Chinese)
Zheng Gang, Cui Tao, Cheng Xue-song, Sun Ji-bin, GE Long-bo, Introduction and analysis of an accident in a shield tunnel, *Chinese Journal of Geotechnical Engineering*, 2017, 39(z2): 132-135. (In Chinese)
Zheng G., Cui T., Cheng X., Zhang T. Safety situation investigation and analysis of the shield tunnels under the condition of the sandy ground leakage, *Journal of Safety and Environment*, 2018: 39-44. (In Chinese)
Zheng G., Cui T., Jiang X. Mechanism of progressive collapse induced by partial failure of shield tunnels in sandy soil, *Chinese Journal of Geotechnical Engineering*, 2015, 37(9): 1556-1571. (In Chinese)
Zheng G., Dai X., Zhang X.. Experimental study and numerical simulation of leaking process of sand and water in underground engineering., *Chinese Journal of Rock Mechanics and Engineering*, 2014, 33(12): 2458–2471. (In Chinese)
Zheng G., Deng C., Cheng X., Huang T., Lei Y. Influence of Space Effect on Progressive Collapse in Excavations Retained by Cantilever Piles, *Journal of Tianjin University*, 2016, 49(10): 1016-1026. (In Chinese)
Zheng G., Gong X., Xie Y., et al. State-of-the-art techniques for ground improvement in China, *China Civil Engineering Journal*, 2012, 45(2): 127–146. (In Chinese)
Zheng G., Lei Y., Cheng X., Li X., Wang R. Influences and mechanisms of anchor failure on anchored pile retaining system of deep excavations, *Chinese Journal of Geotechnical Engineering*, 2020a,42(03): 421-429. (In Chinese)
Zheng G., Lei Y., Cheng X., Li X., Yu D. Experimental study on influences of local failure on steel-strutted pile retaining system of deep excavations, *Chinese Journal of Geotechnical Engineering*, 2020b, 41(08): 1390-1399. (In Chinese)
Zheng G., Li S., Diao Y. Centrifugal model tests on failure mechanisms of embankments on soft ground reinforced by rigid piles. *Chinese Journal of Geotechnical Engineering*, 2012, 34(11): 1977–1989. (In Chinese)
Zheng G., Liu J., Han J. Stability of embankment on soft subgrade reinforced by rigid piles (I)—background and single pile analysis. *Chinese Journal of Geotechnical Engineering*, 2010, 32(11): 1648–1657. (In Chinese)
Zheng G., Liu J., Han J. Stability of embankment on soft subgrade reinforced by rigid inclusions (II) - group piles analysis. *Chinese Journal of Geotechnical Engineering*, 2010, 32(12): 1811–1820. (In Chinese)
Zheng G., Liu S., Chen R. State of advancement of column-type reinforcement element and its application in China. *Proceedings of the US-China Workshop on Ground*
Zheng G., Yang X., Zhou H., Sun J. Researches on stability and control strategy of embankments reinforced by rigid piles based on the concept of progressive failure, *Chinese Journal of Geotechnical Engineering*, 2017, 39(4): 581–591. (In Chinese)
Zheng G., Yi F., Huang T., Cheng X., Yu D., Lei Y., Wang R. Mechanism of overturning progressive collapse of excavations retained by double-row piles induced by over-excavation, *Chinese Journal of Geotechnical Engineering*, 2020b, 43(08): 1373-1381. (In Chinese)
Zheng G., Yu X., Zhou H., Yang X., Guo W., Yang P. Influence of geosynthetic reinforcement on the stability of an embankments with rigid columns embedded in an inclined underlying stratum. *Geotextiles and Geomembranes*, 2021, 49(1): 180-187.
Zheng G., Zhang N., Ba Z., Zhang T. Seismic performance of PHC pipe piles in layered soft soils, *Chinese Journal of Geotechnical Engineering*, 2013, 35(S2): 506-510. (In Chinese)
Zheng G., Zhao J., Zhou H., Yu X., Xiao B., Wang J. State-of-the-art review for techniques on ground improvement of highway and railway, *Chinese Journal of Ground Improvement*, 2021, 3(02): 91-99. (In Chinese)
Zheng G., Zhao J., Cheng X., Yu D., Wang R., Zhu X., Yi F. Mechanism and Control Measures of the Vertical Progressive Collapse of Deep Excavations Retained by a Multilayer Strutting System, *Journal of Tianjin University*, 2021, 54(10): 1025-1038. (In Chinese)
Zheng G., Zhou H. State-of-the-Art Review of Ultimate Bearing Capacity and Stability of Composite Foundations, *Journal of Tianjin University*, 2020, 53(07): 661-673. (In Chinese)
Zheng G., Zhu X., Cheng X., Lei Y., Wang R., Zhao J., Yi F. Study on the control theory and design method of progressive collapse in excavations retained by cantilever piles, *Chinese Journal of Geotechnical Engineering*, 2021, 43(06): 981-990. (In Chinese)
Zheng G., Yu X., Zhou H., Wang S., Zhao J., He X., Yang X. Stability analysis of stone column-supported and geosynthetic-reinforced embankments on soft ground. *Geotextiles and Geomembranes*, 2020, 48(3): 349-356.
Zheng J., Guo J., Lee F. Immune algorithm for reliability analysis of geotechnical engineering, *Chinese Journal of Geotechnical Engineering*, 2007, (05): 785-788. (In Chinese)
Zheng Y. *Engineering Treatment of Slope & Landslide*. China Communications Press, 2007. P817. (In Chinese)
Zheng, Y., Hu, Z., Ren, X., Wang, R., Zhang, E. The influence of partial pile cutting on the pile-anchor retaining system of deep foundation pit in loess area. *Arabian Journal of Geosciences*, 2021, 14(13), 1-12.
Zhou L.. Resilient City: Risk Management and Its Indicators. *Journal of Beijing Administrative College*. 2016(02): 13-20. (In Chinese)
Zhou H, Zheng G, Liu J, et al. Performance of embankments with rigid columns embedded in an inclined underlying stratum: centrifuge and numerical modelling. *Acta Geotechnica*. 2019, 14: 1571–1584
Zuo Y., Zhu H., Li X. An ANN-based four order moments method for geotechnical engineering reliability analysis. *Rock and Soil Mechanics*, 2013, 34(2): 513-518. (In Chinese)

Proceedings of the 20th ICSMGE – Post-conference Volume – Rahman and Jaksa (Eds)
© 2023 Australian Geomechanics Society, Sydney, Australia, ISBN 978-0-9946261-7-2

TC221 Tailings and mine waste – Brief report

TC221 Résidus et déchets miniers – Bref rapporten

Ramon Verdugo
CMGI Ltda. Chile, rverdugo@cmgi.cl

Roberto Cudmani
Zentrum Geotechnik, Technische Universität München, Germany, r.cudmani@tum.de

ABSTRACT: Tailings dams and rock waste dumps are among the largest earth structures that are built these days and require specialized geotechnical engineering to ensure their physical stability. Unfortunately, catastrophic failures of tailings dams have occurred, suggesting that the current state of practice has an insufficient set of regulations, government controls, provisions and geotechnical knowledge that are necessary to address and improve. Accordingly, the TC221 has an important role to play in these issues. In this conference the liquefaction phenomenon, its evaluation, and its use in tailings dam stability, attracted an important number of articles. The consolidation phenomenon was also addressed. The proposals for early warning systems show the use of new technologies in geotechnics. The challenge of characterizing waste rock materials is well addressed in the articles presented. Other important topics such as risk analysis, numerical modeling, filter criteria, liners permeability, arsenic attenuation, strength anisotropy, seasonal weathering, alkalic activation, delayed response time of standpipes piezometers, were also presented.

RÉSUMÉ : Les digues à résidus et les décharges de déchets rocheux comptent parmi les plus grandes structures en terre construites de nos jours et nécessitent une ingénierie géotechnique spécialisée pour assurer leur stabilité physique. Malheureusement, des défaillances catastrophiques de barrages de résidus se sont produites, ce qui suggère que l'état actuel de la pratique comporte un ensemble insuffisant de réglementations, de contrôles gouvernementaux, de dispositions et de connaissances géotechniques nécessaires pour traiter et améliorer. En conséquence, le TC221 a un rôle important à jouer sur ces questions. Dans cette conférence, le phénomène de liquéfaction, son évaluation et son utilisation dans la stabilité des digues à stériles, ont attiré un nombre important d'articles. Le phénomène de consolidation a également été abordé. Les propositions de systèmes d'alerte précoce montrent l'utilisation des nouvelles technologies en géotechnique. Le défi de caractériser les matériaux de stériles est bien abordé dans les articles présentés. D'autres sujets importants tels que l'analyse des risques, la modélisation numérique, les critères de filtrage, la perméabilité des revêtements, l'atténuation de l'arsenic, l'anisotropie de résistance, l'altération saisonnière, l'activation alcaline, le temps de réponse différé des piézomètres des bornes-fontaines, ont également été présentés.

KEYWORDS: Tailings, Mine waste, liquefaction, rock dump, mines.

1 INTRODUCTION.

Since the TC221 officially started in March 2021, it was not initially considered in the call for papers. Only in the final phase of paper submission, the inclusion of TC221 in the conference was decided in agreement with other TCs, which had received contributions related to tailings and mine waste. In total, 17 papers were accepted for publication. It is worth noting that several articles related to TC 221 remained in the original TCs or were sent to other sessions.

TC221 on Tailings and Mine Waste began its activities a few months after the start of the pandemic. Due to this situation, so far only online activities have been carried out, this Conference being the first opportunity for hybrid participation.

The technical presentations associated with TC221, as well as that of other TCs, were divided into two sessions: in-person and virtual.

In the format of parallel sessions, on Thursday 5, in Meeting Room C4.5, the in-person session was held from 8:30 to 9:30. The Chairman of this session was Dr. Ramón Verdugo from Chile.

Also, in the format of parallel sessions, on Thursday 5th, the virtual session was held from 09:45 to 10:45. The Chairman of this session was Prof. Roberto Cudmani from Germany.

Most of the presentations had interesting questions and discussions with good interaction between speaker and attendees. Regrettably, due to time constraints the discussions had to be interrupted.

2 PRESENTATIONS

The speakers and technical papers assigned to the in-person session were the following:

- Research results on soil liquefaction of mine dumps in Eastern Germany
  Prof Wolfram Kudla

- A review of liquefaction potential assessment in engineering practice
  Mr Yashay Narainsamy

- Liquefaction of mine dumps in Eastern Germany - new findings after a five-year monitoring research project
  Dr Tino Rosenzweig

- Flow liquefaction triggering analyses of a tailings storage facility by means of a simplified numerical procedure
  Mauro Sottile

- The concept of static liquefaction and its application in stability analysis of tailings dams
  Dr. Ramon Verdugo

The speakers and technical papers of virtual session were the following:

- The implications of transitional behaviour for tailings
  Prof Matthew Coop

- Towards sustainable and safe tailings dams
  Dr Luca Piciullo

- Natural weathering as a surface crusting tool for tailings management
  Ms. Umme Rima

- Shear strength estimation of very coarse mine waste using Barton's empirical criterion and a Bayesian analysis
  Ms. Sandra Linero Molina

3 GENERAL REPORT

The liquefaction phenomenon, its evaluation, and its effect on the stability of tailings dams attracted the interest of the many participants and resulted in a significant number of contributions. A contribution dealt with the spontaneous soil liquefaction events occurring in the dumps of former open-pit lignite mines in Lusatia, Germany. The consolidation of the slimes in TSFs was also in the focus of the contributions. Proposals for early warning systems show the potential of new monitoring technologies in geotechnics. The challenge of characterizing waste rock materials was also addressed in the TC sessions.

Other important topics such as risk analysis, numerical modeling, filter criteria, liners permeability, arsenic attenuation, strength anisotropy, seasonal weathering, alkalic activation, delayed response time of standpipes piezometers, were also discussed in the sessions.

This variety of topics allows us to conclude that the geotechnics associated with tailings and mine waste is extraordinarily broad, probably requiring a sub-specialization in mining geotechnics.

4 ACKNOWLEDGMENTS

The 20th ICSMGE held in Sydney was one of the first in-person events since the begin of the pandemic restrictions in 2020. This undoubtedly meant an additional effort for the organizers, who managed to successfully carry out this event. On behalf of TC221, we want to express our gratitude to the organizers for everything they have done.

Proceedings of the 20th ICSMGE – Post-conference Volume – Rahman and Jaksa (Eds)
© 2023 Australian Geomechanics Society, Sydney, Australia, ISBN 978-0-9946261-7-2

TC 307 – Sustainability in Geotechnical Engineering – Brief report of the 20th International Conference on Soil Mechanics and Geotechnical Engineering

TC 307 – Durabilité en génie géotechnique – Bref rapport de la 20ème Conférence internationale de mécanique des sols et de géotechnique

Anand J. Puppala
Zachry Department of Civil and Environmental Engineering, Texas A&M University, USA

Md Mizanur Rahman
STEM, University of South Australia, Australia

Olivier Cuisinier
LEMTA (UMR 7563) CNRS, Université de Lorraine, Nancy, France

ABSTRACT: Sustainability in Geotechnical Engineering (TC 307 committee of ISSMGE) was proactive in participating in various technical committee activities at the 20th International Conference on Soil Mechanics and Geotechnical Engineerin g in Sydney, Australia from 1 May to 5 May 2022. TC 307 held a few technical committee meetings for preparation of the International conference and organised two sessions, one in-person session and one parallel session, during the confer ence. Both events are reasonably well attended. This article presents a brief report of TC 307's activities during the conf erence along with various initiatives planned for the next four years.

RÉSUMÉ : Le comité 307 - Sustainability in Geotechnical Engineering - a été proactif en participant à la 20ème Conférence internationale de mécanique des sols et de géotechnique qui s'est tenue à Sydney, en Australie, du 1er mai au 5 mai 2022. Le TC 307 a tenu quelques réunions de comité technique pour la préparation, et a organisé deux sessions, une en présentiel et une parallèle, le 5 mai lors de la conférence. Cet article présente un bref rapport sur les activités du TC 307 pendant la conférence ainsi que les actions qui seront entreprises dans les prochaines quatre années .

KEYWORDS: Sustainability, Geotechnical Engineering, Solid waste, Geothermal

1 INTRODUCTION

The 20th International Conference on Soil Mechanics and Geotechnical Engineering (ICSMGE2022) was held in Sydney from 1 May to 5 May 2022. The conference was held in "hybrid mode" combining in-person and virtual (online) sessions.

In preparation for ICSMGE2022, TC 307 - Sustainability in Geotechnical Engineering organised several technical committee meetings and conducted a survey in 2021 among 42 technical committee members for preference on the mode of TC sessions. Both in-person and virtual sessions were suggested due to COVID-19 related travel restrictions and uncertainties. Therefore, TC 307 organised two sessions: in-person session and a virtual session at International Convention Centre in Sydney, Australia on 5 May. In addition, Prof Mizanur Rahman, Secretary of TC307, was extensively involved in the technical activities of the conference as a Technical Program Committee member and the lead editorial member of the conference proceedings.

2 TC 307 SESSIONS IN ICSMGE2022

A total of 22 articles were allocated to the theme of TC 307 - Sustainability in Geotechnical Engineering. The country of origin of these articles and their distribution is shown in Figure 1. Apart from the host country of Australia, China, Portugal, and the USA had higher article contributions to the TC 307's theme. China, Portugal, and the USA also had higher representatives in the TC.

TC 307 could allocate only five articles to present in the in-person session and five articles in the virtual session.

2.1 *In-person sessions*

The in-person session was one hour on 5 May and was chaired by Professor Puppala and Professor Rahman. Each speaker was allocated 10 minutes of presentation and 2 minutes for question and answers.

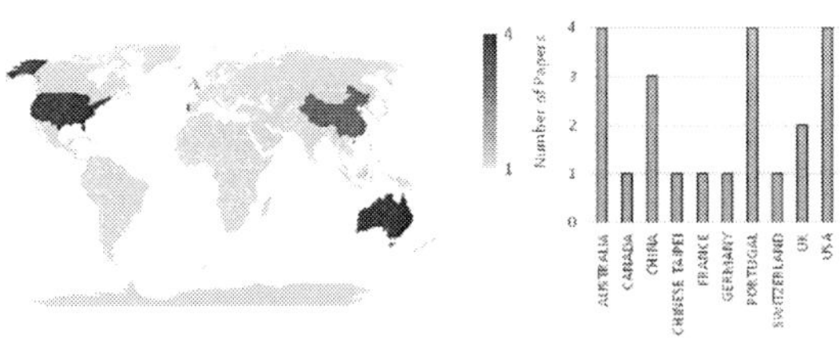

Figure 1. Distribution of TC 307- Sustainability in Geotechnical Engineering Theme articles in the world map.

The title of the contribution and the presenters are listed below:

- Beneficial use of geopolymers in geotechnical engineering applications by Dr Jayawickrama
- Industrial solid wastes acting as barrier material for storing solid wastes (SW) and wastewaters – A critical review by Mr Marchiori
- Geothermal pavements: 3D numerical modelling for the long-term thermal performance by Professor Narsilio
- On the Landslide Hazard and Resilience with the Impact of Climate Change in Taiwan by Professor Shou

- Minimally Invasive Soil Treatment Based on Bio-Grout by Dr Terzis

Figure 2 presents a photograph of all the speakers of the in-person session meeting along with Chair and Secretary of the TC307 committee.

Figure 2. Speakers and TC307 Leadership After the In-Person Session

2.2 *Virtual session*

The virtual session was one hour on 5 May and was chaired by Professor Rahman and Professor Puppala. Each speaker was allocated 10 minutes of presentation/video recording. Then there was 10 minutes of live question and answer. The title of the contribution and the presenters are listed below:

- Recycling of marginal materials to replace traditional capping in track substructure by Dr Tawk
- Performance of recycled aggregates as unbound structural layers for airfield pavements by Dr Yaghoubi
- Enhancing the uniformity and extent of bio-cementation soil improvement: physical experiments and reactive transport modelling by Dr Gomez
- Investigating Water Holding Barriers for Climate Adaptation by Professor Toll
  Effect of wetting-drying cycles on physical and mechanical properties of a recycled construction and demolition material by Professor Vieira

2.3 Committee Meeting

The TC307 meeting was held on 3 May in hybrid mode and we have 20 plus members attended in both modes. TC liaison. Prof Pierre Delage, Former Chair of Technical Oversight Committee and Prof. Marcelo Sanchez, New Incoming Chair of TOC were in attendance. Profs Rahman and Puppala presented TC307 activities including results from Time Capsule project coordinated by Prof Olivier Cuisiner.

Prof Hazarika presented 2023 CREST conference in Japan and invited TC307 to have a major activities as this committee was accepted as the principal TC from ISSMGE. Prof Dipanjan Basu was in-charge for TC307 activities at 17th Asian Regional Conference in Kazakhstan. Committee also discussed about participation at various regional conference events in Europe and Africa.

3 CONCLUSIONS

Two sessions (in-person and virtual) and an hybrid TC307 committee meeting was organised by TC 307: Sustainability in Geotechnical Engineering during the 20th International Conference on Soil Mechanics and Geotechnical Engineering. 22 articles were allocated to TC 307's theme. Five articles were presented during in-person parallel session and five other articles were presented live/recorded video during the virtual session. Committee meeting held at the conference focused on 2022-23 activities, Time Capsule project details, and upcoming conferences where TC307 will be participating. Overall, TC307 was pleased with committee activities in the Sydney conference as it is one of the few ISSMGE committees that were given two sessions and a hybrid committee meeting.

Minutes of ISSMGE Council Meetings

 ISBN 978-0-9946261-7-2

International Society for Soil Mechanics and Geotechnical Engineering

Minutes of the Council Meeting

Held at Century City Conference Centre, Cape Town, South Africa

Sunday, 6th October 2019

PRESENT

Charles Ng	- ISSMGE President
Roger Frank	- Immediate Past President ISSMGE
Etienne-Marcelin Kana	- ISSMGE Vice President Africa
Eun Chul Shin	- ISSMGE Vice President Asia
Philip Robins	- ISSMGE Vice President Australasia
Mario Manassero	- ISSMGE Vice President Europe
Timothy Newson	- ISSMGE Vice-President North America
Alejo Sfriso	- ISSMGE Vice President South America
Mounir Bouassida	- Appointed Board Member
Pedro Sêco e Pinto	- Appointed Board Member
R Neil Taylor	- Secretary General
C Paloma Peers	- ISSMGE Secretariat
Peter Day	- Chair, Corporate Associates Presidential Group
Sukumar Pathmanandavel	- Vice-Chair, Corporate Associates Presidential Group
Lucy Wu	- Chair, Young Member Presidential Group
Dimitrios Zekkos	- Chair, Innovations and Development Committee
Pierre Delage	- Chair, Technical Oversight Committee
Ikuo Towhata	- Chair, Professional Image Committee
Roberto Terzariol	- Chair, Awards Committee

Member Society	**Voting Member**	**Non-Voting Member**
Albania	Professor Mario Manassero *	
Algeria	Professor Bahar Ramdane	
Argentina	Professor Roberto Terzariol	
Australia	Mr Graham Scholey	
Austria	Professor Helmut F. Schweiger	
Bangladesh	---	
Belarus	Professor Askar Zhusupbekov *	
Belgium	---	
Bolivia	Dr Alejo Sfriso*	
Bosnia & Herzegovina	---	
Brazil	Dr Alessander Kormann Bulgaria	
Canada	Professor Tim Newson *	
Chile	Dr Alessander Kormann*	
China	---	
Chinese Taipei	Professor EC Shin *	
Colombia	Dr Alejo Sfriso*	
Costa Rica	---	
Croatia	Professor Pedro Sêco e Pinto *	
CTGA	---	
Cuba	---	
Cyprus	---	
Czech & Slovak Republics	---	
Denmark	---	
Dominican Republic	---	
Ecuador	---	
Egypt	Professor Fatma Baligh	
Estonia	---	
Finland	---	

Member Society	Voting Member	Non-Voting Member
France	Mrs Valérie Bernhardt	
Georgia	---	
Germany	Dr Ing. Wolfgang Sondermann	
Ghana	Professor S.I.K. Ampadu	
Greece	Mrs Valérie Bernhardt *	
Guatemala	---	
Hong Kong	Dr André Archer	
Hungary	---	
Iceland	---	
India	---	
Indonesia	---	
Iran	---	
Iraq	---	
Ireland	---	
Israel	---	
Italy	Mrs Valérie Bernhardt *	
Japan	Professor Toshifumi Mukunoki	Dr Takeshi Katsumi
Kazakhstan	Professor Askar Zhussupbekov	
Korea R	Professor EC Shin *	
Kyrgyzstan	---	
Latvia	---	
Lebanon	---	
Lithuania	---	
Malaysia	Professor EC Shin *	
Mexico	---	
Morocco	---	
Mozambique	Dr Saturnino Diogo	Dr Sidney de Abreu
Nepal	---	
Netherlands	---	
New Zealand	Mr Philip Robins *	
Nigeria	Mr Will Ibim Beresibo	
North Macedonia	Professor Helmut F. Schweiger *	
Norway	---	
Pakistan	---	
Paraguay	Professor Roberto Terzariol*	
Peru	---	
Poland	Professor Mario Manassero *	
Portugal	Professor Pedro Sêco e Pinto	
Romania	Professor Pedro Sêco e Pinto *	
Russia	Prof. Askar Zhusupbekov *	
Serbia	---	
Singapore	Dr Chandrasegaran S Sundararaju	Professor Leung Chun Fai
Slovenia	---	
South Africa	Mr Solly Phalanndwa	Mr Jacobus Breyl
S E Asia	---	
Spain	Professor Antonio Gens	
Sri Lanka	---	
Sudan	Professor Elhussein A. E. Elarabi	Dr Abdelkarim M. Zein
Sweden	---	
Switzerland	Professor Mario Manassero *	
Syria	---	
Tajikistan	---	
Thailand	---	
Tunisia	---	
Turkey	---	
U K	Professor David Toll	
U S A	Professor Domniki Asimaki	
Uzbekistan	---	
Venezuela	---	
Vietnam	---	

* Denotes Proxy vote

Apologies:

Kok Kwang Phoon	- Appointed Board Member
Matthew Coop	- Chair, TC101 – Laboratory Testing
Andy Take	- Chair, TC208 – Slope Stability
Phil Watson	- Chair, TC209 - Offshore
Jianye Ching	- Chair TC304 - Risk
Zhongqiang Liu	- Chair TC309 - Machine Learning

Member Societies

Albania -Luljeta Bozo
Belgium - Noel Huybrechts
Bolivia - Mario Terceros
Croatia - Igor Sokolić
Cuba - Gilberto Quevedo
Cyprus - Dimitrios Loukidis
Denmark - Henrik Mikkelsen
Finland - Leena Korkiala-Tanttu
Hungary - András Mahler
India - GL Sivakumar Babu
Ireland - Ciaran Reilly
Lebanon - Salah Sadek
Norway - Geraldine Sørum
Philippines - Alexis Acacio
Poland - Katarzyna Zabielska-Adamska
Venezuela - Heriberto Echezuria
Vietnam- Phung Duc Long

1 OPENING REMARKS BY PRESIDENT AND AGREEMENT OF THE AGENDA

The President opened the meeting by welcoming all delegates including Member Society representatives, the ISSMGE Board and Chairs of the Board Level Committees to the meeting. He thanked the Geotechnical Division of the South African Institute of Civil Engineers and in particular its Chair, Solly Phalanndwa, and Peter Day for their support and arrangements for this Council Meeting. He went on to ask if there were any changes to be made to the agenda, and there were none.

2 REPORT BY THE PRESIDENT

The President gave a brief report to Council (see Appendix 1). He began by presenting a brief history of ISSMGE, introduced the ISSMGE Board to Council and outlined the ISSMGE organisational structure. He noted the occasions at which the Board had met. He recognised the recent state awards presented to Professor Heinz Brandl and Suzanne Lacasse and congratulated the individuals who had promoted the ISSMGE to the highest national level. He went to describe developments of the ISSMGE Virtual University including additions to the ISSMGE online library and the creation of online courses and further IT developments such as the ISSMGE Conference Paper Review Platform. He acknowledged the great contributions made by Mounir Bouassida, who had been largely responsible for organising the course material and new courses to be added in the future and by Dimitrios Zekkos and Kok Kwang Phoon for their significant developments of the online library and the conference paper review platform.

In summarising key achievements over the last 2 years, the President commented on the new Conference Paper Review Platform that can be accessed from the ISSMGE website, the 3 new Technical Committees on System Performance, Field Monitoring and Machine Learning. A new award for young members had been introduced – the Bright Spark Lecture award – and a number of lectures had now been given. The Corporate Associates Presidential Group led by Sukumar Pathmanandavel and Peter Day had launched a major survey of international practice entitled "Are we overdesigning?" and were also active in developing debate or discussion plenary sessions at all the ISSMGE Regional Conferences in 2019. Finally, he thanked Pedro Sêco e Pinto for the development and delivery of a number of International Seminars that were reaching out to potential new members and strengthening the relationship between ISSMGE and its existing members.

3. LIST OF DELEGATES AND OTHER PERSONS PRESENT

The delegates listed above were noted, as were the voting and proxy vote delegates.

4 CONFIRMATION OF QUORUM (STATUTE 12K)

The Secretary General confirmed that with 39 votes available from 73 Member Societies entitled to the benefits of membership, the meeting was quorate for all business declared on the agenda.

5 CONFIRMATION OF MINUTES OF THE SEOUL COUNCIL MEETING

The minutes of the Council Meeting in Seoul, September 2017 were confirmed as an accurate representation of the meeting.

6 MEMBERSHIP. REPORT BY SECRETARY GENERAL

The Secretary General delivered his report, which is included as Appendix 2. It was noted that although there had been changes in membership of individual Member Societies, the overall trend was of an increase to more than 21,100 members from 89 Member Societies. In addition, there were now 39 Corporate Associates.

The Secretary General reported that an application for ISSMGE membership had been received from the Mongolian Association of Soil Mechanics and Geotechnical Engineering (MASMGE). The Board had approved the application and the MASMGE was welcomed to membership of ISSMGE.

The Secretary General reminded Member Societies that there was regular communication from the Secretariat via the monthly circular and that Member Societies should always forward it to their members. Annual subscription notices were sent out in January each year and these should be settled promptly.

7 AMENDED COOPERATION AGREEMENT FOR THE FEDERATION OF INTERNATIONAL GEO-ENGINEERING SOCIETIES

The President introduced this item stating that during the FedIGS meeting in September 2018, the FedIGS Board had suggested changes to the Cooperation Agreement with respect to the election of its President. They considered that the FedIGS President should serve for one term (of 4 years) only and that the next President should be from a different member organisation than the current president.

In discussion, Professor Ampadu (Ghana) asked why there was not a simple rotation of the member organisation from which the President would be selected, as in his view there is a danger of the presidency moving between the two strongest societies. Roger Frank stated that there would then effectively be no vote. An organisation might not be in a position to provide a candidate and the overall aim was to minimise prescriptive conditions.

There being no further comments, the President called for a vote by a show of hands and the revised Cooperation Agreement, given in Appendix 3, was approved unanimously.

The President continued by reporting on the recent FedIGS meeting held in September 2019. It had been noted that the two joint committees, JTC1: Natural Slopes and Landslides and JTC2: Representation of Geo-engineering Data, were making good progress. During the meeting, the FedIGS Board had suggested that one or two keynote lectures at the International Conference could be given by experts from the other FedIGS Societies. This had met with unanimous approval. The FedIGS Board had also considered that it would be timely for a follow-up conference to the GeoEng2000 event organised in Melbourne in 2000. However, a willing organiser had yet to be found

8 REGIONAL REPORTS BY VICE PRESIDENTS

The President invited the Regional Vice-Presidents in turn to present their reports on regional activities. The written reports are given in Appendix 4.

AFRICA: Etienne Marcelin Kana noted that 10 of the 13 Member Societies from Africa were active and that because of "Le Comité Transnational des Géotechniciens d'Afrique" (CTGA), these societies came from 28 of the 40 prominent countries in Africa. A number of conferences had been held in Africa, the most notable being the 6th African YGEC held in Sudan and the International Conference on Advances in Structural and Geotechnical Engineering (ICASGE19) organised in Egypt. The latter event included 10 keynote lectures, and during the conference, there was a great opportunity to meet with the Egyptian Geotechnical Society to discuss issues of communication, unification of standards etc. Other regional events include the Journées Africaines de la Géotechnique (an annual event), a seminar in Mozambique and various events in South Africa. The membership is distributed evenly between academics and practitioners. While it is encouraging to see a number of upcoming events, there remain many challenges such as communication, transport and poor involvement of academics and practitioners in some countries. For the future, it was noted that there was need to improve training and the range of regional activities including contributions towards Technical Committees.

ASIA: Eun Chul Shin reported that there was considerable activity in the Asia region and a wide range of conferences and other events are listed in the written report. Much of the growth in the region has been due to construction. He was pleased to acknowledge the increase in Member Societies with the addition of the Philippine Society for Soil Mechanics and Geotechnical Engineering and the Mongolian Association of Soil Mechanics and Geotechnical Engineering. The seven regional Technical Committees continued to be very active.

AUSTRALASIA: Philip Robins passed on the apologies from Gavin Alexander who had had to stand down as VP on health grounds. Reflecting on the written report that Gavin had prepared, Philip noted that though the region only had two Member Societies, it had a strong, active and relatively large membership. There were well-established conference series that attracted many delegates and there was good activity with the Young Geotechnical Professionals Conference series. The region had many individuals contributing to Technical Committee activity and the regional membership of the TCs was reviewed regularly. There was an ongoing project looking at the registration of geotechnical engineers and, following the Christchurch earthquake, the New Zealand Geotechnical Society has been active in developing a series of earthquake guidelines.

In discussion, Professor Ampadu enquired how it had been possible to motivate young engineers to be so active. Philip Robins replied that youngsters want to be heard and that it was important to get them to make presentations on their work and for more senior engineers not to judge their work but to be encouraging. Fostering the career development of young engineers had proved very beneficial.

EUROPE: Mario Manassero noted that the European region had 38 Member Societies and was the largest region within ISSMGE in terms of number of societies and number of members. There were very many conferences and events organised throughout the region. These could be national, sub- regional, regional or international events and included Young Geotechnical Engineers conferences. The region was very active within the ISSMGE Technical Committees and there were also 4 Regional TCs.

NORTH AMERICA: Tim Newson commented that although the region had a very lard land mass, it only had three Member Societies. All three societies were very active and organised a number of national and international conferences. Of these, the national conference organised by the Canadian Geotechnical Society is of particular note, as it is the longest running global conference series in geotechnical engineering, having started in 1947. Concerning cross-border awareness, it had been noted that each country had its own geotechnical trade magazine and these had each now agreed to showcase projects from the other two regions. Other cross-border initiatives were geared at ensuring good communication between officers of the Member Societies.

SOUTH AMERICA: Alejo Sfriso noted that in the region, Brazil was by far the largest society but that although the other societies were smaller, they were still very active, organising many events. As VP, he had been engaging with the Member Societies and he was working hard to improve communication and collaboration between the societies. A number of Member Societies had been terminated in 2017. Of these, two, Bolivia and Peru, had restarted with new national society groups and it was hoped that similar relaunching would be possible with Ecuador. A successful series of International Seminars had taken place in Paraguay, Argentina and Mexico in September 2018 and these were well attended. Discussions were ongoing about the creation of a new Technical

Committee on Mine Waste. This could be hosted by Chile and it was hoped to launch early next year.

REPORTS ON ACTIVITIES BY CHAIRS OF THE BOARD LEVEL COMMITTEES

9 INNOVATION AND DEVELOPMENT COMMITTEE (IDC)

Dimitris Zekkos presented his report on the initiatives undertaken by the IDC, which is given in Appendix 5.

(a) *SIDC Survey*

This had been undertaken to assess the past and prioritise possible future developments. The main features appreciated were the ISSMGE website, the open access online library and webinars. Of the possible future developments, the ones receiving most interest were expansion of the open-access publications database and an increase in the technical content and resources.

(b) *ISSMGE website*

The site continues to grow and new features are often added. The aim is to ensure the site is of value both to Member Societies and to geotechnical engineers and receives a continuous growth in traffic. Technical Committees can add news directly to their part of the website and soon they will be able to add new pages with, for example, technical content and achievements of the TC. Anyone can sign up to receive TC news by using the button featured on each TCs main page.

(c) *Open-access library*

The online library remains open access and free to all. It has grown about 30% in the last two years and now holds over 13,000 papers. New conference series are being added which include some Australia New Zealand regional conferences and the latest European and African Regional conferences. Paper downloads are approaching 500,000 each year.

(d) *Conference support*

A new Paper Review Platform has been added via the ISSMGE website, which can be of help to conference organisers in dealing with abstract submission and subsequent paper review. At present, three ISSMGE conferences were making use of the platform and feedback was very encouraging.

Member societies can have free use of the system, but they must be committed to Open Access.

Future developments for supporting conference organisers could include the publication of proceedings. This would require the submission of papers to Scopus and other databases and, depending on the requirements of organisers, may require proceedings to be loaded onto say a USB drive and possibly printed using an on-demand system. The implications of these potential developments were presently being investigated.

(e) *Virtual University Platform*

Progress with this has started with the creation of "courses" from existing and new webinars. A new- look platform will be released in the coming months that utilises OpenEdX and will give the platform an improved "course" feel. There are plans in hand to expand the content in terms of courses, keynote lectures and educational videos and t integrate it with the Online Library. Though not a requirement, a user would be able to register on the platform and would then be able to track the courses followed etc.

Pierre Delage commented that since there are many conferences within the online library then it should be easy to explain to Scopus that the ISSMGE conferences are part of a series. In that way, it should be possible to get a listing on their citation database.

(f) *Communication*

The ISSMGE has a presence on a number of social media sites and the ISSMGE app for mobile phones has over 1000 users who get regular push-in notifications. In addition, the GeoWorld community – a professional network for geotechnical engineers worldwide - continues to grow.

10. TECHNICAL OVERSIGHT COMMITTEE

Pierre Delage presented his report on the initiatives undertaken by the TOC, which is given in Appendix 6.

The Technical Committees are of paramount importance to the ISSMGE. They provide a forum for active participation by the individual members of ISSMGE and are a meeting arena for discussing, developing and applying specialist geotechnical knowledge related to geotechnical engineering. They can also provide good source material for the ISSMGE Virtual University. In order to be of maximum value, it is important that Member Societies are proactive in ensuring their TC representatives are active and committed. It is very important that Member Societies respond with due efficiency and nominate promptly members to new TCs. They should also review periodically their TC members and keep the TC database up to date. A recent comment highlighted the problem with inactive member and Member Societies need to take heed and rectify such problems. All Regional Vice-Presidents are members of the TOC and as well as monitoring TC activity in their region they help in managing the ISSMGE support to the Conferences organised in their region by TCs and ISSMGE members.

In the 2-year period since the last Council Meeting, ten TCs have new Chairs and three new TCs have been created with a fourth in the early stages of development. It is anticipated that the TCs will have excellent opportunities to be involved in the ICSMGE Sydney 2021 via discussion sessions and workshops and, if they wish, in hosting Honours Lectures within their parallel session. TCs have also been asked for proposals of plenary State-of-the-Art lectures for the Sydney conference.

TCs are encouraged to promote talented young geotechnical engineers through Bright Sparks plenary lectures in their conferences. It is also hoped that TCs will publish their conference proceedings using the ISSMGE online library with open access to the content making their work widely available. Their research achievements can thereby be known and transfer to practice. It is envisaged that the TCs will be making many contributions to the ISSMGE Virtual University.

As part of its mission, the TOC undertakes periodic reviews of TC activity and critically reviews the functionality and performance of TCs. The review is generally in the form of analysis of responses to a questionnaire. Where failings are noted, the TOC will act to address the problems and initiate changes that will get the TC back on track.

11. YOUNG MEMBERS' PRESIDENTIAL GROUP

Lucy Wu presented her report on the initiatives undertaken by the TOC, which is given in Appendix 7.

The YMPG presently has a core group of three members (<35 years old) from each region. They have a monthly Skype meeting and activities include developing the Bright Spark Lecture award, collaborating with the IDC and CAPG, contributing articles to the ISSMGE Bulletin and creating a resources page for young members. The YMPG had also been able to hold an in-person meeting in Singapore in March 2019.

In the next term, the YMPG planned to transition to a smaller but tighter group with strong links to Member Societies, TCs and the Board Level Committees. The smaller core group should permit easier dissemination of ISSMGE initiatives and would encourage societies to start young member groups, which could also interact with each other.

The President advised Council that the Bright Spark Lectures were restricted to people under 36 years of age but that they were not limited to Regional or International conferences but could be in TC or other Regional TC conferences; young people should be encouraged not restricted. Sukumar Pathmanandavel asked if Bright Spark lecturers be a source of young members to engage with YMPG and the ISSMGE. Lucy Wu thought this to be a good idea.

12 CORPORATE ASSOCIATES' PRESIDENTIAL GROUP

Peter Day presented the CAPG report, which is given in Appendix 8.

The CAPG is a Board Level Committee of the ISSMGE comprised of various consultants, contractors, researchers, universities, equipment and product suppliers. It developed actions and activities to enhance the commercial sector of the ISSMGE and in the period 2017 – 2021 would address issues of relevance to the Geotechnical Engineering practice. The Committee was established in 2011 and now had 39 members. Benefits of membership of the CAPG included participation in Board-level committee activities with other industry leaders, fostering interaction between practitioners and academics, helping to address important issues in geotechnical engineering practice, having direct interaction with ISSMGE Board members and the President and enjoying the presence of a logo on ISSMGE platforms.

The focus for 2017-2021 was to provide visible and prominent platforms to discuss and promote practical geotechnical engineering issues. The CAPG was involved in a plenary session at all five ISSMGE Regional conferences in 2019. Each session had a different theme, which considered topics from collaboration, implementation, innovation, design and education.

In considering design, the CAPG had initiated a global survey of geotechnical practice. Ten simple design problems were posed that involved sand and clay profiles. A number of responses had been received so far and a preliminary analysis of these would be presented as the Asian Regional Conference. There was a wide range of solutions and practices being followed. The survey would remain live for another year with the closing date for entries being 31 December 2020. A full analysis and key findings from the survey would be presented at the ICSMGE in Sydney 2021.

For the future, the CAPG would be involved with re-imagining geotechnical engineering by practitioners and researchers working in concert to the benefit of the wider community.

13. AWARD COMMITTEE

Roberto Terzariol presented his report, which is given in Appendix 9.

The purpose of the ISSMGE awards was to promote and recognise individual members and societies who made important contributions to our profession. The awards will be presented at the ICSMGE in Sydney 2021. The awards available are:

- Outstanding Geotechnical Project
- Outstanding Innovator
- Outstanding Member Society
- Outstanding Paper Published in the International Journal of Geo-Engineering Case Histories
- Outstanding Public Relations Award
- Outstanding Technical Committee
- Outstanding Young Geotechnical Engineer Award
- ISSMGE Bright Spark Lecture award

Award nominations are open now and will close on 1 July 2020.

14. PROFESSIONAL IMAGE COMMITTEE

Ikuo Towhata presented his report, which is given in Appendix 10.

The Committee had 16 members from 13 different countries. A number of objectives had been defined for the period 2017-2021:

Objective 1: Dissemination of Contribution to the Public Welfare made by Geotechnical Engineering
Objective 2: Mitigation of Geo Risk
Objective 3: Future Possibility of Geotechnical Engineering
Objective 4: Production of Materials
Objective 5: Communication with Public
Objective 6: Communication with Clients
Objective 7: Awards
Objective 8: Other relevant activities to promote the value of geotechnical engineering among public

Regarding dissemination, it was noted that geotechnical engineers have many excellent achievements but these are generally not well known or appreciated. Better methods need to be found to communicate our work to the public. Mitigation of risk is very important. Some studies have demonstrated that increased investment in, for example, site investigation can reduce risk by improved design but importantly also increase profit. Getting clients to understand this requires move effective communication channels than appear to exist at present. A way of improving communication might be by increasing the status of geotechnical engineers so that their opinions carry more weight. Other ways to reach the public could be, for example, by the excellent videos on the theme "Let's talk about geotechnics" produced by the French Geotechnical Society (Comité Français de Mécanique des Sols et de Géotechnique).

The President thanked Ikuo Towhata for his presentation and stated that he was looking forward to hearing about actual achievements of the Professional Image Committee at the Sydney conference.

15 INTERNATIONAL JOURNAL OF GEO-ENGINEERING CASE HISTORIES

Dimitris Zekkos presented his report on the Journal. The activities undertaken in the past two years can be summarised as:

- New Editorial Board Membership
- Established Ombudsman process
- Established TC liaisons
- New Journal website
- New Online Review System
- Statistics on Impact: Paper Downloads
- Established Special issues Policy
- Getting the journal Indexed

The ombudsman process had been introduced such that there was a completely independent and accountable method to deal with any complaints. Establishing liaison with a number of TCs was seen as a good way to both improve the review process and increase submissions to the journal. The new website was seen as a way to rejuvenate the journal and the new online review system had made the review process more efficient and rapid. The journal remained free and open-access and the download rates of papers was very high, averaging 931 downloads per paper per year. The journal was establishing a special issues policy, which would include issues for Technical Committees on specific technical topics, Member Societies for specific regions and other groups that are interested in a special issue. The journal papers are presently indexed in Georef and Google Scholar. The possibility of being indexed in Scopus and ISI is being investigated. The goal is to increase the circulation of the journal papers and to encourage references to the papers in order to achieve wider achieve indexing.

Dimitrios Zekkos thanked the following organisations that had supported the journal and made it sustainable:

Dar Al-Handasah group (with special thanks to the late Dr. Sherif Wissa)

The Shamsher Prakash Foundation (with special thanks to Prof. Shamsher Prakash)

ConeTec (with special thanks to Bruce Miller) Geosyntec (with special thanks to Dr. Rudy Bonaparte

16 THE ISSMGE FOUNDATION

Neil Taylor presented the report on The ISSMGE Foundation, which is given in Appendix 11.

As reported previously, the Foundation was created January 2010 and converted to a UK Charity September 2013. The purpose of the Foundation was to allow geotechnical engineers from around the world to be more able to make use of resources that would otherwise be inaccessible to them. Its stated objective is the advancement of education and educational training in the field of geotechnical engineering and its applications for the public benefit and it achieves this by making awards to individuals who apply for funding. Awards are based on an assessment of the applicants need for an award, the quality of the activity to be undertaken (usually participation in a conference), the extent of cost sharing (100% awards are not made) and the potential impact of the activity on the career progress of the individual and of his or her Member Society. Since August 2010, there have been 154 recipients of an award from 187 applications in 37 countries. The average award is GBP900.

The generosity of the ISSMGE in donating to The ISSMGE Foundation was acknowledged; it is much appreciated. The reserves had reached approximately GBP550000 and the trustees were required by the UK Charities Commission to consider and develop a financial and reserves policy. They had decided that it would be sensible to create an endowment fund to support the objective of the Foundation and had recently invested GBP500000 of the reserves. In the long term, it was hoped that the invested sum could reach GBP2000000 and that annual returns from such an investment would then be sufficient to meet requests for funding.

In discussion, Wolfgang Sondermann (Germany) asked about the finances and in particular, why the Foundation was accumulating money and not simply spending it. Neil Taylor replied that there were very few rejected applications and that despite a high rate of making awards the reserves had nevertheless built up. It seemed to make more sense to invest the money rather than leave it in a low- interest bank account. Helmut Schweiger (Austria) wondered how the figure of GBP2000000 was arrived at. Neil Taylor replied that the trustees thought it a reasonable target.

Hussein Elarabi (Sudan) asked why a recent application from Sudan had been rejected. Neil Taylor explained that banking regulations prevented transfer of money to Sudan. Sudan is one of a number of sanctioned countries and, having taken legal advice it became clear there could be no financial transaction from the UK-based Foundation to Sudan. The problem had also occurred with an applicant from Cuba, which is also a sanctioned country. Regarding who could receive an award from the Foundation, it was noted that there was a wide range of applicants including recently the President of the Venezuelan Geotechnical Society who otherwise would not have been able to attend an important meeting in South America called by their Regional Vice-President Alejo Sfriso. The query was raised if Bright Spark Lecturers could receive an award and it was confirmed that they were eligible to apply.

Philip Robins asked if there was any feedback from the Foundation. Neil Taylor replied that there was an annual report and that if required there could be further reporting to the Board. In addition, the ISSMGE Board can feedback to the Trustees what it would like to see from the Foundation in the future.

Roger Frank volunteered a brief history of the Foundation, stating that the decision about funds being transferred to the Foundation had been taken explicitly by the ISSMGE Council in 2013.

The President commented that the operation of the Foundation lacks of transparency to the ISSMGE Board and the Council. For instance, the recent change of one of Trustees has never been reported to the Board. Neil Taylor replied that the very recent change had occurred after papers for the Council Meeting had been circulated in May 2019. As is normal practice, an updated report will be included with the minutes of this Council Meeting. The President asked for clarifications on who could be a Trustee of the Foundation and the appointment procedure. Neil Taylor stated that the Board of Trustees met at least once per annum and at those meetings considered if the Board of Trustees should be increased in size. If so, they considered who they might approach to join the Board. There was no restriction on who could serve as a trustee other than the UK Charities Commission being averse to the current ISSMGE Board as they could have a conflict of interest when considering awards. However, the Charities Commission had permitted the Secretary General to be (ex officio) a charity trustee since the Secretary General does not represent any ISSMGE Member Society. The Board of Trustees of the ISSMGE Foundation are responsible for maintaining an active group of trustees and consider on an ad-hoc basis when new appointments should be made and whom to approach. Neil Taylor assured the President that the Trustees of the Foundation were all strong supporters of the ethos of ISSMGE and they were committed to making sure that continued to be the case. The President commented that the accountability of the Foundation to the ISSMGE should be strengthened. For example, the Council and the Board, except the Secretary General, cannot decide how to use the funds. This seemed to be a problem. Neil Taylor replied that the Board (2009 – 2013) agreed to create the ISSMGE Foundation as an independent Charitable Incorporated Organisation and that this was formally established at the start of the Board (2013 – 2017). Being independent meant that it was not accountable to the ISSMGE. However, the Board of Trustees were mindful of the generous donations received from the ISSMGE and kept in mind the ethos of the ISSMGE when making awards.

17 THE ISSMGE FOUNDATION

Anthony Leung, Editor-in-Chief of the Bulletin presented his report, which is given in Appendix 12.

Since taking over as editor-in-chief, Anthony Leung had established a new editorial board with good international representation. The Bulletin is publishing bimonthly and is available via the ISSMGE website and the ISSMGE mobile app. Key sections in the Bulletin were research highlights, conference reports, the Young Members Arena, the TC corner and reports from the Board and Member Societies. Anthony Leung was always seeking new material and issued a call for articles on major projects, TC activity and Board Level Committee activity.

In discussion, the President asked the Technical Oversight Committee to request all TCs to produce an article for the Bulletin every four years. Sukumar Pathmanandavel asked how members were notified about the bulletin. Neil Taylor explained that new issues were noted in the monthly circular sent to

Member Societies. It was their commitment and responsibility to forward this information to their membership.

18 PRESENTATION OF AUDITED ACCOUNTS 2017 AND 2018

Neil Taylor presented the accounts, which are given in Appendix 13. He confirmed that the accounts had been independently examined and assessed as giving a true and proper representation of receipts and payments for the given accounting years. It was noted that, as might be expected, there were some variance between the budgeted and actual receipts and expenditure. Subscription fee and Corporate Associate income were both good and the 2017 Seoul conference had been very successful leading to a significant amount remitted to ISSMGE. Costs were generally within budget, which had initiatives relating to the ISSMGE online library and the Conference Paper Review Platform to be funded.

Comments were invited but there was none. Philip Robins proposed that the accounts be accepted and this was seconded by Alejo Sfriso. By a show of hands, the proposal was accepted unanimously.

19 BUDGET 2019 – 2021

Gavin Alexander, former ISSMGE Vice-President for Australasia and Chair of the ISSMGE Finance and Budget, had prepared the budget and circulated this in advance of the meeting. He had recently resigned from the Board on health grounds and Neil Taylor undertook to present the budget on his behalf. He explained that a recent email had advised of increased costs associated with the website and online library. Some small adjustments had been made to account for these increases and were incorporated in the budget now presented. Apart from other generally inflationary changes, the only other notable addition was a biennial travel allowance for an in-person YMPG meeting.

There were no comments on the budget per se but there followed considerable discussion on the ISSMGE financial policy. That policy, agreed in 2013, had been to maintain the ISSMGE end-of-year reserve at GBP500000 adjusted to take into account inflation from 2013 and to transfer any surplus to the ISSMGE Foundation automatically. Assuming an annual inflation rate of 3%, required ISSMGE reserves at the end of 2019 would be approximately GBP581000. The budget forecast then showed that GBP12000 would be donated to the Foundation in 2020 and that there would be no further transfers as the forecast growth in end-of-year balance would be at rate lower than inflation. It was noted that the donation of GBP41000 in 2019, as shown on the budget, had already been paid as this was calculated from the 2018 end-of-year balance.

Graham Scholey (Australia) asked if the cessation of donations from ISSMGE from 2021 onwards would hurt the ISSMGE Foundation. Neil Taylor replied that Trustees of the Foundation would be aware of the situation and would need to investigate fund-raising and to adjust, if necessary, the magnitude of awards made.

Samuel Ampadu (Ghana) asked if the initiatives of IDC outlined by Dimitrios Zekkos were covered by the budget. Neil Taylor replied that the "Future developments" and "Initiatives" components in the budget generally went to cover the programming work undertaken on behalf of the IDC. It was envisaged that the creation of recorded lectures could be achieved with minimal additional costs by not undertaking the same degree of editing as has previously been used for webinars.

Wolfgang Sondermann (Germany) suggested that instead of making automatic annual donations to the Foundation based on a formula there should instead be a proposal within the budget for transferring fixed amounts. These proposals would then have to be agreed by Council every 2 years. Helmut Schweiger (Austria) agreed with this suggestion.

Alejo Sfriso stated that any changes to the ISSMGE financial policy deserved careful consideration and he asked the Secretary General for his views. Neil Taylor replied that it was an issue for Member Societies to consider and debate at their Council Meeting.

Roger Frank confirmed that indeed the policy had been agreed at a previous Council Meeting. If the concern was with the transfer of funds by formula then a change could be to accept the amounts suggested in the forecast budget. Neil Taylor replied that the forecast was just that: a prediction made with generally conservative assumptions. He stated that Council could propose a change in financial policy and this could be decided by voting. If the formula for donating money to the Foundation were removed, then it would be helpful to consider if any transfer to the Foundation should be included in the budget.

During a lengthy discussion, a number of proposals were suggested by the delegates. The President summarised the views expressed and proposed the motion that the financial policy should be changed to remove the formula for making an automatic annual donation to the ISSMGE Foundation. A proposal would be needed and approved by the Council for any future transfer. This was seconded by Philip Robins. Votes were cast as 31 in favour, 0 against with 7 abstentions; the motion was carried.

Neil Taylor asked if anyone wanted to change the budget. Graham Scholey asked if the ISSMGE Foundation could submit a budget request and indicate for what it would be used. Neil Taylor said that the he could not do that now so the next opportunity would be at the Council Meeting in 2021. Helmut Schweiger thought the Foundation already had sufficient funds. Neil Taylor stated that at present, the Trustees of the Foundation thought their current account would be sufficient for the next 2 years though it was difficult to predict future applications especially on the run-up to the International Conference in Sydney. Roger Frank thought it could be sensible for Council to agree a budgeted transfer of funds to the Foundation. The President thought it would be best for the ISSMGE Foundation to submit a request to ISSMGE for donations though he accepted that the first such request could not be considered until 2021. For now, Council should consider what if any donation should be made in the years 2020 and 2021. Philip Robins then proposed that GBP12000 should be donated each year for the next 2 years.

There followed a lively discussion on what amounts might be donated in each of those 2 years. Various amounts were suggested. Neil Taylor said that applications for awards to attend the Sydney conference would gain momentum after the paper acceptance process. Lucy Wu said that word was spreading to young people who may consider applying for an award. Dimitrios Zekkos said it was clear that the trend was an increase in applications to the Foundation for an award. Sukumar Pathmanandavel thought it was the intention of ISSMGE to have more people attend conferences. Wolfgang Sondermann had no problem with supporting young engineers but wanted reassurance that the ISSMGE resources were being used correctly. Roger Frank considered that annual donations to the ISSMGE Foundation could be made using the old "formula" but recognised that such donations would need to be approved each Council Meeting. Neil Taylor replied that having revoked the previous financial policy it would now be simpler to have agreed annual donations.

Finally, three motions were proposed and voted on as follows:

(i) Donation of GBP12000 in each of years 2020 and 2021. Proposed by Philip Robins, seconded by Alejo Sfriso. Votes for 13, against 25, abstentions none: motion lost.

(ii) Donation of GBP12000 in 2020, and no donation in 2021. Proposed by Wolfgang Sondermann, seconded by

Marcelin Kana. Votes for 8, against 23, abstentions 7: motion lost.

(iii) No donation in 2020, donation of GBP40000 in 2021. Proposed by Graham Scholey, seconded by Helmut Schweiger. Votes for 21, against 8, abstentions 9: motion carried.

Neil Taylor undertook to make the necessary changes to the budget. Considering those changes, he asked if there was a proposer and seconder to accept the budget. Philip Robins proposed acceptance of the revised budget and this was seconded by Tim Newson. Voting was unanimously in favour.

The revised budget and the commentary prepared by Gavin Alexander is presented in Appendix 14

20 20ICSMGE, SYDNEY 2021

An oral report on the 20ICSME was given by Graham Scholey on behalf of the Australian Geomechanics Society. Preparations for the conference were progressing well. Mark Jaksa with assistance from Harry Poulos was in charge of the technical program. Graham Scholey had noted with favourable interest the suggestion that experts from other Sister Societies could be invited to give keynote lectures and he would pass this on to the organising committee. He was pleased that the ISSMGE Honours Lectures could now be given in the parallel sessions. The organisers were keen for the conference to be a highly sustainable event with minimal waste. That would mean the conference proceedings would be on-line only with no printed copy or availability on a separate drive. However, he had noted that making the proceedings available using print-on-demand could be investigated.

Abstract submission would be from December 2019 to 15 April 2020. Notification of acceptance would be July 2020 and papers would be required by January 2021. Helmut Schweiger noted that with proceedings available only online, it should be possible to have a significant revision to the normal page allocation. Graham Scholey stated that this was still under discussion. Roger Frank commented that Member Societies must deal with the review of papers they submit and the President agreed, stating that the high quality of papers must be maintained.

Graham Scholey advised that they were working with the Australian Government to make the visa application process as easy as possible. Neil Taylor welcomed this and said that it would nevertheless be important to advertise the deadlines for visa applications.

Anthony Leung asked if there would be an associated International YGEC and Bright Spark lectures. Graham Scholey confirmed that both would take place.

21 ANY OTHER BUSINESS

The President asked if there was any other business and there was none.

22 DATE AND VENUE OF NEXT MEETING

The President noted that the next Council Meeting would be on Sunday 12 September 2021 in Sydney just prior to the 20ICSMGE. The Secretary General commented that Member Societies might need to start planning now to ensure they have a budget available to send delegates to that Council Meeting.

23 THANKS AND CLOSURE

The President thanked all delegates for their thoughtful contributions and then closed the meeting.

Appendix 1: Presidential Report

Charles W. W. Ng
ISSMGE President 2017-2022, cecwwng@ust.hk

Dear colleagues and friends,

It was my great honour to be elected as the 17th President of the ISSMGE at the Council Meeting convened on 17th September 2017, prior to the 19th International Conference on Soil Mechanics and Geotechnical Engineering (ICSMGE) held between 18th and 22nd September 2017 in Seoul, Korea. I am very happy to report you that I have an extremely supportive and outstanding Board, consisting of 11 board members (refer to Plate 1 and Table 1) and seven Board-Level Committee (BLC) chairs/co-chairs (see Table 2). They have supported me in implementing my presidential plan for improving Education, Innovation and Diversity at the ISSMGE, which was announced during my presidential address at the closing ceremony of the 19th ICSMGE (see Plate 1) and published in the 2017 October issue of Bulletin. I am very pleased to say that we have since made significant progress.

Plate 1: Introduction of all Board members and some BLC chairs at the closing ceremony of the 19th ICSMGE (from right to left: Neil Taylor, Pedro Pinto, Kok Kwang Phoon, Mounir Bouassida, Alejo Sfriso, Timothy Newson, Mario Manassero, Gavin Alexander, Eun Chul Shin, Marcelin Kana, Roger Frank, Pierre Delage, Lucy Wu, Sukumar Pathmanandavel, Peter Day and Ikuo Towhata)

Table 1. ISSMGE Board (2017–2021)

Position	Name
President	Charles W.W. Ng (Hong Kong SAR, China)
Immediate Past President	Roger Frank (France)
Vice President for Africa	Marcelin Kana (Cameroon)
Vice President for Asia	Eun Chul Shin (South Korea)
Vice President for Australasia	Gavin Alexander (New Zealand)
Vice President for Europe	Mario Manassero (Italy)
Vice President for North America	Timothy Newson (Canada)
Vice President for South America	Alejo Sfriso (Argentina)
Appointed Board Member	Pedro Sêco E Pinto (Portugal)
Appointed Board Member	Kok Kwang Phoon (Singapore)
Appointed Board Member	Mounir Bouassida (Tunisia)
Secretary General	Neil Taylor (United Kingdom)

Table 2. Board Level Committee Chairs/Co-Chairs

Board-Level Committee	Chair/Co-chair
Innovation and Development Committee	Dimitris Zekkos (USA)
Technical Oversight Committee	Pierre Delage (France)
Young Members' Presidential Group	Lucy Wu (USA)
Corporate Associates' Presidential Group	Sukumar Pathmanandavel (Australia) / Peter Day (South Africa)
Award Committee	Roberto Terzariol (Argentina)
Professional Image Committee	Ikuo Towhata (Japan)

1 EDUCATION – ISSMGE VIRTUAL UNIVERSITY

My proposed "ISSMGE Virtual University" is rapidly taking shape. With the strong support of Professors Mounir Bouassida, Kok Kwang Phoon and Dimitris Zekkos, all existing ISSMGE webinars were reviewed and categorised for the first time into five "themed series of webinars or postgraduate (PG) courses". The first five PG courses are (a) Risk Mitigation, Monitoring & Observational Methods; (b) In Situ Testing; (c) Earthquake Engineering, (d) Foundations and (e) Geo-Engineering Education. Recently, a new purposely designed course on "Unsaturated Soil Mechanics" has been contributed by Professor

Del Fredlund. This makes it the 6th PG course for the University. Each PG course is comprised of three or more webinars giving students and engineers a total of no less than five hours of viewing. Please visit https://www.issmge.org/education/virtual-university for details. More PG courses are being designed and will be produced soon. Other details of the PG courses are provided by Professors Mounir Bouassida in his report (see Appendix A).

2 INNOVATION - ISSMGE VIRTUAL LIBRARY AND PUBLISHER

The open access policy initiated by the immediate Past President, Professor Roger Frank, during his presidential term has been continued and expanded rapidly. Naturally, the open access materials would form the "ISSMGE library" as a part of the "ISSMGE virtual university", for an example, the proceedings of the 7th International Conference on Unsaturated Soils held at the Hong Kong University of Science and Technology in August 2018. Please visit https://www.issmge.org/publications/online-library for more information.

Under the leadership of Professors Dimitrios Zekkos and Kok Kwang Phoon, the Innovations and Development Committee has been exploring and developing our own publisher - the "ISSMGE virtual publisher" - with the intention of providing any organiser of a conference to review and publish conference papers online via the ISSMGE virtual publisher platform.

Following on my presidential plan announced in September 2017, I am also very pleased to inform you that a new technical committee (TC) called TC309 "Machine Learning and Big Data" has been created. To learn more about this new TC, please visit https://www.issmge.org/committees/technical-committees/impact-on-society/machine-learning. On top of that, two other new TCs have been established very recently: TC219 "System Performance of Geotechnical Structures" and TC220 "Field Measurement in Geomechanics". I am grateful to the chair of the Technical Oversight Committee, Professor Pierre Delage, and his committee for reviewing the three new TCs.

3 DIVERSITY

3.1 Recruiting new Corporate Associates to narrow the# gap between academics and practitioners

Under the leadership of the co-chair of the Corporate Associates Presidential Group (CAPG), Mr Sukumar Pathmanandavel, significant progress has been made since September 2017 in narrowing the gap between academics and practitioners. We have recruited four new Corporate Associates: Geoharbour Group from China, GDS Instruments from Malaysia, and AOSA SA and SRK Consulting from Argentina.

In addition, the CAPG together with the Young Members Presidential Group (YMPG) have been working closely with the organising committees of various conferences such as the 16th Pan-American Conference on Soil Mechanics and Geotechnical Engineering (SMGE), the 16th Asian Regional Conference on SMGE and the 17th European Conference on SMGE to create a dialog among geotechnical engineers around the world on matters directly affecting the welfare of the profession. The main purpose of the dialog is to bridge the gap between academics and practitioners so that they may find a common understanding and take collective actions.

3.2 Creating Bright Spark Lecture series for young members

As our young members (less than 36 years of age) have been under-represented in major international lectures for many years, I have created the ISSMGE Bright Spark Lecture series to enable more young members to take the main stage at conferences. I would like to thank the YMPG under the leadership of Lucy Wu for coming up with the wonderful title of this young members' lecture series. Guidelines on organising these lectures can be found at https://www.issmge.org/the-society/issmge-awards.

The first Bright Spark Lecture was delivered by Dr Lin Dong at the 7th Technical Conference in Eastern Asia on Geo-Natural Disasters held in Chengdu, China (see Plate 2). The conference was organised by Asian Technical Committee 3. I was very pleased to attend that lecture and to present a certificate to the lecturer.

Plate 2. Dr Lin Dong from the Lanzhou Institute of Seismology of China Earthquake Administration receives the first "ISSMGE Bright Spark Lecturer" certificate.

Another four ISSMGE Bright Spark Lectures were delivered at the 7th International Conference on Unsaturated Soils held between 3rd and 5th August 2018 at the Hong Kong University of Science and Technology. The four lecturers were also presented with ISSMGE certificates (see Plates 3 to 6).

Plate 3. Dr Amin Askarinejad from TU Delft, the Netherlands receives the certificate.

Plate 4. Dr Majid Ghayoomi from the University of New Hampshire, US receives the certificate.

Plate 5. Dr Annan Zhou from RMIT University, Australia, receives the certificate.

Plate 6. Dr Chao Zhou from Hong Kong University of Science and Technology, HKSAR, receives the certificate

Other ISSMGE Bright Spark Lectures are being planned by the YMPG in collaboration with various local organising committees of different conferences such as the 16th Asian Regional Conference on SMGE, the 16th Pan-American Conference on SMGE, the 17th European Conference on SMGE, and the 17th African Regional Conference on SMGE.

3.3 Connecting with the forgotten societies and exploring new ones

Promoting soil mechanics and geotechnical engineering in different parts of the world, engaging with our existing members and reaching out to new ones are all central to my presidential plan. One way to achieve these goals is through ISSMGE international seminars. Under the outstanding leadership of Professor Pedro Seco e Pinto (Portugal), one ISSMGE International Seminar was held in Brest (Belarus) in April 2018 and another four in November 2018 in Asunción (Paraguay), Buenos Aires (Argentina), Mexico City (Mexico) and León (Mexico). I would like to thank Pedro Seco e Pinto, Roger Frank (France), Liudas Furmanovich (Lithuania), Nikitenko Mikhail (Belarus), Sernov Vyacheslav (Belarus), Roberto Terzariol (Argentina), Alejo Sfrisio (Argentina), Tim Newson (Canada), Sukumar Pathmanandavel (Australia), Eun Shin (South Korea) and Gabriel Auvinet (Mexico) for their outstanding contributions to these seminars. More details of ISSMGE seminars carried out by Professor Pedro Seco e Pinto are given in Appendix B.

4 CONGRATULATIONS TO PROFESSOR HEINZ BRANDL

I would like to congratulate Professor Heinz Brandl on receiving the Medal of Merit for Macedonia. The Board was invited by the then President of the former Republic of Macedonia, Dr Gjorge Ivanov, to join the award ceremony held at the presidential palace (see Plate 7). The Board was also invited by Professor Milorad Jovanovski, the President of Macedonian Association for Geotechnics, to attend the XVI Danube-European Conference on Geotechnical Engineering held between 6th and 8th June 2018 in Skopie and to convene our Board meeting there. I presented an ISSMGE tie to Dr Gjorge Ivanov, the then president of the former Republic of Macedonia at the opening ceremony of the XVI Danube-European Conference (refer to Plate 8) to promote ISSMGE to him.

I would like to thank Professor Brandl for promoting the public and professional image of ISSMGE to the highest state level and also to thank Professor Jovanovski for providing excellent support to our Board meeting.

Plate 7. Professor Heinz Brandl receives the Medal of Merit for Macedonia at the presidential palace.

Plate 8. The President of ISSMGE presents an ISSMGE tie to Dr Gjorge Ivanov at the opening ceremony of the XVI Danube-European Conference.

Looking ahead, I am confident that the ISSMGE will continue to flourish and reach new heights with the help and support from all of you.

APPENDIX A

ISSMGE VIRTUAL UNIVERSITY PROGRAMME

Report by Mounir Bouassida

The President, Professor Charles Ng, announced his plan on Education, Innovation and Diversity in September 2017 at the ICSMGE in Seoul. The proposed "ISSMGE virtual university" has begun to take shape. It is the intention for the virtual university to provide free online structured postgraduate courses for anyone to use. There is no assessment and no credit will be given. ISSMGE will not be held responsible for the use of any of the course materials.

Existing ISSMGE webinars have been reviewed and categorized into themed series of webinars or postgraduate courses. Each series comprises three to four webinars giving students and engineers up to five hours of viewing. It is agreed that this viewing, by login on ISSMGE website, is equivalent to a fifteen hours course in classroom.

To date there are six available courses titled:

1 Risk-Mitigation, monitoring & Observational methods
2 In situ testing:
3 Earthquake Engineering
4 Foundations
5 Soil Characterization
6 Geo-Engineering Education

Those courses were identified from the most available list of delivered webinars. This updated list includes two added webinars delivered in French and two remaining webinars not yet included in identified course, so far.

The first course of ISSMGE virtual university was experienced on February 06-09, 2018 at the National Engineering School of Tunis for 33 PhD candidates in civil engineering, mechanical engineering, electrical and industrial engineering. The viewed course was on "Risk Mitigation, Monitoring and Observational methods". Detailed report on this first experience including: organization of webinar sessions and evaluation of the synthesis of viewed webinars by the candidates was forwarded to the ISSMGE board members.

Incoming webinars
Other potential courses have been identified but require some additional material to achieve a minimum of five hours of viewing. Themes of webinars and speakers are selected by the chairs of TCs and, then, approved the ISSMGE board members. Soil behavior, numerical modeling in geomechanics, ground improvement and Landslides-slope stability are among the forthcoming courses of the ISSMGE virtual university program.

Fully formatted courses
This new option provides much more benefits for students and engineers to follow a complete course on a specific theme composed by five to six well linked lectures. Very recently, the first course on "Unsaturated soil Mechanics" prepared by Prof. Delwyn G. Fredlund (Canada) has been launched publicly on May 15th 2019. Seven PhD candidates, from the National Engineering School of Tunis, viewed six lectures, at rate of two lectures per day, on May 15th, 16th and 17th 2019.

New formatted courses are also scheduled.

It is, then, concluded that the new ISSMGE virtual university program is quite valuable in providing postgraduate courses which are needed for PhD candidates to fulfill required credit by their institutions. Also, it provides for a very helpful learning tool to solve quite challenging geotechnical engineering problems.

The updated list of formatted courses can be viewed on:
https://www.issmge.org/education/virtual-university

26th May 2019
Professor Mounir Bouassida
University of Tunis El Manar- National Engineering School of Tunis
Head of Research Laboratory in Geotechnical Engineering and Geo-risks
Mounir.bouassida@enit.utm.tn
Tél. +216.26.408.666 - +216.98.666.408

APPENDIX B

ISSMGE INTERNATIONAL SEMINARS

Report by Pedro Sêco e Pinto

1. Background

The Touring Lectures Program was planned by ISSMGE to disseminate the current state-of-the art/practice amongst geotechnical engineers involved in the design and practice in developing countries.

ISSMGE International Seminars (Touring Lectures) have the following purposes:

- To address a topic selected by the Host Society;
- To address the interests of practitioner engineers;
- To attract young geotechnical engineers;
- To incorporate a good mix of routine and modern soil mechanics and to cover case histories;
- To allocate after each lecture a period for discussions, in order to stimulate questions and to share experiences from the participants;
- To have a strong local support;
- To have the support of the Industry;
- 2 days of lectures delivered by 4 international experts appointed by the Co-ordinator and 2 lecturers appointed by the Host Society;
- In the 3rd day a technical visit (optional);

All the written versions of the lectures should be sent to the Host Society with 2 months in advance in order to be prepared a Seminar Volume/CD-Rom.

The first Touring Lecture took place in Lagos, 2001 and the second one in St. Petersburg, in 2004.

During the period (2005-2009) 21 Touring Lectures – International Seminars were organized namely in Chile, in Paraguay, in Tunisia, in Vietnam, in Albania, in Croatia, in Costa Rica, in El Salvador, in Indonesia, in China, in Sri Lanka, in Nigeria, in Ecuador, in Pakistan, in Makedonia, in Cambodia, in Laos, in Myanmar, in Hungary, in Ghana and in Mozambique.

During the period (2009-2013) 17 Touring Lectures – International Seminars were organized namely in:

- In 2010 in Dominican Republic, Philippines (Manila), South Africa (Pretoria), Argentine (Cordoba) and Bangladesh (Dacca)
- In 2011 in Peru (Lima), Malaysia (Kuala Lumpur), Laos (Vientiane), Cambodia (Phnom Penh) and Myanmar (Yangoon and Mandalay)
- In 2012 in Singapore, Malaysia (Kuala Lumpur), Philippines (Los Angeles and Puerto Princesa), Laos (Vientiane) and Cambodia (Phnom Penh)
- In 2013 in North Sudan (Khartoum)

From the collected experience to organize a Touring Lecture/International Seminar there is a need of 4-6 months of preparation and to exchange around 60 e-mails with the Host Society and the Lecturers.

ISSMGE President Prof. Charles Ng has proposed for his programme "Education, Innovation and Diversity", to be implemented during his tenure 2017-2021, 10 components. To me was allocated component 7 "Board –level Outreach Committee".

2. ISSMGE Seminars organized in 2018

2.1 Seminar in Belarus, 27-28 April 2018

The theme of the International Seminar that took place in Belarus has covered "Foundations Design-Theory and Practice of implementing Eurocodes in European countries and Republic of Belarus".

The idea was given the utmost attention by Belarus Society to develop geotechnical activities in Belarus.

The Seminar held during 27 and 28 April 2018 and took place at the University of Brest.

The following lecturers have participated: (i) from overseas: Prof. Roger Frank, Dr. Liudas Furmanovich and Prof. Pedro Seco e Pinto; (ii) from Belarus: Prof. Nikitenko Mikihail and Prof. Sernov Vyacheslav.

The lectures have covered Piles Design based in Eurocode 7 "Geotechnical Design" and Eurocode 8 "Design Structures for Earthquakes" and applications for Baltic countries and Belarus.

The written versions of the lectures were collected in a Volume that was distributed to all the participants.

Each lecture had a duration of 45 minutes followed by a period of discussion of 20-30 minutes. The participants were actively engaged in questions, answers and comments for the content of the materials presented during the seminar.

Special thanks are due to Dr. Tamara Ulasik, Secretary of BELGS and to Prof. Petr Poita, Rector of Brest State Technical University, for their great support for the organization of the Seminar and for the warm hospitality.

The co-operation of the lecturers was highly appreciated.

Plate 1. View of seminar audience.

Plate 2. Main table.

2.2. Seminar in Asuncion, 12-13 November 2018

The theme of the International Seminar that took place in Asuncion has covered "Active Design by Codes ".

The Seminar held on 12-13 November 2018.

The following lecturers have participated: (i) Prof. Roger Frank, Prof. Alejo Sfrisio, Prof. Roberto Terzariol and Prof. Pedro Seco e Pinto.

The written versions of the lectures were distributed to all the participants.

Each lecture had a duration of 45 minutes followed by a period of discussion of 20-30 minutes. The participants were actively engaged in questions, answers and comments for the content of the materials presented during the seminar.

Plate 3. View of seminar audience.

Plate 4. General view.

Special thanks are due to Eng. Jose Pavon and Organizing Committee for their great support for the organization of the Seminar and for the warm hospitality.

The co-operation of the lecturers was highly appreciated.

2.3. Seminar in Buenos Aires, 15-16 November 2018

The theme of the International Seminar that took place in Asuncion has covered "Active Design by Codes ".

The Seminar held on 15-16 November 2018.

The following lecturers have participated:(i) Prof. Roger Frank, Prof. Alejo Sfrisio, Prof. Roberto Terzariol and Prof. Pedro Seco e Pinto;

The written versions of the lectures were distributed to all the participants.

Each lecture had a duration of 45 minutes followed by a period of discussion of 20-30 minutes. The participants were actively engaged in questions, answers and comments for the content of the materials presented during the seminar.

Plate 5. Main table.

Plate 6. View of seminar audience.

2.4. Seminar in Mexico, 19-20 November 2018

The theme of the International Seminar that took place in Leon has covered "Foundations Design".

The Seminar held on 19-20 November 2018.

The following lecturers have participated:(i) Prof. Tim Newson, Prof. Roger Frank, Prof. Alejo Sfrisio Dr. Sukumar Pathmanandavel, Prof. Eun Schin, Prof. Gabriel Auvinet and Prof. Pedro Seco e Pinto;

The written versions of the lectures were distributed to all the participants.

Each lecture had a duration of 45 minutes followed by a period of discussion of 20-30 minutes. The participants were actively engaged in questions, answers and comments for the content of the materials presented during the seminar.

2.5. Seminar in Leon, 21 November 2018

The theme of the International Seminar that took place in Leon has covered "Dam Engineering".

The Seminar held on 21 November 2018.

The following lecturers have participated: (i) Prof. Tim Newson, Prof. Roger Frank, Prof. Alejo Sfrisio Prof. Roberto Terzariol and Prof. Pedro Seco e Pinto;

The written versions of the lectures were distributed to all the participants.

Each lecture had a duration of 45 minutes followed by a period of discussion of 20-30 minutes. The participants were actively engaged in questions, answers and comments for the content of the materials presented during the seminar.

2.6 Seminar in Maputo in 2, 3 October 2019

The theme of the International Seminar that will take place in Maputo will covered "Geotechnical Engineering ".

The Seminar will occur on 2-3 November 2018.

The following lecturers will participate:(i) Dr. Etienne Kana, Prof. Roger Frank, Dr. Sukumar Pathmanandavel, Prof. Ikuo Towhata, Prof. Mounir Bouassida, Prof. Roberto Terzariol and Prof. Pedro Seco e Pinto.

The written versions of the lectures will be distributed to all the participants.

Each lecture will have a duration of 45 minutes, followed by a period of discussion of 20-30 minutes. Some short presentations of local colleagues are expected.

3. Final Comments

I would like to take this opportunity to thank the President, the Regional Vice Presidents, Host Member Societies, the Technical Committees, the Invited Lecturers and the Sponsors for their contributions and support.

I will summarize this rich experience in the following words:

- We want all of you to make part of our community, to promote your growth and to make you feel that you are contributing for our dream come true.
- Interaction with people is important, because I believe this experience will help us with new ideas and interesting insights.

Last but not least I would like to address to all of you a word of praise and gratitude for your contributions and a message of hope that these Seminars will allow us to develop a feeling of universal responsibility and to create the ambition to serve better our Society,

Appendix 2: ISSMGE Membership – Member Societies, Corporate Associates

Neil Taylor
Secretary General, ISSMGE

1 MEMBER SOCIETIES

The accompanying Table indicates that the present ISSMGE membership is just over 21,100 in 89 Member Societies, (information correct as of 24th September 2019). Though most member societies have shown a small but steady increase in membership numbers, the following have reported significant changes compared to the report produced for the Seoul Council Meeting: Australia (up to 1133 from 955, Egypt (up to 103 from 75), France (up to 636 from 541), India (down to 193 from 236), Iraq (up to 144 from 42), Korea (down to 585 from 714), Lebanon (up to 85 from 66), Malaysia (down to 69 from 100), New Zealand (up to 775 from 683), Norway (up to 530 from 435), South Africa (up to 659 from 547), Sweden (up to 1301 from 1076), and USA (down to 2975 from 3184).

The applications for membership from the Philippine Society for Soil Mechanics and Geotechnical Engineering (PSSMGE), Peruvian Geotechnical Engineering Association (APGEO), and the Bolivian Society of Soil Mechanics and Geotechnical Engineering (SBMSIG) have been approved by the Board since the last report to Council.

2 CORPORATE ASSOCIATES

There are currently 39 Corporate Associates (up from 31 reported at the Seoul Council Meeting).

AECOM Asia Company Ltd
S.N. Apageo S.A.S.
AOSA SA
Arup
Aurecon
Centre for Geomechanics and Railway Engineering
Coffey Geotechnics Pty Ltd
Dar Al Handasah Corp
Deltares
Fugro Consult GmbH
Gamuda
GDS instruments
Geomil BV
GHD Pty, Ltd.
Giken Ltd
Golder Associates Inc
Ground Investigation
Huesker Synthetic GMBH
International I.G.M.
Jan De Nul N.V.
Keller Group plc
KGS Ltd
MIC - Maccaferri Innovation Centre Srl
Naue GmbH Co KG
Norwegian Geotechnical Institute
Pagani Geotechnical Equipment
RCF Ltd
Shanghai Geoharbour Construction Group
Siemens Energy
SMEC AUSTRALIA PTY LTD
SOLETANCHE BACHY SA/Soletanche Freyssinet
SRK Consulting Argentina S.A.
Tencate Geosynthetics France SAS
Tensar International Ltd
Terrasol
Terre Armée Internationale
A.P. van den Berg
Wagstaff Piling
Zetas Zemin Teknolojisi AS

MEMBERSHIP INFORMATION AS AT 27th SEPTEMBER 2019

Member Society	Members	Africa	Asia	Australasia	Europe	North America	South America
Albania	40				40		
Algeria	43	43					
Argentina	106						106
Australia	1133			1133			
Austria	108				108		
Bangladesh	33		33				
Belarus	30				30		
Belgium	319				319		

Member Society	Members	Africa	Asia	Australasia	Europe	North America	South America
Bolivia	42						42
Bosnia & Herzegovina	18				18		
Brazil	806						806
Bulgaria	31				31		
Canada	751					751	
Chile	101						101
China	472		472				
Chinese Taipei	68		68				
Colombia	20						20
Costa Rica	31						31
Croatia	104				104		
CTGA	100	100					
Cuba*	30						30
Cyprus	30				30		
Czech & Slovak Reps.	45				45		
Denmark	345				345		
Egypt	103	103					
Estonia	30				30		
Finland	45				45		
France	636				636		
Georgia	38				38		
Germany	492				492		
Ghana	21	21					
Greece	69				69		
Guatemala	32						32
Hong Kong	120		120				
Hungary	65				65		
Iceland	32				32		
India	224		224				
Indonesia	140		140				
Iran	802		802				
Iraq	114		114				
Ireland	24				24		
Israel	32				32		
Italy	269				269		
Japan	1124		1124				
Kazakhstan	40		40				
Korea	519		519				
Kyrgyzstan	30		30				
Latvia	44				44		
Lebanon	85		85				
Lithuania	40				40		
Macedonia, FYR	35				35		
Malaysia	82		82				

Member Society	Members	Africa	Asia	Australasia	Europe	North America	South America
Mexico	260					260	
Morocco	28	28					
Mozambique	37	37					
Nepal	22		22				
Netherlands	719				719		
New Zealand	775			775			
Nigeria	40	40					
Norway	530				530		
Pakistan	85		85				
Paraguay	38						38
Peru	27						27
Philippines	34		34				
Poland	378				378		
Portugal	156				156		
Romania	158				158		
Russia	180				180		
Serbia	40				40		
Singapore	46		46				
South East Asia	124		124				
Slovenia	77				77		
South Africa	659	659					
South East Asia	125		125				
Spain	388				388		
Sri Lanka	36		36				
Sudan	37	37					
Sweden	1301				1301		
Switzerland	173				173		
Syria	17		17				
Tajikistan	23		23				
Thailand	32		32				
Tunisia	40	40					
Turkey	267				267		
Ukraine	30				30		
UK	1173				1173		
USA	2975					2975	
Uzbekistan	33		33				
Venezuela	32						32
Vietnam	50		50				
TOTAL	**21,114**	**1,108**	**4,356**	**1,908**	**8,491**	**3,986**	**1,265**

Appendix 3: Amended cooperation agreement for FedIGS

Federation of International Geo-engineering Societies (FedIGS)

COOPERATION AGREEMENT

This Agreement was approved in the meeting of the Presidents of IAEG, ISRM, and ISSMGE in Amsterdam, The Netherlands on May 12, 2006, modified by the FedIGS Board in the meeting of the Board Members of IAEG, ISRM, ISSMGE, and IGS in Shenyang, China on 4 July 2015, and modified again by the FedIGS Board in the meeting of the Board Members of IAEG, ISRM, ISSMGE, and IGS in San Francisco, USA on Sept. 24, 2018.

1 NAME AND CONSTITUTION

1.1 The Name of the Federation is: Federation of International Geo-engineering Societies, abbreviated as FedIGS. It is hereinafter referred to as "FedIGS".
1.2 This Agreement shall formalize and regulate the cooperation between the Member Organizations.

2 DEFINITION OF GEO-ENGINEERING

2.1 Geo-engineering is engineering with, on, or in geological materials.

3 LANGUAGE

3.1 The official language of FedIGS is English.

4 AIMS, OBJECTIVES

4.1 FedIGS is founded on the principle that each of the participating Member Organizations will retain its identity and autonomy. The role of FedIGS is to carry out functions for the international geo-engineering community that could not be carried out with the same effectiveness and efficiency by the Member Organizations individually.

4.2 The aims and responsibilities of FedIGS are to:

- raise the awareness within the geo-engineering profession of the inevitable and essential interaction between geo-engineering practitioners from the Members;
- coordinate the scientific and technical activities in areas with overlapping interest between the Members and
- raise the public awareness and profile of the geo-engineering profession.

by:

- promotion of interaction among geo-professionals within various facets of geo-engineering;
- coordination of research efforts in areas of common interest to the Member Organizations;
- development of a FedIGS website;
- coordination of the scheduling, promotion, and delivery of conferences and symposia of the Member Organizations;
- establishment of Joint Technical Committees on topics and interests common to the Member Organizations;
- stimulation of the development of educational curricula for higher education and for continuing education purposes;
- promotion of interaction with industry and policy making organizations and
- development of policies that will advance the learned and professional functions of the geo-engineering community.

5 MEMBERSHIP

5.1 FedIGS is founded by three member organizations: The International Association of Engineering Geology and the Environment (IAEG), the International Society for Rock Mechanics (ISRM) and the International Society for Soil Mechanics and Geotechnical Engineering (ISSMGE). These three organizations hereinafter will be referred to as the Founding Members.

5.2 Other organisations, which fully subscribe to the FedIGS Cooperation Agreement and in particular the Aims and Objectives of FedIGS as defined in Clause 4 may be invited to join FedIGS as Member Organizations. Acceptance of new Member Organizations will require approval by all existing Member Organizations.

5.3 Conditions for admission to membership in FedIGS are as follows. Organizations invited by the Board to become member organizations, or those applying for membership of FedIGS shall submit a written letter of application to the President of FedIGS which explains the reasons and arguments for the application. Furthermore this letter of application shall confirm that the applying organization subscribes to the aims and objectives of FedIGS, as outlined in Clause 4 and describe how the organization will contribute to the implementation of these aims and objectives. This letter shall be accompanied by the constitution of the organization, the name and contact information of its officers, and a summary of recent activities.

5.4 A member organization shall fulfil its obligations to FedIGS, which includes to inform the President of FedIGS of the address and contact information of its secretariat, as well as the names and contact information of its officers and the address of its web site.

5.5 Termination of membership of FedIGS can be the result of

a. Decision of the Member Organisation itself. Such a decision shall be communicated immediately to the FedIGS President and can become effective only after the organisation has complied with all administrative obligations to FedIGS which result from this Agreement.
b. Decision by the FedIGS Board in case the concerned Member Organisation acts against the aims, objectives, and/or interests of FedIGS. Such a decision is taken in two steps: a preliminary Board decision on suspension of membership is communicated to the concerned Member Organisation, that is given appropriate time to correct its position and comply with all its obligations. If such corrective measures are not taken within a period of six months then the definite decision to terminate the membership can be taken by the FedIGS Board.

6 FINANCES

6.1 FedIGS is a non-profit organisation.

6.2 There are no dues for Membership.

6.3 No Board member shall receive any remuneration from FedIGS.

7 ADMINISTRATION

7.1 FedIGS shall be governed and administered by the Board.

8 BOARD

8.1 The Board of FedIGS is the exclusive decision making body of FedIGS. It is composed as follows:

- the FedIGS president;
- the presidents of all Member Organizations,
- the immediate past presidents of all Member Organizations, and
- the secretaries general of each Member Organzation.

8.2 The Board is chaired by the President of FedIGS.

8.3 In the event that one of the presidents or immediate past presidents cannot fulfil the obligations, its Member Organization should appoint a replacement.

9 PRESIDENT

9.1 The prime duty of the President shall be to foster the aims and objectives of the Federation, thereby serving the interest of all Member Organizations.

9.2 The FedIGS Board elects the President. This election needs the formal approval by the presidents of all Member Organizations.

9.3 The President shall be one of the immediate-past-presidents or another co-opted individual as the Board sees fit. Candidates for the presidency shall be nominated by members of the Board. One year before the expiry of the term of office of the President, the current President shall invite each Member to submit a nomination for the next presidency while noting that the FedIGS President cannot come from the same Member Organisation two times in a row.

9.4 The term of office of the FedIGS President is 4 years.

9.5 The FedIGS President may be substituted by another voting Board member, in any and all of the President's duties. The substitute President is appointed by the FedIGS Board. In the event of the resignation or death of the President, the Board will have to arrange the President's substitution.

10 SECRETARIAT

10.1 A Secretariat is installed for the period coinciding with the term of office of the President.

10.2 The Secretariat will handle the administrative matters for FedIGS, under the responsibility of the President.

10.3 The FedIGS Secretariat will be located and handled by the President.

10.4 The official headquarters of FedIGS is at the office of the President.

10.5 The cost related with the functioning of the FedIGS Secretariat will be covered by the President.

11 MEETINGS, DECISION MAKING

11.1 Meetings of the Board will be held at least once a year.

11.2 Decisions in the FedIGS Board are taken by simple majority. All Board members except the Secretaries General of the Member Organizations have voting rights in the Board meetings. The Secretaries General of the Member Organizations are advisory Board members without voting rights. Furthermore, the President has an additional casting vote in case of a tied vote.

11.3 A voting Board member who is absent at a Board meeting can assign another voting Board member as proxy.

11.4 The presidents of the three Founding Members acting unanimously will have a veto right on all decisions of the Board.

11.5 Voting will take place openly, unless one of the Board members requests a secret ballot. Voting on persons will always take place in a secret ballot.

11.6 Minutes of the Board meetings will be prepared by the designated note taker. A list of key decisions and resulting actions will be circulated within two weeks after a Board meeting.

12 JOINT TECHNICAL COMMITTEES

12.1 In order to further the aims, responsibilities, and objectives of FedIGS, the Board may appoint Joint Technical Committees (hereinafter referred to as JTCs), composed of a responsible chairperson and individual members from the Member Organizations, to study scientific and/or technical matters of interest to the international geo-engineering community.

12.2 In general such JTCs are appointed for a 4-years period, after which the Board will decide about continuation or discontinuation of the JTC and its membership.

12.3 Proposals for the establishment of Joint Technical Committees shall be formulated by Board Members and shall be presented in writing to the President.

A proposal for the installation of a Joint Technical Committee shall contain the following items:

- The proposed title;
- The technical issues to be addressed;
- The terms of reference: definition of the broad terms of reference, with a list of the tasks to be performed, reports to be presented and a well-defined time schedule, and
- Suggestions of one or more names for the chairperson of the JTC.

The proposal shall be included in the agenda for the next Board meeting.

12.4 The Board shall take a decision about the installation of the proposed JTC, the title, issues to be addressed and terms of reference for the JTC.

The FedIGS Board appoints the chairperson for the JTC who is personally responsible for the successful operation of the JTC.

Each Member Organisation will select up to three members in a JTC. These members may be supplemented by up to 6 members selected by the chair. The membership as a whole shall be composed in such a way that it represents a good coverage of the knowledge fields relevant for the JTC theme and also a reasonable regional spreading.

Corresponding members can be nominated by National Groups/Member Societies of the FedIGS Member Organisations to the JTC chairperson who makes the decision. Corresponding members do not have to be members of Member Organizations and do not have a vote.

12. 5 After completion of its term of office, the JTC may be reinstalled with a well-defined term of office of a maximum of 4 years. In this case the procedures for appointment of the JTC chair and all individual members will have to be repeated.

The FedIGS Board has the right to terminate the existence of a JTC or dismiss the chairperson and appoint a new chairperson for lack of progress or when it becomes clear that the goals cannot be achieved for whatever reason. Such decisions can only be taken on the basis of a proposal submitted to the Board.

12.6 The JTC shall hold at least one meeting per year.

JTCs can arrange their own symposia or other forms of organised discussion and do so at their own financial risks. If such meetings generate funds, the JTC and the organising committee shall retain any surplus to be divided as per prior arrangement between them. No members of JTCs or their sub-committees shall be paid for their work for the JTC.

12.7 The chairperson of a JTC shall send a written annual report to the FedIGS President about the progress of the JTC. This report shall reach the President at least 2 weeks before the annual Board meeting of FedIGS. The Board shall discuss the annual JTC reports and inform the JTC chairperson about the comments and suggestions.

All reports produced by the JTCs involving matters of policy shall be sent to the FedIGS Board prior to their publication.

12.8 The creation of Awards and Honour Lectures by JTCs is not encouraged. However if a JTC feels strongly about proposing such a lecture or award, a proposal must be presented to the FedIGS Board for its consideration. This proposal must contain

a. The name of the engineer/scientist who is honoured.
b. The accomplishments of this person demonstrating outstanding and sustained contributions to geo-engineering
c. The record of the vote of the JTC members regarding this Award or Honour Lecture.
d. The procedure to be followed to select the recipient of the Award or Honour Lecture
e. The frequency of the recognition and the typical location of the delivery of the Award or Honour lecture.

The Board will vote on the proposal and a two third majority is required. If the Board approves the proposal, the JTC may proceed with the Award or the Honour Lecture for a period of 8 years. After 8 years, the proposal must be resubmitted for consideration by the FedIGS Board.

13 AMENDMENTS TO THE AGREEMENT

13.1 Amendments to the Agreement of FedIGS may be proposed by any Board Member. These amendments, after unanimous approval by the Board of FedIGS, shall require the written approval of each Member Organization according to its approval procedure.

14 DISSOLUTION

14.1 The dissolution of FedIGS will take place after a decision of the FedIGS Board. A two third majority is required.

Appendix 4: Regional Report – Africa

Etienne Marcelin Kana
ISSMGE Vice-President, Africa

INTRODUCTION

Currently, we have ten active ISSMGE transnational or national societies in the Africa region, namely: Société Algérienne de Géotechnique (ALGEOS), Egyptian Geotechnical Society (EGS), Ghanaian Geotechnical Society GGS (Ghana), Comité Marocain de la Mécanique des sols et des roches (CMMSR), Sociedade Moçambicana de Geotecnica (SMG), Nigerian Geotechnical Association (NGA), South African Institute for Civil Engineer-Geotechnical Division (SAICE), Sudanese Society of Soil Mechanics and Geotechnical Engineering (SSSMGE), Association Tunisienne de Mécanique des Sols (ATMS), and "le Comité Transnational des Géotechniciens d'Afrique" (CTGA). The last cited regroups the geotechnical community of 15 African French speaking countries: Benin, Burundi, Cameroon, Congo, Djibouti, Comoro's Islands, Ivory Coast, Gabon, Mali, RD Congo, Chad, Centrafrique, Niger, Senegal and Mauritania.

Several attempts have been made, and are still on going, to form new national societies in Africa, some faded out due to political unrest in their countries or other reasons, e.g., Angola, Libya, Tanzania and Uganda.

Other attempts are being also made towards the revival of some dormant societies that existed previously, e.g., Zimbabwe and Kenya, Ethiopia. Communications with interesting geotechnical engineers in these countries are established and plans are ongoing to help these societies resolve their internal challenges.

A task force conducted by Professor Fatma Baligh former ISSMGE VP for Africa, is devoted to this duty.

This report comprises firstly, a general overview of the main activities noted around the region since the last council meeting in Seoul on September 2017, and the details of events or activities in the various member societies during the same period.

GENERAL OVERVIEW

According to the decision taken to hold three African Young Geotechnical Engineers' Conferences (AYGEC) between successive Young International Geotechnical Engineers' Conferences (YIGEC), the Sixth Young African Geotechnical Engineers Conference (6thAYGEC) was successfully organized by the Sudanese Society for Soil Mechanics and Geotechnical Engineering (SSSMGE), in collaboration with the Building and Road Research Institute (BRRI), University of Khartoum and the ISSMGE from 25th to 26th November 2018, in Khartoum (Sudan).

This conference was preceded by a workshop on Problematic Soils held at BRRI.

The event was honored by the attendance of young researchers from African countries, and of prominent geotechnical experts from the ISSMGE's Board: the V.P. of the ISSMGE for Africa Prof. E. Marcelin KANA, Prof. Ikuo Towhata and Prof. Mounir Bouassida from the ISSMGE's Board, Prof. Fatma Baligh former V.P. of ISSMGE for Africa. The participation of university staff members from Egypt and Sudan, and geotechnical engineers from SSSMGE enriched the discussions. The overall participants were over seventy (70) Engineers.

Plate 1: 6th YAGEC Opening Day at the Ministry of Higher Education and Scientific Research

During the period under review, changes of officers occurred as follows in some member society in Africa:

Geotechnical Division of the South African Institution of Civil Engineering (SAICE)

From November 2017, 02nd the following candidates were elected for the 2018 committee:

President Solly Phalanndwa
Secretary Jacobus Breyl

Nigerian Geotechnical Association (NGA)

From May 2018, 12th the new officers elected to run the affairs of the Association for the next two years (2018-2020) are as follows:

President Prof Y. A. Jimoh
Secretary Eng. Sebastian C. Ozoamalu

Comité Transnational des Géotechniciens d'Afrique (CTGA)

From November 2017, 17th the new officers elected to run the affairs of the CTGA for the next years are as follows:

President Prof. Ibrahima Khalil CISSE
Secretary Eng. Jean Pondy

A geotechnical workshop, for the Portuguese speaking countries in Maputo-Mozambique, has been scheduled from 1st to 3rd October 2019. It aims to bring together the "Community of Portuguese Language Countries ": (Mozambique, Angola, Equatorial Guinea, Guinea-Bissau and Cape Verde). Some core members of the ISSMGE's Board will attend the workshop and present lectures.

Professor Heinz Brandl and Professor Charles Ng has been invited by the CTGA, in collaboration with the National Civil Engineering Laboratory of Cameroon (LABOGENIE), to deliver lectures during a two days' workshop, scheduled from 29th to 30th July 2019, in the University of Yaounde 1 (Cameroon), on the themes:

for Professor Brandl

1- Long-term settlements of soft soils and ground improvement
2- Flood protection / defense (Stability and failure of dykes; Emergency measures and restoration)

for Professor Charles Ng

1- Plant-Soil slope Interaction
2- Soil-structure Engineering of Deep Foundations, Excavations and Tunnels

After the huge success of the 9th and the 10th edition held respectively in Lome (TOGO) and Abidjan (Ivory Coast), the 11th edition of one of the main geotechnical event for the French speaking countries in Africa "11ème edition des Journées Africaines de la Géotechnique (11ème JAG 2019)" (The 11th edition of the African Geotechnical Engineering days) is scheduled from 21st to 24 October 2019 in Niamey (Niger). It is organized by the CTGA, in collaboration with the Association of African civil engineering laboratories for buildings and publics works (ALBTP), under the main theme: Geotechnical engineering and sustainable socio-economic development strategies of countries in the intertropical region of Africa (Géotechnique et stratégies de developpement socio-économique durable des pays d'Afrique intertropicale).

To enable in some way, for African French and English speaking ISSMGE individual's members, an honorable participation to main geotechnical engineering events in Africa, the availability of translation services, English to French and vice versa, should be assured. Each of them is representing approximately 45% of African ISSMGE's membership.

It has been noted that the open access policy is being very well appreciated in the Africa region. It surely needs to be more widely advertised in that region.

In collaboration with the Young Member Presidential Group (YMPG), two young engineers has been selected, from the continent. They will have the opportunity to present their work as the recipient of the ISSMGE Bright Spark Lecture award, at a special lecture session at the next XVII African Regional Conference on Soil Mechanics and Geotechnical Engineering Cape Town, South Africa.

The Bright Spark award is dedicated to the young geotechnical engineer that exemplifies mature research or practice and epitomizes what the ISSMGE is looking for in engineers of tomorrow. The selected are:

- Mohamed Morsy (Egypt): He has published a papers on the topic of contaminant transport and geosynthetics.
- Talia Da Silva (RSA): Her topic is focused on granular soils and their behavior during/after the formation of voids.

EVENTS OR ACTIVITIES IN THE VARIOUS MEMBER SOCIETIES

The events or activities in the various member societies, since the last Council Meeting held in Seoul September 2017, are summarized subsequently.

ALGEOS (Algeria):

The Algerian Geotechnical Society (ALGEOS) has organized:

From 6th to 7th November2017, in the Auditorium of STHB of Algiers, the second edition of the ALGEO'S annual technical colloquium entitled "Congrès Algérien de Géotechnique" (CAG).

In the town Oran, from 30th to 31st October 2018, with the collaboration of the Oran's University of sciences and technologies, the 4th edition of the UNSAT.dz group's colloquium on unsaturated soils and sustainable constructions titled: "UNSATOran 2018". The presentations and the discussions during this colloquium were focused on four themes:

1- Mechanical behavior of unsaturated soils;

2- Modelisation and tests of the transfer in unsaturated soils;

3- The use unsaturated soils as geo materials for buildings and roads constructions

4- Pathologies related to the non-saturation of soils

CTGA

The 9th edition of the CTGA's Annual Technical Colloquium (9ème JAG 2017), was held on the main theme: "Relevance of innovative geotechnical engineering practices on the growth strategies of African countries", in Lomé (Togo), from 14th to 17th November, 2017.

Figure 1 Visit of stands of exhibition by ISSMGE VP for Africa and Some Ministers in charge of Public Works of ECOWAS

A workshop on the project of the revision of civil engineering codes and guidelines (P2RT project) was carried out and widely appreciated.

Under the patronage of the Togolese Minister of Public Works, this colloquium was attended by almost 225 delegates from 16 African countries: Benin, Burundi, Burkina Faso, Cameroon, Ivory Coast, Congo, Democratic Republic of Congo, Gabon, Niger, Mali,

Three main lectures where given in plenary sessions. There are:

- Growth of concrete pavement construction in the intertropical region of Africa, by Engineer Jean PONDY
- Prevention and attenuation of the risk of hydric erosion of soils occurrence for civil engineering constructions, by Prof. Etienne-Marcelin KANA
- Tropical soils specificities in road constructions in African intertropical regions, by Prof. CAMAPUM DE CARVALHO and Prof. Etienne Marcelin KANA

During this colloquium, four workshops were also organized. And, in conformity with the ISSMGE's open access policy, the various presentation done can be downloaded on the website of CTGA: www.ctgaafrique.org,

A parallel session also took place, with some courses given under the theme: Constructions on problematic soils in Africa.

The 10th edition of "the African Geotechnical Engineering Days (10ème JAG 2018)", took place under the main theme: "Review and prospect of Geotechnical engineering in intertropical region of Africa", was held in Abidjan (Ivory Coast), from 22nd to 25th October, 2018.

That colloquium was attended by more than 250 delegates from 18 countries: Benin, Burundi, Burkina Faso, Cameroon, Ivory Coast, Congo, Democratic Republic of Congo, Gabon,

Niger, Nigeria, Ghana, Mali, Senegal, Togo, Mauritania, Cap Verde, Morocco, Comoro's islands, Madagascar, and France.

In perspective, the 11th edition of "the African Geotechnical Engineering days" that will be hold in Nyamey (NIGER), is scheduled from 21st to 24 October 2019 in the Congress Hall of Niamey (Niger), under the main theme: Geotechnical engineering and sustainable socio-economic development strategies of countries in the intertropical region of Africa (Géotechnique et stratégies de developpement socio-économique durable des pays d'Afrique intertropicale). Translation from French to English language and vice-versa will be provided.

EGS (Egypt):

The EGS has organized:

On the 28th July 2017, a workshop entitled: A Geotechnical Day, hosted by the College of Engineering and Technology, Arab Academy for Science, Technology and Maritime Transport (AAST) in Abo Qir, Alexandria. A lecture on Collapsible Soils: Great Challenge was delivered by Prof. Dr. Tarek Mostafa, Egypt.

During this event Prof. Fathi AbdRabbo of Alexandria University was honored for his contribution to Geotechnical Engineering in Egypt, and Eng. Mohamed Salah Morsy was nominated for the Outstanding Young Geotechnical Engineer Award

An international Congress and Exhibition, in collaboration with Egyptian government, the Housing and Building Research Center (HBRC), and the Soil-Structure Interaction Group in Egypt (SSIGE), The GeoMEast2018. Under the main theme : "Sustainable Civil Infrastructures", It was held from November 24th to 28th, 2018 in Giza, Egypt at Marriott Mena House hotel, in front of the Great Pyramids.

Over 900 experts from 71 countries representing the globe attended the conference. Thirty (30) companies from 20 countries took part to the exhibition. 570 full-papers were received, and 300 accepted and published in 16 Edited Volumes with Springer in SUCI book series.

GeoMEast 2018 organized five official lectures delivered by the current Presidents of ASCE, CSCE and ISSMGE, the former President of IGS, and the General Manager of GMA-IFAI. Twelve (12) Workshops/Courses and over 250 presentations were delivered.

GeoMEast2019 is scheduled during the period of 10-14 November 2019.

An International conference on Advances in structural and Geotechnical Engineering ICASGE'19 was held in HURGADA, under the main theme: "Together we can construct promising future", from 24th to 27th March 2019. It was organized by the Structural Engineering Dept., Tanta University, and the Egyptian Geotechnical Society (EGS). There were 50 papers accepted by the Geotechnical Division. Among them were 15 papers from Algeria, Nigeria, Sudan, Japan, Cameroon, Kuwait, USA Saudi Arabia, and the rest was from Egypt.

Figure 2: Participants to the ICASGE, HURGADA- MARCH 2019

Among others, two main lectures for the geotechnical engineering science were given by the Vice President for Africa Prof. E. Marcelin KANA, on Investigation and classification of African intertropical Soils, and the Prof. Ahmed Elsharief (SSSMGE-Sudan), on collapsible soils of Sudan

Figure 3: Participant to the meeting between ISSMGE VP and the Egyptian Geotechnical Society (EGS), 27th March 2019, HURGADA-EGYPT

A meeting between EGS and the ISSMGE VP for Africa, was organized at the end of this event to seek for solutions to the eventual problems encountered by EGS. Among the main solutions taken was the enhancement of the communication between African individual members and the Board of ISSMGE, via at least, the regional VP or the SG of ISSMGE.

GGS (Ghana)):

An annual general meeting (AGM) involving both Ghana Geosynthetic Society and the Ghana Geotechnical Society (GGS) was organized on 24th February 2018. Ms. Samuella Denise Boadu was selected as the coordinator for the student chapters of the two societies. Due to a constitutional difficulty, the election of new officers could not come on as scheduled.

Being part of the Civil Engineering technical group of the Ghana Institution of Engineering (GhIE), the Ghana Geotechnical Society also sponsors activities under the GhIE Evening lecture series. Some of the events organized under the evening lecture series include

Presentation on Geotechnical Design and Construction held on 25th October 2018 at the Engineers Center in Accra and it involved engineering experiences on geotechnical designs and construction and lessons learnt. It involved on how multidisciplinary engineers work together to deliver successful projects. The presentation was given by Monica Ansah-Osam, a senior geotechnical and tailings engineer with an Oil and Gas company.

Presentations and discussions on recent earthquakes in Ghana was held on 25th April 2019 at the Kumasi National University of Technology (KNUST). There were four short presentations which were then followed by a discussion. The presentations and discussions were streamed online and were followed simultaneously by the northern group based in KNUST

Following the strategies at the AGM, GGS started a presentation series known as the Y-GeoPresentation which is given by the young geotechnical engineers. The first presentation was made on 27th April 2019 on the theme. "Expansive Soils of Ghana". Two presentations were made. The first presentation on "stabilizing expansive soils using geogrid" was presented by Ms Samuella Denise Boadu while Mrs Rosemary Kweisi made a presentation on the "expansive soils of Pomadze". The presentations were based on the MSc thesis research of the presenters. The presentations were followed by discussions of the topic. This was attended by 15 members and was chaired by Prof. S.I.K. Ampadu.

SMG (Mozambique):

Sociedade Moçambicana de Geotecnica (SMG) intend to organize a geotechnical workshop, for the Portuguese speaking countries in Maputo-Mozambique, from 1st to 3rd October 2019. It aims to bring together the "Community of Portuguese Language Countries": (Mozambique, Angola, Equatorial Guinea, Guinea-Bissau and Cape Verde). Some core members of the ISSMGE's Board are invited to attend the workshop and to deliver some lectures.

CMMSR (Morocco), NGS (Nigeria) and ATMS (Tunisia):

Nothing has been noticed so far from these NS during the period under review.

SAICE (South Africa)

The Geotechnical Division of the South Africa Institute of Civil Engineers (SAICE), held an AGM on 02 November 2017, the following candidates were elected for the 2018 committee:

No.	Portfolio	Person
1.	Chairperson/President	Solly Phalanndwa
2.	Secretary	Jacobus Breyl
3.	Treasurer	Heather Davis
4.	Awards & YMPG	Dan Avutia
5.	Evening Lectures	Gabi Wojtowitz
6.	Codes and YGE Representative	Frans van der Merwe
7.	Marketing and Publications	Jurie Engelbrecht
8.	Courses & CPD	John Pavlakis
9.	Conferences and Jennings Lecture	Scott Gover
10.	Past Chairperson	Trevor Green
11.	Durban Representative	Bazi Dukhan
12	Cape Town Representative	Denis Kalumba

At present the Geotechnical Division has 551 members as of 25th January 2018

In that AGM, discussions were held on the establishment of the South African Geotechnical Design code, Lateral Support Code, & Piling Code. It was agreed that the Geotechnical Society will adopt a Eurocode but have with South African annexures.

The following events were successfully organized by the Geotechnical Division committee in 2018:

Date	Nature of event & Comments
15 February 2018	Geotechnical Investigation course for no-geotechnical engineers at Central University of Technology presented by Peter Day
06, 12, & 14 March 2018	16th Memorial Jennings Lecture by Mike Jefferies in Pretoria, Durban and Cape Town
07, 08, & 09 March 2018	Critical Soil Mechanics workshop presented by Prof. Mike Jefferies and Dr Dawn Shuttle
16-18 July 2018	3 Days Plaxis Numerical Modeling course by Denis Waterman
28 August 2018	1 Day Finite Element Modelling Geosynthetics
28 February 2019	The Latest Development in Soil Improvement presented by Prof. Dr.-Ing. M.Sc, Wolfgang Jimmy Wehr

The following events are planned by the Geotechnical Division committee for 2019:

Date	Nature of event & Comments
30 May 2019	Evening Lecture: Terzaghi Oration Lecture by Peter Day, Focus Rooms in Sunninghill
06 October 2019	African Young Geotechnical Engineering Conference (AYGEC) at the University of Cape Town
07-09 October 2019	African Regional Conference (ARC) of the ISSMGE Conference in Cape Town
07/11/2019	Three awards will be awarded during our AGM 2019 • Barry van Wyk Award for the best final undergraduate thesis in Geotechnical engineering • Jennings Award- best technical paper published in Southern Africa during the 2018 year • Gold Medal his outstanding work in Geotechnical industry

The division has presented a course "Geotechnical Investigation course for non-geotechnical engineers" at the Central University of Technology in Free State aimed at educating students about the importance of conducting geotechnical investigation for all civil engineering projects.

For the student outreach program, the division presented two courses aimed at educating students about the importance of conducting geotechnical investigation for all civil engineering projects on the respectives themes

-"The importance of Geotechnics in Civil Engineering "at the University of the Witwatersrand;

-"Geotechnical Investigation course for non-geotechnical engineers" at the Central University of Technology in Free State

SSSMGE (Sudan)

The Sudanese Society for Soil Mechanics and Geotechnical Engineering (SSSMGE) in collaboration with Building and Road Research Institute (BRRI), University of Khartoum and the ISSMGE organized the 6th AYGEC, from 25th to 26th November 2018 in Khartoum, Sudan.

This conference was preceded by a workshop on Problematic Soils held at BRRI. A group of young geotechnical engineers from the SSSMGE headed by Prof. Ahmed M. Elsharief worked hard to make this event a successful one.

Appendix 4: Regional Report – Asia

Eun Chul Shin
ISSMGE Vice-President, Asia

OUTLINE

The Asian Region of ISSMGE has 25 Member societies which are situated in geotechnical conditions that consist of good ground condition or difficult ground conditions like soft ground, earthquake prone area. The Asian Member Societies represent the following nations and areas: Bangladesh, China, Chinese Taipei, Hong Kong, India, Indonesia, Iran, Iraq, Japan, Kazakhstan, South Korea, Kyrgyzstan, Lebanon, Malaysia, Nepal, Pakistan, Philippines, Singapore, Southeast Asia, Sri Lanka, Syria, Tajikistan, Thailand, Uzbekistan and Vietnam. The Philippines became a new member of ISSMGE in 2018. Geotechnical Society of Singapore (GeoSS) was founded in 2008 and one of active societies in Asian region. The GeoSS hosted the ISSMGE Board Meeting in Singapore on March 10, 2019.

The previous Asian Council Meeting of ISSMGE was` held in September 2017, during the 19th ICSMGE, Seoul, Korea. The Asian Council Meeting will be held on October 15, 2019 during the 16th Asian Regional Conference in Taipei, Taiwan.

1 ON ASIA

There are 25 member Societies. The interim report (Oct. 2017 – Sept. 2017) has been received from individual Society and summarized. The activity report for 7 Asian Technical Committees are summarized. In the past 20 years, many geotechnical related the social infrastructures such as highway, airport, port and harbor, subway, industrial complex have been constructed in the Asian region.

The new technology and construction methodology were developed for solving difficult problems in various ground conditions. Some of these findings are presented in the event of international conference and symposium as well as workshops.

(1) *General scope*

The construction practice is related the political situation in the region. Instability of government in the western part of Asia is caused problem to the development of the country. Therefore, the countries in the region has constrained their ISSMGE activity.

The ISSMGE member societies in Asian region demonstrate a good collaboration model to the other part of member society throughout the joint conference and symposium as well as workshop together, because many societies are involved collaboration agreement between societies, particularly Southeast Asia, far eastern counties and central part of Asian countries, or neighboring countries. For example, the GeoSS and Malaysian Geotechnical Society could help to promote ISSMGE in neighboring countries in the region such as Cambodia, Laos, and Myanmar by organizing international seminars.

In this respect, the ISSMGE could support the outreach program to establish their own member society.

(2) *Climate change and natural disaster*

The two-thirds of entire population on the earth are lived in the Asian region. The large population can contribute to the emission of carbon dioxide due to the land cultivation and also the shortage of useful land in the urban area. Therefore, the living environmental of people are getting worse.

The great amount emission of carbon dioxide is contributed the temperature rise around the living space of earth. The coverage land of glacier in the cold region is dwindling and hence the sea level is rising. We should prepare the certain countermeasure against the inundation of low-land area. The irregular heavy rainfalls are caused of the landslides and debris flow in the mountain area.

The earthquake and tsunami are the common natural disasters in Japan, China and recently in Palu and Sulawesi, Indonesia (Sept. 2018). The liquefaction of ground, tsunami of water as well as mud, landslides are the deadly force to the human life and the property of government and private. Several earthquakes were occurred in India (2019), Pakistan (2013), Nepal (2015) at the Himalayan Mountains range. The ISSMGE would support to dispatch the geotechnical expert to the place where needs a help and mitigation of problem.

(3) *Low cost conference*

The registration fee of the international conference is relatively high these days. The people from low income countries have a problem to attend he international conference. The former Vice President for Asia, Ikuo Towhata initiated the Low Cost Conference (LCC). He actively involved the LCC policy to the Asian Regional Conference in Fukuoka, Japan in November, 2015 (USD 407). This LCC program is also adopted for the 16th AEC in Taiwan, Taipei, October 2019 (USD 450 for member of ISSMGE). The registration fee for the Southeast Asian Geotechnical Conference which was held in Jakarta, November 2018 was about USD 400 for the ISSMGE member. I would like to emphasize the LCC policy for the forth coming international conference and symposium in Asian region.

2 ASIAN REGIONAL CONFERENCE

The 16th Asian Regional Conference will be held in Taipei, Taiwan from October 14-18, 2018.
(1) There are 4 keynote lectures, 6 invited lectures, 4 them lectures, Corporate Associate Lecture, and CAPG session
(2) 7 Asian Regional Technical Committee sessions and 2 ISSMGE. TC sessions will be held on October 14-15. Three YMPG sessions will be held in the afternoon of October 14, 2019.

The venue of 17th Asian Regional Conference in the year of 2023 will be decided by the ballot voting during the Asian Council Meeting on October 15, 2019 in Taipei, Taiwan.

Currently the Kazakhstan Geotechnical Society submitted a proposal and the Nur-Sultan city, the capital of Kazakhstan would be the venue of the 17th ARC in 2023.

The Thai Geotechnical Society is also submitted a proposal and the venue of the 17th ARC would be the Bangkok city, Thailand in 2023.

3 ASIAN YOUNG GEOTECHNICAL ENGINEERS CONFERENCE

The Kazakhstan Geotechnical Society hosted the 8th Asian Young Geotechnical Engineer Conference (8th AYGEC) in the capital city of Astana in August 5-7, 2016. Now, the 9th AYGEC is hosted by the Pakistan Geotechnical Engineering Society and will be take place in the city of Lahore, December 5-7, 2019.

4 ASIAN REGIONAL TECHNICAL COMMITTEES

(1) AsRTC-1: Geotechnical Mitigation and Adaption to Climate Change-induced Geo-disaster in Asia-Pacific Regions, Chair: Prof. Bergado, Dennes T. (Formerly AIT, Thailand) and Prof. Yasuhara, Kazuya (Japan), Secretary Prof. Hazarika Hemanta(Japan).

(2) AsRTC-3: Geotechnical for natural hazards Chair: Prof. Kazama Mokito (Japan). Secretary Prof. Hazarika, Hemanta (Japan).

(3) AsRTC-6: Urban Geoengineering, Chair Prof. OU Chang-Yu (Chinese Taipei), Secretary Bin-Chen Benson Hsiung (Chinese Taipei). 43 members from 13 member societies in Asia, 2nd ATC6 symposium in Changsha, Hunan Province in China, 24th and 25th of November, 2017 and 100 researchers participate, more than 50 papers. The 3rd ATC6 symposium will be held in 14th to 18th of October, 2019 in Taipei, Taiwan
16th Asian Regional Conference

(4) AsRTC-7: Thick Deltaic Deposits, Chair Prof. Kim, Jiman (South Korea), Secretary Prof. Ahn, Jaehun (South Korea).
(5) AsRTC-8 on Groundwater Environmental and Quality Management is going to stop because of no activity.

(6) AsRTC-10: Urban Geo-informatics, Chair: Dr. Mimura, Mamoru (Japan), Secretary: Ms. Kitada, Naoko (Japan).

(7) AsRTC-18: Mega Foundation, Chair Jeong, Sang-Seom (South Korea). ATC-18 workshop was held under Prof. Myoungmo Kim with Six presentation, 15th Asian Regional Conference in Fukuoka Japan. 2017 International Symposium on Design and Analysis of Piled Raft Foundations, 12-13 Sep. 2017. Supporting "16th Asian Regional Conference on Soil Mechanics and Geotechnical Engineering (16ARC)" (Oct. 2019).

(8) AsRTC-19: Geo-engineering for conservation of heritage monuments and historical sites Chair: Prof. Iwasaki, Yoshinori (Japan), Secretary: Prof. Mimura, Mamoru (Japan). Special Workshop of ATC19 at Yunlin University, Taiwan ,16ARC, 2019. Sakura Workshop 2020 in Japan, April, 2020 at Kyoto or Osaka

5 ACTIVITY OF VP OF ISSMGE FOR ASIA

The term of my Vice President of ISSMGE for Asia was began since the September 2017, the 19th ICSMGE, Seoul, Korea. I have been traveled extensively for participating ISSMGE Board Meeting, ISSMGE TC and ATC conference which were held in Asian region. As a Director of Creative Educational Program for Future City Design which is supported by the Ministry of Education in Korea, the extensive overseas volunteer works have been carried out with the Incheon National University student delegation in Mongolia for planting trees in desert area, Cambodia for road construction and environmental clean-up activities and in Nepal for reconstruction of earthquake damaged water supply pipe line.

Summary of Attending the International Event and Delivering Lectures

2017
Sept. 20 - Organized ISSMGE Asian Council Meeting as the Vice President.
Oct. 31 - Keynote Lecture, Indonesia Geotechnical Engineering Conference in Jakarta.
Nov. 20 - Keynote lecture, Singapore Geotechnical Engineering Conference and attended GeoSS General Assembly Meeting, election new president of GeoSS.
Dec. 5 - Delivered lecture and attended Indian Geotechnical Engineering National Conference at IIT Delhi and attended IGS General Assembly Meeting, elected new President of IGS (Prof. S. Babu, IIS).

2018
Jan. 23 - Zurich Institute of Switzerland Geotechnical Laboratory (Prof. Anastaspoulos Ioannis).
Jan. 26 - Visiting Vienna Technical University Geotechnical Laboratory (Prof. H. Brandle).
Feb. 5 - Special invitation lecture for Singapore Geotechnical Society at NUS. Attended GeoSS Board Meeting
April 6 - Invited speaker at the ISSMGE TC 202 Transportation Geotechnics at IIT Delhi.
April 4 - Keynote Lecture at International Disaster Management Conference at Padang, Indonesia.
June 5 - Attending ISSMGE Board Meeting at Skopje, Macedonia.
July 3 - Keynote Lecture at International Disaster Reduction and Mitigation Conference at Sakhalin, Russia, organized by Russian Geotechnical Society, and Far Eastern Transportation University.
Nov. 5 - Invited speaker, Southeast Geotechnical Engineering Conference at Jakarta, Indonesia.
Nov. 16 - Attending ISSMGE Board Meeting and delivering Touring Lecture at Mexico City.
Dec. 11 - Attending ISSMGE TC 207, Soil-Structure Interaction and Retaining Walls and delivered invited lecture at Indian Geotechnical Engineering Conference at IIS, Bangalore, India. Attended IGS General Assembly Meeting and elected IGS president (Prof. S. Babu at IIS)

2019
Jan. 28 - Round Table Meeting with the Board Member of Vietnamese Geotechnical Engineering Society at Hanoi City. Discuss about Geotec Hanoi 2019 Conference in Nov. with Dr. Phung Duc Long (President of VSSMGE).
Feb. 11 - Delivering Special Invitation Lecture at IGS-Chennai Chapter IIT Madras, India (Coordinated by Prof. Buminantan). Discussion about ISSMGE TC 301 Preservation of Historic Sites Conference in September, Chennai, India.
Feb. 15 - Delivering Special Invitation Lecture at IGS-Cochin Chapter, Cochin, India (Coordinate by Dr. Anil Joseph)
Feb. 23 - Keynote Lecture, ISSMGE TC 305 Geotechnical Infrastructures for Mega Cities and New Capitals. Symposium at Astana, Kazakhstan, Kazakhstan Geotechnical Society (Prof. Askar Zhussupbekov).
March 10 - Attending ISSMGE Board Meeting and delivering Touring Lectures at Singapore. Round Table Board Meeting with GeoSS.

May 18 - Memorial Lecture, Silver Jubilee Celebration Conference, 25th Anniversary of Nepal Geotechnical Society, Kathmandu, Nepal. Discussion about April 2020 International Geotechnical Engineering Conference in Nepal.
May 21 - Invited Lecture, TRANSOILCOLD 2019 Conference, Transportation Soil Engineering in Cold Regions, Russian Geotechnical Society, St. Petersburg, Russia.
June 17 - Keynote Speaker at ISCORD 2019, Cold Region Development Association Oulu, Finland.
June 26 - Invited Lecture, ISSMGE TC 203 earthquake at Chania-Crete, Greece.
Aug. 29 – Attend Unsaturated Soil Symposium at Okayama, Japan
Sept. 17 - Invited Lecture, ISSMGE TC 301 Preservation of Historic Sites, Chennai, India, IIT Madras.
Sept. 20 - Invited Lecture, ISSMGE TC 107 Laterites and Lateritic Soils, Goa, India.
Oct. 5 - Attending ISSMGE Council Meeting, Cape Town, South Africa.
Oct. 14 – Invited Lecture and Asian Council Meeting, ISSMGE, 16th ARC, Taipei, Taiwan.
Nov. 28 – Invited Lecture, 4th International Conference on Geotechnics, Geotec Hanoi, Vietnam.
Dec. 5 -Keynote Lecture, 9th Asian Young Geotechnical Engineering Conference, Lahore, Pakistan.
Dec. 19 -Invited Lecture, Indian Geotechnical Conference, SVNIT Surat, Gujarat, India.

Appendix 4: Regional Report – Australasia

Gavin Alexander
ISSMGE Vice-President, Australasia

1 MEMBER SOCIETIES AND MEMBERSHIP

The Australasian region of the ISSMGE, comprises two Member Societies, those of Australia and New Zealand. While it is the smallest ISSMGE region in terms of the number of Member Societies (2% of the total), its individual membership, at 1908, represents approximately 8% of the ISSMGE membership. Australia and New Zealand are modestly populated counties and the current membership in the region represents 80 ISSMGE members per million inhabitants. This compares to an overall average of 3 ISSMGE members per million inhabitants for the Society as a whole. These statistics reflect both the high level of engagement with the ISSMGE by engineers in the region and the level of geotechnical engineering activity taking place in Australia and New Zealand.

The AGS (www.australiangeomechanics.org) is a joint technical society of the Institution of Engineers, Australia and the Australasian Institute of Mining and Metallurgy. The number of members has increased significantly over the last 4 years, and this is consistent with a general rebound in the infrastructure market in Australia during this period. At the time of writing this report the total AGS membership was 1999, which is 15% up on the 1,726 members reported in 2017. Of these, 999 (50%) are members of the ISSMGE.

The NZGS (www.nzgs.org) is a Collaborating Technical Society of Engineering New Zealand (previously the Institution of Professional Engineers NZ). It was recognised as the outstanding ISSMGE member society at the 2013 ICSMGE in Paris, and remains extremely proud of this achievement. At the time of writing this report, the NZGS has 1,282 members (also up 15% on the 2017 membership of 1,110). Of these, 775 (60%) are ISSMGE members. The NZGS works closely with the national structural engineering (SESOC) and earthquake engineering (NZSEE) societies to develop and deliver technical guidance and training. It also works closely with Engineering New Zealand and with Government in the development of policy and professional guidance.

2 AUSTRALASIAN REGIONAL CONFERENCES

The conference series that serves the ISSMGE region is the Australia New Zealand (ANZ) Conference on Geomechanics, which is held once every 4 years and involves a rotation cycle, over a 12-year period, where two are held in Australia and one in New Zealand. During the period 2017–2019, one ANZ Conference was held. The 13th Australia New Zealand Conference on Geomechanics, was held in Perth from 1 to 3 April 2019, and was hosted by the AGS. More than 450 registrants participated in the event. The conference was led by Michael Smith, Barry Lehane and Hugo Acosta and was very successful, both technically and socially. The next in the series of ANZ Conferences is to be held in 2023 in Cairns.

3 AUSTRALASIAN SUB- REGIONAL CONFERENCES

The NZGS holds a symposium once every two years. From November 24-26, 2017 the 20th NZGS Symposium was convened in Napier with the theme "What in Earth is Going on? Balancing Risk, Reward, Regulation and Reality." Approximately 250 delegates attended, with papers prepared by a mix of consultants, contractors and academics. These were summarised by theme reporters, with authors available to respond to questions from the floor. An invited session provided in depth insights into various aspects of the 2016 Kaikoura Earthquake. The 2017 symposium was preceded by a one day seminar of the recently released Earthquake Geotechnical Engineering Guidance series. No national NZ conference was held in 2019, due to the 13th ANZ Conference on Geomechanics. In Dunedin, in October 2019, the NZGS will host the next in this series of symposia, with this one entitled "Good grounds for the future".

Unlike the NZGS, the AGS holds no national conferences. Instead, most of the AGS chapters hold an annual or biennial seminar on a wide variety of geotechnical engineering related themes. Details of these are available from the AGS website.

4 AUSTRALASIAN YOUNG GEOTECHNICAL ENGINEERS CONFERENCES

The NZGS and AGS also share responsibility for organising and hosting the biennial Young Geotechnical Professionals Conference series in the region. The conferences are intended for young professionals, 35 years old or younger, with a maximum of 10 years experience. The number of attendees is generally capped at around 50 delegates, so as to facilitate an intimate and nurturing conference experience. Usually 3 or 4 senior geotechnical engineers from industry and academia are actively involved in the conference and provide mentorship to the delegates. From November 7-9, 2018, the AGS organised the 12th Young Geotechnical Professionals Conference in Hobart, Tasmania. These conferences continue to be well attended, well organised and enjoyed by all delegates.

The level of over-subscription of these events has led to the introduction of local one-day events around New Zealand for young professionals who have not been able to attend the regional conferences. These local events provide similar opportunities to present to one's peers in a supportive environment, to develop professional networks, and to be mentored by more experienced practitioners.

5 20TH ICSMGE, SYDNEY 2021

The Australian Geomechanics Society was delighted to have been unanimously awarded hosting rights for the 2021 ICSMGE.

This conference is the highlight of the Presidential term, and Sydney is a truly stunning location for it. Preparations for Sydney 2021 are well in hand, with the venue confirmed and the PCO appointed. The Local Organising Committee, led by John Carter, with Graham Scholey's support, has been formed and has had useful discussions with members of the ISSMGE Board to start shaping this as an outstanding event. GeoEng2000 is recognised as the Gold Standard, and John and his team are seeking to create an event that is at least as successful. Details can be obtained from the conference website www.ICSMGE2021.org

6 TECHNICAL COMMITTEES AND BOARD LEVEL COMMITTEES

Both the AGS and NZGS have strong participation in the ISSMGE's technical committees. The Region's membership of the Technical Committees has been comprehensively reviewed and refreshed through 2018 and 2019.

Sukumar Pathmandanaval of the AGS has been an extremely active Chair of the Corporate Associates Presidential Group (CAPG) since Professor Roger Frank's presidential term. Sukumar has worked tirelessly in this role to increase the CAPG membership, and to foster collaboration between research and practice. Highlights within our region have been a number of well attended and lively panel discussions that have been held in Australia and New Zealand on collaboration in geotechnical engineering.

Ashe Cooper of the NZGS has been active on the Young Members Presidential Group, coordinating the CAPG panel discussion in Auckland in 2019 and attending the inaugural YMPG face to face workshop in conjunction with the ISSMGE Board Meeting in Singapore in March 2019.

7 INTERACTION WITH MEMBER SOCIETIES

The NZGS and AGS have a strong working relationship. They interact regularly, with the Chairs of both societies attending one another's committee meetings at least once each year. Moreover, the Vice Presidents of the 3 sister societies (ISSMGE, IAEG and ISRM) regularly attend both societies' committee meetings, and work closely with local liaison officers from the other Society. In addition, both societies collaborate on the development of joint codes of practice and guidelines, and the AGS and NZGS are in regular communication regarding visiting overseas speakers and visitors so that both societies can exploit opportunities for presentations in both countries. An example of this is the Rankine Downunder series of presentations, where the Rankine Lecturer from the previous year is invited to tour both Australia and New Zealand in order to present their lecture to all chapters and regions of the AGS and NZGS.

8 GENERAL REMARKS AND OUTLOOK

The Australasian region is vibrant, active and its members are engaged with the profession and the ISSMGE.

In collaboration with Engineers Australia, the AGS has been developing a framework for professional registration of geotechnical engineers and engineering geologists that is informed by the UK's Registration of Geotechnical Engineering Professionals (ROGEP) and the NZGS's existing registration model for engineers and for professional engineering geologists. This project is ongoing.

In the aftermath of the 2011 Christchurch Earthquake, the NZGS has been very active in developing a series of earthquake guidelines with modules on liquefaction, foundations, retaining walls, site investigation practice, ground improvement design, ground improvement specifications guidance, scoping geotechnical investigation programmes, rockfall design and post disaster geotechnical response. Each of the guidelines publications is available on the NZGS's website, and a project has recently been confirmed to finalise these documents over the next two years.

The most important and exciting opportunity for the region in the next Presidential term is Australia's hosting of the 20th ICSMGE in Sydney, which I have discussed above.

I am extremely grateful and humbled to have been given the opportunity to serve both the AGS and NZGS, as Vice President for Australasia and Treasurer on the ISSMGE Board over these last two years. I have thoroughly enjoyed this experience but, unfortunately, have to stand down at this point due to a sudden health issue which prevents me from fulfilling my duties. My successor, for the remainder of this term, is being confirmed as I write this, and will be attending the mid-term Council meeting in Cape Town. I am sure he will continue the energetic and engaged tradition of Vice Presidents from our region.

I wish the Society and all its members every success for the future.

Appendix 4: Regional Report – Europe

Mario Manasserio
ISSMGE Vice-President, Europe

This mid-term report contains an overview of the state of the ISSMGE in Europe and a summary of the highlights of the activities in the region since the Council meeting in Seoul in September 2017.

1 MEMBER SOCIETIES

Today, there are 38 European Member Societies (out of a total of 89), involving 39 countries, since the Czech and Slovak Republics joined as a single society. As reported in the membership list submitted at the Seoul Council of September 2017, it should be noted that in the year 2017, under the mandates of the President, Roger Frank, and the Vice-President for Europe, Antonio Gens, Cyprus became a new Member Society with the name of the Cypriot Society for Soil Mechanics and Geotechnical Engineering (CSSMGE).

I am also happy to anticipate that an active group of Moldavian geo-technicians, with the help and support of the Romanian Geotechnical Society, is going to establish the Moldavia society of soil mechanics and geotechnical engineering. Thereafter, their intention, in the near future, is to apply to ISSMGE for membership, in accordance with the established procedures. In this desired case, Moldavia would become the 40th European Country Member of ISSMGE.

2 REGIONAL CONFERENCES

Obviously, the main Regional Conference event is the quadrennial European Conference which will be held in Reykjavik from the 1st to the 6th September 2019. At the time of writing this report (May 2019), the organisation is well under way to deliver what promises to be an outstanding Conference.

The XVII European Conference on Soil Mechanics and Geotechnical Engineering has as overall subject: Geotechnical Engineering, foundation of the future and covers a wide range of topics that have been classified in 6 themes: A: Modelling and experimental assessment of geomaterials, B: Geotechnical construction and soil improvement, C: Geohazards, earthquakes and mitigation, D: Environment, water and energy, E: Historical heritage preservation, F: Special and specific Issues. The Conference has been organised by the Icelandic Geotechnical Society through a Conference Organising Committee (COC) chaired by Mr. Haraldur Sigursteinsson. The Conference has been structured over four days and the technical programme includes: 5 Keynote lectures, 5 Invited lectures, 8 Main Sessions and around 50 Discussion Sessions. In addition, there will be a number of Workshops and Meetings organised by various Technical Committees. As in the case of the previous edition of the ECSMGE, held in Edinburgh (2015), no limits were set for the number of papers that may have been submitted by each Member Society. Out of the 1,050 Abstracts initially proposed, 720 papers were finally accepted for publishing in the corresponding printed and digital Proceedings. As many as 500 papers will be presented orally, and the remaining ones have been offered poster presentations. It is worth noting that all the papers will become available for downloading from the conference website from the very beginning of the conference in order to allow the audience to follow all the presentations on both the main screens and/or on their computers, tablets or smart phones.

In addition, a series of Technical Tours and an extensive Social Programme will complete the Conference activities.

More than 35 Technical Exhibition stands have been sold and a number of Sponsors have been secured, although additional stands and sponsorship opportunities are still available for any organisation that may be interested.

Just two weeks after the Reykjavik Conference, where great visibility will be provided to young members through a joined main session in cooperation with Corporate Members dealing with Bridging the Gap Between Academics and Practitioners and through the Bright Spark Awards, the 27th European Young Geotechnical Engineers Conference will take place in Bodrum, Muğla-Turkey, on 26th and 27th September 2019.

Moreover, the dates of the latter conference offer the attendees a special chance to follow another international event in the same week devoted to young ISSMGE members i.e. the 1st Mediterranean Young Geotechnical Engineers Conference which will take place in the same venue from the 23rd to the 24th September, 2019. In this way, a blended atmosphere, a bigger get together and a larger platform of collaboration will be provided for the young attendees, together with some invited keynote speakers who will be, let's say, "differently young". Hopefully, this initiative should give rise to a new conference series that will be able to promote a lasting and profitable cooperation among all the participants for the common technical and scientific progress of the Mediterranean Regions within the Geotechnical Engineering discipline.

In addition to the aforementioned general European Conference, there are also a number of European Conferences characterized by a long and outstanding tradition and involving sub-regional groups of Member Societies: The Danube-European Conference on Geotechnical Engineering, The Nordic Geotechnical Meeting and the Baltic Sea Geotechnical Conference.

The 16th Danube-European Conference on Geotechnical Engineering, organised by the Macedonian Association for Geotechnics, was held at the Hotel Aleksandar Palace of Skopje on 7-9 June 2018. The Organising Committee was chaired by Professor Milorad Jovanovski and co-chaired by Professor Ljupčo Dimitrievski. The conference theme was: Geotechnical Hazards and Risks: Experiences and Practices (Geotechnische Gefahren und Risiken: Erfahrungen und Praxis). Participants from 46 countries took part, and there were 10 invited lecturers and more than 150 papers. During this event, the President of the Republic of Macedonia awarded prof. Heinz Brandl the Medal for Merit for his outstanding contribution to the establishment and to general increase in the Macedonian geotechnical community. Moreover, it was decided that the next XVII Danube Geotechnical Conference will take place in Bucharest, Romania, in 2022.

Both the 18th Nordic Geotechnical Meeting and the 14th Baltic Sea Geotechnical Conference will take place in Helsinki (Finland) on 25-27 May 2020. They are being organised by the Finnish Geotechnical Society under the chairmanships of Leena Korkiala-Tanttu and Juho Mansikkamäki.

One of the yearly highlights of the European meeting calendar is the European Young Geotechnical Engineers Conference (EYGEC). The 26th European Young Geotechnical Engineers Conference was held at Hotel Klugbauer in Reinischkogel near Graz in Austria from the 11th to the 14th September 2018 and was organised by the Austria Member Society. It was attended by 45 delegates nominated by 26 different countries. Four keynote lectures were delivered by Janko Logar (the University of Ljubljana, Slovenia), Clemens Kummerer (Keller Kolding, Germany), Dietmar Adam (Vienna University of Technology, Austria) and Helmut F. Schweiger (Graz University of Technology, Austria). Overall, the presentations were of a high level and well prepared by the young geotechnical engineers, and they covered a wide range of geotechnical problems. The participants represented a well-balanced group of academics and practitioners and, therefore, not only were high level research results discussed, but also interesting problems that arose from practice. As noted above, the 27th EYGEC will be held in Bodrum, Muğla-Turkey on 26th and 27th September 2019, whereas the 28th EYGEC will be organised by the Russian Member Society for the year 2020.

3 OTHER INTERNATIONAL CONFERENCES

In addition to the above, a number of ISSMGE Conferences have been held in Europe and have often been organised by Technical Committees or by Member Societies under the auspices of ISSMGE:

- 2nd International Symposium on Coupled Phenomena in Environmental Geotechnics (CPEG2). Leeds, The UK, 06-08 September 2017. (the University of Leeds and endorsed by TC215).
- 3rd International Soil-Structure Interaction Symposium. Izmir, Turkey, 18-20 October 2017. (Turkish Society for Soil Mechanics and Geotechnical Engineering).
- Geo-Expo 2017 Scientific and Expert Conference. Sarajevo, Bosnia and Herzegovina, 26-27 October 2017. (Geotechnical Society of Bosnia and Herzegovina).
- 2nd International Conference "Challenges in Geotechnical Engineering". Kiev, Ukraine, 20-23 November 2017. (Kyiv National University of Building and Architecture, the University of Zielona Góra, the Ukrainian Society for Soil Mechanics, Geotechnics and Foundation Engineering).
- Granular Matter Workshop - CEGD Workshop. Budapest, Hungary, 12-13 June 2018. (Óbuda University, the Hungarian Academy of Sciences and the BME Morphodynamics Research Group).
- 4th International Symposium on Cone Penetration Testing (CPT'18). Delft, The Netherlands, 21-22 June 2018. (Delft University of Technology and endorsed by TC102).
- 9th European Conference on Numerical Methods in Geotechnical Engineering. Porto, Portugal, 25-27 June 2018. (the University of Porto and endorsed by ERTC7).
- 9th International Conference on Physical Modelling in Geotechnics. London, The UK, 17-20 July 2018. (City, the University of London and endorsed by TC104).
- China – Europe Conference on Geotechnical Engineering. Vienna, Austria, 13-16 August 2018. (the Institute of Geotechnical Engineering, the University of Natural Resources and Life Sciences Vienna (BOKU) and the University of Leeds).
- Urban Planning Below the Ground Level: Architecture And Geotechnics. Saint Petersburg, Russia, 19-21 September 2018. (St. Petersburg Council of Architects and endorsed by TC207).
- International Symposium on Energy Geotechnics. Lausanne, Switzerland, 26-28 September 2018. (the Swiss Federal Institute of Technology in Lausanne and endorsed by TC308).
- Geo-Expo 2018 Scientific and Expert Conference. Neum, Bosnia and Herzegovina, 18-19 October 2018. (the Geotechnical Society of Bosnia and Herzegovina).
- International Scientific-Technical Conference "Geotechnics of Belarus: Science and Practice". Minsk, The Republic of Belarus, 23-26 October 2018. (Belarusian National Technical University and Belarusian Geotechnical Society).
- Geotechnical Challenges in Karst. Omiš, Croatia, 11-13 April 2019. (the Croatian Geotechnical Society).
- Prague Geotechnical Days 2019. Prague, The Czech Republic, 13-14 May 2019. (the Czech and Slovak Society for Soil Mechanics and Geotechnical Engineering, SG Geotechnika, Charles University - Faculty of Science).
- 4th International Conference "Transportation Soil Engineering in Cold Regions". Saint Petersburg, Russia, 20-23 May 2019. (Emperor Alexander I Petersburg State Transport University and endorsed by TC202, TC215 and TC216).
- Construction in Historical Cities: Problems and Solutions. Saint Petersburg, Russia, 06-07 June 2019. (Institute "Georeconstruction" and endorsed by TC207).
- 7th ICEGE 2019 - International Conference on Earthquake Geotechnical Engineering. Rome, Italy, 17-20 June 2019. (the Italian Geotechnical Society and endorsed by TC203).
- 7th International Symposium on Deformation Characteristics of Geomaterials. Glasgow, UK, 26-29 June 2019. (the University of Strathclyde, the University of Bristol and endorsed by TC101).
- 3rd International Conference "Challenges in Geotechnical Engineering". Zielona Góra, Poland, 10-13 September 2019. (the University of Zielona Góra, Kyiv National University of Building and Architecture, the Polish Committee on Geotechnics).
- 1st Mediterranean Young Geotechnical Engineers Conference. Bodrum, Turkey, 23- 24 September 2019. (the Turkish Society for ISSMGE - ZMGM).
- 27th European Young Geotechnical Engineers Conference. Bodrum, Turkey, 26-27 September 2019. (the Turkish Society for ISSMGE - ZMGM).
- 3rd International Conference on Information Technology in Geo-Engineering (3rd ICITG 2019). Guimarães, Portugal, 29 September – 02 October 2019. (the University of Minho, the Portuguese Geotechnical Society and endorsed by JTC2 of FedIGS)

Future Conferences

- International Conference on Geotechnical Engineering Education. Athens, Greece, 24- 25 June 2020 (TC306).
- 4th European Conference on Unsaturated Soils - Unsaturated Horizons. Lisbon, Portugal, 24-26 June 2020. (Instituto Superior Técnico, Delft University of Technology, Universitat Politècnica de Catalunya and endorsed by TC106).
- TC204: Geotechnical Aspects of Underground Construction in Soft Ground - TC204 Cambridge 2020. Cambridge, The UK, 29 June – 01 July 2020. (the University of Cambridge and endorsed by TC204).
- 6th International Conference on Geotechnical and Geophysical Site Characterization.

4 NATIONAL MEETINGS

Many of the Member Societies have organised National meetings, but they are too numerous to be listed here. For instance, I have attended the following ones:

- - 16° Congresso Nacional de Geotecnia on Riscos Naturais e Geotécnicos e Sustentabilidade. 27-30 May 2018, Ponta Delgada, S. Miguel, Açores, Portugal (Sociedade Portuguesa de Geotecnia - SPG, Laboratório Regional de Engenharia Civil - LREC, Sociedad Española de Mecánica del Suelo e Ingeniería Geotécnica SEMSIG).
- First National Symposium on Geo-environmental Engineering. 2-3 July 2018, Bucharest, Romania (the Rumanian Society of Geotechnical and Foundation Engineering).
- XXV Geotechnical Conference of Torino on Seismic Analysis and Design of Geotechnical Works. 8-9 November 2018 (the Italian Geotechnical Society).
- XVI Croce Lecture on Environmental Geotechnics: Looking Back, Looking Forward. Italian Geotechnical Society, December 13, 2017, Rome (Italy).
- XXV Széchy Károly memorial session, 8th of February 2019, Budapest, Hungary (the Hungarian Geotechnical Society).

5 TECHNICAL COMMITTEES

Thirteen ISSMGE Technical Committees (TCs) are under the care of European Member Societies:

- TC-101: Laboratory Stress Strain Strength Testing of Geomaterials. The United Kingdom, Chair: Matthew Richard Coop.
- TC-102: Ground Property Characterization from In-Situ Tests. Portugal, Chair: Antonio Viana da Fonseca.
- TC-106: Unsaturated Soils. The United Kingdom, Chair: David Toll.
- TC-201: Geotechnical Aspects of Dykes and Levees and Shore Protection. The Netherlands, Chair: Cor Zwanenburg.
- TC-204: Underground Construction in Soft Ground. The Netherlands, Chair: Adam Bezuijen.
- TC-207: Soil-Structure Interaction and Retaining Walls. Russia, Chair: Michael Lisyuk.
- TC-211: Ground Improvement. Belgium, Chair: Noel Huybrechts.
- TC-212: Deep Foundations. Italy, Chair: Alessandro Mandolini.
- TC-213: Scour and Erosion. The United Kingdom, Chair: Richard Whitehouse.
- TC-220: Field Measurement in Geomechanics. The United Kingdom, Chair: Andrew Ridley.
- TC-301: Preservation of Historic Sites. Italy, Chair: Renato Lancellotta.
- TC-306: Geo-engineering Education. Greece, Chair: Marina Pantazidou.
- TC-309: Machine Learning and Big Data. Norway, Chair: Zhongqiang Liu.

European Regional Technical Committees.

In addition to the ISSMGE TCs, there are several European Regional Technical Committees focused on issues that are particularly relevant for European Geotechnics.

- ERTC-3: Piles. Chair: Maurice Bottiau (Belgium),
  This is a very active Committee with a steadily expanding membership. The Committee has held a number of online and face-to-face meetings. In particular, a very successful meeting, hosted by the Belgian Member Society, was organised during the 19th ICSMGE in Seoul, on 20th September 2017. Moreover, an information session of a technical approval system of pile foundations was held in Brussels on 22nd May 2018.
- ERTC-7: Numerical Methods. Chair: Helmut Schweiger (Austria)
  The Committee organised a very successful 9th European Conference on Numerical Methods in Geotechnical Engineering (NUMGE) in Porto (Portugal) on 25-27 June 2018. The Committee is also going to hold a workshop in the XVII ECSMGE, Reykjavik, Iceland on Surveying, Compare and Discuss the Development and Applications of Numerical Methods in Geotechnical Engineering, 1st September 2019.
- ERTC-10 Evaluation of Eurocode 7. Chair: Trevor Orr (Ireland).
  The main activity of ETC 10 was the planning and preparation, together with TC205 and TC304, of several workshops and face to face meetings. Following the resignation of Professor Trevor Orr, a new chairman is going to be officially appointed at the XVII ECSMGE in Reykjavik, Iceland, September 2019.
- ERTC-12: Evaluation of Eurocode 8. Chair: Christos Vrettos
  The main purpose of ERTC-12 is to provide general, as well as specific comments on the EC8 code from the earthquake geotechnical engineering practice point of view, and in the light of the latest research. Moreover, it intends to conduct comparisons with other national or international codes and to propose improvement and modifications to EC 8. The new chairman, Professor Christos Vrettos, in the role of DIN delegate of Eurocode 7 and Eurocode 8 and as a member of several DIN Committees on calculation methods in geotechnics and seismic design of buildings, is in the ideal position to re-assemble the committee and to start with a fruitful phase of activities.

6 ADDITIONAL ITEMS

The selection of the Member Society that will be in charge of the next XVIII European Conference on Soil Mechanics and Geotechnical Engineering, to be held in 2022, will be decided during the meeting of the European Member Societies that will be held in Reykjavik on 3rd September 2019. So far, three Member Societies have formally expressed their intention of

submitting a bid at the meeting: The Belgian Society for Soil Mechanics and Geotechnical Engineering, BGGG-GBMS (Belgium), The Portuguese Geotechnical Society (Portugal) and the Turkish Geotechnical Society (Turkey).
the annual "Carlos Oteo Award" was created in 2018 to award the best geotechnical paper published in Spanish in either the SEMSIG Bulletin or in the Geotecnia Journal and in order to promote the publication of technical papers in the Spanish language. The first Carlos Oteo award, for 2018, was awarded in February 2019 to the paper entitled "Modelización numérica de los drenes verticales" (Numerical modeling of vertical drains), published in edition 144 of the Geotecnia Journal and written by Svetlana Melenjitevic, Javier Moreno Robles and Pablo Martín Blanco.

It is now worth mentioning a much appreciated initiative that was promoted and headed by the French Committee of Soil Mechanics and Geotechnical Engineering (CFMS) in cooperation with a number of supporters and sponsors: the Virtual Geotechnical Museum, which was completed and presented, in website form, at the beginning of 2019. This initiative is based on the "Les dessous des grands travaux" exhibition hosted by the Museum of Arts and Crafts in Paris, and which was first inaugurated at the 18th ICSMGE held in 2013. The virtual visit through the website of the exhibition is available in French and English (http://expo.geotechnique.org). This installation is a first in France and abroad, and it illustrates the know-how of French engineering and construction companies in the geotechnical field.

7 DEPARTED

I am very sad to report that two key people of the European and International Geotechnical community have recently passed away: Professor Iacint Manoliu (Founding father of the Rumanian Society of Geotechnical and Foundation Engineering and Chair of the ISSMGE ERTC-16 on Education and Training for 2007-2011) and Professor Sven Hansbo (president of the Swedish Geotechnical Society for 8 years and awarded the Kevin Nash Gold Medal by ISSMGE in the year 2009).

APPENDIX

REPORTS SUBMITTED BY INDIVIDUAL MEMBER SOCIETIES IN EUROPE

ALBANIA

Year 2017
We have published the annual magazine "Geotechnics Nr. 19", and the book "Tunnels" by our member PhD Ervin Paci (for students and engineers).
We have organized the training course with engineers and technicians who worked in geotechnical laboratories.
12 of our members participated in international conferences in Prague, San Petersburg, Seoul, Tirana.
President of AGS was member of the scientific committees in San Petersburg, Vienna, Skopje, Tirana.

Year 2018
We have published the annual magazine "Geotechnics Nr. 20", and the book "Stability of the artificial and natural slopes, engineering measures" by Professor Emeritus Luljeta Bozo (for students and engineers).
We have organized the open lecture about the "Modern construction of the railway" which was held by PhD Fitim Shala with participation of over 80 engineers and students in M.Sc.
The first geotechnical engineer graduated from Polis University.
6 of our members participated in international conferences in Skopje, Graz, Vienna.
President of AGS was member of the scientific committees in Island, Poland, Turkey, Austria.

AUSTRIA

Conferences and Symposia

- 11th Austrian Geotechnical Conference (ÖGT) & VÖBU Fair, 02.-03.02.2017, Vienna, about 350 participants and almost 1000 visitors (fair), Vienna Terzaghi Lecture delivered by Prof. Sarah Springman (ETH Zürich, Switzerland).
- 32nd Christian Veder Kolloquium (CVK), 19.-20.04.2017, Graz, about 400 participants.
- 33rd Christian Veder Kolloquium (CVK), 05.04.-06.04.2018, Graz, about 400 participants.
- 26th European Young Geotechnical Engineers Conference (EYGEC), Klugbauer, Austria, 11.-14.09.2018 (organized by TU Graz).
- 40 Years of Roller Integrated Continuous Compaction Control (CCC) - Anniversary Symposium & Exhibition, 29.11.2018, Vienna, about 180 participants (organized by TU Wien in cooperation with KTH Stockholm).
- 12th Austrian Geotechnical Conference (ÖGT) & VÖBU Fair, 31.01.-01.02.2019, Vienna, about 800 participants including fair, Vienna Terzaghi Lecture delivered by Prof. Jürgen Grabe (TU Hamburg, Germany).
- 34th Christian Veder Kolloquium (CVK), 25.04.-26.04.2019, Graz, about 400 participants.
- Close cooperation and exchanges with colleagues from neighboring societies, e.g. Prof. Luis Ibarra (The University of Utah, USA), visiting professor at TU Wien, 10/2017-06/2018.
- Seminars and presentations given by colleagues from abroad.

Keynote, Invited and Special Lectures (selection)

Prof. Heinz Brandl
- Near-surface geothermal geotechnics for heating and cooling - World Multidisciplinary Earth Sciences Symposium – WMESS 2017, Prague, 11.09.2017 (Opening-Key Speaker).
- Stability and failures of dykes/levees; Flood protection, defense, restoration - XVI Danube-European Conference, Skopje, 07.06.2018 (Invited Lecture).
- Thermo-active ground structures for future cities with renewable energy systems - 2018 Future City Forum – Design and Construction of Urban Development, Incheon, Republic of Korea, 13.09.2018 (Invited Lecture).
- Box-shaped deep foundations (piles, etc.) improving bearing-settlement and seismic behaviour of structures - 1st Dr. Clement and Carrie Chang Chair Lecture, Taipei, 23.10.2018 (Invited Lecture).
- Structures and retaining measures along unstable slopes - 1st G. Stefanoff Memorial Lecture, Sofia, 22.05.2019.

Prof. Helmut Schweiger
- Advanced finite element modelling for practical geotechnical problems - XXIVth Széchy Károly Memorial Lecture, Budapest, 09.02.2018.
- On the merits of numerical modelling in practical geotechnical engineering - 19th Šuklje´s Lecture, Maribor, Slovenia, 16.11.2018.
- Examples of successful numerical modelling of complex geotechnical problems GeoMEast 2018, Cairo, Egypt, 26.11.2018 (Invited Lecture).

Prof. Dietmar Adam
- Piles as Retaining Structures - Prague Geotechnical Days 2017, Academy of Science, Prague, Czech Republic, 09.05.2017 (Invited Introductory Lecture).
- 25 Jahre Geotechnik-Kolloquium an der TU Darmstadt und 25 Jahre Bodendynamik an der TU Wien – Zwei besondere Geotechnik-Jubiläen - 25. Darmstädter Geotechnik- Kolloquium mit Fachausstellung – Anniversary Colloquium, Darmstadt, Germany, 08.03.2018 (Invited Anniversary Lecture).
- The role of engineering geologists in geotechnical engineering - 5th Slovenian Geological Congress / 5. Slovenski geološki kongres (5 SGK), Velenje, Slovenia, 03.10.2018 (Invited Lecture).

BELGIUM

Activities Belgian Member Society of ISSMGE (BGGG-GBMS)

- ICSMGE Seoul September 2017
- Participation of 2 young professionals to the iYGEC
- Participation of the Chairman of BGGG-GBMS (M. Bottiau) to the ISSMGE Council meeting
- Organisation of a meeting of ETC3-Piles (ETC 3 is hosted by the Belgian Member Society, Chairman ETC3 is M. Bottiau)
- Organisation of a workshop TC 211 – Ground Improvement (TC 211 is hosted by the Belgian Member Society, Chairman TC 211 is N. Huybrechts)
- Organisation of 1/2-day conference "La pratique du pressiomètre en Belgique" - in memoriam prof. Van Wambeke, 28 September 2017, Brussels (100 participants)
- Organisation of 4 evening sessions on "Deformations and geotechnical structures", 22 February, 6 March, 19 and 26 April 2018, Brussels (125, 80, 125 and 80 participants respectively)
- Organisation of 1 day conference in collaboration with CFMS "Journée Franco-Belge – Dimensionnement sismique des fondations", 15 March 2018, Brussels (100 participants)
- Organisation of the General Assembly and Award ceremony of the De Beer Award to ir. F. De Cock + Lecture by the laureate, 14 June 2018, Waterloo
- Selection and participation of 2 young professionals to the YGEC Reinigskögel, September 2018
- Organisation of an evening Lecture by C. Treve (CFE) on "Experiences in geotechnics", 10 December 2018, Brussels (125 participants)
- 3 Meetings of the BGGG-GBMS Task Force 2 on "Quality of soil Investigation" (12 December 2017, 16 February 2018 and 3 October 2018). This task force is elaborating procedures for a registration/qualification system for geotechnical engineers and geologists dealing with the soil investigation and/or geotechnical design.

Planned activities in the near future

- Information session of a technical approval system of pile foundations, 22 May 2018, Brussels
- Evening session on Jet Grouting, 28 May 2018, Brussels
- General Assembly, End of June 2018.

HUNGARY

The Hungarian Geotechnical Society has experienced an active period since the 19th International Conference on Soil Mechanics and Geotechnical Engineering, Seoul, Korea.

The annual Széchy Károly Memorial Lectures were held at the Hungarian Academic Sciences in Budapest. According to tradition, three speakers were invited: – a well-known International, a well-known Hungarian, and a young Hungarian engineer. In February 2018 the 24th Széchy Károly Memorial Lecture focused on the topic of numerical modelling. The agenda was as follows:
1) prof. Helmut Schweiger, Graz University, Austria - Advanced finite element modelling for practical geotechnical problems,
2) Manninger Marcell, Bohn Ltd., Hungary - Design of special deep foundation,
3) Szilvágyi Zsolt, Geoplan Ltd., Hungary as a young engineer - Dynamic parameters of Danube sands.

In February 2019 the 25th Széchy Károly Memorial Lecture emphasized the topic of geotechnical testing including physical, laboratory and field tests. The international speaker for the Memorial Lecture was Charles W.W. Ng, the president of the ISSMGE. The agenda was as follows:
1) prof. Charles W.W. Ng, HKUST, Hong Kong - Understanding geotechnical phenomena to advance engineering design through centrifuge modelling,
2) Móczár Balázs, BME, Hungary – Contributions to the interpretation of shear strength based on triaxial tests,
3) Nagy Péter, TU Wien, Austria - Experimental and theoretical investigation of vibroflotation.

The Hungarian Geotechnical Society organized two scientific symposiums in 2018:
1) In June 2018. prof. Gyan N. Pande, Swansea University, United Kingdom, gave a short course on "Particle/pore size distribution & microstructure of saturation – key elements for rational description of mechanical behaviour of unsaturated soils",
2) In November 2018. prof. Ikuo Towhata, University of Tokyo, Japan, shared his presentation titled "Activities to solve the nuclear accident problems at Fukushima No.1 Power Station during the 2011 earthquake and tsunami disaster" with the members of the HNC.

The International Conference on Geotechnical and Geophysical Site Characterization Conference (ISC'6) will be held in Budapest, Hungary in 2020. About 500 abstracts were submitted and the Organizing Committee has been working on the arrangements.

MACEDONIA

Name of the president, international secretary and their contact e-mail address:

- Jovan Br. Papić, papic@gf.ukim.edu.mk;
- Igor Peševski, pesevski@gf.ukim.edu.mk.

Approximate number of members (both domestic and international): 60 Approximate number of student members: 10 Organizing International Conferences, Workshops, Seminars etc.:

- 16th DECGE in June 2018 (participants from 46 countries, 10 invited lecturers, more than 150 papers, published via Ernst&Sohn publishing house, including e-journal);

- International Tunneling and Underground Space Association (ITA) Seminar in June 2018 (about 130 participants);
- Meetings of Board Members of ISSMGE and Executive Board of ITA.

Organizing National Conferences, Workshops, Seminars etc.:

- Three annual meetings of MAG (with presentations given by experienced engineers on geotechnical problems and young researchers on results from scientific projects);
- Several invited foreign lecturers (e.g., prof. Ahmet Saglamer, prof. Ernest Olinic);
- Workshops (realized by MOBA and TENSAR);
- Round tables (on National Annexes to Eurocode 7 and CEN/ISO technical standards).

Ongoing MOU with other member societies of ISSMGE:

- Mutual with Serbian, Bosnian, Croatian and Slovenian geotechnical societies.

Current hosting of ISSMGE TC and ETC: No

Periodical journals and magazines:

- special edition of Scientific Journal of Civil Engineering dedicated to 16th DECGE;
- planned bulletin for 2019.

Awards:

- Nominating and awarding of prof. Heinz Brandl by the President of the R. Macedonia (State Recognition: Medal for Merit).

Other important achievements:

- Financial support (covering scholarship) to students at first cycle of studies at geotechnical department;
- Memorandum for cooperation with local companies;
- Established prize for extraordinary dedication in developing geotechnics in the country;
- Established league of young geotechnical engineers;
- Membership in the Engineering Institution of Macedonia;
- Membership in the International Society for Rock Mechanics and Rock Engineering (ISRM);
- Published dictionary of rock mechanics terms at ISRM.net;
- Honorary members: prof. Heinz Brandl and prof. Ahmet Saglamer;
- Participation at regional and international conferences and events, both for young and senior engineers, as well as visiting other international institutions;
- Development of Facebook profile GF.geotehnika;
- Close cooperation with national sister-associations (covering dam engineering, roads, structural engineering etc.).

Planned future activities (in 2019):

- Continuing tradition of inviting eminent scientists to deliver lectures to MAG's members (e.g., prof. Dietmar Adam);
- Membership in the Balkan geophysical association;
- Organizing seminar on landslides' hazards and risks (in cooperation with national Directorate for protection and rescue);
- Promotion of new web-page;
- Round table on implementation of common prices for geotechnical services;
- Participation in preparation of new law for civil engineering works from geotechnical point of view.

POLAND

Country of National Society: POLISH COMMITTEE ON GEOTECHNICS

Principal office bearers President: Prof. A. Szymański
Honorary President: Prof. E. Dembicki
Vice Presidents: Prof. Z. Lechowicz, Prof. K. Gwizdała
Secretary: Prof. K. Zabielska-Adamska
Treasurer: Prof. Joanna Bzówka

In the frame of Polish Committee on Geotechnics there are two subcommittees:
- Underground Constructions Subcommittee
- Engineering Geology Subcommittee

Member of the Awards Committee ISSMGE – prof. E. Dembicki. Number of ISSMGE members – 371 (2017), 378 (2018), 378 (2019).
In period 2017-2019 the Certificate of the Polish Committee on Geotechnics for the recognition of professional qualifications and experience in the geotechnical engineering was granted for 13 members.

Number of National Corporate members – 9.
Organization of Conferences and Workshops
- 6th International Workshop on "In situ and laboratory characterisation of OC subsoil", Poznań, 26-27.06.2017, Chairperson of Organizing Committee – prof. Z. Młynarek, Invited Speakers: dr. M. Long, dr. T. Lunne, D. Marchetti, dr. P. Monaco, dr. R. Massarsch, prof. F. Danzinger, prof. H. Tanaka, dr. Knut Schjetne, J-B Clausen, (125 participants), Proceedings 203 pp.
- XVIII National Conference on Soil Mechanics and Geotechnical Engineering with VII Young Geotechnical Engineers Conference, Warsaw, Poland, 04-07.09.2018, Chairperson of Organizing Committee – prof. K. Garbulewska, Chairperson of Scientific Committee – prof. Z. Lechowicz, 195 participants, 76 papers were published in Polish in Journal "Maritime Engineering and Geotechnics", Acta Scientiarum Polonorum. Architectura as well as in English in Journal, Studia Geotechnica et Mechanica" and Annals of Warsaw University of Life Sciences-SGGW, Land Reclamation.
- Third International Conference "Challenges in Geotechnical Engineering" CGE-2019, Zielona Góra, 10-13.08.2019, Co-Chairs of Organizing Committee: V. Sakharov and W. Szajna, Key note lectures: prof. Rolf Katzenbach, Germany, prof. Roberto Nascimbene, Italy, prof. Shahab Yasrobi, Iran, prof. Ivan Vaníček, Czech Republic.

Participation in International Conferences

- 19th International Conference on Soil Mechanics and Geotechnical Engineering, 17- 22.09.2017, Seoul, Korea, 8 papers published (4 oral presentations), 12 participants.
- Transportation Geotechnics and Geoecology, ISSMGE – Emperor Alexander I Petersburg State Transport University, 17-19.05.2017, St. Petersburg, Russia, 6 papers published and oral-presented, 6 participants, member of Scientific Committee – prof. K. Zabielska-Adamska, chairman of session – dr W. Gosk.
- XVI Danube-European Conference on Geotechnical Engineering, 07-09.06.2018, Skopje, Macedonia, co-chairman of session F – prof. Z. Lechowicz, 5 papers published, 8 participants.
- Transportation Soil Engineering in Cold Regions (Transoilcold 2019), ISSMGE – Emperor Alexander I Petersburg State Transport University, 20-23.05.2019, St. Petersburg, Russia, members of Scientific Committee – prof. K. Zabielska-Adamska, dr J. Kawalec.

Participation in ISSMGE Technical Committees

TC 101 – A. Szymański, M. Lipiński; TC 102 – Z. Młynarek, J. Wierzbicki; TC 103 –

M. Cudny, A. Truty; TC 105 – D. Łydżba, M. Lefik; TC 106 – K. Garbulewski, Z. Skutnik; TC 202 – L. Rafalski, K. Zabielska-Adamska; TC 203 – Z. Bednarczyk, W. Świdziński; TC 204 – A. Siemińska-Lewandowska, M. Mitew-Czajewska; TC 205 – L. Bałachowski; TC 206 – P. Popielski; TC 207 – G. Horodecki, G. Kasprzyk; TC 208 – P. Srokosz, TC 209 – R. Coufal; TC 211 – M. Topolnicki, J. Bzówka; TC 212 – K. Gwizdała, A. Krasiński; TC 214 – Z. Lechowicz; TC 215 – E. Koda; TC 216 – T. Kozłowski; TC 217 – W. Cichy; TC 218 – A. Duszyńska, J. Kawalec; TC 302 – W. Cichy; TC 304 – W. Puła; TC 305 – E. Dembicki; TC 307 – W. Sas.

Main activities for young geotechnical engineers - Participation in International and European Young Geotechnical Engineers' Conferences:

- 6th International Young Geotechnical Engineer's Conference, 6th IYGEC, Seoul, Korea, 31.08-01.09.2017: M. Wasil and E. Soból.
- 26th European Young Geotechnical Engineer's Conference, 26th EYGEC, Graz, Austria, 11-14.09.2018: M. Kreczmar, M. Chwała.

Specify journals, in which geotechnical engineers can publish

- Special Sections – "Geotechnical Engineering" in the Polish Journal "Maritime Engineering and Geotechnics" Polish and English http://www.imig.pl/
- Journal, Studia Geotechnica et Mechanica" (English) http://sgem.pwr.wroc.pl/
- Annals of Warsaw University of Life Sciences – SGGW Land Reclamation (English) http://ann_landreclam.sggw.pl/

SPAIN

Number of ISSMGE members 2019: 388 members
The activities carried out each year by the Spanish Society of Soil Mechanics and Geotechnical Engineering (SEMSIG) from September 2017 up to now are the following:
- 2 seminars about geotechnical works carried out by Spanish engineering firms, organized by SEMSIG, and presented by the invited companies took place in Madrid; a short volume is published for each of them, printed and uploaded in SEMSIG website, as well as the video recording of the event (in Spanish language, available to SEMSIG members):
- 51st Seminar, October the 25th 2017: SENER
- 52nd Seminar, November the 7th 2018: ARUP
- Annual workshops, jointly with the Spanish Association of Geotechnical Contractors (AETESS), about theoretical and practical aspects of special foundations:
- 18th SEMSIG-AETESS Workshop, 22/02/2018: "Control and Instrumentation in Geotechnical Works", Colegio de Ingenieros de Caminos, Madrid.
- 18th SEMSIG-AETESS Workshop, second edition, 18/10/2018: "Control and Instrumentation in Geotechnical Works", Escuela Politécnica de Mieres, Universidad de Oviedo, Asturias.
- 19th SEMSIG-AETESS Workshop, 21/02/2019: "Advances in Soil improvement Techniques and Special Foundations"; CEDEX, Madrid.
- Annual SEMSIG Lectures (in Spanish, also recorded and available in our website for SEMSIG members):
- 2018's SEMSIG Lecture: "Effects of soil-Structure Interaction during excavation on tunnels in clays"; Prof. Davor Simic Sureda; 15/03/2018, Madrid.
- 2019's SEMSIG Lecture: "Geotechnical Seismic Engineering: Past, Present and Future"; Eng. Alberto Bernal Riosalido; 13/03/2019.
- Every four years, normally the year before the International Conference, we celebrate the National Congress on Geotechnical Engineering. Preparations for next one (XI Congress) have already started. It will be held in October 2020 in Asturias.
- The Bulletin of SEMSIG is published with four issues per year with technical papers.
Since September 2017, issues N. 190 to N. 195 have been published.
- Together with the Portuguese Geotechnical Society (SPG) and with the Brazilian ABSM, SEMSIG co-edits GEOTECNIA journal, in Portuguese and Spanish languages, three issues per year. Since September 2017 issues N. 141, 142, 143, 144 and 145 have been published.
- As to Young Geotechnical Engineering Conferences, SEMSIG has sent two delegates to the 6th International Conference in Seoul, South Korea, in September 2017 and other two to the 26th European Conference in Graz, Austria, in September 2018.
- In 2018 the annual "Carlos Oteo Award" has been created to reward the best geotechnical paper published in Spanish either in the SEMSIG Bulletin or in GEOTECNIA Journal, and economically endowed, in order to promote publication of technical papers in our language. The first Carlos Oteo award, corresponding to 2018, was given in February 2019.
- SEMSIG's Internet website has been remodelled and a profile in Linkedin has been created, presently with more than 1200 followers.
In 2018 elections to SEMSIG board took place, to renew half of its members.
SEMSIG delegates to the different ISSMGE Technical committees have been renewed at the end of 2018.

SWITZERLAND

We organized annual conferences in spring and autumn every year which lasted one or two days. The conferences included presentations and discussions. The conference presentations are published in brochures and distributed to all our members and to the participants of the conference in question. The extra copies can be purchased by anyone who is interested in the subject and two years after the conference they are placed on our website. Please find the details hereafter.

Past Conferences

Year	Title of the Conference	City	Particip.	Publication Mitteilungen der Geotechnik Schweiz" no.	Available on the website
2018	Forschung und Innovation in der Geotechnik, 2 Tage Recherche et innovation en géotechnique, 2 jours	Lausanne	98	177	fall 2020
2018	Beobachtungs methode und Überwachung in der Geotechnik Méthode observationnell e et surveillance en géo-technique	Yverdon	148	176	spring 2020
2017	Stützmauern Murs de soutènement	Bern	250	175	fall 2019

Future Conferences

Year	Title of the Conference	City	Particip.	Publication Mitteilungen der Geotechnik Schweiz" no.	Available on the website
2019	Vereisungsmassnahmen in der Geotechnik Congélation en géotechnique	Bern		178	Spring 2021
2019	Technische Exkursion Grand-Paris, 2 Tage Excursion technique Grand- Paris, 2 jours	Parisn	8	None	

PORTUGAL

1. Country of National society:
National Society: Sociedade Portuguesa de Geotecnia SPG (Portuguese Geotechnical Society)
Country: PORTUGAL
2. Number of SPG members – evolution between 2017 – 2018
December 2017: 567
December 2018: 561
3. Number of Corporate SPG members – evolution between 2017 – 2018
December 2017: 36
December 2018: 37
4. Number of ISSMGE members – evolution between 2017 – 2018:
December 2017: 156
December 2018: 155
5. Number of Corporate ISSMGE members – evolution between 2017 – 2018:
December 2017: 3
December 2018: 3
6. Main international activities:
- VIth Portuguese and Spanish Geotechnical Journeys, in association with SEMSIG- Spain, held at Ponta Delgada, Azores, May, 28, 2018, with the theme: Geotechnical Works in Volcanic Rocks.
- IXth Portuguese-Brazilian Geotechnical Congress, in association with ABMS- Brazil, held at Salvador - Bahia, Brazil, August 30, 2018, with the theme: Safety of Dams.
Proceedings published in CD-ROM. Papers are in Portuguese.
- VIth Victor de Mello Lecture (in association with ABMS-Brazil):
Lecturer: Norbert R. Morgenstern, Title: Geotechnical Risk, Regulation and Public Policy, Salvador, Brazil, August 30, 2018.
Victor de Mello Lectures are published in Journal Soils and Rocks, in English.
- Since 2017, SPG has been preparing a candidacy to organize and host the 2023 ECSMGE; this candidacy will be presented at the meeting of the Geotechnical European Societies that will occur in Reykjavik next September.
- Other international SPG Activities during 2017-2019
Workshop on Shallow Geothermal Systems: Applications and Perspectives, held at University of Aveiro, Portugal, May 8, 2017, under the framework of the SUCCESS Project.
António Gomes Correia, Past-President of SPG, taught the 2nd Proctor Lecture: From fundamentals to applications in compaction: recent developments in embankments and structural layers of pavements and railways, during the 19th ICSMGE, Seoul, South Korea, September 18, 2017;
6th Conference of Young Geotechnical Engineers (iYGEC6), integrated in 19th ICSMGE, Seoul, South Korea, , September, 16-17, 2017; SPG was represented by Daniel Ribeiro, with the oral communication: Modelling the mechanical behaviour of sand-cement mixtures;

9th European Conference on Numerical Methods in Geotechnical Engineering - NUMGE2018, held at Porto, Portugal, June, 25-27, 2018, with the joint organization of the Univ. of Porto and SPG, under the auspices of ISSMGE. The conference had 240 participants, from which 200 came from 38 countries outside Portugal and from the 5 continents. The conference had 5 Keynote Lectures (in plenary session) and 204 communications, all presented orally (in 4 parallel sessions). The proceedings were published by Taylors & Francis;
26th European Young Geotechnical Engineers Conference (EYGEC2018), held in Austria, September 11-14, 2018; SPG was represented by Manuel Parente, with the oral communication: Intelligent earthworks optimization system;
7. Main national activities:
- Manuel Rocha Lecture (yearly lecture):
XXXIV Lecture, Lecturer: Paul W. Mayne, Title: Stress History of Soils from Cone Penetration Tests, November 11, 2017, Lisbon;
XXXV Lecture, Lecturer: Laura Caldeira, Title: Internal Erosion and Safety of Dams, October 9, 2018, Lisbon.
The lectures are published in English in the journal Soils and Rocks.
- XVIth Geotechnical National Conference (Congresso Nacional de Geotecnia), May 27 - 30, 2018, Ponta Delgada, Azores, with 266 reviewed papers and 330 participants. The Proceedings are published in CD-ROM and will be published in an ebook. Papers are either in Portuguese or in English.
- Yearly National Meetings (includes an invited talk and the General Assembly of SPG):
13th SPG Annual Meeting, Lisbon, March 29, 2017, Lecturer: Paulo Matias Ramos,
Tittle: Foundations of fluvial and marine structures;
14th SPG Annual Meeting, Lisbon, March 28, 2018, Lecturer: Jaime Santos Tittle: The role of dynamic testing in geotechnical characterization;
15th SPG Annual Meeting, Lisbon, March 27, 2019, Lecturer: Paulo da Venda Oliveira Tittle: Embankments on soft soils. From research to practice.
- Other national SPG Activities during 2017-2019
Technical Field Visit to the Águas Santas and Marão Tunnels (in Portugal), February 14, 2017. The event was attended by over 60 participants.
1st Technical Session on Geosynthetics, LNEC, Lisbon, June 6, 2017. The event was attended by 104 participants.
1st Portuguese Seminar on Transportation Geotechnics, LNEC, Lisbon, October 12-13, 2017. Proceedings published in electronic version, in Portuguese. The event was attended by 240 participants.
8th Meeting of Young Geotechnical Engineers, held in Lisbon, October 13, 2017.

Workshop on Circular Economy in Environmental Geotechnics, held at University of Coimbra, November 22, 2017. The event was attended by 150 participants.
Workshop on Execution of Tunnels and Complex Geotechnical Works. What changes with the revised CCP (Portuguese Official Code for Contracts of Public Works)? International Good Practices, LNEC, Lisbon, February 10, 2018. The event was attended by 110 participants.

2nd Technical Session on Geosynthetics - Solutions of Geosynthetics for Reinforcement of Soft Soils, LNEC, Lisbon, July 6, 2018. The event was attended by 54 participants.
Cycle of Geotechnical Lectures Thinking about the Environment - 4th Lecture: The environmental recovery of degraded mining areas: performance and future perspectives, by Edgar Carvalho, LNEC, Lisbon, July 10, 2018. The event was attended by 36 participants.
2nd Edition of the Seminar on Terrain Anchorages, Porto, October 25-26, 2018. The event was attended by 129 participants.
2nd Portuguese Seminar on Transportation Geotechnics, held in Vila Franca de Xira, January 28-29, 2019. Proceedings published in electronic version, in English. The event was attended by 165 participants.
Cycle of Geotechnical Lectures Thinking about the Environment - 5th Lecture: Surface Geothermal: Geothermal Heat Capture / Rejection System - Visit to the Ombria Resort, February 12, 2019.
8. Prizes:
- Manuel Rocha Prize for the best PhD thesis in a 4 year-period. The prize for the 2012-2015 quadrennium was attributed ex-aqueo to Luís Manuel Araújo Santos, for the thesis Characterization of the Mechanical Behavior of the Sand of Coimbra under Generalized Actions, and Ricardo Neves Correia dos Santos, for the thesis Experimental Investigation on Limitation of Progression of Internal Erosion in Zoned Dams. An Honorable Mention was given to Ana Teresa Pereira de Carvalho, for Seismic Response of Anchored Curtains, Physical and Numerical Modeling.
- Yearly the José Folque -Young Geotechnical Engineer Prizes (including two awards, one for papers in English language and another for papers in Portuguese) are attributed. The prizes consist of a diploma and a cash value, for the author of the best paper in Portuguese, and of a diploma and funding for the author of the best paper in English. The funding permits the participation of this author at the Young Geotechnical Engineer Conference of ISSMGE.
The José Folque - Young Geotechnical Engineer Prize in Portuguese Language
2017 awarded, Catarina Fartaria, with the paper Landfills of Approach to Integrated Bridges: Mitigation of Settlements through Reinforcement with Geogrids. Two Honorable Mentions were given to Ana Sofia Saldanha (Microzoning of susceptibility to liquefaction applied to a case study of the Lisbon region) and Pedro Patrício (Influence of resistance to the bending of containment walls in the bottom breaking of excavations in sandy soils).
The Prize in English Language awarded, on 2017, Manuel Parente, with the work Intelligent Eartworks Optimization System.
The José Folque - Young Geotechnical Engineer Prize in Portuguese Language 2018 awarded Martim Rodrigues de Matos, with the work Excavation of vertical wells in rock through the method of Suspended Loads at Gouvães Hydroelectric Power Plant.
The Prize in English Language awarded in 2018 Luís Miranda, with the work Laboratory Testing and Numerical Modeling of the Dynamic Behavior of Tagus River Sand.
- Ricardo Esquível Teixeira Duarte: Prize for the Best Geotechnical Project.
The 1st edition of the prize took place in 2018. The two winner projects in ex-aequo were:
Project Venda Nova III. Reinforcement of Venda Nova Hydroelectric Power, authored by Celso Lima and EDP Production Project Team.
Algiers Metro Project, authored by Miguel Conceição and Carlos Baião (CENOR - TPF Planege).
Both projects were presented at the first plenary session of the XVIth Geotechnical National Conference, Ponta Delgada, Azores.
9. Journals:
SPG publishes two journals: GEOTECNIA (in Portuguese, in association with ABMS-Brazil and SEMSIG-Spain) and SOILS & ROCKS (in English, in association with ABMS-Brazil). All papers are peer-reviewed.
In 2017 were published 3 issues of the following SPG journals: i) GEOTECNIA (139 to 141); ii) SOILS & ROCKS (Volume 40 – issues 1, 2 and 3);
In 2018 were published 3 issues of the following SPG journals: i) GEOTECNIA (142 to 144); ii) SOILS & ROCKS (Volume 41 – issues 1, 2 and 3);
10. Newsletters:
SPG provides to all active members an electronic Newsletter bimonthly. In 2017 were published Issues 28 to 32.
In 2018 were published Issues 33 to 37.
In 2019, till April 30th, was published Issue 3

Appendix 4: Regional Report – North America

Tim Newson
VP ISSMGE North America

1 INTRODUCTION

The international geotechnical activities in North America are primarily organized by three professional groups. These are the Mexican Geotechnical Society (SMIG), Canadian Geotechnical Society (CGS) and the Geo-Institute (G-I) of the American Society of Civil Engineers (ASCE). A number of other prominent organizations also host and sponsor technical and professional activites, including the Deep Foundations Institute (DFI), Association of Drilled Shaft Contractors (ADSC), Pile Drivers Contractors Association (PDCA), and Geosynthetics Institute (GSI). A brief summary of the activities held by the SMIG, CGS, and G-I are described in the next few sections.

2 MEMBER SOCIETIES

2.1 *Mexican Society of Geotechnical Engineering (SMIG)*

The current membership for the society is 523. Activities on geotechnical engineering are held every two years by the Mexican Society of Geotechnical Engineering and are posted at: www.smig.org.mx. Since the 19th ICSMGE in Seoul, one national conference has occurred:

- XXIX National Meeting of Geotechnical Engineering and XX National Meeting of Professors of Geotechnical Engineering was held from November 21st to 24th, 2018 in León, Guanajuato, Mexico. www.rnig2018.com.mx.

The SMIG has three invited lectures that are given at these national events:

- The Nabor Carrillo Lecture;
- The Leonardo Zeevaert Lecture;
- The Raúl J. Marsal Lecture.

Other major events that have been held are:

- 5th Colloquium of Young Geotechnicians and 2nd Meeting of Teachers: Challenges in the 60th Anniversary. Held from October 27th- 28th, 2017 in Puebla.
- 4th DFI-SMIG-GI-ISSMGE International Deep Foundations Symposium from November 15th-16th, 2017 in Mexico City.
- 5th International Symposium on Tunnels and Shafts in Soils and Rock. Held from August 16th-18th, 2018.

An important upcoming event is the XVI Pan American Conference on Soil Mechanics & Geotechnical Engineering (PCSMGE) in Cancún from November 17th-20th 2019. This is the quadrennial regional conference that is alternately shared between North America and South America. There are a number of prestigious invited lectures associated with this conference.

The Casagrande Lecture will be given by Professor Gabriel Auvinet and the Plenary Lecture will be given by Professor Charles Ng. Seventeen keynote lectures will be provided by global expert speakers. Three rising stars will give Bright Spark Lectures. Seven hundred and fifty-five abstracts were submitted from forty-one countries.

2.2 *Canadian Geotechnical Society (CGS)*

The current membership for the society is 1,190. National activities on geotechnical engineering are held every year by the Canadian Geotechnical Society. The national CGS conferences generally alternate between the eastern and western sides of the country, with details found at: www.cgs.ca. Over the past two years, the national conferences have included:

- GeoOttawa 2017: 70 Years of Geotechnics & GeoSciences. This was the 70th Canadian Geotechnical Conference and the 12th Joint CGS/IAH-CNC Groundwater Conference and was held at the Shaw Centre in Ottawa, Ontario from 1st-4th October 2017. http://www.geoottawa2017.ca.
- GeoEdmonton 2018: Transportation Geotechnique - Moving Forward. This was the 71st Canadian Geotechnical Conference and the 13th Joint CGS/IAH-CNC Groundwater Conference and was held in Edmonton, Alberta in September 23-26th, 2018 at the Shaw Conference Centre. http://www.geoedmonton2018.ca.

The CGS has two invited lectures that are given at these national events:

- The R.M. Hardy Address;
- The CGS Colloquium.

Upcoming national conference events of the CGS in the next two years include:

- GeoSt.John's 2019: Under Land and Sea. This will be the 72nd CGS annual conference, in St. John's, Newfoundland, September 29th-October 2nd, 2019. www.geostjohns2019.ca.
- 6th cYGEGC, September 26th-28th, 2019, in St. John's (NL). www.cygegc2019.com.
- 73rd CGS annual conference: GeoCalgary 2020. www.geocalgary2020.ca.

2.3 *ASCE GeoInstitute*

The current membership for the society is 3,184. The ASCE Geo-Institute (www.geoinstitute.org) holds a national congress once a year, plus on occasion, additional specialty conferences may be held. Since the Seoul conference, the G-I congresses have included:

- March 5th-10th, IFCEE 2018, Orlando, Florida: 3547 attendees, 333 final papers, 198 exhibitors, approximately 25 sponsors.
- March 24th-27th, GeoCongress 2019 (Geotechnical Frontiers – joint conference with IFAI), Philadelphia, Pennsylvania: 1375 attendees, 405 final papers, 114 exhibitors, 15 sponsors.

The ASCE G-I has a number of invited lectures that are given at these national events including:

- The Ralph B. Peck Medal Lecture;
- The H. Bolton Seed Medal Lecture;
- The Karl Terzaghi Lecture;
- The Mercer Award Lecture.

The next GeoCongress will be held from February 25th-28th, 2020 (partnered with COPRI, CGS, and MGS) in Minneapolis, Minnesota.

Further conferences and events that have occurred include:

- PanAm UNSAT, November 12th-15th, 2017, Dallas, Texas.
- GEESD V, June 10th-13th, 2018, Austin, Texas.
- EMI/G-I joint conference, June 18th-21st, 2019, Pasadena, California

3 TECHNICAL COMMITTEES

The North American region of the ISSMGE hosts several technical committees (TC) and their associated chairs are shown below:

TC 105 - Geomechanics from Micro to Macro: Kenichi Soga (United States)
TC 202 - Transportation Geotechnics: Erol Tutumluer (United States)
TC 203 - Geotechnical Earthquake Engineering & Associated Problems: Ross Boulanger (United States)
TC 205 - Safety & Serviceability: Rodrigo Salgado (United States)
TC 208 - Slope Stability in Engineering Practice: Andy Take (Canada)
TC 218 - Reinforced Fill Structures: John Sankey (United States)
TC 307 - Sustainability in Geotechnics: Dipanjan Basu (Canada)
TC 308 - Energy Geotechnics: Marcelo Sanchez (United States)

These TCs have undertaken a significant number of activities in this period that are reviewed in the mid-term report of the Technical Oversight Committee (TOC).

In addition, the Young Member Presidential Group (YMPG) is chaired by Lucy Wu (United States).

4 CROSS-BORDER INITIATIVES AND INTERACTIONS BETWEEN SOCIETIES

In the United States, a new policy is in place, endorsed and supported through the International Activities Council of G-I, to ensure that invitations will be sent to the leadership of the CGS and SMIG for attendance at the annual conferences. The policy ensures that registration costs will be covered for at least one representative from each of Canada and Mexico. This policy also ensures that travel funds will be available for G-I representatives to attend cross-border events, when appropriate.

During the period of reporting, Mario Ruel, as President of CGS, attended the Terzaghi Dinner and the International Activities Council Meeting in Philadelphia in March 2019. Prof. Youssef Hashash, as President of G-I, and Prof. Jim Hanson, as Chair of the G-I International Activities Council, attended GeoEdmonton (September 2018) and each briefly presented context on the status and potential for cross-border collaborations during plenary sessions. Prof. Jim Hanson and Dr. Nazli Yesiller attended the Legget Dinner at GeoEdmonton in September 2018.

The strategic plan for the International Activities Council of the G-I has been formalized and endorsed by the G-I. Specific cross-border collaborations and activities with both Canada and Mexico have been highlighted as priorities. As part of this initiative, the CGS annual conference in 2021 GeoNiagara will have the IAH-CNC and ASCE Buffalo as participating partners.

5 GENERAL REMARKS

It is gratifying to see that there is more cooperation and shared activities are beginning to occur in the region, including workshops, symposia, and conferences between the various North American professional organizations. I have taken advantage of the National Meetings and the International Conferences that I have attended to interact with officers and members of the North American member societies. The following items have been raised for discussion: membership, links with neighbouring and other countries, participation in International Conferences and Technical Committees, North American code provisions, how to raise the profile and visibility of our profession and the relationships between the national societies and the ISSMGE.

I have also been trying to expose members of the national societies to research and professional practice occurring in the other countries of the region through the technical publications of the societies (i.e. GeoStrata, Geotechnia and Geotechnical News). Special issues of these publications expounding geotechnical projects from the other regions have been published (http://www.readgeo.com/geostrata/mar_apr_2019) or are in press. I hope to continue this work over the remainder of my term. Finally, serious consideration is being given by the ASCE G-I to host an ICSMGE in the USA, with the 100-year anniversary of the First ICSMGE being a distinct possibility. This would certainly give North American the opportunity to showcase the considerable geotechnical expertise that exists in this region.

Appendix 4: Regional Report – South America

Alejo Sfriso
ISSMGE Vice-President, South America

1 INTRODUCTION

This is a report of activities of the South America and the Caribbean region of ISSMGE in the period Sep 2017 – Sep 2019 presented at the Council Meeting of ISSMGE in Cape Town, October 2019.

The region has been fairly active in the period, with a large number of events being held in Brazil, Argentina and, to a lesser extent, Chile. A few of the societies show little activity or are largely dormant.

2 ACTIVITIES OF THE VP

2.1 *Bringing national societies back home*

Ecuador, Bolivia, Peru and Dominican Republic were terminated by ISSMGE in 2017 due to lack of activity, fees in arrears and no communication with ISSMGE. A new National Society was formed in Peru -Asociación Peruana de Ingeniería Geotécnica- and admitted in 2017. A new National Society was also formed in Bolivia -Sociedad Boliviana de Mecánica de Suelos e Ingeniería Geotécnica- and admitted in late 2018.

Sfriso is now actively working in resuscitating the national society of Ecuador. He attended an international event in August 2019 and met prospective candidates for seeding this new society.

2.2 *Collaboration and communication*

Efforts aiming at improving the collaboration among sister societies in the region were productive: two in-person meetings in Brazil – 18/09/17 @ ICSMGE in Seoul and 31/08/18 @ COBRAMSEG - and five meetings using Skype were held among the presidents and secretary generals of the various societies in the region. In these meetings, Sfriso briefed the authorities of the national societies on the activities of ISSMGE and the decisions made at ISSMGE board meetings.

Participation of regional attendees in events organized by sister societies was also encouraged and, to some extent, successful. A strong and sustained effort was carried out to encourage the participation of professionals from South America in the upcoming Panamerican Conference on Soil Mechanics and Geotechnical Engineering, to be held in Cancun, Mexico, in November 2019.

2.3 *Corporate associates*

Two new CAPG members from South America – SRK Consulting and AOSA - were admitted into ISSMGE. Two more companies are now in informal talks and might join in the coming months.

2.4 *Young Members Presidential Group*

A large number of candidates showed up to form the regional chapter of YMPG. Three were formally appointed: Mauro Sottile, Vitor Pereiro and Mario Terceros Arce. There is a large spool of other young geo-professionals willing to participate, an encouraging fact which guarantees sustainability of the society in the region.

2.5 *International Seminars*

An ISSMGE International Seminar Series -Former Touring Lecture- was held in Paraguay, Argentina and Mexico in September 2018, and was well attended. The subject selected for this Seminar was "Codes for Foundations". Lectures were delivered by Roger Frank, Roberto Terzariol and Alejo Sfriso.

2.6 *Visits to National Societies*

Sfriso visited several national societies and met in person with their senior officers. Various aspects of the above-described initiatives were discussed, and actions were taken to promote their advance. The countries visited in the period are Argentina, Bolivia, Brazil, Chile, Cuba, Mexico, Paraguay, Peru. Not being (yet) member societies, Ecuador and Uruguay were also visited, and meetings with prominent geotechnical engineers of both countries were held. During these visits, Sfriso delivered a few lectures where the mission of ISSMGE was communicated to attendees:

- Impacts of Geotechnical Engineering. SAIG Seminar, Buenos Aires, Argentina (2019).
- LRFD foundation design technicques and reliabiltiy design. Seminar of Foundations ASCE, Guayaquil, Ecuador (2019).
- Large excavations in urban environments. National Civil Engineering Student's Conference, Puno, Perú (2019).
- Validation of construction procedures for tunnels using numerical methods. 2nd International scientific convention. Santa María, Cuba (2019).
- Recent experiences with ground retaining systems in Buenos Aires, Argentina. IV International Conference on Deep Foundations in Bolivia (2019).
- Experiences with the Buenos Aires Metro in the last 20 years. Workshop on Urban Geotechnics, Singapore (2019).
- Seismic analysis of earth and tailings dams. SMIG, Leon, Mexico (2018).
- Building a code of foundations for Argentina. ISSMGE Touring Lecture, Paraguay, Argentina (2018).
- LRFD and numerical modeling of foundations. ISSMGE Touring Lecture, Paraguay, Argentina, México (2018).
- Elements for a (first draft of) code of foundations for Argentina. XXIV CAMSIG Salta, Argentina (2018).
- Large excavations and their support systems. UTN Concordia, Entre Ríos, Argentina (2018).

- Constitutive models for mining geotechnics. 1st International Seminar on Mine Waste Disposal, Arequipa, Peru (2017).
- Seismic design of waste dumps. 1st International Seminar on Mine Waste Disposal, Arequipa, Peru (2017).
- Modelling of construction procedures for underground projects. 2nd Intl Tunneling ITA Conference, Cochabamba, Bolivia (2017).
- Evolution of analysis tools for the seismic response of earth dams. Hong Kong Univ of Science and Technology (2017).
- El Paseo del Bajo. Seminar on Excavations SAIG, Córdoba (2017).

This VP apologizes for not visiting Guatemala, Costa Rica and Venezuela. These visits are expected to occur at the earliest opportunity.

2.7 *Technical Committee on Tailings and Mine Wastes*

The convenience and opportunity for creating a TC on Mine Waste were discussed in several Board Meetings of ISSMGE. Sfriso proposed that the TC be chaired by a representative of South America -as the region has no chairs at the moment- and this was accepted.

In informal conversations with colleagues of the countries in the region where the mining industry is large -Chile, Peru, Brazil- it was found that the best way forward was to ask the Chilean Society to start planning for creating such TC, which was accepted. Dr. Ramón Verdugo submitted a letter to the TOC Chair, Dr. Pierre Delage, containing the basic elements for starting this TC. It is expected that the TC will start no later than end 2019 and be in full position to host a session on mine waste by the next ICSMGE in 2021.

3 UPDATE ON MEMBER SOCIETIES

There are eleven active national societies in the region. Four societies were terminated – Bolivia, Dominican Republic, Ecuador and Peru – but two of them rejoined ISSMGE in the period: Bolivia and Peru. A brief description of each society follows.

3.1 *Argentina*

The national society of Argentina is "Sociedad Argentina de Ingeniería Geotécnica", SAIG (www.saig.org.ar). Currently, SAIG has 106 members. It's president is Dr. Diego Manzanal (diego.manzanal@gmail.com). SAIG held the following activities in the period:

- Workshop for faculty members on teaching CSSM in undergrad courses, Buenos Aires, 2017
- Technical Seminar on Urban Excavations, Cordoba, 2017
- 24th National Conference, Salta, 2018
- Workshop for faculty members on teaching shear strength and retaining structures, Salta, 2018
- ISSMGE Seminar on Codes for Foundations, 2018
- Workshop on Foundations and Excavations, Santa Fe, 2018
- Support to the 14th International Conference on Rock Mechanics, Foz de Iguazu, 2019
- Seminar on Geotechnical Challenges, Buenos Aires, 2019

3.2 *Bolivia*

The former national society of Bolivia was terminated in 2017. A new national society was formed and admitted by ISSMGE in December 2018, the "Sociedad Boliviana de Mecánica de Suelos e Ingeniería Geotécnica" with 43 members. It's president is Mr. Mario Terceros Herrera (math@incotec.cc). They are organizing the Fourth International Conference on Deep Foundations (www.cfpb4.com).

3.3 *Brazil*

The "Associação Brasileira de Mecânica dos Solos e Engenharia Geotécnica", ABMS, is the largest national society in the region, with 1.296 members (www.abms.com.br). It's president is Dr. Alexandre Duarte Gusmão (alexandreduartegusmao@gmail.com). The society publishes two journals, Soils and Rocks and Geotecnia. ABMS performed the following events in the period, some of them attracting +1000 professionals:

- I GeoBase - Seminário Geotécnico da Bahia/Sergipe, 2017
- I GeoMin - Geotecnia Aplicada à Mineração, B.Horizonte, 2017
- VII Conferência Brasileira sobre Estabilidade de Encostas, Florianópolis, 2017
- 4º Simpósio da Prática da Engenharia Geotécnica na Região Centro-Oeste, Brasília, 2017
- 6º Simpósio de Geotecnia do Nordeste, Maceió, 2017
- II Simpósio de Engenharia Geotécnica do Rio de Janeiro, Rio de Janeiro, 2017
- Palestra Milton Vargas da ABMS, itinerante, 2017
- 19º Congresso Brasileiro de Mecânica dos Solos e Engenharia Geotécnica, Salvador, 2018
- 14a Palestra Milton Vargas da ABMS, itinerante, 2018
- Geocaruru, Salvador, 2018
- FloripaTUN – Seminário Internacional de Túneis, Florianópolis, 2019
- Conferência Tecnologia de Fundações – CTF'2019, Vitória, 2019
- Seminário de Engenharia de Fundações Especiais e Geotecnia, São Paulo, 2019
- GeoRS – Seminário de Engenharia Geotécnica do Rio Grande do Sul, Ijuí, 2019
- IX Congresso Brasileiro de Geotecnia Ambiental, VIII Congresso Brasileiro de Geossintéticos, São Carlos, 2019
- International Society for Rock Mechanics and Rock Engineering - ISRM'2019, Foz do Iguaçú, 2019
- GeoMin – Geotecnia Aplicada à Mineração, Ouro Preto, 2019
- XII Simpósio de Práticas de Engenharia Geotécnica da Região Sul, Joinville, 2019
- 5º Simpósio de Prática de Engenharia Geotécnica na Região Centro-Oeste, 2019
- 6º Simpósio de Geotecnia do Nordeste, Recife, 2019
- 2º Simpósio de Prática de Engenharia Geotécnica na Região Norte, 2019
- 3º Simpósio de Prática de Engenharia Geotécnica no Rio de Janeiro, 2019
- 15a Palestra Milton Vargas da ABMS, itinerante, 2019

3.4 *Chile*

The national society of Chile is "Sociedad Chilena de Geotecnia", SOCHIGE (www.sochige.cl). Currently, SOCHIGE has 92 members. It's president is Mrs. Daniela Pollak (dpollak@gmail.com). SOCHIGE held the following activities in the period:

- Geotechnical aspects of Puebla Earthquakes, Santiago, 2017
- Technical meeting: Performance of a trial embankment at the Ballina Soft Soil Field Testing Facility, Santiago, 2018
- Seminario de Ingeniería Geotécnica: Avances teóricos y prácticos para su aplicación en Obras Civiles, Concepción, 2018
- Seminario Excavaciones Profundas en zonas urbanas, Concepción, 2018
- Challenge for Young Geotechnical Students, Concepción, 2018
- X Congreso Nacional de Geotecnia, Valparaíso, 2018
- Conversatorio: Tranques de Relaves: Diseño, Operación y Plan de Cierre, Santiago, 2019

• Seminario Desafíos en el Diseño, Construcción y Operación de Presas de Relave, Santiago, 2019
• Seminario SOCHIGE: Primer encuentro de Geotecnia en la Región del Maule, Curicó, 2019
• Geotechnical Earthquake Engineering, Santiago, 2019

Also, a vast amount of geotechnical seminars, lectures and workshops were organized by Universities and sister societies related to the vibrant mining industry in Chile.

3.5 *Colombia*

The national society of Colombia is "Sociedad Colombiana de Geotecnia", SCG (www.scg.org.co) with 24 members. It's president is Edgar Eduardo Rodriguez Granados (presidenciascg@gmail.com). SCG held the following activities in the period:
• Curso Integral de Diseño Sismo-Geotécnico de Cimentaciones, Bogotá, 2018
• XVI Congreso Colombiano de Geotecnia, Paipa, 2018
• III Curso de Exploración y Caracterización del Subsuelo y Diseño de Excavaciones, Bogota, 2019
• Geotechnical Earthquake Enginnering, Bogota, 2019

SCG is organizing a major regional event, XIII International Symposium on Landslides (http://www.scg.org.co/xiii-isl/), to be held in Cartagena in 2020.

3.6 *Costa Rica*

The national society of Costa Rica is "Asociación Costarricence de Geotecnia", ACG (www.geotecniacr.com) with 31 international members. It's president is Mr. Oldemar Bermudez (obermudez@maccaferri.co.cr). ACG held the following activities in the period:
• Cimentaciones intermedias de vanguardia y su aplicación práctica en proyectos de Costa Rica, San José, 2017
• Congreso de Geotecnia, San José, 2018

3.7 *Cuba*

The national society of Cuba is "Comisión Técnica Nacional de Mecánica de Suelos e Ingeniería Geotécnica", CNTMSIG with 31 international members. It's president is Mr. Rolando Armas Novoa (rarmasn@yahoo.com). The society is organizing a multi-disciplinary event, "XII Simposio Internacional de Estructuras y Geotecnia 2019", in Cayo Santa María, 2019.

3.8 *Guatemala*

The national society of Guatemala is "Asociación Guatemalteca de Mecánica de Suelos e Ingeniería Geotécnica" AMSIG with 32 members. It's president is Dr. Rodolfo Semrau (dr.rodolfosemrau@gmail.com). AMSIG held their first national conference, "Primer Congreso Nacional de Ingenierías Sísmica y Geotécnica" in 2019.

3.9 *Paraguay*

The national society of Paraguay is "Sociedad Paraguaya de Geotecnia" SPY with 38 members (www.spg.org.py). It's president is José Pavón(presidente@spg.org.py). SPG held a Symposium on earth and rockfill works and hosted an ISSMGE seminar in 2018. It also plans to hold a Seminar on Dispersive soil in 2019.

3.10 *Peru*

The former national society of Peru was terminated in 2017. A new national society was formed and admitted by ISSMGE in December 2017, the "Asociación Peruana de Ingeniería Geotécnica", APGEO, with 27 members. It's president is Dr. Jorge Zegarra (jorge.zegarra@pucp.edu.pe).

3.11 *Venezuela*

The national society of Venezuela is "Sociedad Venezolana de Geotecnia" SVG with 32 members. It's president is Heriberto Echezuria (hechezuria51@gmail.com).

In-person meeting of Presidents of National Societies in COBRAMSEG.

Appendix 5: Innovation and Development Committee

Dimitrios Zekkos
Chair, Innovation and Development Committee

1 INTRODUCTION

This report describes the IDC activities carried out since the last ICSMGE held in Seoul in September 2017, during which the new President Charles Ng has requested to continuation of IDC activities for the 2017 – 2021 term.

2 MEMBERSHIP

As part of the new appointment of the IDC, we decided it was beneficial to have a smaller task-driven IDC core membership that can be active, flexible and easy to meet, and then engage active members of the geo-community in specific IDC Activities. Thus for the 2017-2021 term, the core IDC committee consists of the following individuals:

- Chair: Dimitrios Zekkos (USA), Associate Professor, University of Michigan
- Vice Chair: Kok-Kwang Phoon (Singapore), Professor, National University of Singapore
- Secretary : Jennifer Nicks (USA), Research Geotechnical Engineer, Federal Highway Administration
- Roger Estephan (Lebanon), Director, Dar Al-Handasah (Shair & Partners), Beirut

The IDC has also put a together a succession plan, and it is expected that Jennifer Nicks will Chair the IDC committee starting in 2021 (subject to the new President's approval).

3 IDC INITIATIVES FOR 2017-2021

For the 2017-2021, in collaboration with the President of ISSMGE, and with feedback from the Board, the IDC has decided to focus on the following activities:

Initiative #1: Conduct a survey to assess the opinion of the geo-community on past, and future IDC activities.

Initiative #2: Expand Open Access through the ISSMGE Online Library

Initiative #3: Support open-Access Publications by developing a Conference Paper Review Platform

Initiative #4: Establish ISSMGE Virtual University

Initiative #5: Upgrade ISSMGE website to increase its functionality, usage and impact

Initiative #6: Continue to support networking opportunities, and expand ISSMGE's reach to the geo-community

Initiative #1: IDC Survey

Since its establishment in 2009, IDC has been leading various initiatives. Among those, major initiatives include the launch of a new ISSMGE website, the establishment of a webinar series, the launch of an open access library, the launch of GeoWorld, the development of an online Lexicon, as well as the development of a mobile app. It was, thus, deemed useful to "take stock" and assess past initiatives before setting new ones. Thus, the IDC conducted a survey with the objective to have the geo-community review previous IDC activities and their impact and solicit feedback on what should be the priorities of the IDC for the 2017-2021 period.

The survey was conducted from March 6th 2018 to May 1st 2018 with 701 participants and included 13 questions. The results of the survey have been analyzed and a public report on the survey can be found on the ISSMGE website.

Briefly, the survey indicated that all initiatives undertaken where generally valuable, but especially the ones that related to the ISSMGE website, open-access and educational resources (webinars, online lectures, etc.) were particularly appreciated. The survey respondents also provided feedback on proposed activities, and an outcome of this feedback are the six initiatives of IDC for the 2017-2021 period.

Initiative #2: Expand Open Access Library

During the 2013-2017 tenure, the IDC established within the ISSMGE website the Online Library (URL: https://www.issmge.org/publications/online-library). The online library provides open access to papers published in conferences organized under the Auspices of ISSMGE. The IDC has been since working with many member societies, Technical Committees and other groups to expand the database. As of May 2019, the online library has 11,865 papers from 42 conferences. From November 2017, when the platform was launched, to November 2018 there were 305,980 downloads of the papers included in the Open Access Library.

The following proceedings have been added to the online library since September 2017:

- 17th Nordic Geotechnical Meeting (NGM2016); May 25-28, 2016, Reykjavik, Iceland (139 papers added)
- 6th International Conference on Earthquake Geotechnical (6ICEGE), November 1-4, 2015, Christchurch, New Zealand (371 papers added)
- 5th International Conference on Earthquake Geotechnical (5ICEGE), 2011, Santiago, Chile (182 papers added)
- Performance Based Design in Earthquake Geotechnical Engineering III (PBD-III), July 16-19, 2017, Vancouver, British Columbia (238 papers added)
- ISSMGE 2011 PanAmerican conference, Toronto, Canada, (503 papers added)
- EUROFUGE 2016, Nantes, France (58 papers added)

A listing of the conferences in the ISSMGE Online library is provided below in Appendix A.

The IDC remains committed to expand the content of the open access library.

Initiative #3: Support open-Access Publication: Develop a Conference Paper Review Platform

The IDC has also been working towards supporting the geo-community in expanding open-access publication for future events. As part of that initiative, the IDC has developed, following authorization by President Ng and the Board, a Conference Paper Review Platform (URL: https://www.issmge.org/publications/review-platform/ongoing-conferences). The platform is part of a broader initiative to support open access of the papers published in ISSMGE conferences.

The ISSMGE Conference Review Platform is part of ISSMGE's cyber-infrastructure and aims to support the review of abstracts and papers for Conferences organized by ISSMGE Member Societies, Technical Committees and other groups affiliated with ISSMGE.

This platform is provided at no cost with the requirement that published papers will become available in open access through the ISSMGE Online Library. It is intended to support the management of conference papers for ISSMGE and ISSMGE-affiliated events and allows also for the seamless integration between this paper review platform and the ISSMGE Online Library. The platform can be used by conference organizers of the ICSMGE, ISSMGE sponsored events, ISSMGE regional conferences, or conferences organized or sponsored by Technical Committees. Member Societies can also directly request to use the platform.

The platform is presently operational and the first conference that is supported by it is the 6th International Conference on Geotechnical and Geophysical Site Characterization, "Toward synergy at site characterisation", that will be held on 7-11 September, 2020 in Budapest, Hungary.

IDC sees this platform as the first step towards a more integrated approach to open access of publications in the field of geotechnics. The IDC is also envisioning that in the future, ISSMGE may have the capacity to also print and publish electronic proceedings of these conferences and even support conference organizers in indexing conference proceedings in databases such as Scopus.

Initiative 4: Establish ISSMGE Virtual University

A major initiative of the President of ISSMGE, Charles Ng, is the development of additional educational material through the ISSMGE to professionals globally. The initiative, titled "Virtual University" is a step beyond the webinars that ISSMGE has launched in the last decade and is presently managed by Board Member Prof. Bouassida.

The purpose of the platform is to educate geoprofessionals at a global scale, and support Universities that may not have the resources to provide detailed educational material on specific topics. A powerful platform is currently under development and is expected to be launched in 2019. The platform, which will be hosted under the issmge.org domain name, has the capacity to deliver educational content such as educational videos, webinars, keynote lectures, or short courses freely to anyone. An ISSMGE account will be needed to login. The platform will also bundle videos and provide recommendations, based on previous videos watched, and will also have the capacity to organize content based on topics

Initiative 5: Upgrade ISSMGE website to increase its functionality, usage and impact.

IDC has been continuously upgrading ISSMGE's online presence, with the goal to provide relevant and current content to ISSMGE membership. In the first two years, the following upgrades have been completed:

- Visitors of the ISSMGE website have now the capability to "sign-up" to receive updates from Technical Committees that fit their interests. Once content is posted by the TCs, it is automatically promoted through the website, but also, by integrating with GeoWorld, an e-mail will be sent to the signed member with the posted content. This feature allows ISSMGE to directly reach out to members and provide valuable content to those that need it.
- Emphasis was also placed to further increase the use of ISSMGE website by TCs and easily assess level of activity. Towards this objective, the current website platform was expanded to allow TC Chairs to upload their progress reports online and make them accessible to TOC leadership.

Initiative 6: Continue to support networking opportunities, and expand ISSMGE's reach

The opportunities for international networking that ISSMGE provides are valued by ISSMGE members. Beyond the ISSMGE website, IDC has been trying to positively affect the careers of geoprofessional by expanding ISSMGE's messaging and reach, through the following means:

GeoWorld: Since its launch in 2011, with the support of ISSMGE, GeoWorld has become the largest global network of geoprofessionals. It presently includes 13690 professional from 145 countries and is supporting many aspects of ISSMGE activities. Through GeoWorld, ISSMGE delivers content and information posted on the ISSMGE website.

Social Media Presence: ISSMGE's social media presence has also been increasing. Currently, there are 3850, 606, and 2754 members on ISSMGE's Facebook, Twitter and LinkedIn account respectively.

Mobile Platform: ISSMGE's mobile platform, launched in September 2017, is presently used by 523 iOS and 450 Android users. Individuals who download the ISSMGE app, receive push notifications of ISSMGE news.

APPENDIX A:

CONFERENCE PROCEEDINGS AVAILABLE IN OPEN-ACCESS VIA THE ISSMGE ONLINE LIBRARY

1. 18th International Conference on Soil Mechanics and Geotechnical Engineering (Paris)
2. 17th International Conference on Soil Mechanics and Geotechnical Engineering (Alexandria)
3. 16th International Conference on Soil Mechanics and Geotechnical Engineering (Osaka)
4. 15th International Conference on Soil Mechanics and Foundation Engineering (Istanbul)
5. 14th International Conference on Soil Mechanics and Foundation Engineering (Hamburg)
6. 13th International Conference on Soil Mechanics and Foundation Engineering (New Delhi)
7. 12th International Conference on Soil Mechanics and Foundation Engineering (Rio De Janeiro)
8. 11th International Conference on Soil Mechanics and Foundation Engineering (San Francisco)
9. 10th International Conference on Soil Mechanics and Foundation Engineering (Stockholm)
10. 9th International Conference on Soil Mechanics and Foundation Engineering (Tokyo)
11. 8th International Conference on Soil Mechanics and Foundation Engineering (Moscow)
12. 7th International Conference on Soil Mechanics and Foundation Engineering (Mexico)
13. 6th International Conference on Soil Mechanics and Foundation Engineering (Montréal)
14. 5th International Conference on Soil Mechanics and Foundation Engineering (Paris)
15. 4th International Conference on Soil Mechanics and Foundation Engineering (London)
16. 3rd International Conference on Soil Mechanics and Foundation Engineering (Switzerland)
17. 2nd International Conference on Soil Mechanics and Foundation Engineering (Rotterdam)

18. 1st International Conference on Soil Mechanics and Foundation Engineering
19. 19th International Conference on Soil Mechanics and Geotechnical Engineering (Seoul)
20. ICETGES - 2008
21. SFGE - 2012
22. SFGE - 2016
23. IS NEW DELHI '94
24. IS LONDON '96
25. IS TOKYO '99
26. IS TOULOUSE '02
27. IS AMSTERDAM '05
28. IS SHANGHAI '08
29. IS ROME '11
30. IS SEOUL '14
31. ICGE Colombo 2015
32. Fifth International Conference on Geotechnical and Geophysical Site Characterization (ISC'5)
33. 5th International Conference on Forensic Geotechnical Engineering
34. 6th International Young Geotechnical Engineers' Conference (iYGEC6)
35. 14th Pan-American Conference on Soil Mechanics and Geotechnical Engineering 64th Canadian Geotechnical Conference, October 2-6, 2011, Toronto, Ontario, Canada
36. 26th European Young Geotechnical Engineers Conference
37. 17th Nordic Geotechnical Meeting (NGM-2016)
38. EUROFUGE 2016 (Nantes, France)
39. 5th International Conference on Earthquake Geotechnical Engineering (Santiago, Chile)
40. 3rd International Conference on Performance-based Design in Earthquake Geotechnical Engineering (Vancouver 2017)
41. 6th International Conference on Earthquake Geotechnical Engineering (Christchurch, New Zealand)
42. 7th International Conference on Unsaturated Soils 2018 (Hong Kong)

Appendix 6: Technical Oversight Committee

Pierre Delage
Chair, Technical Oversight Committee

1 INTRODUCTION

This report describes the TC and TOC activities carried out at mid-term (May 2019), since the last ICSMGE hold in Seoul in September 2017, during which the new President Charles Ng extended my mission as TOC Chair for the 2017 – 2021 term. Besides the various changes that occurred during this half term, the report also provides a feedback on the activities carried out by the TCs, based on their mid-term report. The questions asked for elaborating these mid-term reports are given in Annex 1, and the answers had to be uploaded on line by the TC officers.

The new TOC members, the Vice-presidents of the ISSMGE regions, are given below, together with the number of TCs of their region:

- Africa: Etienne-Marcellin Kana - 1 TC
- Asia: Eung Shul Shin - 11 TCs
- Australasia: Alexander Gavin, 2 TCs
- Europe: Mario Manassero, 13 TCs
- North America: Tim Newson, 9 TCs
- South America: Alejo Sfriso

The idea is to have TOC members fully aware of the activities of the TC hosted by their region, given that TC activities are an important part of the ISSMGE activities, in particular in terms of organising specialised Conferences and Workshops and providing State of the art, State of practice reports and Guidelines in their field. The input of VPs and other TOC members for evaluating the activity of the TCs hosted by their region is appreciated.

The TOC secretary from September 2017 to 5 March 2019 was Kenichi Soga, who decided to resign once he has been nominated Chair of TC105 Micro-macro. The appointment of a new secretary is pending.

Note that the option adopted during the previous term of having another member from the same region has not been kept for this term, given its limited efficiency. One also observe that the dispatching of TC leaders among the regions is not fully satisfactory, with 1 TC107 (Lateritic soils) in Africa, whereas the number of TC in South America was previously of 2 (TC214 Soft soils and TC 306 Geo-Education). It is true that the main criterion of selecting TC Chairs is the expertise, regardless of any geographical consideration. But a minimum re-equilibration would be welcome. The possible creation of a new TC on Tailing Dams is considered in this context.

2 CHANGE IN TC CHAIRS

The rule of having no more than two terms of four years for TC Chairs has been quite well followed, and the nominations of new TC Chairs have all been made with the agreement of TC members, as recommended in the Guidelines. The following changes in TC Chairs occurred since 2017: Matthew Coop (UK) for TC101 Lab testing (20/10/2017), Akira Murakami (Japan) for Numerical methods (3/5/2017), Dong-Soo Kim (Korea) for TC104 Physical modelling, Kenichi Soga (USA) for TC105 micro-macro (5/3/2019), Rod Salgado (USA) for TC205 Safety-serviceability (9/2017), Phil Watson (Australia) for TC209 Offshore (9/2017), Limin Zhang (HK) for TC210 Dams (18/6/2018), Hanlong Liu (China) for TC303 Floods (26/9/2017), Marina Pantazidou (Greece) for TC306 Geo-Education (20/10/2017). It is believed that these numerous changes in TC Chair indicate a good health of the TC system, and helped maintaining the dynamism of TC activities.

3 NEW TCS

As a part of the ISSMGE's President initiatives, a new TC on Machine learning and big data was proposed by NGI. Together with other two TCs, all proposals and potential new Chairs were approved and presented by TOC to the Board for discussion and final approval. Among the various proposals received, the following ones have been accepted:

- TC309: Machine learning (25/1/2018), by Dr. Zhongqiang Liu from NGI (Norway)
- TC219: System Performance (2/4/2019), by Prof. Gang Zheng from Tianjin University (China)
- TC220: Field instrumentation (2/4/2019), by Dr. Andrew Ridley from Geotechnical Observations (UK)

Note that TC309 has already started to set its member list and plan activities, whereas the two other more recent TCs are still in the process of constitution. In this regard, it is recalled that the Officers of new TCs will contact all the Member Societies to set up their members' list, and Member Societies are hence invited to react rapidly to allow for a rapid start of these new TCs.

A proposal of a TC on Water geotechnics (Prof. Sitharam, IIS Bangalore, India) is still being discussed.

A proposal on a new TC on Geo-Multiphysics has been made in 2018, but various discussions made with some participants to this proposal with the TOC and some officers of other TCs finally reached the conclusion that there would perhaps be not enough available experts, given that many of them are already mobilised in other TCs.

The possibility of creating a TC on "Mine Wastes" that would include the hot topic of "Tailings Dams" has been considered by the VPs of both North and South America, with some contacts taken in countries from both regions, including Chile (a TC on Tailing dams had been run by Chile in the past).

There is present number of active TCs is 35.

4 FUNCTIONALITY AND PERFORMANCE OF TCS

The monitoring of the functionality and performance of TCs is part of the ISSMGE's President initiatives, it has still been a priority for the TOC. An assessment of TC activities has been done based on a certain number of indicators coming from the TC mid-term reports (see Annex 1), that include the functionality of the TC website, the organisation of regular TC Conferences or Workshops at ICSMGEs, the delivery of Bright Spark lectures, the regular organisation of administrative meetings (either in person or by teleconference/e-mail exchanges), the publication of Conference Proceedings in Open Access, the existence of operational Task Force, the production of State of the Art or State of Practice (SoA – SoP) documents, the interactions with other TCs and the production (or intended production) of papers for the International Journal of Geoengineering Case Histories. The aim of this assessment is not to rank TCs in any order of merit. It is more used as an indicator of which TCs has significant activity, and which TCs need to be considered more closely in an aim to improve their level of activity, if the field still appears to be of interest for the ISSMGE and geotechnical engineering.

It is good to observe that most TCs now have an active website (with updated Terms of Reference), most often based on the facilities proposed in the last version of the ISSMGE website (2016). In this regard, the TOC and the ISSMGE are most grateful to the excellent efficiency and availability of the ISSMGE Webmaster (webmaster@issmge.org). Some exceptions are with most active TCs that have their own website, sometimes most sophisticated and with more possibilities than through the ISSMGE website. It is now well admitted among TC officers that they have to feed and update their websites so as to let geotechnical engineers informed about what they do in their field of expertise (TCs are not clubs) and to make their technical and scientific production easily available (Open access proceedings of Conferences and Workshops, newsletters, minutes of TC meetings,…).

14 TCs have an Honour Lecture, given either during an International or TC specialised Conferences. It is recalled that the presentation of Honour Lectures at ICSMGEs depends on the options taken by the Conference organisers, and what has been done in the past (Paris 2013, Seoul 2017) will not necessarily be extended in the forthcoming ICSMGEs.

An important aspect of TC activities is related to the organisation of a specific TC Conference, which is the case of most of the active TCs. In this regard, the importance of producing Open access proceedings has been recalled at various occasions to TC officers and will, hopefully, concern all the TC Conference proceedings. Actually, some active TCs don't have their own Conference series, but actively participate to a series of regular existing Conferences, either by participating to the organisation or by giving keynote lectures.

5 CONCLUDING REMARKS

As illustrated in the Table, most TCs perform quite well to support the ISSMGE in their objective of improving the technical and scientific level of geotechnical practice over the world. This is made possible by the commitment of many experts who accept to serve and hardly work to reach this objective on a volunteer basis. Once again, the TOC and the ISSMGE would like to thank the TC officers and members for their dedication. On the other hand, for those few not well-performed TCs, the TOC will keep closely monitor their performance.

As commented in this report, the role of the TC websites to communicate and to promote their activity for the benefit of geotechnical practice is most important, and it has been observed that this tool has been more and more used, which is quite satisfactory. The creation of new TCs on topics of growing importance and the collaboration between TCs are another indication of the dynamic of the system. Another important contribution is the production of Open Access Proceedings of Conference and Workshops, a practice highly recommended by the ISSMGE that has been increasingly adopted by most TCs, to make even stronger the impressive on-line library of the ISSMGE.

The TOC and the TCs had during the previous term a significant collaboration with the Corporate Associates Presidential Group (CAPG, led by Sukumar Pathmanandavel) in order to reduce the gap between State of the Art and State of Practice. It seems that these actions increased the awareness of TCs about their mission of improving the general technical and scientific of the geotechnical practice all over the world, and the TOC will look forward to encouraging TCs in this direction.

Besides their own activities, the contribution of TCs has been quite important in the running of the two previous ICSMGEs of Paris (2013) and Seoul (2017), and it seems that their role will also be important for the success of the forthcoming Sydney (2021) ICSMGE.

Annex 1: mid-term report form, filled online

International Society for Soil Mechanics and Geotechnical Engineering

Société Internationale de Mécanique des Sols et de la Géotechnique

MID-TERM TC REPORT

October 2017 – April 2019

This questionnaire, finalised and approved during the Singapore Board meeting of 10 March 2019, is presented here as a word file to let TC officers aware of the questions asked to complete the mid-term report.

The questionnaire will have to be filled directly on the ISSMGE website, *that will be set up accordingly by mid-April 2019.*

The synthesis of all TC reports will be presented during the forthcoming Council meeting of 6 October 2019 in Cape Town (17th African Regional CSMGE). To cope with this deadline, TC officers are kindly asked to fill it on line by ***1st May 2019***

Many thanks to TC officers for their contribution to the ISSMGE's technical activities and objectives!

Name of the Technical Committee:	
Chair:	
Vice-Chair (if any):	
Secretary:	
Form filled by:	
Number of TC members:	
Number of chair nominated members	
Website:	
Date of last website update:	

	Query	Yes Please provide details	No Why?
1	Are you happy with the Terms of Reference of your TC in the ISSMGE website?		
2	Is your List of Members satisfactory?		
3	Do you have an Honour lecture? If yes, where and when has it been given last time?		
4	Did you or do you plan to select Bright Spark lecturers during your TC Conferences? see https://www.issmge.org/filemanager/article/48/ Criteria_Bright_Spark_Lecture.pdf		
5	Did you hold TC administrative meetings during the period? If yes, please give details (date, location, participation,…)		
6	Do you plan to organise a TC administrative meeting or Workshop during a forthcoming SMGE Regional Conference or at another occasion?		
7	If any, have you been happy with your contribution to the organisation of a Discussion session on your topic for the 19th ICSMGE in Seoul?		
8	Did you organise a TC workshop on your topic during the Seoul 2017 ICSMGE?		
9	Did you organise any other event (including conference, workshop, short course,…) during the period? If yes, please give details (date, location, participation,…)		
10	Were the proceedings of your events published in Open Access?		
11	Did you assign some tasks to TC members or Task force group? If yes, what has been produced?		
12	Did you produce any deliverable (State of the art report, State of practice report, Guidelines,…) ?		
13	Did you have any interactions with other TCs? If yes, which TC?		
14	Are there any future activities being considered?		
15	Did you or do you plan to have some contributions to any of the ISSMGE President's initiatives? (see list below)		
16	Do you plan to provide a case study for the ISSMGE Case Histories Journal? (cf e-mail D. Zekkos, 11/02/2019)		
17	Any further comment		

Appendix 7: Young Members' Presidential Group

Lucy Wu
Chair, Young Member's Presidential Group

1 INTRODUCTION

The Young Members' Presidential Group (YMPG) is a Board Level Committee within ISSMGE working to engage the next generation of geotechnical engineers. This is accomplished through mobilising students and younger members from across the world to share knowledge and ideas and being a nexus of opportunities for younger members in the international geotechnical community. This report details the continuing work of the YMPG from 2017 to 2019.

2 MEMBERSHIP

Consisting of a Chair, appointed for a 4-year term by the ISSMGE President, and 3 members from each region, nominated by the Regional Vice Presidents, the YMPG has 19 members total representing a broad global reach. Membership is restricted to ISSMGE members 35 years of age or younger at the time of appointment, and members serve 2-year terms, renewable for an additional 2 years depending on interest, activity level, and age eligibility.

Full membership and biographies of each member can be found on the ISSMGE website. The YPMG is led by the current Chair, Lucy Wu (USA). The Chair is supported by the Vice-chair, Jean Potgieter (South Africa) and the Secretary, Ceres Chung (Hong Kong).

Despite the formalised 2-year terms, due to the voluntary nature of the group and the changing circumstances for young members, membership to the group remains fluid. Thus, in some circumstances, members are added or may exit during the middle of a term. Currently, there are also two members who are joining the group on a transitional basis from May 2019 and intending to fully join the group for the 2019-2021 term. In the past, members have expressed difficulties in being able to quickly come onboard and thus are not being able to fully participate. This is an attempt to shorten the learning process by bringing new members on an audit basis.

3 CORRESPONDING MEMBERSHIP

In addition to the committee members of the YMPG, the Corresponding Member (CM) group has grown to 854 members as of May 2019 since its inception in 2013. The CM group signs up to receive emails related to the ISSMGE, conferences, post-doc or entry level positions, etc. This allows the ISSMGE, the Society's other Board Level Committees (BLCs) and Technical Committees (TCs) to reach younger members directly.

As of May 2019, the YMPG LinkedIn page has grown to 470 members and the GeoWorld group has grown to 112 members. Previous terms have driven the membership numbers and achieved a critical mass such that the groups continues to grow with little direct recruiting effort. The focus of this term has switched to providing more opportunities for these corresponding members as well as to respond to their needs and requests.

4 INITIATIVES

At the start of the term, Task Forces similar to previous term was set up, each with many objectives and initiatives. However, due to the difficulties of setting up regular meetings across different time zones and the amount of new initiatives, it was decided that the committee would work on initiatives as a group rather splitting the group into different groups. Larger initiatives also benefited from the entire group's participation rather than a specified task force.

4.1 *Bright Spark Lectures*

The Bright Spark Lecture award is an initiative started by the current ISSMGE president, Charles Ng to foster geotechnical talent and accomplishment within the geotechnical community. The purpose of the Bright Spark Lecture award is to promote young and promising geotechnical engineers/academics by giving them a platform to deliver keynote and invited lectures at these conferences.

With the support of Regional Vice Presidents and Conference Organising Committees, the YMPG advertised and vetted applications for three of the 2019 Regional Conferences: Europe, Africa, and Pan-America. The evaluation was based mainly in three criterias: Relevance of the subject, Quality of the application, and Personal background (CV and recommendation letters). The applications were impressive and the winners are all very accomplished, Winners have a combination of interesting and relevant subjects, good experience, and have reputable journal publications, even at a young age.

The committee began its efforts with the PanAmerican Conference. Announcements and applications were sent out to all of the North and South American societies. The societies were asked to forward their best applications. A total of 10 applications were received. A sub-committee consisting of YMPG, ISSMGE Vice Presidents, and representatives from the conference organiser evaluated the applications. A discussion session was held to discuss the applicants. The selected winners were presented to the President, who approved the selection. A similar process was followed for the African and European conferences. A total of 6 applications were received for Africa and 8 for Europe.

The committee feels that this initiative has proven to be successful and well received. Due to the fact that the Regional Conferences are only held every four years and the large size of the pool of potential applicants, there is some inherent luck with the applicants. Thus the Regional Conference and the ICSMGE Bright Spark awards should be utilised as platforms to encourage Member Societies and Technical Committees to host Bright Spark Lectures at their own conferences. The frequency of these conferences and the pool of applicants provides a much more sustainable platform to develop young engineers while providing value for conference organisers.

The YMPG is currently collaborating with the organising committee of the ICSMGE 2021 to host a Bright Spark lecture. An initial call was conducted between the YMPG Chair and Professor Mark Jaksa, the Technical Committee Chair.

4.2 *Joint Workshops with CAPG*

The Corporate Associates Presidential Group (CAPG) invited the YMPG to jointly host workshops and plenary sessions at the 2019 Regional Conferences. The collaboration has taken many different forms, from conference call discussions, panelist selection, to organising pre-conference workshops.

A series of pre-conference workshops supporting the theme of Collaboration have been held in preparation for the Australasia conference. The workshop in Auckland with support from the New Zealand Geotechnical Society was largely organised and executed by the YMPG. The evening session consisted of invited panelists with an audience of 40 to 50 people, of which around 10 to 15 were younger members. An instant polling tool was used to get quick feedback from the audience. The session generated excellent discussion and debate.

For the Asia conference, the group has helped the CAPG with choosing and inviting speakers and moderators for the conference session. Work is underway to plan the logistics for the day of the conference. The committee is also helping to disseminate a design survey to collect data points on the overall theme of Are We Overdesigning.

In the upcoming months, the group plans to continue working with CAPG to prepare for the conference sessions by co-hosting local workshops, disseminating the design survey, and preparing material for the conference sessions.

4.3 *Other Initiatives*

Several other initiatives and ongoing efforts were also areas of focus for the committee. Particularly, the YMPG supported the Innovation and Development Committee (IDC) in their initial efforts to establish the ISSMGE University by doing a review of current massive online open courses (MOOC). The group reviewed the functionality of popular websites and summarised findings for the IDC. The YMPG was contacted by TC304 to help advertise their student competition. This has evolved into an ongoing relationship with TC304 taking up the Bright Spark Lectures for their committee's conference. Other ongoing work includes maintaining the relevant ISSMGE webpages, posting news through various channels, and continuing to solicit articles for the YM Arena portion of the ISSMGE Bulletin.

5 MEETINGS

The YMPG typically meets via Skype conference call once every month. At the beginning of the term, a series of three rotating times were selected to give everyone the chance to call in two out of every three months. However this was too complicated so the strategy switched to having two meetings at 12 hours apart. This gave everyone a reasonable chance to join every meeting. The downside is that the same group of people always join each of the meetings and could never talk to the other half of the team. It is very difficult to operate a volunteer group across international and cultural boundaries and having remote meetings makes it even harder. Therefore the group began discussing the possibility of having an in-person meeting to foster better teamwork and collaboration.

With the generous support of the ISSMGE, the YMPG held its first ever in-person meeting in Singapore on 9-10 March 2019. During these two days, YMPG representatives met to discuss current and upcoming initiatives, foster networks, plan future initiatives, and propose a strategy for the future of the YMPG. Meeting in person, sharing meals, and connecting face to face allowed the group to connect more deeply, discuss topics in depth, and foster new ideas to successfully advance the YPMG.

Since it was held at the same time as the Board Meeting, the members who attended were able to meet the Board Members and present at the Board Meeting. The ISSMGE board is in agreement with the value of the YMPG meeting in-person. It was agreed that representatives of the YPMG should aim to meet every two years in the beginning of each term and that the ISSMGE will assist financially.

During the Singapore visit, the YMPG was also invited by the Student and Young Wing of Geotechnical Society of Singapore (SYGeoSS) to co-host a networking event for young members in Singapore. The event was held with generous support from the local professional community. It was well attended with over 40 SYGeoSS members. Lucy Wu presented opportunities and initiatives of the ISSMGE and YMPG.

One of the most important conclusions of the meeting is the recognition that the YMPG's core strength is to serve as a connector of groups worldwide. The structure and longevity of the YMPG team is not the best setup to create initiatives. However, given its place within the ISSMGE BLC, international connections of the members, and the group's increasing reach, the committee is a natural place for other young member groups to look to for assistance, resources, and leadership. Since establishing the ympg.issmge@gmail.com email account, the group has received emails from several young member groups looking to find out more about the international community, wanting to connect.

6 TRANSITION TO NEXT TERM

As part of the outcome of the extensive discussions at the March meeting, the group will concentrate efforts on a few new initiatives. The remaining few months of this term will focus on establishing the way forward and transitioning the majority of the effort to the next term.

1. Bright Spark Lectures. Encouraging member societies and technical committees to host Bright Spark Lectures for their conferences, using the Regional Conferences as demonstration cases.
2. Member Society Liaisons. Establishing a network of young member representatives from each of the member society. This network will allow the YMPG to find out about young member activities around the world and offer societies a way to connect with each other.
3. YGEC Network. Establish a platform for YGEC organisers past and present to share ideas, know-how, and lessons learned.
4. Alumni Group. Establish a YMPG alumni network. Former YMPG members are in an unique position through their understanding of the society and committee to help with future initiatives.

7 ACKNOWLEDGEMENTS

The group would like to thank the efforts of other Board Level Committees, Board Members, and Technical Committees to collaborate and support our initiatives. Special thanks goes to the CAPG, IDC, and TC304 for their enthusiasm and unwavering support.

The group would also like to applaud the individual efforts of the YMPG members, who operate on a voluntary basis in hope to advance the field of geotechnical engineering for younger members around the world.

Appendix 8: Corporate Associates' Presidential Group

Sukumar Pathmanadavel, Peter Day
Co-Chairs, Corporate Associates' Presidential Group

The Corporate Associates Presidential Group (CAPG) is a Board-level committee established by the President of the ISSMGE to develop actions and activities to enhance the commercial sector of the ISSMGE. It provides a home and a voice within the ISSMGE for the commercial sector of the international geotechnical community.

1 MEMBERSHIP

There are currently 36 Corporate Associates, including contractors, consultants, equipment and material suppliers, research institutes and universities. A list of CAPG members is given on https://www.issmge.org/corporate-associates/listing. The aim of the ISSMGE is to grow the number of corporate associates to 50 or more. Further information on the benefits of membership and membership applications are available at www.issmge.org/corporate-associates/why-how-to-join.

2 FOCUS 2017-2021

In this 2017 – 2021 Presidential term, the CAPG is focussing on providing visible and prominent platforms to discuss and promote geotechnical engineering issues that have a significant impact on the geotechnical engineering industry. The ISSMGE Regional Conferences play a significant role in achieving this objective. The issues identified are collaboration, implementation, over-design, innovation and education

3 2019 REGIONAL CONFERENCES

Plenary sessions have been organised for all Regional Conferences as outline below.

3.1 *XIII Australasian Conference, 1-3 April: Theme – Collaboration*

A report on the highly successful session titled "Collaboration in geotechnical engineering – Impact on research and project delivery" in Perth will appear in the June issue of the ISSMGE newsletter. Opportunities identified include a CAPG sponsored session on the management of tailings storage facilities involving the major mining houses, a 2-minute video competition for young engineers and a special edition of the Australian Geomechanics Society journal on Industry Collaboration.

3.2 *XVII European Conference, 1-6 September: Theme – Design Implementation*

On 2nd April, a CAPG coffee morning was hosted by Siemens (a CAPG member) in Offenbach, Germany, to lay the groundwork for a plenary panel discussion to be held in Reykjavik on "Bridging the gap between designers and contractors". This discussion will explore how we can ensure effective transfer of design intentions into construction realities.

3.3 *XVII African Conference, 7-9 October: Theme – Geotechnical Innovation*

Many of our geotechnical conferences focus on research with less attention being paid to the considerable achievements of geotechnical practitioners. The CAPG session in Cape Town will provide Corporate Associates an opportunity to showcase innovations in the geotechnical industry. Abstracts have been invited from all CAPG members and selected papers will be presented in a 90-minute plenary session in Cape Town.

3.4 *XVI Asian Conference, 14 -18 October: Theme – Are we overdesigning?*

At the Danube Sub-regional Conference in Skopje in 2017, the Asian Region posed the question "are we overdesigning? In response, the CAPG is conducting an international survey comprising 10 common geotechnical design problems on both clay and sand profiles. The results will be presented in a plenary session in Taipei, alongside a CAPG panel discussion. For more information see https://www.issmge.org/news/are-we-overdesigning-a-survey-of-international-practice.

3.5 *XVI Pan American Conference, 17-20 November: Theme – Geo-Education*

The face of education is changing worldwide, and geo-education is no exception. In addition, the availability of both knowledge and data is on the increase. The CAPG panel discussion in Cancun will concentrate on "Geo-education over the next 30 years" including discussion on machine learning and big data.

4 LOOKING TO THE FUTURE

We look to achieve increase membership of the Corporate Associates through visible and compelling case for the value of being a Corporate Associate of the ISSMGE.

We plan to continue to create interesting and engaging platforms where geotechnical engineering can be reimagined by academia and practitioners working in concert, for the benefit of the wider community.

CAPG will contribute to the success of the ICSMGE 2021 in Sydney by strong engagement and unique offerings to the attendees.

5 CAPG PARTNERS

The success of the 2017 Global state of the art and state of practice in geotechnical engineering survey was due largely to excellent cooperation between the CAPG, the Technical

Oversight Committee and the Technical committees of the ISSMGE. The "overdesigning" survey currently underway is being undertaken in collaboration with the YMPG, TC 205 and TC304.

In April 2019, the CAPG held a panel discussion session at the DFI conference in Dubai. Many valuable outcomes emerged including the need for improved databases, communication between industry and researchers, formalisation of local knowledge and practices, and greater involvement of geotechnical specialists in major projects.

Cooperation with other committees within the ISSMGE and related institutions is a key part of the CAPG's strategy.

6 MEETINGS

Representatives of CAPG members meet every six weeks by telephone. These hour-long meetings are used to review achievements, formulate strategies and plan activities. All interested CAPG member representatives are welcome to join these meetings.

More face-to-face meetings of CAPG representatives, such as the "Siemens Coffee Morning", should be planned. The CAPG sessions at ISSMGE International and Regional Conferences also provide opportunities for us to meet.

7 ACKNOWLEDGEMENTS

The CAPG expresses its thanks to the ISSMGE President and Board for their enthusiastic and continued support.

We also thank all the Corporate Associates who have chosen to show visible alignment and financial support to the ISSMGE.

Thanks also to the many colleagues who have assisted with the organisation of recent CAPG activities including Kim Chan, Yuli Doulala-Rigby, Joachim Müller, Finlay Leibrick, Hugo Acosta-Martinez, Lucy Wu and so many others in the past and present who have contributed to making CAPG stronger and vibrant.

Appendix 9: Awards Committee

Roberto Terzariol
Chair, Awards Committee

1 INTRODUCTION

Professor Charles W.W. Ng, president effective from September 2017, appointed Roberto Terzariol as the new chairman of the Awards Committee (AWAC). In order to maintain continuity, Esve Jacobsz was invited to continue to serve in the same role, until he resigned in March 2019. A new secretary, André Archer from South Africa, was nominated and his position as secretary was approved at the ISSMGE Board Meeting in Singapore on 10 March 2019. Other memberships to this committee were formed through consultation and nominations by regional vice-presidents. The composition of this new AWAC consists of representatives from six regions. This report outlines AWACs actives and changes from September 2017 to June 2019 and acts as a guide for the award currently on offer by the ISSMGE.

2 MEMBERSHIP AND RESPONSIBILITIES OF AWARDS COMMITTEE

Committee members

Chairman: Roberto Terzariol (Argentina)

Secretary: André Archer (South Africa)

Regional representatives:

- Lidija Zdravkovic (Europe)
- Werner Bilfinger (South America)
- Hugo Acosta (Australasia)
- Teik Aun Ooi (Asia)
- Kerry Rowe (North America)
- Carlos Quadros (Africa)

Max Ervin retired and as such are no longer a member of AWAC. Gavin Alexander, VP for Australasia, appointed Hugo Acosta as replacement and as representative for the Australasian region, effective from April 2019.

The main responsibilities are to work with the ISSMGE Board to devise and improve awards guidelines and nomination procedures, to coordinate review processes, to conduct review of nominations and to make recommendations to the Board for consideration and approval. Ad hoc members may be recruited to serve on the AWAC, if necessary

3 TYPES OF ISSMGE AWARDS

The ISSMGE offers the following awards to recognise those individuals and bodies who have made important contributions to our profession, society and the world:

1. ISSMGE Outstanding Geotechnical Project Award
2. ISSMGE Outstanding Innovator Award (individual or team)
3. ISSMGE Outstanding Member Society Award
4. ISSMGE Outstanding Paper Published in the International Journal of Geo-Engineering Case Histories Award
5. ISSMGE Outstanding Professional Image Award
6. ISSMGE Outstanding Technical Committee Award
7. ISSMGE Outstanding Young Geotechnical Engineer Awards (up to three awards)
8. ISSMGE Bright Spark Lecture award

Awards will be given once every four years, except for the Bright Spark lecture award, which is given yearly at various conferences.

Guidelines for the nomination of the various awards follow below. All nominations must be made through the local ISSMGE Member Society (except the Outstanding Paper Award) and must reach the ISSMGE Secretariat by the deadline, which will be announced approximately one year before the International Conference on Soil Mechanics and Geotechnical Engineering (ICSMGE) to be held in Sydney, Australia in 2021.

4 PURPOSE AND NOMINATION PROCEDURES FOR EACH TYPE OF AWARDS

1. ISSMGE Outstanding Geotechnical Project Award

Purpose

To recognize projects that best illustrate superior geotechnical engineering skills and represent significant contributions to geotechnical engineering progress and society. Criteria reflect the importance placed on sustainability while continuing to recognize innovative design and construction. Any geotechnical engineering project, regardless of its location, is eligible, e.g. projects such as tunnels, dams, innovative foundation engineering solutions and military projects.

Nomination of candidates

Nominations for the award are invited from ISSMGE Member Societies and must reach the ISSMGE Secretariat by the announced date. Member societies are encouraged to arrange their own competitions to identify an outstanding geotechnical project which can then be nominated for the ISSMGE award.

The nomination package should include:

- A cover letter, signed by the nominator (Chairman of the ISSMGE Member Society), introducing the nominated project.
- A written motivation, not more than five pages in length, to address the following:

(i) Originality and innovation – new or innovative application of technology, design, materials, process/methods and construction;

(ii) Resourcefulness in planning and solving design challenges – (a) complexity of the problem or situation addressed, (b) creativity in solutions;

(iii) Sustainability considerations – (a) environmental, (b) social, (c) economic;

(iv) Project planning and delivery – (a) financing, budget and schedule, (b) meeting the client's needs particularly related to budget and schedule adherence;

(v) Contribution to the well-being of people and communities, including aesthetic value;
(vi) Photographs, design drawings, figures, sketches as applicable.

2. ISSMGE Outstanding Innovator Award (individual or team)

Purpose
The ISSMGE Award for Innovation in Geo-engineering is awarded to an individual ISSMGE member or groups of ISSMGE members (researchers, consultants and contractors) in recognition of innovations in Geo-engineering that have a pronounced impact on geo-engineering practice, research and education. The term "innovation" is used broadly to describe any major, unprecedented achievements that led to a major advancement in our profession. Examples include:
- The development of a theory or a design approach that has pronouncedly affected our understanding of earth materials or design methods for geotechnical infrastructure.
- The invention of a technique or a technology that has significant applications in field or laboratory investigations or measurements.
- The successful implementation of an unprecedented construction technique that has a sustained impact in geotechnical construction.
- Other outstanding innovations or approaches (e.g. development of software, internet technologies) that impact the practice or education of the Geo-engineering profession worldwide.

Nomination of candidates
ISSMGE Member Societies are invited to nominate important contributors in their respective fields. Nominations are limited to one per Member Society for the award. The nomination package should include the following:
- A cover letter, signed by the nominator, indicating the person/parties nominated and the name of the award (Outstanding Innovator).
- A written motivation, not more than two pages in length, which should include reference to the supporting evidence of distinguished accomplishments (technical papers, etc.) by the nominee.
- The nominee's CV (up to 5-pages) or biographical information. In case of a group of individuals, CVs for all individuals should be provided if judged applicable by the nominator.
- Three letters of recommendation.

3. ISSMGE Outstanding Member Society Award

Purpose
To recognise the most outstanding ISSMGE Member Society. Criteria for evaluation relate to extent of the society's activities, given the number of members and the rate of progress and advancement. These activities could include:
- Conferences and workshops supported by the local Society and/or endorsed by ISSMGE in the past 4 years.
- Encouragement given by the Member Society and recognitions for its members in terms of awards and honours.
- Contributions to ISSMGE Technical Societies through member participation.
- Contributions to the advancement of science and technology through publications by the members in scholarly journals and conferences.
- Any other information representing the contributions by the Society, including those that have made an impact on community welfare and safety (e.g. affected by natural hazards).
- In judging the nominations, emphasis will be placed on growth in achievements or new and recent developments.

Nomination of member societies
Nominees for these awards are individual member societies. Member Societies are invited to nominate themselves by submitting a brief submission (not more than two pages) addressing the above. The necessary supporting documentation should accompany the nomination. Nominations must reach the ISSMGE Secretariat by the announced date.

4. ISSMGE Outstanding Paper in The International Journal Of Geo-Engineering Case Histories Award

Purpose
To recognise the most outstanding paper published by an ISSMGE member in the International Journal of Geo-Engineering Case Histories.

Nomination of candidates
The two best papers published in the International Journal of Geo-Engineering Case Histories (IJGCH) during the past four years (2013-2016) are selected by the IJGCH Editorial Board. The IJGCH will be requested to nominate two papers and send the nominations to the ISSMGE Secretariat who will relay them to the ISSMGE Awards Committee.

5. ISSMGE Outstanding Professional Image Award

Purpose
The ISSMGE Outstanding Professional Image Award in Geo-engineering is awarded to any individual or organization, preferably from outside the ISSMGE Society, who contributed outstandingly in promoting geotechnical engineering and showing its importance as a profession in life.

Nomination of candidates
ISSMGE Member Societies are invited to nominate candidates for this Award. Candidates would be public relations individuals or organization, author of any geotechnical contribution that has made an impact within or through the news, media, internet, and/or the general public.

Nominations are limited to one per Member Society. Nominations must reach the ISSMGE Secretariat by the announced date.

The nomination package should include the following:
- A cover letter, signed by the nominator, indicating the person/organization/parties nominated and the name of the award (Outstanding Professional Image).
- A written motivation, not more than two pages in length, which should include reference to the supporting evidence of distinguished accomplishments (advertisement, movie, promotion, documentary, activity or innovative approach) by the nominee, with a copy of that accomplishment if applicable (printout, CD, USB, etc.).
- The nominee's CV (up to 5-pages) or biographical information. In case of a group of individuals, CVs for all individuals should be provided if judged applicable by the nominator. An organization can submit one qualification document that describes it and its activities with the CVs of key personnel as deemed applicable by the nominator.

6. ISSMGE Outstanding Technical Committee Award

Purpose
To reward an ISSMGE Technical Committee in recognition of one or more of the following achievements:
- Conducting work in solving geo-engineering problems that are within the Technical Committee's sphere of activity.

- Finding effective solutions to important theoretical geo-engineering problems.
- Effectively enhance current geo-engineering practice through up to date and innovative solutions concerning guidelines, standards, technologies, laboratory methods or recommended laboratory apparatus.
- Providing education and training with emphasis on current geo-engineering issues.

Nomination of technical committees
Award nominations for particular committees are to be submitted to the ISSMGE Secretariat by the announced date by the chairs of the ISSMGE Member Societies who wish to nominate for this award. The nomination package should include the following:
- A written motivation, not more than two pages in length, which should include relevant attachments (publications, patents, reviews, confirmations of implementation, etc.) concerning the subject of the award.
- Information regarding active participation in conferences, seminars, workshops as well as other meetings concerning the Technical Committee's work which is presented in such a way as to document and confirm that the Technical Committee's outstanding activities entitle it to an award.
- Information as regards to the organising of conferences, seminars and other types of meetings concerning the Technical Committee's activities since the last quadrennial ISSMGE conference (regular meetings, internet conferences, etc.).
- Demonstration of the effectiveness of organisational meetings for scientific, engineering, educational and training purposes.

7. ISSMGE Outstanding Young Geotechnical Engineer Awards (up to three awards)

Purpose
The aim of the ISSMGE Outstanding Young Geotechnical Engineer Award is to acknowledge the contribution of young members to the profession and to the society. The award will be given in recognition of achievement during the four-year period preceding the year of the award. The ISSMGE Outstanding Young Geotechnical Engineer Award will be presented up to three individuals who have made outstanding contributions to the practice or research in geotechnical engineering through their contribution to one or more research or industrial projects. There would be a maximum of 3 awards made every 4 years, which are presented at the ICSMGE. The awards each amount to a cash prize of £1,000. The nominations would be evaluated holistically based on contributions to any or several of the following criteria:
- Geotechnical project development and construction;
- Research contributions in the geotechnical community;
- Involvement in national and international geotechnical societies; and
- Publications and education in the geotechnical discipline.

Nomination of candidates
The awards will be given to ISSMGE young members who are less than 36 years of age on 31 December of the year of the quadrennial ICSMGE conference. One nomination may be made by each member society, which should include the following:
- A written motivation, not more than two pages in length, clearly explaining why the candidate is deserving of this award, provided by the candidate's current organization and endorsed by the member society.
- The nominee's one-page CV.

Nominations must reach the Secretary General by the announced date.

Candidates will be evaluated by the ISSMGE Awards Committee in consultation with the ISSMGE Young Members Presidential Group (YMPG) and the finalists will then be recommended to the Board for final decision.

8. ISSMGE Bright Spark Lecture Award

Purpose
The aim of the ISSMGE Bright Spark Lecture Award is to promote young members of ISSMGE to play a major role in various international and regional conferences. The award offers promising young geotechnical engineers/academics to deliver keynote and invited lectures at these conferences and was the creation of the ISSMGE president Professor Charles W.W. Ng.

Eligibility and basic requirements for the award are:
- Outstanding and promising young members at age 36 or under at the last day of the conference they are nominated for.
- The member shall be nominated by member societies and the YMPG.
- Limit of 3 nominees per member national society

Nomination of candidates and approval procedure
Each local conference organising committee will be responsible for the preliminary selection of Bright Spark lecturers, in collaboration with the YMPG. The preliminarily selected lecturers together with their CVs will be submitted to the President of ISSMGE for final approval. Although it is intended to provide young members more opportunities, no more than four lecturers should be selected and approved in any single conference.

Each awardee will be invited to give a 20-30 minute presentation at the conference and write a plenary paper. The Bright Spark award will go to young geotechnical engineers that exemplifies mature research or practice and epitomizes what the ISSMGE is looking for in engineers of tomorrow. An "ISSMGE BRIGHT SPARK LECTURER" certificate will be given either by the President or his/her representative.

Assessment Procedures

To avoid any potential conflict of interests, five sub-committees responsible for different types of awards were formed to assess the suitability and quality of nominations and to make recommendations to the AWAC for considerations. The five sub-committees are listed as follows:

Sub-committee Chairs:
Dr Hugo Acosta (Australia)
Dr André Archer (South Africa)
Professor Kerry Rowe (Canada)
Professor Lidija Zdravkovic (UK)
Professor Dimitrios Zekkos (USA)

Committee members:
Dr Dennis Becker (Canada)
Dr Werner Bilfinger (Brazil)
Professor Jon Bray (USA)
Dr Johnny Cheuk (Hong Kong)
Dr Peter Day (South Africa)
Professor Russel Green (USA)
Professor Mark Jaksa (Australia)
Professor Mike Jamiolkowski (Italy)
Dr Zachieh Moh (Chinese Taipei)
Dr TA Ooi (Malaysia)
Professor K.K. Phoon (Singapore)

Professor Pedro Pinto (Portugal)
Professor Carlos Quadros (Mozambique)

The Bright Spark Lecture Award procedure is independent of AWAC with nominations sent directly to the Young Members Presidential Group (YMPG). The YMPG will assess and short-list candidates, with the selected lecturers sent to the president for final approval.

5 AWARDS APPROVED BY THE BOARD

Based on the recommendations made by AWAC, the Board approved the following awards at the Board meeting held on 10 March 2019 in Singapore:

1. One ISSMGE Outstanding Member Society award
2. One ISSMGE Outstanding Paper in the International Journal of Geo-Engineering Case Histories award
3. One ISSMGE Outstanding Technical Committee award
4. Three ISSMGE Outstanding Young Geotechnical Engineer awards
5. ISSMGE Bright Spark Lecture award

In general, AWAC believes the variety of rewards presently on offer meets the needs of the ISSMGE.

The award ceremony will be held during a plenary session at the 20th International Conference on Soil Mechanics and Geotechnical Engineering in September 2021 in Sydney, Australia.

6 FUTURE WORK

The nomination process for the 20th ICSMGE will start on 1 October 2019 with a closing date for nomination of 1 July 2020. Awards will be promoted at ISSMGE sponsored events, especially at the regional conferences leading up to the 20th ICSMGE. As such, a short presentation will be prepared by AWAC that will be circulated to VPs and conference organisers. At the board meeting in Singapore on 10 March 2019, it was suggested to create a new award category for short (30 sec) videos promoting geotechnical engineering and the profession. AWAC is in the process of developing clear guidelines, descriptions and legalities for such an award, which would be sent for board approval once it is finalised. Should this be finalised, and approved, before the start of nominations on 1 October 2019, the award will be included at the 20th ICSMGE.

Appendix 10. Professional Image Committee (PIC)

Ikuo Towhata
Chair, Professional Image Committee

1 HISTORY

During the previous term of 2013-2017, the PIC committee was led by Dr. Sherif Wissa. I found in those days that promotion of our profession among public is a difficult task.

I decided to take over the Chair of this committee in September, 2017, and drafted a new Terms of Reference in order to start activities. That draft was basically approved through Skype Board meeting and e-mail discussion in February-March period in 2018. The activities were not able to start until another Skype meeting at the end August, 2018, because of unclear administrative situations. In the meantime, I organized a domestic supporting group in Tokyo and initiated preliminary works. So, the net activity period of PIC is officially 8 months which is from September 2018 to early May, 2019, when this report was prepared.

2 CURRENT SITUATION IN AND AROUND OUR PROFESSION

The activity should start with the recognition of the reality. Dream is important but dream without knowing reality is illusion.

Pictures in Fig. 1 were drawn by young fellows (elementary and junior high school students up to 15 years old) when the Kanto Chapter of the Japanese Geotechnical Society organized a contest of pictures on young people's image of geosphere. Young people have a good image of the ground. This point should be borne in our mind.

Fig. 1 Young fellow's image of geosphere

The current situation is unsatisfactory for us in many countries. People enjoy many benefits from the products of geotechnical engineering. Despite that, people do not recognize the importance of underground works and take everything for granted. Difficulties in deep excavation without disturbing surrounding structures, tunneling through fault fractures, and protection of infrastructures from severe natural disasters are not known to the public. On the other hand, once any problem occurs and the progress of a project is delayed or hindered, the public and the media blame engineers. It is noteworthy that the advanced astronomy observes galaxies at twenty billion light years (=1.9*1026 m) while geotechnical engineering cannot see soil at 0.5 m below the ground surface, unless special attention is paid. This fact implies the difficulty in underground works.

It is unfortunate that the public do not know the difference between purchasing PCs of mass production in a market and working on a "single-item production" in an invisible geo-space. There is no recipe for a single-item production. Because of invisibility, information on the geological and hydrological conditions in situ is limited. The associating risk is profoundly higher than the risk in purchasing a new PC. Furthermore, it is easy to buy a new PC if a previous purchase is unsatisfactory, whereas it is costly and difficult to re-start an underground construction if any difficulty is encountered during the work.

Traditionally, "dirt" is a different name of soil and its adjective form is "dirty". Dirty people's dirty work is not highly recognized by the public. Even in research world, big funding tends to go to advanced science and technology which appears substantially more beautiful than dirt. We should be recalled that people of advanced science and technology are making efforts to develop new ideas, draw pictures of better future, and appeal to the public. Examples are found in flying cars, quantum computers, immunotherapy of cancer, etc. Although some of the ideas appear impossible, they do appeal to the public. Geotechnical engineering did not do efforts in this direction. For promoting our image, we need to compare our profession with others. Otherwise, our efforts for promotion will be self-satisfaction.

3 MEMBERS OF PIC

Currently, PIC consists of the following people. Please point out any mistakes herein.

Name	MS	Affiliation
Ikuo TOWHATA	Chair	Director, Tohata Architects and Engineers
François DEPARDON	France (Vice Chair)	Egis Geotechnique
Sam MACKENZIE	Australia	Former Chair of Australian Society
Alessander KORMANN	Brazil	Prof and President of Society
Tony SANGIULIANO	Canada	Ministry of Transport in Ontario
Wenjie XU	China	Associate Professor, Tsinghua University

Bruno DEMAY	France	Antea Group
Wolfgang SONDERMANN	Germany	President of German Society
Hemanta HAZARIKA	India	Professor, Kyushu University
Tatsuya NOJIMA	Japan	Engineer, JGC Corporation Nikki, Tokyo
Jin Man KIM	Korea	Professor, Pusan National University
Stephen CRAWFORD	New Zealand	Tonkin and Taylor
Adam LATIMER	UK	Ian Farmer Associates
Donald A. BRUCE	USA	President, Geosystems
Rick DESCHAMPS	USA	VP, Nicholson Construction Company
Trevor GREEN	South America	Department Head, Jones & Wagener

4 OBJECTIVES OF PIC

The objectives of PIC were finally decided as what follows.

Objective 1: Dissemination of Contribution to the Public Welfare made by Geotechnical Engineering

This the most traditional activity for promotion. There have been an uncountable number of significant geotechnical contributions to the human welfare, among which are transportation/communication infrastructures, reliable building foundations and disaster mitigation structures. Geotechnical engineering has been appealing those achievements to the public for years. However, other fields of science and technology have been the same efforts as well and, frankly, have more successful than us.

The present PIC pays attention to urban water systems that supply good drinking water to people (Fig. 2) and treat waste water/sewage (Fig. 3). It is said that the installation of modern water supply/sewerage system has saved more number of lives from epidemics than medical doctors. It is important to compare us and others in dissemination.

Fig. 2 Aswan High Dam of Egypt that provides water and electricity

Fig. 3 One of the earliest sewerage systems in Paris

Objective 2: Mitigation of Georisk

Working on an invisible world is not easy. There is always a possibility to encounter unexpected problems and a project is delayed or becomes more expensive. Once such a trouble is found, our image becomes worse among the public. Fig. 4 shows a ground depression in the center of Fukuoka City, Japan, where ongoing subway tunnel construction triggered depression at the surface. It was officially declared later that insufficient ground investigation missed locally thin part of an impervious layer that collapsed when tunnel excavation reached underneath.

Fig. 4 Collapse of subway tunnel under construction (https://www.nikkei.com/content/pic/20161115/96958A9E93819891E3E79AE2868DE3E7E3E3E0E2E3E4E2E2E2E2E2E2-DSXMZO0954116015112016MM0001-PB1-9.jpg)

Georisk means unexpected difficulty, delay and loss of resources in construction projects (both before and after initiation of construction works). Efforts are necessary to reduce the georisk by conducting good risk management. PIC is interested in two kinds of georisk management which are promotion of more ground investigation and development of open-access geotechnical data base.

Inappropriate and/or insufficient ground investigation leads to georisk. Poor maintenance of testing device is not rare, unfortunately. The quality of field works has to be improved continuously. Another issue is that many clients are not aware of the problems in the ground. When nature created the ground, no care was taken of the uniformity. No quality control was made either. It is very possible that the ground characteristics change drastically over a short distance of several meters. Despite this, many clients are not aware of this risk and tries to save money by reducing the number of subsoil investigations. It is essential, therefore, to let clients know the difficulty of geotechnical engineering and the importance of detailed subsoil investigations.

Increasing the number of ground investigation has another benefit to the geotechnical community. That is increasing more business and more income to our discipline, which will make geotechnical engineers more wealthy, although its extent is still uncertain. Generally, people do not respect a community of poor people; please do not misunderstand my intention in this sentence. Job of low income cannot attract a good number of brilliant young people.

Objective 3: Future Possibility Of Geotechnical Engineering

Geotechnical engineering has a wide range of activity. The target issues are so many as foundation, excavation, underground structure, slope stabilization, lifelines, disaster mitigation, ground investigation, numerical analyses, etc. Noteworthy is that those issues concern the demands from clients only and have little to do with the general public. Accordingly, public do not care geotechnical engineering. This situation is quite different from those of smart phones in which people are keen to learn about and purchase the most advanced version of devices. The underlying problem is that geotechnical engineering has been interested too much in satisfaction of clients and did not care the general public. Certainly, it is important to satisfy clients because geotechnical engineers have to do business, get income and feed their family. However, other fields of science and technology have been doing something more and have been recognized by the public more successfully. Thus, it is stressed again that we should always compare ourselves with others. We need to develop a new idea on what activity is appreciated by the public in general.

Objective 4: Production of Materials ((PPT, Texts etc. to promote the activity)

Materials to support the above-mentioned activities should be prepared and used. Although PPT and academic papers are traditional ways for dissemination of the output, they are appreciated only among academicians. We need to prepare a shorter document and use it for dissemination among non-engineering communities. Website or other IT communication channels will be useful as well. This activity will start after September, 2019.

Objective 5: Communication with Public

Care is needed in choosing the target of communication. Who can understand correctly the products of Objectives 1, 2 and 3? It is thought that the important targets for objective 2 are law makers, governmental body, and private sectors who have their own construction projects. Media willingly understands us only after big and scandalous accidents/disasters. Otherwise, media will be biased to consider us "dirty" businessmen.

Objective 6: Communication with Clients: Importance of subsurface investigation in avoiding georisk

Idea is same as for Objective 5.

Objective 7: Awards and Objective 8: Other relevant activities:

To be considered later.

5 GEORISK MANAGEMENT

PIC's activities are supported by a special committee that was organized by the Japanese Geotechnical Society which collaborates with the Georisk Society and the Japan Geotechnical Consultants Association. In this collaboration, the most important issue is georisk management/mitigation by promoting ground investigation. It is believed that real ground is too heterogeneous and complicated for probabilistic theory to handle. Hence, more ground investigation is the choice.

Georisk Society has been organizing annual symposia since 2010 in which experiences of successful and unsuccessful use of ground investigation have been reported on a variety of construction projects (Fig. 5). Those experiences have been classified into four groups that are namely

A: risk was anticipated well in advance and additional ground investigation helped avoid the risk (successful management),

B: risk occurred during projects and construction cost/period increased,

C: risk occurred in the early stage of projects and additional ground investigation partially mitigated the risk, and

D: others.

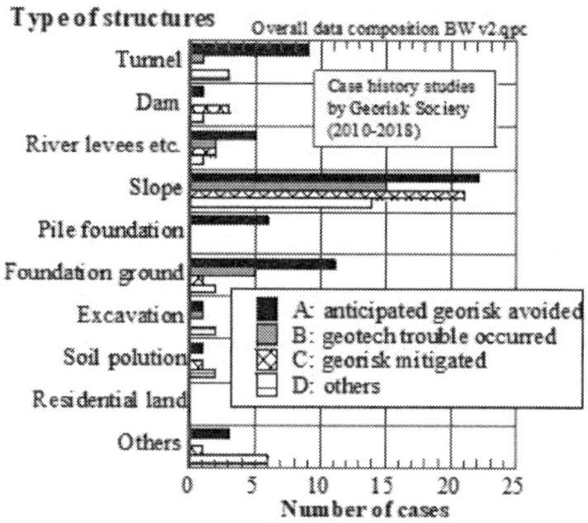

Fig. 5 Composition of 143 studied projects

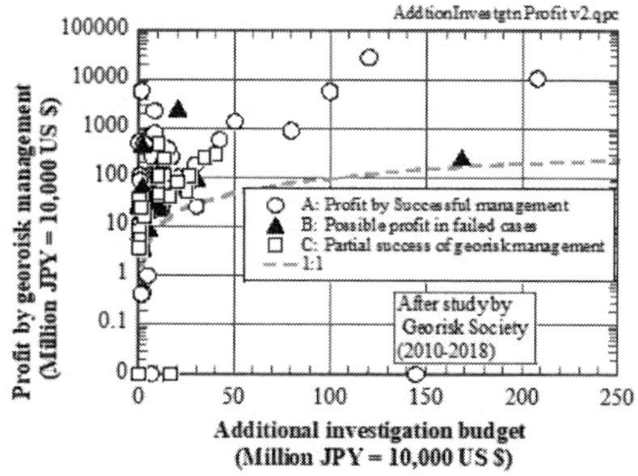

Fig. 6 Cost reduction achieved by extra expenditure on ground investigation

PIC Chair reinterpreted the symposium publications and analyzed the presented data. It was found that many problems occurred as instability of cut slopes where geological structures, material strength and ground water flow unexpectedly affected the stability. One of the outputs from this study is illustrated in Fig. 6 where a small amount of extra expenditure on ground investigation brought about good reduction of the total construction cost. Thus, ground investigation deserves attention.

It is desirable that more data is collected internationally on the value of ground investigation. I would be delighted to communicate with voluntary people who can contribute PIC with more info. It is certainly possible to keep names and sites anonymous. Details of desired data are shown at the end of Chapter 7 of this report.

6 FUTURE TOPICS FOR GEOTECHNICAL ENGINEERING

There are many topics to be studied that will be important in future. Among them, this report stresses the development of new water resources in arid and densely-populated regions. It is intended that water is an urgently needed resource for people and that working on water will easily attract people's interest in geotechnical engineering. I would say that geotechnical community should show up in front of people in place of staying under the umbrella of clients. Another good point is that development of new water resource is a friendly activity for people and give them an impression that geotechnical engineering is working on "mother earth."

Figure 7 illustrates the idea of underground dam that stops water flow below the ground surface and provides water to a local community. The advantage of the underground dam in coastal regions is as what follows:

- Underground water flows into sea and nobody claims rights to its use. In contrast, surface water is under strict control of resource management.
- The underground dam and reservoir do not intervene the land use at the surface.
- The dam needs just to be impervious. Sandwiched between soil on both up- and downstream sides, it does not need mechanical strength.
- Underground water is less prone to pollution than surface reservoir water.
- Underground dams of small scale have been already constructed in remote islands where soil is highly pervious and surface stream cannot be maintained.

In contrast, detailed subsoil investigation is necessary in order to determine the size of an impervious wall (dam).

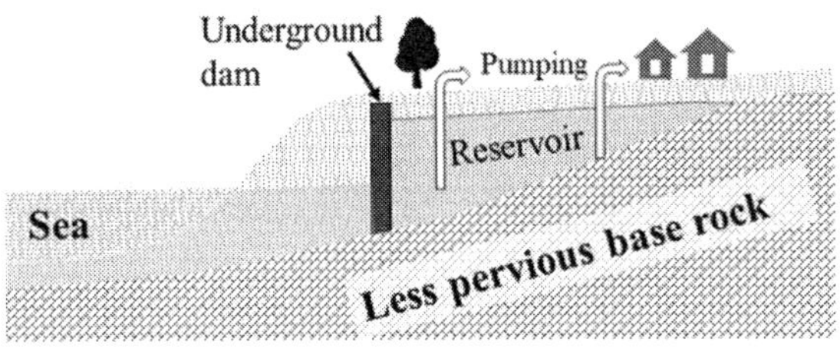

Fig. 7 Conceptual sketch of an underground dam

PIC is expecting establishment of one new Technical Committee on Water Geotechnics. The proposal of this TC describes a similar idea that river water during floods flows into the sea without use and that this surplus water should be stored in a coastal reservoir and used later. Collaboration with this group will be essential in PIC activities for better image of geotechnical engineering.

7 COLLABORATION WITH TECHNICAL COMMMITTEE AND MEMBER SOCIETIES

There is an agreement with TC102 on In-situ Testing. Presentation is planned on the detailed contents of Chapter 5 in this report during the TC102's upcoming conference in Budapest in September 2020.

I visited Paris in February, 2019, in order to visit the French Member Society and talk about further development of the PIC activity. Upon this occasion I was able to see promotion videos for our international community in which attractive points of our discipline are demonstrated in a very relaxed way. The French Society kindly allows other member societies to use them for domestic promotion. They can be downloaded freely from the following sites:

- https://www.youtube.com/playlist?list=PLQg3J2byMVWKFzavNVh1br8Kre7Y8Oceu
  vocation of geotechnical engineer
  geotechnical engineering by construction industry stakeholders
- https://youtu.be/mSMl17vqh8A?list=PLQg3J2byMVWL8tMegRgGC-E30UPWzj-dN
  interview to people in the street

To summarize the international situation of georisk management, the following data is desired from member societies and other institutions. For details, please contact myself at towhata.ikuo.ikuo@gmail.com.

Table 1 List of desired data on georisk management in practice

	A: successful management	B: failed management	C: partial success
Important data	<u>General info on project</u> **Original construction plan:** total budget by local currency and total construction period (months) **How was the risk detected and when? What mitigation was taken?** **Actions (additional ground investigation or else) taken for georisk management:** Type of investigation, cost & period **Cost and period (months) for improved construction:** New budget after reduction and/or shortened construction period **Note: info on periods is less important.**	<u>General info on project</u> **Original construction plan:** total budget by local currency and total construction period (months) **How was the risk detected and when?** **Consequence of georisk:** Increased total budget and elongated total construction period, including what were spent on emergency actions **If the georisk had been captured earlier, what would have been** type of investigation, cost and period for georisk management, total cost after reduction and shortened construction period?	<u>General info on project</u> **Original construction plan:** total budget by local currency and total construction period (months) **How was the risk detected and when? What mitigation was taken?** **Actions (additional ground investigation or else) taken for georisk management:** Type of investigation, cost & period **Cost and period (months) for improved construction:** New budget after reduction and/or shortened construction period **Note: info on periods is less important.**
Optional	PDF copy of related publication	PDF copy of related publication	PDF copy of related publication

Appendix 11: ISSMGE FOUNDATION

Neil Taylor
Secretary General, ISSMGE

REPORT TO DONORS

The ISSMGE Foundation

A charitable incorporated organisation (CIO) associated with the International Society for Soil Mechanics and Geotechnical Engineering

Charity Commission of England and Wales - No. 1153795

Trustees:

Professor Jean-Louis Briaud,
Professor and Holder of the Buchanan Chair, Texas A&M University, USA

Professor Michael C.R. Davies (Chair)
Pro Vice Chancellor, University of Sussex, UK

Professor R. Neil Taylor
Professor in Geotechnical Engineering, City, University of London, UK

Professor Mark B. Jaksa
Professor of Geotechnical Engineering, University of Adelaide, Australia

BACKGROUND

The concept of the ISSMGE Foundation dates back to September 2009, when Professor Jean-Louis Briaud, the newly-elected President of the International Society for Soil Mechanics and Geotechnical Engineering (ISSMGE) in his acceptance speech to the ISSMGE Council Meeting (Alexandria, Egypt), introduced his vision of a foundation that would enable geotechnical engineers from around the world to make better use of resources that may otherwise be inaccessible to them. These resources would include participation in conferences, attending knowledge development courses, etc., with a view to enhancing the awardees own education and professional formation and, as a result, their contribution to society at large. In short, and as stated in the Foundation constitution, its objective is "the advancement of education and educational training in the field of geotechnical engineering and its applications for the public benefit".

In September 2013, The ISSMGE Foundation achieved charitable status and was entered onto the Register of Charities of the Charity Commission of England and Wales, with the Registered Charity Number 1153795.

STRUCTURE AND MANAGEMENT

The ISSMGE Foundation currently has four trustees: Professor Jean-Louis Briaud, Professor Michael C.R. Davies (Chair), Professor R. Neil Taylor, and Professor Mark Jaksa, all of whom contribute to decisions concerning the granting of awards. In the future and as the Foundation develops, the Trustees may find it necessary to name additional trustees. The new trustees may be recruited from the ISSMGE Membership, from its own (past) beneficiaries, or from other independent institutions, depending on the needs arising and skills and experience required to consolidate the activities of the Foundation.

Professor Taylor is also in charge of the day-to-day management; the ISSMGE Foundation does not have salaried staff members. Administrative support is provided by the ISSMGE.

ACTIVITIES

The purpose of the Foundation is to award grants to applicants who fulfil the requirements as specified below.

PROCEDURE FOR AWARDING GRANTS

The financial assistance is in the form of a bursary which the successful applicant can use to pay registration costs, travel and/or accommodation expenses. Although membership of the ISSMGE is not a prerequisite for receiving support from the Foundation, applicants are strongly encouraged to become members of their national geotechnical societies, and subsequently of the ISSMGE.

The following procedure has been developed to assess applications for ISSMGE Foundation bursaries:

a. Applications can be submitted four times yearly (at 3-month intervals) on a standard form, to the Secretary General of ISSMGE;
b. These applications are sent to the trustees (Professor Jean-Louis Briaud, Professor Michael Davies, Professor Neil Taylor, and Professor Mark Jaksa);

c. The trustees assess the applications of the basis of the following criteria:

i. Demonstration of need;
ii. Quality of the activity;
iii. Cost-sharing;
iv. Potential impact of the activity on the career progress of the individual and of his or her Member Society.

d. Decisions are made and communicated normally within three weeks following the application round.

e. After the successful applicant has completed his/her nominated activity, they provide a written report to the Trustees on the benefits that they and their Society have gained from the award of the bursary.

The criteria and procedure described above were established in February 2014, though may be modified when considered necessary.

To date (May 2019), there have been a total of 187 applicants with 154 awards being approved at a success rate of 82%. Of these, 130 bursaries (in total £115,394) have been paid with a further 24 due when awardees have completed their activities. Some of this funding took place in the early days through the ISSMGE, prior to the Foundation achieving charitable status.

The reports submitted by the awardees are available to read on the ISSMGE website: https://www.issmge.org/issmge-foundation/recipients-reports.

AMERICAN FUND FOR CHARITIES

In April 2014, the ISSMGE Foundation successfully applied to become a member of the America Fund for Charities (AFC - http://www.americanfund.info). The AFC is a US 501(c)(3) non-profit organisation which allows donations from US taxpayers to be tax deductible to the extent allowed by US Law.

FINANCES AND FUND RAISING

Funding for the ISSMGE Foundation is generated from companies and individuals who concur with the objectives of the Foundation. The funding levels are as follows:

- Diamond: $50,000 and above
- Platinum: $25,000 to $49,999
- Gold: $10,000 to $24,999
- Silver: $1000 to $9,999
- Bronze: $0 to $999

The names of the donors are recognised with their corresponding levels on the ISSMGE web site (https://www.issmge.org/issmge-foundation/donors-list) and in the ISSMGE Bulletin on the page dedicated to the ISSMGE Foundation. They are also be recognised at International Conferences on Soil Mechanics and Geotechnical Engineering.

As reported previously at its Council Meeting in Paris in September 2013, the ISSMGE Council agreed that the annual surplus over the ISSMGE's cash balance of £500,000 (allowing for suitable index linking) should be transferred to the Foundation. Accordingly, when the ISSMGE Accounts for 2016 and 2017 were completed and inspected (in May 2017 and June 2018 respectively) donations of £70,000 and £110,000 were approved by the ISSMGE President, the Chair of the Finance and Budget Committee, and the Secretary General.

ISSMGE Foundation cash reserves at the end of 2018 were £537,664 (at the end of 2017 they were £440,043), as per audited accounts, extracts attached. Currently, the Board is considering awards to the value of approximately US $2000 (c. £1,630).

The policy of the ISSMGE Foundation trustees is, in the long-term, to build up the reserves to establish an endowment fund and to make suitable investments to enable the Foundation to make awards from the income earned by this fund. With this in mind, the trustees are in the process of appointing Tilney Ltd to help the Foundation achieve its investment objectives.

IMPACT OF ISSMGE FOUNDATION ACTIVITIES

It is clear that the financial assistance provided by the ISSMGE Foundation has started to have a very positive impact on the individuals receiving the bursaries and consequently on their Member Societies.

As awareness of the awards becomes more widespread among our Member Societies, the number of applications has increased. We expect the number of applications to rise with the approach of the next International Conference, and the associated Young Engineers' Geotechnical Conference (both in Sydney, in September 2021).

The ISSMGE Foundation is fulfilling its objective of providing financial help to geotechnical engineers throughout the world who wish to further their knowledge and practice through activities which they could not otherwise afford.

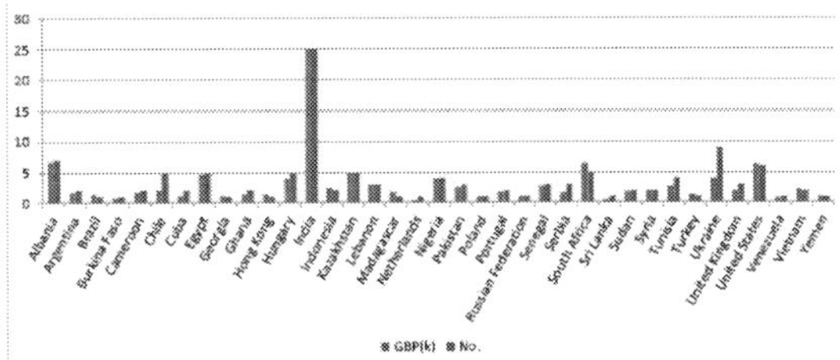

Fig 1: Number and value of awards (GBP k) per country

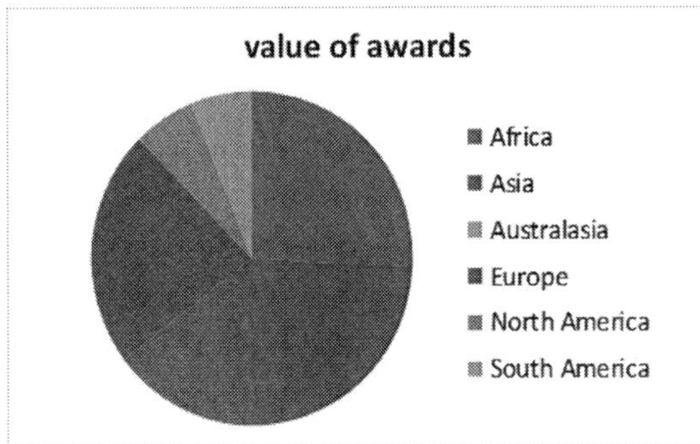

ISSMGE Foundation Accounts
For the year ended 31 December 2017
(As submitted to the Charity Commission)

		Funds
RECEIPTS		
Donations		70,353
Interest		229
	Total	**70,582**
EXPENDITURE		
Awards		36,453
Bank charges		781
	Total	**37,234**
SURPLUS/(DEFICIT)		**33,348**
Cash balance at 31st December 2016		**406,69**
Cash balance at 31 December 2017		**440,04**

Appendix 12: The ISSMGE Bulletin

Anthony Leung
Editor-in-Chief, ISSMGE Bulletin

Dr Anthony Leung (formerly University of Dundee, UK; now HKUST, Hong Kong) was appointed as the Editor-in-Chief (EiC) of the ISSMGE Bulletin since Oct 2017, after he served the Editorial Board (Europe) for more than 3 years. Dr Leung would like to thank Prof. Charles Ng, who undertook the role of EiC from Aug 2014 to Aug 2017. Under his leadership, 19 issues have been published over the last 3 years. I am grateful for his great contribution to the bulletin and dedication to the role. My warmest thanks also go to the retiring members, who served the Editorial Board during the last 3 years, including Profs Antonio Gomez, Andy Take, Alejo Sfriso, Teik Aun Ooi and Marawan Shahien. Since Dr Leung's new appointment, a new Editorial Board has been formed:

Editors for Asia:
- Prof. al-Damluji, Omar
- Prof. Chang, Der-Wen

Editors for Africa:
- Prof. Bouassida, Mounir
- Prof. Heather, Davis

Editors for Europe:
- Prof. Cazzuffi, Daniele
- Prof. Jefferies, Stephan

Editor for Australasia:
- Prof. Rujikiatkamjorn, Cholachat

Editor for North America:
- Prof. Sanchez, Marcelo
- Prof. Siemens, Greg

Editor for South America:
- Prof. Gonzalez, Marcelo

Editor for YMPG (Young Member Presidential Group):
- Prof. Baser, Tugce

Ex-officio:
- Prof. Ng, Charles Wang Wai
- Prof. Taylor, Neil

The bulletin is published bimonthly. A total of 10 issues have been published since October 2017. Each issue of the bulletin is open-accessed and can be downloaded freely from the following website: http://www.issmge.org/publications/issmge-bulletin.

The 10 issues have reported "Research Highlights" for the ISSMGE members to get to know more about the research activities from reputed geotechnical engineering research groups around the world, including University of Alberta (Canada; Vol 11, issue 5), Tsinghua University (China; Vol 11, issue 6), Korea Advanced Institute of Science and Technology (KAIST; South Korea; Vol 12, issue 1), Delft University of Technology (Netherlands; Vol 12, issue 2), Pontificia Universidad Catolica de Chile (Chile) and the Universidad de Chile (Chile; Vol 12, issue 3), University of Texas Arlington (U. S.; Vol 12, issue 4) and University of Southampton (U. K.; Vol 13, issue 2).

Thanks to the close liaison of the editorial board members with conference organisers, the Bulletin also serves as an effective platform for sharing highlights of some major international conferences in "Conference reports". The content includes conference programme, keynote/invited speakers, conference tour, social events and the awardees of ISSMGE Bright Spark Lecture, a new series created by the ISSMGE President for enabling young members to take the main stage at conference.

The Bulletin also has a section called "Young Member's Arena", for allowing young members to share their latest research activities. With the recent appointment of the new editor of YMPG, Prof. Tugce Baser introduced the 17 members of the YMPG, representing different member societies around the world, for the term 2018 – 2019, in Vol 12, issue 1, published in February 2018. During this report period, activities about a student contest on Data Analytics, organised by TC304, including the context question, logistics and awards are reported (Vol 12, issue 5).

Activities organised by TCs of ISSMGE are reported under the section "TC Corner" in the bulletin. This includes the geotechnical engineering competition of college students in China, organised by the TC on Geotechnical Engineering Education (GEE) under Chinese Institution of Soil Mechanics and Geotechnical Engineering (Vol 11, issue 6), CPT data sharing and statistical interpretation jointly organised by TCs 304 and 309 (Vol 12, issue 1) and the 1st ENU-HIT Scientific forum organised by TC305 (Vol 12, issue 4).

The Bulletin also publishes "Board-Level Committee Activity", such as the one reported in Vol(12), issue 3, written by the Corporative Associates Presidential Group (CAPG). This is a very comprehensive report that outlines the contribution of Corporate Associates (CA) to the ISSMGE and also offers various opportunities that can boost the CA to narrow the gaps between academics and practitioners.

The Bulletin also receives a couple of "Reports from Member Society", including two reports from Canadian Geotechnical Society (Vol 12, issue 3 and Vol 12, issue 6) and one from the Kazakhstan Geotechnical Society (Vol 12, issue 2).

In every issue, an updated "Event Diary" is provided closer to the end of the bulletin to advertise all ISSMGE- and non-ISSMGE-sponsored events (mainly conferences and workshops) that are registered in the ISSMGE website. Updated lists of "Corporate Associates" and "Foundation donors" are also given in every issue to acknowledge their financial support to the ISSMGE.

On behalf of the Editorial Board, I cordially invite you to contact your own Member Society and share any interesting events/activities related to soil mechanics and geotechnical engineering in the Bulletin. I am sure our members will be interested in hearing your exciting news over the world. Feel free to get in touch via my email: ceanthony@ust.hk

Appendix 13: Presentation of audited accounts 2017, 2018

Neil Taylor
Secretary General, ISSMGE

INDEPENDENT EXAMINERS' REPORT TO THE COUNCIL MEMBERS OF THE
INTERNATIONAL SOCIETY FOR SOIL MECHANICS AND GEOTECHNICAL ENGINEERING

I have examined the receipts and payments account of the International Society for Soil Mechanics and Geotechnical Engineering (the Society) for the year ended 31 December 2017 and the related information.

This report is made solely to the Council members, as a body, in accordance with our engagement letter. Our examination work has been undertaken so that we might state to the Council members those matters we are required to state to them in an examiners' report and for no other purpose. To the fullest extent permitted by law, we do not accept or assume responsibility to anyone other than Council members as a body, for our work, for this report, or for the opinions we have formed.

Respective responsibilities of Secretary General and Independent Examiner

The Society's Secretary General is responsible for the preparation of the receipts and payments account, which has been prepared under the Society's statutes and By-Laws.

Our responsibility is to examine the receipts and payments account and related information.

We report to you our opinion as to whether the receipts and payments account properly presents the receipts and payments of the Society.

Basis of Independent Examiners Report

We conducted our examination in accordance with the ISSMGE's Statutes and By-Laws An examination includes a review of the accounting records kept by the Society and a comparison of the accounts presented with those records. It also includes consideration of any unusual items or disclosures in the accounts and seeking of explanations from you as Council Members concerning any such matters. The procedures undertaken do not provide all the evidence that would be required in an audit and, consequently no opinion is given as to whether the accounts present a 'true and fair view' and the report is limited to those matters set out in the statement below.

Independent Examiners Statement

In my opinion the accounts:

- properly present the receipts and payments of the Society for the year ended 31 December 2017.

Alison Sargent FCA
Sargent & Co
Chartered Accountants
219 Croydon Road
Caterham
Surrey CR3 6PH

Date: 24 May 2018

INTERNATIONAL SOCIETY FOR SOIL MECHANICS AND GEOTECHNICAL ENGINEERING

RECEIPTS AND PAYMENTS ACCOUNT
FOR THE YEAR ENDED 31 DECEMBER 2017

	Barclays Bank Plc SFr	Barclays Bank Plc £	Investment Accounts £	City University £
RECEIPTS				
Member Society Subscriptions	-	250,407	-	-
Corporate Member Subscriptions	-	40,063	-	-
Donations and Other Income	-	-	-	716
Interest	-	728	683	-
Conference Income	-	33,221	-	-
	-	**324,419**	**683**	**716**
EXPENDITURE				
Secretariat				
President part time support		9,400	-	-
Secretary general support	-	-	-	48,761
Office Expenses - President	-	5,397	-	-
Office Expenses Secretariat	-	3,969	-	829
Secretary General	-	24,800	-	-
Conference support	-	20,000	-	-
Kevin Nash Gold Medal	-	3,087	-	-
Prizes/Awards	-	3,000	-	-
	-	**69,653**	-	**49,590**
Travel, accommodation etc				
President	-	16,003	-	-
Secretary General		6,898	-	-
Board	-	31,490	-	-
	-	**54,391**	-	-
Independent Examiners Fees & Bank Charges	1	3,566	-	-
Corporation Tax	-	451	-	-
	1	**4,017**	-	-
Information Technology (website)				
Licensing Fee	-	10,338	-	-
Future Developments	-	-	-	-
	-	**10,338**	-	-
Other Initiatives				
International Seminars (e.g. travel)	-	-	-	-
Travel of ISSMGE Officers to FedIGS Meetings	-	2,424	-	-
Webinars	-	5,026	-	-
Initiatives	-	11,979	-	-
Contingency	-	-	-	-
ISSMGE Foundation	-	70,000	-	-
		89,429	-	-
Total Expenditure	**1**	**227,828**	-	**49,590**
SURPLUS/(DEFICIT)	**(1)**	**96,591**	**683**	**(48,874)**
Transfers from other accounts	-	(49,000)	-	49,000
Cash balance at 1 January 2016	**106,468**	**383,720**	**145,124**	**194**
Cash balance at 31 December 2017	**106,467**	**431,311**	**145,807**	**320**

Information December 2017 - £1 = SFr 1.31

Secretary General Prof. R N Taylor **Date** 24 **May 2018**

INTERNATIONAL SOCIETY FOR SOIL MECHANICS AND GEOTECHNICAL ENGINEERING

SUBSCRIPTIONS RECEIVED IN THE YEAR ENDED 31 DECEMBER 2017

Member Society	Number of members	Barclays Bank £
Albania	40	393
Algeria*	43	-
Argentina	106	2,002
Australia	979	10,747
Austria	108	2,146
Bangladesh	33	478
Belarus	30	1,009
Belgium	319	5,356
Bosnia & Herzegovinia	19	372
Brazil	889	7,797
Bulgaria	31	2,446
Canada	744	8,891
Chile	101	3,126
China	419	4,915
Chinese Taipei	68	1,339
Colombia	26	431
Costa Rica	31	452
Croatia	108	2,146
CTGA	100	-
Cuba*	30	-
Cyprus	36	417
Czech & Slovak Republics	49	974
Denmark*	345	-
Egypt	103	2,216
Estonia*	21	172
Finland	46	914
France	541	7,282
Georgia	45	518
Germany	540	7,274
Ghana	21	600
Greece	87	1,865
Guatemala	32	326
Hong Kong	129	2,163
Hungary	66	-
Iceland*	32	-
India	251	2,274
Indonesia	140	1,902
Iran*	802	-
Iraq	42	698
Ireland	25	596
Israel	32	1,174
Italy	296	5,334
Japan	1,162	12,213
Kazakhstan	40	780
Korea	714	8,654
Kyrgyzstan*	30	-
Latvia	44	1,620
Lebanon	66	954
Lithuania	40	-

INTERNATIONAL SOCIETY FOR SOIL MECHANICS AND GEOTECHNICAL ENGINEERING
SUBSCRIPTIONS RECEIVED (cont.)
IN THE YEAR ENDED 31 DECEMBER 2017

Member Society	Number of members	Barclays Bank £
Macedonia, FYR	35	1,032
Malaysia	100	1,987
Mexico	265	4,322
Morocco	28	804
Mozambique	37	199
Nepal	22	389
Netherlands	718	8,671
New Zealand	683	8,403
Nigeria	43	885
Norway	435	6,438
Pakistan	85	736
Paraguay	30	763
Poland	371	5,929
Portugal	155	3,080
Romania	156	5,549
Russia	202	4,374
Serbia	37	519
Singapore	53	1,053
Slovenia	77	3,060
South Africa	547	4,903
South East Asia	119	1,091
Spain	368	5,906
Sri Lanka	36	910
Sudan*	37	-
Sweden	1,076	11,535
Switzerland	173	3,435
Syria*	17	-
Tajikistan	23	608
Thailand	32	1,043
Tunisia	40	1,054
Turkey	236	4,713
Ukraine	30	-
UK	1,196	12,486
USA	3,208	28,475
Uzbekistan*	33	558
Venezuela*	32	-
Vietnam	30	531
Total	20,766	250,407

* Membership number is from the last payment received

INTERNATIONAL SOCIETY FOR SOIL MECHANICS AND GEOTECHNICAL ENGINEERING

SUBSCRIPTIONS RECEIVED (cont.) IN THE YEAR ENDED 31 DECEMBER 2017

Corporate Member	US Dollars $	Bank Account £
AECOM Asia Company Ltd	1,500	1,195
Ove Arup & Partners Ltd	1,500	1,182
Aurecon	1,500	1,188
AP van den Berg Ingenieursburo	1,500	1,158
CGRE (UNIC WOLLONGONG)	1,500	1,133
Coffey Geotechnics	-	-
Dar	1,500	1,130
Deltares	1,500	1,176
Fugro GmbH	1,500	1,126
GDS Instruments	3,000	2,203
Geomil	1,500	1,092
GHD Pty Ltd	1,475	1,159
Golder Associates	1,490	1,172
Huesker Synthetic GmbH	1,480	1,157
International I.G.M. S.A.R.L.	1,500	1,109
Jan de Nul NV	1,500	1,162
KellerGround Engineering	1,475	1,160
KGS Ltd	1,500	1,166
LLC Geoizol	-	-
Maccaferri Innovation Centre SRL	1,500	1,094
Naue	1,500	1,182
Norwegian Geotechnical Institute	1,500	1,145
Pagani Geotechnical Equipment	1,460	1,136
RCF Ltd	1,475	1,150
Shanghai Geoharbour Construction Group	3,000	2,199
Siemens Energy	1,500	1,167
SMEC Australia	1,500	1,119
Soletanche Bachy SA	1,480	1,163
S N Apageo S.A.S., France	1,500	1,148
Tencate Geosynthetics France SAS	1,500	1,182
Tensar International Ltd	1,463	1,142
Terrasol	1,485	1,118
Terre Armee Internationale	1,480	1,163
Wagstaff Piling	1,500	1,112
Zetas Zemin Teknolojisi AS	1,500	1,175
	52,263	40,063

INTERNATIONAL SOCIETY FOR SOIL MECHANICS AND GEOTECHNICAL ENGINEERING

NOTES TO THE FINANCIAL STATEMENTS YEAR ENDED 31 DECEMBER 2017

ACCOUNTING POLICIES

(a) Income
Income has been taken to the credit of the revenue account on a cash received basis.

(b) Expenditure
Expenditure, inclusive of VAT, has been charged to the revenue account on a cash paid basis.

RELATED PARTIES

Included in expenditure are certain governance costs incurred by The ISSMGE Foundation which the Society has agreed to pay on behalf of the Foundation to ensure that as much as possible of the income of the Foundation is used for awards. The total amount paid on behalf of the Foundation in 2017 was £1,950 (2016 - £1,831). The payments in 2017 related to the Independent Examination costs for the 2016 accounts of the ISSMGE Foundation, the ISSMGE Foundation Trustees Insurance and the subscription for the American Fund for Charities.

INDEPENDENT EXAMINERS' REPORT TO THE COUNCIL MEMBERS OF THE

INTERNATIONAL SOCIETY FOR SOIL MECHANICS AND GEOTECHNICAL ENGINEERING

I report to the society's Secretary General and society council members on my examination of the receipts and payments accounts of the society for the year ended 31 December 2018 which are set out on pages 2 to 6.

Responsibilities and basis of Report

The society's Secretary General is responsible for the preparation of the receipts and payments account, which has been prepared under the Society's statutes and By-Laws.

My responsibility is to examine the receipts and payments account and related information in accordance with the society's Statutes and By-Laws.

I report in respect of my examination of the society's receipts and payments account.

Independent Examiners Statement

I have completed my examination. I confirm that no material matters have come to my attention in connection with the examination giving me cause to believe that in any material respect:

1. accounting records were not kept in respect of the society as required under the society's Statutes and By-Laws; or
2. the accounts do not accord with those records.

I have no concerns and have come across no other matters in connection with the examination to which attention should be drawn in this report in order to enable a proper understanding of the accounts to be reached.

Alison Sargent FCA
Sargent & Co
Chartered Accountants and Registered Auditors
219 Croydon Road
Caterham
Surrey CR3 6PH

Date: 3 May 2019

INTERNATIONAL SOCIETY FOR SOIL MECHANICS AND GEOTECHNICAL ENGINEERING

RECEIPTS AND PAYMENTS ACCOUNT FOR THE YEAR ENDED 31 DECEMBER 2018

	Barclays Bank Plc SFr	Barclays Bank Plc £	Investment Accounts £	City University £
RECEIPTS				
Member Society Subscriptions	-	229,034	-	-
Corporate Member Subscriptions	-	35,044	-	-
Donations and Other Income	-	-	-	27
Interest	-	905	1,057	-
Conference Income	-	1,731	-	-
	-	266,714	1,057	27
EXPENDITURE				
Secretariat				
President part time support		20,600	-	-
Secretary general support	-	-	-	52,727
Office Expenses - President	-	5,500	-	-
Office Expenses Secretariat	-	1,160	-	-
Secretary General	-	25,300	-	-
Conference support	-	-	-	-
Kevin Nash Gold Medal	-	-	-	-
Prizes/Awards	-	-	-	-
	-	52,560	-	52,727
Travel, accommodation etc				
President	-	16,500	-	-
Secretary General		4,509	-	-
Board	-	40,728	-	-
Board Level Chairs	-	10,594	-	-
	-	72,331	-	-
Independent Examiners Fees & Bank Charges	-	3,718	-	-
Corporation Tax	-	272	-	-
	-	3,990	-	-
Information Technology (website)				
Licensing Fee	-	10,388	-	-
Future Developments	-	-	-	-
	-	10,388	-	-
Other Initiatives				
International Seminars (e.g. travel)	-	1,613	-	-
Travel of ISSMGE Officers to FedIGS Meetings	-	2,856	-	-
Webinars	-	1,081	-	-
Initiatives	-	16,472	-	-
Contingency	-	-	-	-
ISSMGE Foundation	-	110,000	-	-
		132,022	-	-
Total Expenditure		271,291	-	52,727
SURPLUS/(DEFICIT)		(4,577)	1,057	(52,700))
Transfers from other accounts	-	(52,330)	-	52,330
Cash balance at 1 January 2018	106,467	431,311	145,807	320
Cash balance at 31 December 2018	106,467	374,404	146,864	(50)

Information December 2018 - £1 = SFr 1.2547

Secretary General Prof. R N Taylor Date 3 **May 2019**

INTERNATIONAL SOCIETY FOR SOIL MECHANICS AND GEOTECHNICAL ENGINEERING

SUBSCRIPTIONS RECEIVED IN THE YEAR ENDED 31 DECEMBER 2018

Member Society	Number of members	Barclays Bank £
Albania	40	594
Algeria*	43	-
Argentina	106	1,831
Australia	998	10,389
Austria	108	2,043
Bangladesh*	33	-
Belarus	30	474
Belgium	319	5,251
Bosnia & Herzegovina	18	380
Brazil	845	6,963
Bulgaria	31	534
Canada	784	8,773
Chile	101	1,911
China	375	4,487
Chinese Taipei	68	1,275
Colombia	21	410
Costa Rica	31	450
Croatia	108	2,043
CTGA	100	893
Cuba*	30	-
Cyprus	30	568
Czech & Slovak Republics	45	845
Denmark	345	11,171
Egypt*	103	-
Estonia	30	568
Finland	46	870
France	595	6,930
Georgia*	45	-
Germany	514	6,722
Ghana	21	-
Greece	87	-
Guatemala*	32	-
Hong Kong	121	2,473
Hungary	70	2,636
Iceland*	32	1,241
India	193	2,116
Indonesia	140	1,673
Iran*	802	-
Iraq	144	2,260
Ireland	25	568
Israel*	32	-
Italy	285	4,996
Japan	1,162	11,448
Kazakhstan	40	742
Korea	585	7,266
Kyrgyzstan*	39	-
Latvia*	44	-
Lebanon	66	910
Lithuania	40	1,552

INTERNATIONAL SOCIETY FOR SOIL MECHANICS AND GEOTECHNICAL ENGINEERING

SUBSCRIPTIONS RECEIVED (cont.)
IN THE YEAR ENDED 31 DECEMBER 2018

Member Society	Number of members	Barclays Bank £
Macedonia, FYR	35	491
Malaysia	69	1,293
Mexico	260	-
Morocco	28	200
Mozambique	37	188
Nepal*	22	-
Netherlands	723	8,324
New Zealand	704	8,166
Nigeria	40	-
Norway	466	6,366
Pakistan*	85	-
Paraguay	38	324
Peru	27	382
Philippines	34	183
Poland	378	5,700
Portugal	156	2,952
Romania*	156	-
Russia	219	3,885
Serbia	37	505
Singapore	65	1,230
Slovenia*	77	-
South Africa	559	5,091
South East Asia	124	1,193
Spain	375	5,676
Sri Lanka*	36	-
Sudan*	37	-
Sweden	1,196	11,888
Switzerland	170	3,214
Syria*	17	-
Tajikistan*	23	-
Thailand	32	482
Tunisia*	40	-
Turkey	272	4,896
Ukraine	30	993
UK	1,187	11,823
USA	3,184	26,925
Uzbekistan*	33	-
Venezuela*	32	-
Vietnam	47	408
Total	20,952	229,034

* Membership number is from the last payment received

INTERNATIONAL SOCIETY FOR SOIL MECHANICS AND GEOTECHNICAL ENGINEERING

SUBSCRIPTIONS RECEIVED (cont.)
IN THE YEAR ENDED 31 DECEMBER 2018

Corporate Member	US Dollars $	Bank Account £
AECOM Asia Company Ltd	1,500	1,041
Ove Arup & Partners Ltd	1,500	1,064
AOSA SA	1,500	1,017
Aurecon	1,500	1,045
AP van den Berg Ingenieursburo	1,500	1,046
CGRE (UNIV WOLLONGONG)	1,500	1,121
Coffey Geotechnics	1,500	1,029
Dar Al Handasah	-	1,036
Deltares	-	1,044
Fugro GmbH	1,500	1,031
GDS Instruments	-	-
Geomil	-	-
GHD Pty Ltd	1,475	1,025
Golder Associates	-	1,111
Huesker Synthetic GmbH	1,500	1,102
International I.G.M. S.A.R.L.	1,500	1,013
Jan de Nul NV	-	1,045
KellerGround Engineering	-	1,095
KGS Ltd	-	1,148
Maccaferri Innovation Centre SRL	1,500	1,010
Naue	-	1,056
Norwegian Geotechnical Institute	1,500	1,053
Pagani Geotechnical Equipment	1,500	1,061
RCF Ltd	1,500	1,096
Shanghai Geoharbour Construction Group	-	-
Siemens Energy	1,500	1,038
SMEC Australia	1,500	1,020
Soletanche Bachy SA	1,500	1,025
S N Apageo S.A.S., France	-	1,119
SRK Consulting	1,500	1,061
Tencate Geosynthetics France SAS	-	1,121
Tensar International Ltd	1,500	1,056
Terrasol	1,500	1,023
Terre Armee Internationale	-	1,143
Wagstaff Piling	1,475	1,042
Zetas Zemin Teknolojisi AS	1,500	1,107
	34,450	35,044

INTERNATIONAL SOCIETY FOR SOIL MECHANICS AND GEOTECHNICAL ENGINEERING

NOTES TO THE FINANCIAL STATEMENTS
YEAR ENDED 31 DECEMBER 2018

ACCOUNTING POLICIES

(a) Income
Income has been taken to the credit of the revenue account on a cash received basis.

(b) Expenditure
Expenditure, inclusive of VAT, has been charged to the revenue account on a cash paid basis.

RELATED PARTIES

Included in expenditure are certain governance costs incurred by The ISSMGE Foundation which the Society has agreed to pay on behalf of the Foundation to ensure that as much as possible of the income of the Foundation is used for awards. The total amount paid on behalf of the Foundation in 2018 was £1,964 (2017 - £1,950). The payments in 2018 related to the Independent Examination costs for the 2017 accounts of the ISSMGE Foundation, the ISSMGE Foundation Trustees Insurance and the subscription for the American Fund for Charities.

Appendix 14: ISSMGE Budget – 2019–2023

Gavin Alexander, Phil Robins
ISSMGE Vice-President Australasia

1 GENERAL

The ISSMGE budget for the period 2019–2023 is presented in the attached table and is based on the 2017–2021 budget approved at the Seoul Council Meeting in September 2017. The actual amounts for 2017 and 2018 from the audited accounts are also presented to inform the budget forecasts for 2019–2023. To maintain consistency with previous budgets and with the independent auditor's reports, UK pounds (£) are used.

As with previous budgets, the overriding rationale for the budget is to permit the Society to function, both effectively and professionally. Therefore, the budget should facilitate the strategies developed by the President and the Board, together with the operational activities proposed by the board level and presidential committees.

2 RECEIPTS (INCOME)

2.1 *Member Society Annual Subscriptions (MSASs):*

Since 2007 the fee structure has been based on a Basic Fee per Capita, discounted to allow for low Purchasing Power Parity (PPP) and for large Member Societies. At the 2015 Edinburgh Council meeting annual subscriptions were budgeted to increase modestly to keep pace with a predicted inflation of 2%. However, the Council voted not to increase MSASs. As a result, the MSASs in the 2019–2023 period are also budgeted to remain static. At the Seoul Council Meeting, the forecast amounts in 2019 and 2021 were set to reflect the likelihood that a greater proportion of fees are received in years in which a Council meeting is held. This variance was perhaps on the high side and has now been adjusted to reflect a consistent annual income.

2.2 *Corporate Associate Subscriptions:*

In 2011 the fee for corporate membership of the Society was set at US$1,500. It is assumed that this fee level will not increase. There has been an increase in CAs for which the CAPG is thanked. The forecast subscription income has been adjusted accordingly.

2.3 *Interest:*

The predicted revenue from interest remains modest.

2.4 *Conference Income:*

This item accounts for revenues from registration fees from conferences organised under the auspices of the ISSMGE. As a consequence of a Board decision in 2012, the ISSMGE mandates that the Member Society organising the International Conference (ICSMGE) transfer 5% of the registration fee income to the ISSMGE and 3% of the Regional Conference registration fee income. Conference income is, therefore, only expected in the years when the International and Regional Conferences take place.

3 EXPENDITURE

3.1 *Secretariat*

3.1.1 *Secretary General*

Support for the Secretary General is based on the May 2010 and April 2018 proposals from the British Geotechnical Association to host the secretariat. A modest increase is forecast.

3.1.2 *President*

Administrative support and associated expenses are allocated to the President. Please note that this budget does not include the costs of additional support that might be provided to the President by their host Member Society or from other sources.

3.1.3 *Conference Support*

Includes support to Regional Young Geotechnical engineering conferences. Additional funding to support the 7th International Young Geotechnical Engineers' Conferences is included in 2021.

3.1.4 *List of members*

The compilation of an annual list of members has proved to be problematic and it has been decided that the list will no longer be produced. In the proposed budget, this entry has been removed.

3.1.5 *Kevin Nash Gold Medal and Prizes*

Provision is made for the award of the Society's Gold Medal and Prizes at the 20th International Conference for Soil Mechanics and Geotechnical Engineering in 2021.

3.2 *Travel expenses*

Provision is made for a modest annual increase in the normal budget for Board travel to account for inflation. At the Board meeting held in Beirut in March 2017, it was agreed that an additional allowance be allocated to the Chairs of Board level committees (i.e. AC, CAPG, TOC, IDC, PIC, YMPG) to assist them to defray some of their travel expenses. An amount of £18,000 p.a. was allocated for this purpose.

The Board agreed, at the meeting in Singapore, in March 2019, to provide a 50% travel subsidy for the Editor of the Bulletin to attend Council meetings to report on that publication. The Board also agreed to contribute to the cost of biennial face-to-face meetings of key representatives of the YMPG, to be held in conjunction with Board meetings to facilitate interaction between the two groups. Allowance for these items has been included from 2019.

3.3 *Information Technology (website)*

3.3.1 Licensing fee/web maintenance:
In 2012 hosting of the website was transferred to Geoengineer.org from Webforum. Provision for the maintenance fee payable to Geoengineer.org is included in the budget.

3.3.2 *Future Developments*
Provision for upgrades to the website (or other IT developments) is made to fund initiatives proposed by the President and Board.

It was agreed at the March 2019 Singapore Board Meeting, to fund the initial development of a structured education platform to host the ISSMGE Virtual University. Open edX is proposed to be used, with the initial budget derived from sums for Future (IT) developments and Other initiatives. This may lead to a shortfall for other initiatives in the short term, which will be covered from contingency.

A Conference Paper Review Platform has been developed by the IDC and is currently being used for several events. Allowance has been included for nominal support of the present system Future, as yet uncommitted, development of this platform to include compilation and publication of proceedings will need to be funded by others.

3.4 *Other Activities*

3.4.1 *International Seminars*
An amount is provided to contribute to the travel for ISSMGE Board members and speakers participating in these outreach activities.

3.4.2 *FedIGS Meeting Travel*
Provision is made to support travel expenses for the President (or Immediate Past President) and Secretary General to attend meetings of the FedIGS Board.

3.4.3 *Initiatives*
Provision is made for unspecified initiatives that may be developed by the Board. This has included, for example, a modest financial contribution in 2019 towards the cost of migrating a physical exhibition on geotechnical engineering held in Paris from 2013 to 2018 to a virtual exhibition.

3.4.4 Webinars:
An annual fee for the WISTIA video conferencing service and webinar preparation is included. The previous Board decided that these costs should be absorbed by the Society rather than passed on to members. Therefore the budget does not contain a credit in respect of fees for members to access this service.

3.4.5 Contingency:
Provision is made for a modest contingency to account for unforeseen circumstances.

4 TRANSFER TO ISSMGE FOUNDATION AND CASH BALANCE

Consistent with the practice of the previous Board, it is considered prudent to maintain the Society's reserves (i.e. Cash Balance) at £500,000, as at 2013 and adjusted annually for inflation. Consistent with previous budgets and in order to maximise benefit to ISSMGE members, annual surpluses in excess of the adjusted £500,000 are proposed to be transferred to the ISSMGE Foundation. The amount transferred to the Foundation in any given year is contingent on the financial outcome of the Society in that year. On this basis, a sum of £110,000 was transferred to the Foundation in 2018. Further transfers are expected in 2019 and 2020, but surpluses are not expected or allowed for in subsequent years. The arrangements will need to be considered by the Board and the Foundation in 2020 and any revised proposal put to the 2021 Council meeting.

ISSMGE BUDGET 2017–2019 APPROVED AT SEOUL COUNCIL WITH ACTUALS

	2017		2018		2019	2020	2021
	Budget £	Actual £	Budget £	Actual £	Budget £	Budget £	Budget £
RECEIPTS (INCOME)							
Member Society Subscriptions	240,000	250,407	200,000	229,034	240,000	200,000	240,00
Corporate Associate Subscriptions	28,000	40,063	30,000	35,044	32,000	30,000	32,000
Interest	2,500	1,411	2,200	1,962	2,500	2,200	2,200
Conference Income	20,000	33,221	20,000	1,731	10,000	10,000	10,000
Donations and other income		716					
TOTAL INCOME	290,500	325,818	252,200	267,771	284,500	242,200	284,200
EXPENDITURE							
Secretariat							
President part time support	20,200	9,400	20,600	20,600	21,000	21,000	21,00
Secretary General support	43,800	48,761	48,000	52,727	49,000	49,000	49,000
Office expenses - President	5,400	5,397	5,500	5,500	5,600	5,700	5,800
Office expenses - Secretariat	3,500	4,798	3,000	1,160	3,000	3,200	3,200
Secretary General	24,800	24,800	25,300	25,300	25,800	25,800	25,800
Conference support	20,000	20,000	6,000	-	6,00	6,000	20,000
List of Members	-	-	-	-	-	-	-
Kevin Nash Gold Medal	3,500	3,087	-	-	-	-	4,000
Prizes	5,000	3,000	-	-	-	-	5,000
Sub total	126,200	119,243	108,400	105,287	110,400	110,700	133,800
Travel Expenses							
President	16,000	16,003	16,500	16,500	17,000	17,000	17,000
Secretary General	12,000	6,898	12,500	4,509	13,000	13,000	13,000
Board members	40,000	31,490	40,000	40,728	40,000	42,000	42,000
Board Level Chairs			18,000	10,594	18,000	18,000	18,000
Other (YMPG, Bulletin Editor)					12,000		12,000
Sub total	68,000	54,391	87,000	72,331	100,000	90,000	102,000
Audit Fees & Bank Charges	3,000	3,566	3,000	3,718	3,000	3,000	3,000
Corporation Tax	1,000	451	1,000	272	1,000	1,000	1,000
Information Technology (web site)							
Licensing fee/web maintenance	13,000	10,338	9,000	10,388	9,000	10,000	10,000
Future developments	5,000	-	5,000	-	5,000	5,000	5,000
Sub total	22,000	14,355	18,000	14,378	18,000	19,000	19,000
Other Activities							
International Seminars (e.g. travel)	5,000	-	-	1,613	-	-	-
FedIGS meetings travel	2,500	2,424	3,000	2,856	3,000	3,000	3,000
Initiatives	5,000	5,026	8,000	16,472	8,000	8,000	8,000
Webinars	6,000	11,979	7,000	1,081	7,000	7,000	7,000
Contingency	10,000	-	3,000	-	5,000	3,000	5,000
Sub total	28,500	19,429	21,000	22,022	23,000	21,000	23,000
TOTAL EXPENDITURE	244,700	207,418	234,400	214,018	251,400	240,700	277,800
SURPLUS/(DEFICIT)	**45,800**	**118,400**	**17,800**	**53,753**	**33,100**	**1,500**	**6,400**
TRANSFER TO ISSMGE FOUNDATION		70,000		110,000	41,000	28,000	-
CASH BALANCE							
Cash balance at 1 January		613,738		658,711	606,073	598,173	571,673
Cash balance at 31 December (incl. CPI inc.)		658,711		606,073	598,173	571,673	578,073

ISSMGE PROPOSED BUDGET 2019–2021 WITH FORECASTS TO 2023

	2019	2020	2021	2022	2023
	Budget	**Budget**	**Budget**	**Budget**	**Budget**
	£	**£**	**£**	**£**	**£**
RECEIPTS (INCOME)					
Member Society Subscriptions	240,000	230,000	230,000	230,000	230,000
Corporate Associate Subscriptions	32,000	35,000	35,000	35,000	35,000
Interest	2,500	2,000	2,000	2,000	2,000
Conference Income	10,000	-	20,000	-	10,000
Donations and other income	-	-	-	-	-
TOTAL INCOME	284,500	267,000	287,000	267,000	277,000
EXPENDITURE					
Secretariat					
President part time support	21,000	21,000	21,000	21,000	21,000
Secretary General support	54,000	55,000	56,000	57,000	58,000
Office expenses - President	5,600	5,700	5,800	5,800	5,800
Office expenses - Secretariat	3,000	3,000	5,000	3,000	3,000
Secretary General	25,800	26,300	26,800	27,300	27,800
Conference support	6,000	6,000	20,000	6,000	6,000
Kevin Nash Gold Medal	-	-	4,000	-	-
Prizes	-	-	5,000	-	-
Sub total	115,400	117,000	143,600	120,100	121,600
Travel Expenses					
President	17,000	17,000	17,000	17,000	17,000
Secretary General	13,000	13,000	13,000	13,000	13,000
Board members	40,000	42,000	44,000	44,000	44,000
Board Level Chairs	18,000	18,000	18,000	18,000	18,000
Other (YMPG, Bulletin Editor)	12,000		12,000		12,000
Sub total	100,000	90,000	104,000	92,000	104,000
Audit Fees & Bank Charges	3,000	4,000	4,000	4,000	4,000
Corporation Tax	1,000	1,000	1,000	1,000	1,000
Information Technology (web site)					
Licensing fee/web maintenance	13,500	14,000	15,000	15,000	15,000
Future developments	3,000	5,000	5,000	5,000	5,000
Sub total	20,500	24,000	25,000	25,000	25,000
Other Activities					
International Seminars (e.g. travel)	-	2,000	2,000	2,000	2,000
Travel of ISSMGE officers to FedIGS meetings	3,000	3,000	3,000	3,000	3,000
Initiatives	8,000	8,000	8,000	8,000	8,000
Webinars	6,000	7,000	7,000	7,000	7,000
Contingency	3,000	3,000	3,000	5,000	5,000
Sub total	20,000	23,000	23,000	25,000	25,000
DONATION TO ISSMGE FOUNDATION		-	40,000		
TOTAL EXPENDITURE	255,900	254,000	335,600	262,100	275,600
SURPLUS/(DEFICIT)	**28,600**	**13,000**	- 48,600	**4,900**	**1,400**
TRANSFER TO ISSMGE FOUNDATION	41,000	-	-	-	-
CASH BALANCE					
Cash balance at 1 January	606,073	593,673	606,673	558,073	562,973
Cash balance at 31 December	593,673	606,673	558,073	562,973	564,373

 ISBN 978-0-9946261-7-2

International Society for Soil Mechanics and Geotechnical Engineering

Minutes of the Council Meeting

Held at International Convention Centre Sydney, Australia, in hybrid mode

Sunday, 1st May 2022

PRESENT

Charles Ng	- ISSMGE President
Roger Frank+	- Immediate Past President ISSMGE
Eun Chul Shin	- ISSMGE Vice President Asia
Philip Robins	- ISSMGE Vice President Australasia
Mario Manassero	- ISSMGE Vice President Europe
Timothy Newson	- ISSMGE Vice-President North America
Alejo Sfriso	- ISSMGE Vice President South America
Mounir Bouassida +	- Appointed Board Member
Kok Kwang Phoon +	- Appointed Board Member
Pedro Sêco e Pinto +	- Appointed Board Member
R Neil Taylor	- Secretary General
C Paloma Peers +	- ISSMGE Secretariat
Peter Day	- Chair, Corporate Associates Presidential Group
Sukumar Pathmanandavel	- Vice-Chair, Corporate Associates Presidential Group
Lucy Wu+	- Chair, Young Member Presidential Group
Pierre Delage	- Chair, Technical Oversight Committee
Professor William Van Impe +	- ISSMGE Past President
Professor Jean-Louis Briaud	- ISSMGE, Past President
Professor Matthew Coop +	- Chair, TC Laboratory Testing
Dr Jason DeJong	- Chair, TC In-situ Testing
Professor Akira Murakami +	- Chair, TC Numerical Methods
Professor Ahmed Chraibi +	- Chair, TC Tropical Residual Soils
Professor Erol Tutumluer	- Chair, TC Transportation Geotechnics
Professor Philip Watson	- Chair, TC Offshore Geotechnics
Professor Alessandro Mandolini +	- Chair, TC Deep Foundations
Professor Marina Pantazidou +	- Chair, TC Geo-Engineering Education
Professor Zhongquiang Liu +	- Chair, TC Machine Learning
Dr Roger Estephan	- Dar Al Handasah Corp
Dr Ang Koh An +	- GDS Instruments
Mr Karel Allaert +	- Jan de Nul NV
Dr Yuli Doulala- Rigby	- Tensar International Ltd
Dr Marc Ballouz	- ISSMGE Presidential Candidate
Professor Malek Bouazza	- ISSMGE Presidential Candidate
Professor Michael Davies	- ISSMGE Presidential Candidate

Member Society	**Voting Member**	**Non-Voting Member**
Albania	--	
Algeria	Professor Bahar Ramdane +	Professor Abdelmalek Bekkouche +
Argentina	Diego Manzanal +	
Australia	Dr David Lacey	Mr Graham Scholey
Austria	Professor Helmut F. Schweiger	Professor Dietmar Adam +
Bangladesh	Dr Sarwad Yasin +	
Belarus	Ms Tatiana Tronda	
Belgium	--	
Bolivia	Mario Terceros	
Bosnia & Herzegovina	Sabid Zekan +	
Brazil	Professor Fernando Schnaid +	

Member Society	Voting Member	Non-Voting Member
Bulgaria	---	
Canada	Dr Ian Moore +	
Chile	Ms Daniela Pollak	
China	Professor Jianhong Zang +	
Chinese Taipei	Professor Tai-Tien Wang +	Professor Benson Hsiung +
Colombia	Dr Miguel Cabrera +	
Costa Rica	Adrián Fernandez Castro +	
Croatia	Igor Sokolić	
CTGA		Dr Billang Nyogog Serge Aurelien+
Cuba	---	
Cyprus	Dimitrios Loukidis +	
Czech & Slovak Republics	Dr Jana Frankovska +	
Denmark	Dr Varvara Zania	
Egypt	Dr Marawan Shahin +	
Estonia	Lehar Leetsaar +	
Finland	Assoc. Prof. Wojtek Sołowski +	
France	Nicolas Utter +	
Georgia	--	
Germany	Dr Ing. Wolfgang Sondermann +	
Ghana	Dr Frederick Owusu-Nimo +	
Greece	Dr Michael Bardanis +	
Guatemala	José Alberto Perez Zarco +	
Hong Kong	Dr Andy Leung +	
Hungary	Dr Andras Mahler +	Dr. Edina Koch +
Iceland	Mr Haraldur Sigursteinsson +	
India	Prof Narendra Kumar Samadhiya +	
Indonesia	Prof. Widjoyo Prakoso +	
Iran	Prof Fardin Jafarzadeh	
Iraq	--	
Ireland	Dr Carl Bragan +	
Israel	Professor Mario Manassero*	
Italy	Dr Daniele Cazzuffi	
Japan	Professor Toshifumi Mukunoki +	Dr Takeshi Katsumi
Kazakhstan	Professor Askar Zhussupbekov	Dr Assel Sarsaembayeva
Korea R	Dr Moonkyung Chung	Professor Jong-Sub Lee
Kyrgyzstan	Gulzat Kadyralieva	
Latvia	Kaspars Bondars +	
Lebanon	Salah Sadek +	Dr Grace Abou-Jaoude
Lithuania	---	
Malaysia	Dr Dominic Ong +	
Mexico	Walter Paniagua	
Mongolia	Nyamdorj Setev +	
Morocco	---	
Mozambique	Hans Dete +	
Nepal	--	
Netherlands	--	
New Zealand	Mr Philip Robins *	
Nigeria	Mr Will Ibim Beresibo +	
North Macedonia	Assoc. Prof Igor Peshevski +	
Norway	--	
Pakistan	Sohail Kibria	
Paraguay	Nicolas Zabrodiec	Roberto Andrada +
Peru	--	
Philippines	Dr Mark Albert Zarco	Engr. Roy Anthony Luna
Poland	Prof. Katarzyna Zabielska- Adamska+	
Portugal	Professor Alexandre Pinto +	Dr Isabel Fernandes +
Romania	Professor Loretta Batali +	
Russia	Professor Vyacheslav Ilyichev +	Elena Bragar
Serbia	--	
Singapore	Professor Colin Chun Fai Leung	Assoc Prof Darren Siau Chen Chian
Slovenia	Ms Suzana Svetlicic +	
South Africa	Dr John Pavlakis	Dr André Archer
South East Asia GS (SEAGS)	Dr. Suttisak Soralump +	Dr. Geoff Chao +
Spain	Professor Fernando Pardo de Santayana +	
Sri Lanka	Professor Athula Kulathilaka	

Member Society	Voting Member	Non-Voting Member
Sudan	---	
Sweden	Mr Håkan Garin	Dr Gunilla Franzén
Switzerland	Professor Lyesse Laloui +	
Syria	Dr Talal Awwad +	
Tajikistan	Professor Pulod Aminzoda +	
Thailand	Dr Suttisak Soralump +	
Tunisia	---	
Turkey	Professor S. Feyza Cinicioglu +	
Ukraine		Professor Yuriy Kirichek +
U K	Professor David Toll +	Dr Andrew Ridley +
U S A	Professor James Hanson	Professor Anand Puppala
Uzbekistan	Professor Askar Khasanov	Dr Zokhir Khasanov
Venezuela	Professor Wagdi Naime +	
Vietnam	Professor Duc Long Phung +	

+ Denotes Virtual attendance
* Denotes Proxy vote

Apologies:

Etienne-Marcelin Kana	- ISSMGE Vice President Africa
Dimitrios Zekkos	- Chair, Innovations and Development Committee
Ikuo Towhata	- Chair, Professional Image Committee
Roberto Terzariol	- Chair, Awards Committee

1 OPENING REMARKS BY PRESIDENT AND AGREEMENT OF THE AGENDA

The President welcomed delegates to the meeting and explained that the Council Meeting would be very different to the normal format and procedure and would be held as a hybrid event with in-person and online delegates. Similarly, the global pandemic had led to the 20ICSMGE being postponed by over 7 months and would also be held as a hybrid event. With some delegates attending in-person but many more from around the world, the start time of 20:30 AEST was chosen and the meeting would need to finish in about 3.5 hours. The agenda includes three items requiring anonymous voting. An online voting system (ElectionBuddy.com) was chosen and to minimise the risk of poor internet connections, voting opened 48 hours before the Council Meeting started. Voting would close at midnight, i.e. the scheduled end of this meeting and results declared as soon as possible thereafter.

2 LIST OF DELEGATES AND OTHER PERSONS PRESENT - APOLOGIES FOR ABSENCE

The Secretary General conducted a roll call to confirm that the voting delegates above were present either in-person or online.

3 CONFIRMATION OF QUORUM (STAUTE 12K

The Secretary General confirmed that with 71 votes available from 85 Member Societies entitled to the benefits of membership, the meeting was quorate for all business declared on the agenda.

4 CONFIRMATION OF MINUTES OF CAPE TOWN COUNCIL MEETING

The minutes of the Council Meeting in Cape Town, October 2019 were confirmed as an accurate representation of the meeting.

5 MEMBERSHIP

The Secretary General delivered his report, which is included as Appendix 1. It was noted that there had been some changes in membership of individual Member Societies and there were now 20,800 members from 90 Member Societies. Of these societies, 5 were in financial arrears. In addition, there were now 41 Corporate Associates.

6 ELECTION OF REGIONAL VICE-PRESIDENTS

It was noted that the elected Vice-Presidents for the period 2022 to 2026 were:

- Africa – Professor Marawan Shahien (Egypt)
- Asia – Professor Keh-Jian Shou (Chinese Taipei)
- Australasia – Mr Graham Scholey (Australia)
- Europe – Professor Lyesse Laloui (Switzerland)
- North America – Professor Walter Paniagua (Mexico)
- South America – Professor André Assis (Brazil)

Section B – Items requiring a vote

7 ELECTION OF PRESIDENT FOR THE PERIOD 2022-2026

The President reiterated that the votes needed to be submitted using the electronic system that had been circulated to all nominated voting delegates 48 hours before the Council Meeting. There were five candidates for President and each would be given the opportunity to make a 5-minute pitch. The candidates were:

Marc Ballouz
Mounir Bouassida
Malek Bouazza
Michael Davies
Askar Zhussupbekov

The online voting system closed at midnight and the following results declared:

First round:	**Votes**
Marc Ballouz	19
Mounir Bouassida	7
Malek Bouazza	12
Michael Davies	21
Askar Zhussupbekov	21

Mounir Bouassida, having received the fewest votes was eliminated and his ballots reallocated using the single transferable vote system.

Second round:	**Votes**
Marc Ballouz	22
Malek Bouazza	13
Michael Davies	21
Askar Zhussupbekov	24

Malek Bouazza, having received the fewest votes was eliminated and his ballots reallocated.

Third round:	**Votes**
Marc Ballouz	27
Michael Davies	27
Askar Zhussupbekov	25

Askar Zhussupbekov, having received the fewest votes was eliminated and his ballots reallocated.

Fourth round:	**Votes**
Marc Ballouz	40
Michael Davies	39

Marc Ballouz, having received a majority of the votes cast, was duly elected President of ISSMGE for the period 2022 – 2026.

8 VENUE FOR 2023/24 BOARD AND COUNCIL MEETINGS.

The Secretary General reported that in accordance with the statutes, he had written to organisers of regional conferences in 2023/4 in Asia, Australasia and South America. Only one offer to host the Council Meeting had been received from Asia as follows:

- 17th Asian Regional Geotechnical Engineering Conference, Nur-Sultan, Kazakhstan. 14-18 August 2023

Professor Zhussupbekov gave a brief presentation about the conference and the venue. Member Societies were asked to vote on accepting this offer to host the next Council Meeting. The results of the anonymous online ballot were as follows:

For	48 votes
Against	7 votes
Abstentions	25 votes

With the majority vote, it was confirmed that the next Council Meeting will be held on Sunday 13 August 2023 on the occasion of the 17th Asian Regional Geotechnical Engineering Conference, Nur-Sultan, Kazakhstan.

9 PRESENTATIONS OF INVITATIONS TO HOST XXI ICSMGE 2026.

Two proposals for hosting the next international conference had been received from the geotechnical societies in Austria and USA. Representatives from these societies were invited to make a short presentation to Council giving details of their bids. The result of the anonymous online vote was as follows:

Austria	60 votes
USA	20 votes

The 21ICSMGE will therefore be held in Vienna, Austria in June/July 2026.

10 PRESENTATION OF ACCOUNTS 2019 AND 2020 AND DRAFT ACCOUNTS FOR 2021

The inspected accounts for 2019 and 2020 are given in Appendix 2, along with a financial statement for 2021 that compares the draft accounts with the budget for 2021. In general terms, income was a little greater than forecast in the budget. Travel expenditure in 2020 and 2021 was much lower than predicted due to the pandemic. Initiatives was largely focussed on improving functionality of the virtual university which now includes 143 videos. The conference income for the 20ICSMGE had been budgeted for 2021 but postponement of the event means that the income would be received in 2022.

Delegates were asked if there were any comment or questions on the accounts; there were none. Philip Robins proposed the motion to accept the accounts, which was seconded by Alejo Sfriso. Voting by a show of hands indicated 57 in favour with none against and no abstentions, confirming acceptance of the accounts.

11. BUDGET 2022 – 2024 AND FORECAST BUDGET TO 2026

Philip Robins, Chair of the Finance Sub-Committee, presented the budget proposals for 2022 – 2024 with forecast budget to 2026 (see Appendix 3).

The budget income for 2022 to 2026 remains consistent with previous years and again, it is proposed that there is no increase in the basic annual subscription fee per member. Interest rates are forecast to remain low. The main Regional and International Conferences should generate income with a return to significant in-person participation.

The present Secretary General comes to the end of his term of office at the next Council Meeting, and he has advised the ISSMGE Board that he will not seek another appointment. Consequently, the proposed budget includes provision for anticipated costs involved in the search, hand-over and embedding of the new Secretary General.

Budget expenditure on travel allows for small inflationary increases plus an allowance for the Young Members Presidential Group to hold a biennial meeting in-person. Expenditure on IT and related activities is anticipated to remain at current levels tough an allowance has been made to support the Internal Journal on Geoengineering Case Histories.

At the Council Meeting in Paris, 2013, it was proposed to maintain the ISSMGE reserves at about GBP500,000 with any surplus transferred to the ISSMGE Foundation. The new budget proposes to return to that policy.

In discussion, David Toll (UK) queried the large travel expenditure and wondered if there would be moves to reduce the carbon footprint of Board meetings by holding more of these online. The Secretary General replied that the Board had held all its meetings online during the pandemic. While these had been adequate, he felt that in-person meetings were generally more successful. So, although online meetings would no doubt continue, allowing for some in-person meetings was, in his opinion, a good idea.

There were no further questions and the motion to accept the budget was proposed by Walter Paniagua (Mexico) and seconded by Mario Manassero. Voting by a show of hands indicated 63 in favour with none against and 4 abstentions, confirming acceptance of the budget.

Section C: Activities and Achievements

12 PRESENTATION BY THE PRESIDENT SUMMARISING ACTIVITIES AND ACHIEVEMENTS

The President summarised his report (Appendix 4), which gave an overview of his presidential term in office. The key theme of this period was to improve the ISSMGE commitment to education, innovation and diversity. The online ISSMGE Virtual University had been created with the collation of existing and new webinars into a number of Postgraduate-level teaching courses. A conference paper review platform had been created as a benefit to conference organisers and as a means to easily transfer papers to the ISSMGE online library. The increased activity in big data and machine learning had led to new Technical Committees being created. Engagement with Member Societies had increased and the Corporate Associates Presidential Group had organised special sessions at the 2019 Regional Conferences that were aimed at narrowing the gap between academics and practitioners. The creation of Bright-Spark lectures has promoted younger members to deliver keynotes in plenary sessions in conferences and symposia.

The President then continued with summary presentations from other Board members and Chairs of the Board level Committees.

12.1 Regional Reports By Vice-Presidents Etienne-Marcelin Kana, Eun Chul Shin, Philip Robins, Mario Manassero, Timothy Newson, Alejo Sfriso On Regional Activities (See Appendix 5)

Asia (Eun Chul Shin): All 9 Asian Regional Technical Committees were very active and had organised several events. Other major events had been the Asian Regional Conference in Taipei and the 9th Asian Young Geotechnical Engineers Conference in Lahore.

Australasia (Philip Robins): All past proceedings of the Australia New Zealand Geotechnical Conferences had been digitised and uploaded to the ISSMGE online library. Successful and well-attended events had been organised and the Young Geotechnical Professional events continued to be popular.

Europe (Mario Manassero): With 38 Member Societies, the region continues to be very active with many events organised and with significant contribution to Technical Committee activity and output.

North America: (Timothy Newson): Despite the pandemic, many successful events had been organised and there continued to be important contributions by members of Technical Committees.

South America (Alejo Sfriso): Over 300 events had been organized in the past 4 years. A new TC on Mine Waste was hosted by the Chilean Geotechnical Society and there were 3 new Corporate Associate from the region.

12.2 ISSMGE Virtual University (Mounir Bouassida, see Appendix 6)

Existing and new webinars have been used to create 13 identifiable Master's level courses.

12.3 Innovation and Development Committee (Dimitrios Zekkos, see Appendix 7)

The IDC continues to oversee developments of the ISSMGE website which provides a central location for news and technical resources and sub-websites to support Technical and Board Level Committees. The online library supports the Virtual University with well over 16,000 technical papers. The library can now issue DOIs for papers and ISSMGE reports. The conference review platform can support new events and can facilitate direct publishing in the online library. Developing the ISSMGE Virtual University has allowed the various webinars and other videos to be categorised and arranged in courses with individual themes. GeoWorld is a professional network connecting geotechnical engineers and is also used to promote ISSMGE content.

12.4 Technical Oversight Committee (Pierre Delage, see Appendix 8)

There are 38 Technical Committees providing forums for knowledge in geotechnical engineering. They develop and disseminate knowledge and geotechnical practice, establish guidelines and technical recommendations, assist with technical programs of international and regional conferences organised by the ISSMGE and interact with industry and overlapping groups working in specialist areas. TC activity is assessed at least biennially and action taken if reactivation is needed. New TCs are create in response to developments in geotechnical engineering.

12.5 Young Members' Presidential Group (Lucy Wu, see Appendix 9)

The key aims of the YMPG are enabling the sharing of knowledge, connecting groups with each other and facilitating communications between younger members and the ISSMGE leadership. During the past 4 years, there has been a steady growth in corresponding members of YMPG. Major accomplishments have been participation the Reginal Conferences, delivering Bright Spark Lectures, holding joint meetings with the CAPG and participating in the design challenge, hosting an in-person meeting and organising the Future of Geotechnics event in October 2021.

12.6 Corporate Associates' Presidential Group (Peter Day, see Appendix 10)

The CAPG organised a major session at each of the Regional Conferences and will present their findings on their "Are we overdesigning?" initiative.

12.7 Award Committee (Roberto Terzariol, see Appendix 11)

It should be noted that nominations were received for awards in all categories, i.e. Outstanding Technical Committee, Geotechnical Project, Innovator, Member Society, Paper Published in the International Journal of Geo-Engineering Case Histories, Public Relations and Young Geotechnical Engineer. The awardees will be announced during a special session of the 20ICSMGE, Sydney.

12.8 Professional Image Committee (Ikuo Towhata, see Appendix 12)

A major report has been prepared showing the contribution of geotechnical engineering to human society. Geotechnical engineering activity is often invisible and geotechnical engineers often do not command high salaries which can lead to poor respect. Subsurface investigation is essential to avoid or at least minimise risks that can lead to damage, death and financial loss.

12.9 International Journal of Geo-Engineering Case Histories (Dimitrios Zekkos, see Appendix 13)

The IJGCH is an ISSMGE journal and the only diamond open access journal in the geotechnical engineering field. The journal focuses on geotechnical practice though the careful documentation of case histories. In 2021, the 124 papers published in the journal were downloaded 84,000+ times. Efforts are underway to better integrate the journal with the International Society by establishing Technical Committee Liaisons. Special Issues organised by Technical Committees have now been published.

12.10 The ISSMGE Foundation (Chair of the Trustees, Michael Davies, see Appendix 14)

The ISSMGE Foundation was established as an independent charity in 2013 and has been supported by several private donations and substantial donations from ISSMGE. The Trustees hope to eventually establish it as an endowment fund to secure the future of the Foundation. The Foundation provides financial support to individuals participating in technical and professional activities such as conferences and training courses. To date, about 180 awards have been made totalling over £150,000.

12.11 Time Capsule Project (Sukumar Pathmanandavel, see Appendix 15)

The Time Capsule Project (TCP) is an initiative designed to bring the ISSMGE individual members closer together, remember the past and look to the future. Contributions have been made by Past-Presidents, geotechnics champions, Member Societies and Technical Committees and access to these is via the ISSMGE website.

12.12 The ISSMGE Bulletin (Anthony Leung, Editor-in-Chief, see Appendix 16)

The Bulletin continues to be published bimonthly with major articles being research highlights, TC Corner, major projects, the president's messages, reports from member societies, Board Level Committee activity, Time Capsule Project Blogs, conference reports, and the event diary.

Section D: Other business

13 MOTION PROPOSED BY UKRANIAN SOCIETY FOR SOIL MECHANICS, GEOTECHNICS AND FOUNDATION ENGINEERING

In response to the attack by Russia against Ukraine, the Ukrainian Geotechnical Society had written to the Russian Geotechnical Society seeking solidarity and asking Russian scientists to oppose the war. No response had been received and consequently, the Ukrainian Society for Soil Mechanics, Geotechnics and Foundation Engineering proposed the motion that the Russian Society for Soil Mechanics, Geotechnics and Foundation Engineering should be excluded from ISSMGE. The Russian Geotechnical Society had responded to the ISSMGE Secretariat with the view that the motion by the Ukrainian Society was inconsistent with the aim of the International Society, which is the promotion of international cooperation amongst engineers and scientists for the advancement of knowledge in the field of geotechnics and its engineering applications.

Professor Shin stated that other military events had occurred in the past without proposals to exclude Member Societies. The Secretary General stated that while that was the case, this is the first time the ISSMGE had received such a proposal and that it should therefore be brought to Council. Mario Manassero seconded the motion, which then allowed it to be discussed.

In the discussion that followed, Professor Kirichek (Ukraine) reiterated the first motion that had been submitted and Professor Ilyichev (Russia) voiced their counter argument. A number of views were expressed by delegates including Duc Long Phung (Vietnam) essentially arguing that political issues were not the business of ISSMGE. Mr Garin (Sweden) thought it would be wrong to terminate the membership of the Russian Geotechnical Society and proposed an alternative motion that their membership should be suspended. This second motion was seconded by Dr Cazzuffi (Italy). It was followed by a third motion proposed by Mr Pardo De Santayana (Spain), that the first two motions be abandoned. This was seconded by Mr Robins (New Zealand).

The President noted that several views had been expressed but since there were no new arguments, the motions could be voted upon. Noting that many Societies might feel uncomfortable with voting by a show of hands, the President proposed that voting should be by email to the Secretary General with copy to the President. The Secretary General undertook to organise this postal ballot in which the voting delegates would be given 24 hours to cast their vote.

Following normal practice, the motions are considered in reverse order and after the 24hr period, the ballot for Motion 3 had 40 votes for the motion, 16 against and 3 abstentions. The motion is therefore carried and the other motions concerning suspension or termination of the Russian Society for Soil Mechanics, Geotechnics and Foundation Engineering are abandoned.

14 ANY OTHER BUSINESS

The President asked if there was any other business, and there was none.

15 DATE AND VENUE OF NEXT MEETING

The next Council Meeting will be on Sunday 13 August 2023 in Nur-Sultan, Kazakhstan.

16 THANKS AND CLOSURE

The President thanked all those attending the meeting, either in-person or online, for their considered deliberation of the various discussion items. Thanks were also expressed to the Sydney conference organising teams and to the staff of ICC Sydney for their help and support in making the meeting possible. The meeting closed at 00:30.

Appendix 1: Membership

Neil Taylor
Secretary General, ISSMGE

1 MEMBER SOCIETIES

The accompanying Table indicates that the present ISSMGE membership is just over 20,800 in 90 Member Societies, (information correct as of 29th April 2022). The membership of the majority of the societies has remained static during the pandemic, though a few have reported some changes compared to the report produced for the Cape Town Council Meeting: Canada (down to 624 from 751), Germany down to 432 from 492), Indonesia (up to 200 from 140), Iraq (down to 50 from 114), Korea (down to 450 from 519), Philippines (up to 79 from 34), Romania (down to 119 from 158), Singapore (up to 64 from 46), South Africa (down to 532 from 659), South East Asia (down to 95 from 124), USA (down to 2607 from 2975), and Vietnam (up to 70 from 50).

The application for membership from the Mongolian Association of Soil Mechanics and Geotechnics was approved by the Board since the last report to Council.

2 CORPORATE ASSOCIATES

AECOM Asia Company Ltd
AOSA SA
S.N. Apageo S.A.S.
Ove Arup & Partners Ltd
Aurecon
Bentley Systems (UK) Limited
Dar Al Handasah Corp
Deltares
EPB SA
CCC Fourth Harbor Engineering Company
Fugro Australia Marine Pty Ltd
Gamuda (Australia) Branch
GDS instruments
Geomil BV
GHD Pty, Ltd.
Giken Ltd
Ground Investigation Ltd
Huesker Synthetic GMBH
Inora sp. z.o.o.
Jan De Nul N.V.
Keller Group plc
KGS Ltd
Naue GmbH & Co KG
Norwegian Geotechnical Institute
Pagani Geotechnical Equipment
RCF Ltd
Rocscience Inc
Shanghai Geoharbour Construction Group
Siemens Energy Global GmbH & Co. KG;
SMEC AUSTRALIA PTY LTD
SRK Consulting Argentina S.A.
Teclab Limited
Tencate Geosynthetics France SAS
Tensar International Ltd
Terrasol
Terre Armée (Soletanche Freyssinet Group)
Tetra Tech Coffey Pty Ltd
Transport Research Centre
A.P. van den Berg
Wagstaff Piling
WSP - Golder Associates Inc

MEMBERSHIP INFORMATION AS AT 29th APRIL 2022

MEMBER SOCIETY	Membership	Africa	Asia	Australasia	Europe	North America	South America
ALBANIA	30				30		
ALGERIA	30	30					
ARGENTINA	106						106
AUSTRALIA	1218			1218			
AUSTRIA	108				108		
BANGLADESH	33		33				
BELARUS	30				30		
BELGIUM	319				319		
BOLIVIA	42						42
BOSNIA & HERZEGOVINA	16				16		
BRAZIL	779						779

MEMBER SOCIETY	Membership	Africa	Asia	Australasia	Europe	North America	South America
BULGARIA	32				32		
CANADA	624					624	
CHILE	100						100
CHINA	511		511				
CHINESE TAIPEI	68		68				
COLOMBIA	22						22
COSTA RICA	31						31
CROATIA	102				102		
CTGA	100	100					
CUBA	30						30
CYPRUS	30				30		
CZECH & SLOVAK REPS.	45				45		
DENMARK	594				594		
EGYPT	30	30					
ESTONIA	30				30		
FINLAND	45				45		
FRANCE	636				636		
GEORGIA	30				30		
GERMANY	432				432		
GHANA	21	21					
GREECE	69				69		
GUATEMALA	32						32
HONG KONG	138		138				
HUNGARY	65				65		
ICELAND	32				32		
INDIA	141		141				
INDONESIA	200		200				
IRAN	802		802				
IRAQ	50		50				
IRELAND	24				24		
ISRAEL	32				32		
ITALY	292				292		
JAPAN	1066		1066				
KAZAKHSTAN	40		40				
KOREA	450		450				
KYRGYZSTAN	30		30				
LATVIA	44				44		
LEBANON	30		30				
LITHUANIA	40				40		
MALAYSIA	108		108				
MEXICO	260					260	
MONGOLIA	47		47				
MOROCCO	28	28					

MEMBER SOCIETY	Membership	Africa	Asia	Australasia	Europe	North America	South America
MOZAMBIQUE	37	37					
NEPAL	22		22				
NETHERLANDS	731				731		
NEW ZEALAND	762			762			
NIGERIA	40	40					
NORTH MACEDONIA	35				35		
NORWAY	580				580		
PAKISTAN	85		85				
PARAGUAY	38						38
PERU	27						27
PHILIPPINES	79		79				
POLAND	354				354		
PORTUGAL	150				150		
ROMANIA	119				119		
RUSSIA	165				165		
SERBIA	43				43		
SINGAPORE	64		64				
SLOVENIA	75				75		
SOUTH AFRICA	532	532					
SOUTH EAST ASIA	95		95				
SPAIN	394				394		
SRI LANKA	36		36				
SUDAN	37	37					
SWEDEN	1605				1605		
SWITZERLAND	169				169		
SYRIA	17		17				
TAJIKISTAN	23		23				
THAILAND	32		32				
TUNISIA	40	40					
TURKEY	250				250		
UKRAINE	33				33		
UK	1155				1155		
USA	2607					2607	
UZBEKISTAN	33		33				
VENEZUELA	32						32
VIETNAM	70		70				

Appendix 2: Presentation of Audited Accounts 2019, 2020

Neil Taylor,
Secretary General, ISSMGE

FINANCIAL STATEMENTS FOR THE YEAR ENDED 31 DECEMBER 2019

INDEPENDENT EXAMINERS' REPORT TO THE COUNCIL MEMBERS OF THE INTERNATIONAL SOCIETY FOR SOIL MECHANICS AND GEOTECHNICAL ENGINEERING

I report to the society's Secretary General and society council members on my examination of the receipts and payments accounts of the society for the year ended 31 December 2019 which are set out on pages 2 to 6.

Responsibilities and basis of Report

The society's Secretary General is responsible for the preparation of the receipts and payments account, which has been prepared under the Society's statutes and By-Laws.

My responsibility is to examine the receipts and payments account and related information in accordance with the society's Statutes and By-Laws.

I report in respect of my examination of the society's receipts and payments account.

Independent Examiners Statement

I have completed my examination. I confirm that no material matters have come to my attention in connection with the examination giving me cause to believe that in any material respect:

1. accounting records were not kept in respect of the society as required under the society's Statutes and By-Laws; or
2. the accounts do not accord with those records.

I have no concerns and have come across no other matters in connection with the examination to which attention should be drawn in this report in order to enable a proper understanding of the accounts to be reached.

Alison Sargent

Alison Sargent FCA
Sargent & Co
Chartered Accountants and Registered Auditors
219 Croydon Road
Caterham
Surrey CR3 6PH

Date: 14 September 2020

INTERNATIONAL SOCIETY FOR SOIL MECHANICS AND GEOTECHNICAL ENGINEERING
RECEIPTS AND PAYMENTS ACCOUNT
FOR THE YEAR ENDED 31 DECEMBER 2019

	Barclays Bank Plc SFr	Barclays Bank Plc £	Investment Accounts £	City University £
RECEIPTS				
Member Society Subscriptions	-	239,208	-	-
Corporate Member Subscriptions	-	44,176	-	-
Donations and Other Income	-	-	-	=
Interest	-	1,142	1,175	-
Conference Income	-	22,755	-	-
	-	**307,281**	**1,175**	-
EXPENDITURE				
Secretariat				
President part time support		21,000	-	-
Secretary general support	-	-	-	56,227
Office Expenses - President	-	5,600	-	-
Office Expenses Secretariat	-	1,375	-	297
Secretary General	-	25,800	-	-
Conference support	-	1,500	-	-
Kevin Nash Gold Medal	-	-	-	-
Prizes/Awards	-	-	-	-
	-	**55,275**	-	**56,524**
Travel, accommodation etc				
President	-	17000	-	-
Secretary General		6,132	-	-
Board	-	44,557	-	-
Board Level Chairs	-	18,569		
Other (YPMG, Bulletin editor)	-	5,563	-	-
	-	**91,821**	-	-
Independent Examiners Fees & Bank Charges	-	4,082	-	-
Corporation Tax	-	373	-	-
	-	**4,455**	-	-
Information Technology (website)				
Licensing Fee	-	10,576	-	-
Future Developments	-	-	-	-
	-	**10,576**	-	-
Other Initiatives				
International Seminars (e.g. travel)	-	-	-	-
Travel of ISSMGE Officers to FedIGS Meetings	-	2,206	-	-
Webinars	-	5,071	-	-
Initiatives	-	16,634	-	-
Contingency	-	-	-	-
		23,911	-	-
Total Expenditure		**186,038**	-	**56,524**
SURPLUS/(DEFICIT)		**121,243**	**1,175**	**(56,524)**
Transfer to the ISSMGE Foundation	-	(41,000)	-	-
Transfers from other accounts	-	(56,570)	-	56,570
Cash balance at 1 January 2019	**106,467**	**374,404**	**146,864**	**(50**
Cash balance at 31 December 2019	**106,467**	**398,077**	**148,039**	**(4)**

Information December 2019 - £1 = SFr 1.29

Secretary General Prof. R N Taylor **Date** (**September 2020**

INTERNATIONAL SOCIETY FOR SOIL MECHANICS AND GEOTECHNICAL ENGINEERING SUBSCRIPTIONS RECEIVED IN THE YEAR ENDED 31 DECEMBER 2019

Member Society	Number of members	Barclays Bank £
Albania	40	524
Algeria*	43	-
Argentina	106	1,986
Australia	1133	12,043
Austria	108	2,158
Bangladesh*	33	-
Belarus	30	519
Belgium	319	4,623
Bolivia	42	419
Bosnia & Herzegovina	18	412
Brazil	806	7,222
Bulgaria*	31	-
Canada	751	8,997
Chile	101	2,018
China	472	5,617
Chinese Taipei	68	1,369
Colombia	20	438
Costa Rica	32	493
Croatia	104	2,078
CTGA	100	-
Cuba*	30	-
Cyprus	30	599
Czech & Slovak Republics	45	899
Denmark	345	5,755
Egypt*	103	-
Estonia	30	599
Finland	45	899
France	636	8,489
Georgia	38	804
Germany	492	6,927
Ghana	21	323
Greece	69	4,062
Guatemala*	32	-
Hong Kong	120	2,383
Hungary	65	1,299
Iceland	32	639
India	224	1,505
Indonesia	140	1,830
Iran*	802	-
Iraq	114	1,859
Ireland	24	599
Israel*	32	-
Italy	269	5,146
Japan	1,124	11,982
Kazakhstan	40	784
Korea	519	7,147
Kyrgyzstan	30	639
Latvia*	44	-
Lebanon	85	1272
Lithuania*	40	-

INTERNATIONAL SOCIETY FOR SOIL MECHANICS AND GEOTECHNICAL ENGINEERING SUBSCRIPTIONS RECEIVED (cont.) IN THE YEAR ENDED 31 DECEMBER 2019

Member Society	Number of members	Barclays Bank £
Macedonia, FYR	35	521
Malaysia	82	1,638
Mexico*	260	-
Mongolia	47	778
Morocco*	28	-
Mozambique	37	206
Nepal*	22	-
Netherlands	719	-
New Zealand	775	9,192
Nigeria*	40	-
Norway	530	7,235
Pakistan	85	1,442
Paraguay	38	410
Peru	27	397
Philippines	34	401
Poland	378	6,018
Portugal	156	3,117
Romania	158	5,635
Russia	180	3,596
Serbia	40	581
Singapore	46	919
Slovenia	77	1,465
South Africa	659	5,740
South East Asia	125	1,087
Spain	388	6,099
Sri Lanka	36	932
Sudan*	37	-
Sweden	1,301	13,400
Switzerland	173	3,455
Syria*	17	-
Tajikistan*	23	-
Thailand	32	531
Tunisia	40	476
Turkey	267	5,132
Ukraine	30	942
UK	1,173	12,375
USA	2,975	26,763
Uzbekistan	33	903
Venezuela*	32	-
Vietnam	50	466
Total	**21,162**	**239,208**

* Membership number is from the last payment received

INTERNATIONAL SOCIETY FOR SOIL MECHANICS AND GEOTECHNICAL ENGINEERING SUBSCRIPTIONS RECEIVED (cont.) IN THE YEAR ENDED 31 DECEMBER 2019

Corporate Member	Bank Account £
AECOM Asia Company Ltd	-
Ove Arup & Partners Ltd	1,158
AOSA SA	1,089
Aurecon	1,156
AP van den Berg Ingenieursburo	1,115
CGRE (UNIV WOLLONGONG)	1,212
Coffey Geotechnics	1,130
Dar Al Handasah	2,246
Deltares	1,120
EBP SA	1,098
Fugro GmbH	1,124
Gamuda	1,162
GDS Instruments	1,118
Geomil	1,194
GHD Pty Ltd	1,104
Giken Ltf	1,156
Golder Associates	1,191
Ground Investigation	1,174
Huesker Synthetic GmbH	1,128
INDRA	1,123
International I.G.M. S.A.R.L.	1,161
Jan de Nul NV	1,159
KellerGround Engineering	1,096
KGS Ltd	1,178
Maccaferri Innovation Centre SRL	1,099
Naue	1,135
Norwegian Geotechnical Institute	1,111
Pagani Geotechnical Equipment	1,113
RCF Ltd	1,126
Shanghai Geoharbour Construction Group	-
Siemens Energy	1,179
SMEC Australia	1,091
Soletanche Bachy SA	-
S N Apageo S.A.S., France	1,097
SRK Consulting	1,092
Tencate Geosynthetics France SAS	1,109
Tensar International Ltd	1,140
Terrasol	1,112
Terre Armee Internationale	1,140
Wagstaff Piling	1,111
Zetas Zemin Teknolojisi AS	1,125
	44,176

INTERNATIONAL SOCIETY FOR SOIL MECHANICS AND GEOTECHNICAL ENGINEERING
NOTES TO THE FINANCIAL STATEMENTS
YEAR ENDED 31 DECEMBER 2019

ACCOUNTING POLICIES

(a) Income
Income has been taken to the credit of the revenue account on a cash received basis.

(b) Expenditure
Expenditure, inclusive of VAT, has been charged to the revenue account on a cash paid basis.

RELATED PARTIES

Included in expenditure are certain governance costs incurred by The ISSMGE Foundation which the Society has agreed to pay on behalf of the Foundation to ensure that as much as possible of the income of the Foundation is used for awards. The total amount paid on behalf of the Foundation in 2019 was £2,032 (2018 - £1,964). The payments in 2019 related to the Independent Examination costs for the 2018 accounts of the ISSMGE Foundation, the ISSMGE Foundation Trustees Insurance and the subscription for the American Fund for Charities.

FINANCIAL STATEMENTS FOR THE YEAR ENDED 31 DECEMBER 2020

INDEPENDENT EXAMINERS' REPORT TO THE COUNCIL MEMBERS OF THE INTERNATIONAL SOCIETY FOR SOIL MECHANICS AND GEOTECHNICAL ENGINEERING

I report to the society's Secretary General and society council members on my examination of the receipts and payments accounts of the society for the year ended 31 December 2020 which are set out on pages 2 to 6.

Responsibilities and basis of Report

The society's Secretary General is responsible for the preparation of the receipts and payments account, which has been prepared under the Society's statutes and By-Laws.

My responsibility is to examine the receipts and payments account and related information in accordance with the society's Statutes and By-Laws.

I report in respect of my examination of the society's receipts and payments account.

Independent Examiners Statement

I have completed my examination. I confirm that no material matters have come to my attention in connection with the examination giving me cause to believe that in any material respect:

1. accounting records were not kept in respect of the society as required under the society's Statutes and By-Laws; or
2. the accounts do not accord with those records.

I have no concerns and have come across no other matters in connection with the examination to which attention should be drawn in this report in order to enable a proper understanding of the accounts to be reached.

Alison Sargent FCA
Sargent & Co
Chartered Accountants and Registered Auditors
219 Croydon Road
Caterham
Surrey CR3 6PH

Date: 19 May 2021

INTERNATIONAL SOCIETY FOR SOIL MECHANICS AND GEOTECHNICAL ENGINEERING
RECEIPTS AND PAYMENTS ACCOUNT
FOR THE YEAR ENDED 31 DECEMBER 2020

	Barclays Bank Plc SFr	Barclays Bank Plc £	Investment Accounts £	City University £
RECEIPTS				
Member Society Subscriptions	-	237,594	-	-
Corporate Member Subscriptions	-	40,764	-	-
Donations and Other Income	-	-	-	-
Interest	-	516	1,066	-
Conference Income	-	5,000	-	-
	-	**283,874**	**1,066**	-
EXPENDITURE				
Secretariat				
President part time support		21,000	-	-
Secretary general support	-	-	-	57,444
Office Expenses - President	-	5,700	-	-
Office Expenses Secretariat	-	962	-	-
Secretary General	-	26,300	-	-
Conference support	-	-	-	-
Kevin Nash Gold Medal	-	-	-	-
Prizes/Awards	-	-	-	-
	-	**53,962**	-	**57,444**
Travel, accommodation etc				
President	-	17,000	-	-
Secretary General		3,587	-	-
Board	-	3,334	-	-
Board Level Chairs	-	3,632	-	-
	-	**27,553**	-	-
Audit Fees & Bank Charges	-	3,791	-	-
Corporation Tax	-	440	-	-
	-	**4,231**	-	-
Information Technology (website)				
Licensing Fee	-	12,988	-	-
Future Developments	-	15,281	-	-
	-	**28,269**	-	-
Other Initiatives				
International Seminars (e.g. travel)	-	-	-	-
Travel of ISSMGE Officers to FedIGS Meetings	-	-	-	-
Initiatives	-	11,772	-	-
Webinars	-	7,777	-	-
Contingency	-	-	-	-
		19,549	-	-
Total Expenditure		**133,564**	-	**57,444**
SURPLUS/(DEFICIT)		**150,310**	**1,066**	**(57,444)**
Transfers from other accounts	-	(57,276)	-	57,276
Cash balance at 1 January 2020	**106,467**	**398,077**	**148,039**	**(4)**
Cash balance at 31 December 2020	**106,467**	**491,111**	**149,105**	**(172)**

Information December 2020 - £1 = SFr 1.22

Secretary General Prof. R N Taylor **Date** 18 **May 2021**

INTERNATIONAL SOCIETY FOR SOIL MECHANICS AND GEOTECHNICAL ENGINEERING SUBSCRIPTIONS RECEIVED IN THE YEAR ENDED 31 DECEMBER 2020

Member Society	Number of members	Barclays Bank £
Albania	43	556
Algeria*	43	-
Argentina	106	3,629
Australia	1,175	12,020
Austria	108	2,096
Bangladesh*	33	-
Belarus*	30	-
Belgium	319	829
Bolivia*	42	-
Bosnia & Herzegovina	18	412
Brazil	864	7,284
Bulgaria*	32	-
Canada	719	8,491
Chile*	101	-
China	531	5,695
Chinese Taipei	68	1,320
Colombia	21	416
Costa Rica*	31	-
Croatia	101	1,960
CTGA	100	1,140
Cuba*	30	-
Cyprus*	30	-
Czech & Slovak Republics*	45	-
Denmark	345	5,589
Egypt*	50	-
Estonia	30	582
Finland	45	874
France	636	7,848
Georgia*	32	-
Germany	463	6,505
Ghana*	21	-
Greece*	69	-
Guatemala	32	965
Hong Kong	126	2,431
Hungary*	65	-
Iceland	32	621
India	241	2,620
Indonesia	200	2,546
Iran*	802	-
Iraq	90	1,381
Ireland	24	582
Israel*	32	-
Italy	289	5,156
Japan	1,121	11,614
Kazakhstan*	40	-
Korea	450	6,404
Kyrgyzstan*	30	-
Latvia*	44	-
Lebanon*	85	-
Lithuania	40	1,304

INTERNATIONAL SOCIETY FOR SOIL MECHANICS AND GEOTECHNICAL ENGINEERING SUBSCRIPTIONS RECEIVED (cont.) IN THE YEAR ENDED 31 DECEMBER 2020

Member Society	Number of members	Barclays Bank £
Malaysia	91	1,766
Mexico	260	12,735
Mongolia*	47	-
Morocco*	28	-
Mozambique	37	200
Nepal*	22	-
Netherlands	704	17,121
New Zealand	784	8,996
Nigeria*	40	-
North Macedonia	35	511
Norway	550	7,235
Pakistan*	85	-
Paraguay*	38	-
Peru	27	404
Philippines	34	392
Poland	377	5,840
Portugal	151	2,931
Romania*	119	-
Russia	98	1,902
Serbia	40	605
Singapore*	46	-
Slovenia	75	1,440
South Africa	703	5,645
South East Asia	107	920
Spain	395	5,976
Sri Lanka*	36	-
Sudan*	37	-
Sweden	1,403	13,806
Switzerland	172	3,337
Syria*	17	-
Tajikistan*	23	-
Thailand*	32	-
Tunisia*	40	-
Turkey	285	5,112
Ukraine*	33	-
UK	1,158	11,784
USA	2,577	22,897
Uzbekistan	33	297
Venezuela	32	2,383
Vietnam	53	489
Total	**20,848**	**237,594**

* Membership number is from the last payment received

INTERNATIONAL SOCIETY FOR SOIL MECHANICS AND GEOTECHNICAL ENGINEERING SUBSCRIPTIONS RECEIVED (cont.) IN THE YEAR ENDED 31 DECEMBER 2020

Corporate Member	Bank Account £
AECOM Asia Company Ltd	2,231
Ove Arup & Partners Ltd	1,120
AOSA SA	-
Aurecon	1,133
AP van den Berg Ingenieursburo	1,136
CCCC Fourth Harbor Engineering Institute Co.Ltd	1,191
TRU/UTS	1,104
Coffey Geotechnics	1,199
Dar Al Handasah	-
Deltares	1,140
EBP SA	-
Fugro GmbH	-
Gamuda	1,140
GDS Instruments	1,097
Geomil	1,196
GHD Pty Ltd	1,108
Giken Ltf	1,142
Golder Associates	1,102
Ground Investigation	1,136
Huesker Synthetic GmbH	1,107
INORA	-
International I.G.M. S.A.R.L.	-
Jan de Nul NV	1,119
KellerGround Engineering	1,125
KGS Ltd	1,164
Naue	1,130
Norwegian Geotechnical Institute	1,124
Pagani Geotechnical Equipment	1,126
RCF Ltd	1,138
Shanghai Geoharbour Construction Group	1,108
Siemens Energy	1,140
SMEC Australia	1,135
Soletanche Bachy SA	1,119
S N Apageo S.A.S., France	1,121
SRK Consulting	1,119
Techfab India Industries Ltd	1,131
Tencate Geosynthetics France SAS	1,133
Tensar International Ltd	1,139
Terrasol	1,166
Terre Armee Internationale	1,128
Wagstaff Piling	1,117
	40,764

INTERNATIONAL SOCIETY FOR SOIL MECHANICS AND GEOTECHNICAL ENGINEERING NOTES TO THE FINANCIAL STATEMENTS YEAR ENDED 31 DECEMBER 2020

ACCOUNTING POLICIES

(a) **Income**
Income has been taken to the credit of the revenue account on a cash received basis.

(b) **Expenditure**
Expenditure, inclusive of VAT, has been charged to the revenue account on a cash paid basis.

RELATED PARTIES

Included in expenditure are certain governance costs incurred by The ISSMGE Foundation which the Society has agreed to pay on behalf of the Foundation to ensure that as much as possible of the income of the Foundation is used for awards. The total amount paid on behalf of the Foundation in 2020 was £2,161 (2019 - £2,032). The payments in 2020 related to the Independent Examination costs for the 2019 accounts of the ISSMGE Foundation and the ISSMGE Foundation Trustees Insurance.

Appendix 3: Proposed ISSMGE Budget for 2022–2026

Budget proposé de l'ISSMGE pour 2022–2026

Philip Robins
Vice President for Australasia

SUMMARY: This paper outlines the proposed budget for the next Council period (i.e. 2022–2024) and provides budget forecasts to the next ICSMGE in 2026. The 2022–2026 budget is informed by the audited accounts for 2019, 2020 and 2021 included with the meeting papers. The budget again proposes no increase in Member Society Annual Subscriptions.

1 GENERAL

The ISSMGE Budget 2019 - 2021 (with Forecasts to 2023) is presented in the attached table and is based on the budget approved at the Cape Town (mid-term) Council Meeting in October 2019.

The actual amounts for 2017 and 2018 from the audited accounts were presented and approved at the Cape Town meeting. This paper includes the proposed budgets for the presidential next term (2022 to 2026). To maintain consistency with previous budgets and with the independent auditor's reports, UK pounds (£) are used.

As with previous budgets, the overriding rationale for the budget is to permit the Society to function, both effectively and professionally. Therefore, the budget should facilitate the strategies developed by the President and the Board, together with the operational activities proposed by the board level and presidential committees.

Table 1 summarizes the ISSMGE Budget with forecasts to 2023 as approved at the Cape Town Council Meeting in October 2019. The table includes actual values for the years 2019 – 2021.

Table 2 includes the proposed budget for the period 2022 to 2026.

2 RECEIPTS (INCOME)

2.1 Member Society Annual Subscriptions (MSASs):

Since 2007 the fee structure has been based on a Basic Fee per Capita, discounted to allow for low Purchasing Power Parity (PPP) and for large Member Societies. At the 2015 Edinburgh Council meeting annual subscriptions were budgeted to increase modestly, however, the Council voted not to increase MSASs.

The subscriptions in the 2019–2023 period was also budgeted and approved by the Council to remain static. At the mid-term Cape Town Council Meeting. The forecast amounts for the 2022 to 2026 budget is once again not to increase MSASs and keep the income amounts static.

2.2 Corporate Associate Subscriptions:

In 2011 the fee for corporate membership of the Society was set at US$1,500. The 2022 to 2026 budges assumes that this fee level will not increase. The forecast subscription income was adjusted modestly at the mid-term Cape Town Council Meeting in October 2019 to account for an increase in Corporate Associates, for which the CAPG is thanked. For the 2022 to 2026 budget, Corporate Associate Subscriptions have been forecast to be kept static.

2.3 Interest:

The predicted revenue from interest remains modest.

2.4 Conference Income:

This item accounts for revenues from registration fees from conferences organised under the auspices of the ISSMGE. Because of a Board decision in 2012, the ISSMGE mandates that the Member Society organising the International Conference (ICSMGE) transfer 5% of the registration fee income to the ISSMGE and 3% of the Regional Conference registration fee income. Conference income is, therefore, only expected in the years when the International and Regional Conferences take place.

Conference income for the 20th International Conference for Soil Mechanics and Geotechnical Engineering (20ICSMGE) in Sydney was budgeted for year 2021. As this conference was delayed to May 2022, any income will be reflected in the actuals for 2022.

3 EXPENDITURE

3.1 Secretariat:

3.1.1 Secretary General:
Support for the Secretary General is based on the May 2010 and April 2018 proposals from the British Geotechnical Association to host the secretariat. A modest increase is forecast.

The Secretary General has advised the ISSMGE Board that they will be stepping down from this role at the next mid-term Council Meeting in 2024.

As such, the proposed budget forecast for the 2022 to 2026 period includes a single modest increase in year 2024 to allow for costs anticipated as part of the search, hand-over and embedding of the new Secretary General.

3.1.2 President:
Administrative support and associated expenses are allocated to the President. Please note that this budget does not include the costs of additional support that might be provided to the President by their host Member Society or from other sources.

3.1.3 Conference support:
Includes support to Regional Young Geotechnical engineering conferences. Additional funding to support the 8th International Young Geotechnical Engineers' Conference in 2024 has been included in the 2022 to 2026 forecast budget.

3.1.4 List of members:
The compilation of an annual list of members has proved to be problematic and it has been decided that the list will no longer be produced. In the proposed budget, this entry has been removed.

3.1.5 Kevin Nash Gold Medal and Prizes:
Provision is made for the award of the Society's Gold Medal and Prizes at the 21st International Conference for Soil Mechanics and Geotechnical Engineering (Sydney) in 2026.

Table 1. ISSMGE Budget 2019 - 2021 (with Forecasts to 2023) as approved at Cape Town Council Meeting

	2019		2020		2021		2022	2023
	Budget	**Actual**	**Budget**	**Actual**	**Budget**	**Actual**	**Budget**	**Budget**
	£	**£**	**£**	**£**	**£**	**£**	**£**	**£**
RECEIPTS (INCOME)								
Member Society Subscriptions	240,000	239,208	230,000	237,594	230,000	261,901	230,000	230,000
Corporate Associate Subscriptions	32,000	44,176	35,000	40,764	35,000	40,099	35,000	35,000
Interest	2,500	2,313	2,000	1,582	2,000	50*	2,000	2,000
Conference Income	10,000	22,755	-	5,000	20,000	-	-	10,000
Donations and other income	-	-	-		-	-	-	-
TOTAL INCOME	284,500	308,452	267,000	284,940	287,000	302,050	267,000	277,000
EXPENDITURE								
Secretariat								
President part time support	21,000	21,000	21,000	21,000	21,000	19,000	21,000	21,000
Secretary General support	54,000	56,227	55,000	57,444	56,000	57,849	57,000	58,000
Office expenses - President	5,600	5,600	5,700	5,700	5,800	-	5,800	5,800
Office expenses - Secretariat	3,000	3,112	3,000	962	5,000	1,088	3,000	3,000
Secretary General	25,800	25,800	26,300	26,300	26,800	26,800	27,300	27,800
Conference support	6,000	1,500	6,000	-	20,000	-	6,000	6,000
Kevin Nash Gold Medal	-	-	-	-	4,000	3,906	-	-
Prizes	-	-	-	-	5,000	-	-	-
Sub total	115,400	113,239	117,000	111,406	143,600	108,643	120,100	121,600
Travel Expenses								
President	17,000	17,000	17,000	17,000	17,000	-	17,000	17,000
Secretary General	13,000	6,132	13,000	3,587	13,000	-	13,000	13,000
Board members	40,000	44,557	42,000	3,334	44,000	-	44,000	44,000
Board Level Chairs	18,000	18,568	18,000	3,632	18,000	-	18,000	18,000
Other (YMPG, Bulletin Editor)	12,000	5,563	-	-	12,000	-	-	12,000
Sub total	100,000	91,820	90,000	27,553	104,000	-	92,000	104,000
Audit Fees & Bank Charges	3,000	2,642	4,000	3,791	4,000	3,053	4,000	4,000
Corporation Tax	1,000	373	1,000	440	1,000	301	1,000	1,000
Information Technology (web site)								
Licensing fee/web maintenance	13,500	10,576	14,000	12,988	15,000	13,635	15,000	15,000
Future developments	3,000	-	5,000	15,281	5,000	-	5,000	5,000
Sub total	20,500	13,591	24,000	32,500	25,000	16,989	25,000	25,000
Other Activities								
International Seminars (e.g. travel)	-	-	2,000	-	2,000	-	2,000	2,000
Travel of ISSMGE officers to FedIGS	3,000	2,206	3,000	-	3,000	-	3,000	3,000
Initiatives	8,000	5,071	8,000	11,772	8,000	3,086	8,000	8,000
Webinars	6,000	16,634	7,000	7,777	7,000	6,488	7,000	7,000
Contingency	3,000	-	3,000	-	3,000	-	5,000	5,000
Sub total	20,000	23,911	23,000	19,549	23,000	9,574	25,000	25,000
Donation to ISSMGE Foundation					40,000	40,000		
TOTAL EXPENDITURE	255,900	242,561	254,000	191,008	335,600	175,206	262,100	275,600
SURPLUS/(DEFICIT)	**28,600**	**65,891**	**13,000**	**93,932**	**48,600**	**126,844**	**4,900**	**1,400**
TRANSFER TO ISSMGE FOUNDATION	41,000	41,000	-	-			-	-
CASH BALANCE								
Cash balance at 1 January	606,073	606,072	593,673	630,963	606,673	724,895	853,289	858,189
Cash balance at 31 December	593,673	630,963	606,673	724,895	558,073	851,739	858,189	859,589

* estimated

3.2 Travel Expenses:

Provision is made for a modest annual increase in the normal budget for Board travel to account for inflation. Expenses are allocated to the Chairs of Board level committees (i.e. AC, CAPG, TOC, IDC, PIC, YMPG) to assist them to defray some of their travel expenses.

The Board agreed, at the meeting in Singapore, in March 2019, to provide a 50% travel subsidy for the Editor of the Bulletin to attend Council meetings to report on that publication. The Board also agreed to contribute to the cost of biennial face-to-face meetings of key representatives of the YMPG, to be held in conjunction with Board meetings to facilitate interaction between the two groups. Allowance for these items has been included for the budget forecast period 2022 to 2026.

3.3. Information Technology (web site)

3.3.1 Licensing fee/web maintenance:
In 2012 hosting of the website was transferred to Geoengineer.org from Webforum. Provision for the maintenance fee payable to Geoengineer.org is included in the budget.

3.3.2 Future developments:
Provision for upgrades to the website (or other IT developments) is made to fund initiatives proposed by the Chair - Innovations and Development Committee and as agreed by the Board.

It was agreed at the March 2019 Singapore Board Meeting, to fund the initial development of a structured education platform to host the ISSMGE Virtual University. The ISSMGE Virtual University now includes a collection of 143 videos in total! This initiative is included in the 2022 to 2026 Budget at the similar level of funding.

Similarly, allowance for the Conference Paper Review Platform which was developed by the IDC and which is currently being used for several events has been included in the 2022 to 2026 Budget.

3.4 Other Activities

3.4.1 International Seminars:
An amount is provided to contribute to the travel for ISSMGE Board members and speakers participating in these outreach activities.

3.4.2 FedIGS meeting travel:
Provision is made to support travel expenses for the President, Past President and Secretary General to attend meetings of the FedIGS Board.

3.4.3 Initiatives:
Provision is made for unspecified initiatives that may be developed by the Board. This has included, for example, a modest financial contribution in 2019 towards the cost of migrating a physical exhibition on geotechnical engineering held in Paris from 2013 to 2018 to a virtual exhibition.

3.4.4 Webinars:
An annual fee for the WISTIA video conferencing service and webinar preparation is included. The previous Board decided that these costs should be absorbed by the Society rather than passed on to members. Therefore, the 2022 to 2026 budget provides an allowance to host these Webinars.

3.4.5 Contingency:
Provision is made for a modest contingency to account for unforeseen circumstances.

4 CASH BALANCES

The global pandemic over the past two years and meant that most conferences and meeting have either been cancelled (or postponed) or gone online.

As a result, allowances for international travel have not been realized and the Society's reserves (i.e.: Cash Balance) on December 2021, were about £850,000.

Consistent with the practice of the previous Board, it is considered prudent to maintain the Society's reserves (i.e. Cash Balance) at about £500,000.

5 ISSMGE FOUNDATION

Consistent with previous budgets and to maximize benefit to ISSMGE members, annual surpluses more than the adjusted £500,000 are proposed to be transferred to the ISSMGE Foundation.

At the Council Meeting in Cape Town, October 2019, it was agreed to suspend the "automatic" transfer to the Foundation and distribute £40,000 in 2021 in part to encourage and support students from Member Societies to travel internationally to attend the 20ICSMGE in Sydney.

Typically, the amount transferred to the Foundation in any given year is contingent on the financial outcome of the Society in that year and adjusted annually for inflation.

In previous years sums in the order of £130,000 (in 2015) to £70,000 (in 2017) were transferred to the Foundation. A sum of £110,000 was transferred to the Foundation in 2018 while maintaining the Society's reserves (i.e:. Cash Balance) above about £500,000

The 2022 to 2026 budget includes the transfers to the Foundation, as previously, that is before the Council Meeting in Cape Town.

Further transfers are expected in subsequent years, and these arrangements will need to be considered by the Board and the Foundation in 2024 and any revised proposal put to the 2024 Council meeting.

Table 2. ISSMGE Proposed Budget 2022 -2026

	2022 Budget (£)	2023 Budget (£)	2024 Budget (£)	2025 Budget (£)	2026 Budget (£)
RECEIPTS (INCOME)					
Member Society Subscriptions	230,000	230,000	230,000	230,000	230,000
Corporate Associate Subscriptions	35,000	35,000	35,000	35,000	35,000
Interest	1,000	1,000	1,000	1,000	1,000
Conference Income	10,000	5,000	5,000	-	20,000
Donations and other income	-	-	-	-	-
TOTAL INCOME	276,000	271,000	271,000	266,000	286,000
EXPENDITURE					
Secretariat					
President part time support	21,000	21,000	21,000	21,000	21,000
Secretary General support	57,000	58,000	70,000	58,000	58,000
Office expenses - President	5,800	5,800	6,000	6,000	6,000
Office expenses - Secretariat	5,000	3,000	5,000	3,000	5,000
Secretary General	27,300	27,800	31,000	28,000	28,000
Conference support	6,000	6,000	20,000	6,000	6,000
Kevin Nash Gold Medal	-	-	-	-	4,000
Prizes	5,000	-	-	-	5,000
Sub total	127,100	121,600	153,000	122,000	133,000
Travel Expenses					
President	17,000	17,000	17,000	17,000	17,000
Secretary General	13,000	13,000	13,000	13,000	13,000
Board members	44,000	44,000	44,000	44,000	44,000
Board Level Chairs	18,000	18,000	18,000	18,000	18,000
Other (YMPG, Bulletin Editor)	-	-	12,000	-	12,000
Sub total	92,000	92,000	104,000	92,000	104,000
Audit Fees & Bank Charges	4,000	4,000	4,000	4,000	4,000
Corporation Tax	1,000	1,000	1,000	1,000	1,000
Information Technology (web site)					
Licensing fee/web maintenance	15,000	15,000	15,000	15,000	15,000
Future developments	5,000	5,000	5,000	5,000	5,000
Sub total	25,000	25,000	25,000	25,000	25,000
Other Activities					
International Seminars (e.g. travel)	2,000	2,000	2,000	2,000	2,000
Travel of ISSMGE officers to FedIGS meetings	3,000	3,000	3,000	3,000	3,000
Initiatives	5,000	6,000	7,000	8,000	8,000
Webinars + Virtual University	9,000	9,000	9,000	9,000	9,000
Int'l Journal on Geoengineering Case Histories	8,000	8,000	8,000	8,000	8,000
Contingency	3,000	3,000	3,000	3,000	3,000
Sub total	30,000	31,000	32,000	33,000	33,000
Donation to ISSMGE Foundation	100,000	100,000	75,000	75,000	50,000
TOTAL EXPENDITURE	374,100	369,600	389,000	347,000	345,000
SURPLUS/(DEFICIT)	**(98,100)**	**(98,600)**	**(118,000)**	**(81,000)**	**(59,000)**
CASH BALANCE					
Cash balance at 1 January	853,000	754,900	656,300	538,300	457,300
Cash balance at 31 December	754,900	656,300	538,300	457,300	398,300

Appendix 4: Presidential Report (2017–2022)

Charles W. W. Ng
ISSMGE President 2017-2022, cecwwng@ust.hk

1 INTRODUCTION AND VISION

It was my great honour to be elected as the 17th President of the International Society for Soil Mechanics and Geotechnical Engineering (ISSMGE) for a four-year term at the Council meeting convened on 17th September 2017, prior to the 19th International Conference on Soil Mechanics and Geotechnical Engineering (ICSMGE) held between 18th and 22nd September 2017 in Seoul, Korea (see Plates 1 and 2). The names of all the ISSMGE presidents since the establishment of the professional body in 1936 are shown in Table 1. Due to the COVID-19 pandemic, my term of office has been extended to 1st May 2022 when the 18th President will be elected in Sydney, Australia.

Plate 1. Former UN Secretary General Ban Ki-moon delivering the opening address at the 19th ICSMGE held on 18th September 2017 in Seoul.

Plate 2. Prof. Roger Frank (France), immediate past President, introducing the incoming President of ISSMGE at the closing ceremony of the 19th ICSMGE held on 21st September 2017.

During my presidential address at the closing ceremony of the 19th ICSMGE (see Plate 3), I shared my vision and plan to enhance "Education, Innovation and Diversity" at ISSMGE. In implementing this vision and plan, I am delighted to have received the full support of an outstanding Board, consisting of 11 board members (see Table 2) and 7 Board-Level Committee (BLC) chairs/co-chairs (see Table 3). I am very pleased to report to you that we have made significant progress.

Plate 3. Introduction of all Board members and some of the Board-level Committee (BLC) chairs at the closing ceremony of the 19th ICSMGE (from right to left: Neil Taylor, Pedro Pinto, Kok Kwang Phoon, Mounir Bouassida, Alejo Sfriso, Timothy Newson, Mario Manassero, Gavin Alexander, Eun Chul Shin, Marcelin Kana, Roger Frank, Pierre Delage, Lucy Wu, Sukumar Pathmanandavel, Peter Day and Ikuo Towhata).

Table 1. ISSMGE Presidents since the Establishment of the Professional Body in 1936.

	Year of Service	Name	Member Society
1	1936–1957	K. Terzaghi	USA
2	1957–1961	A. W. Skempton	UK
3	1961–1965	A. Casagrande	USA
4	1965–1969	L. Bjerrum	Norway
5	1969–1973	R. B. Peck	USA
6	1973–1977	J. Kerisel	France
7	1977–1981	M. Fukuoka	Japan
8	1981–1985	V. F. B. de Mello	Brazil
9	1985–1989	B. B. Broms	Sweden
10	1989–1994	N. R. Morgenstern	Canada
11	1994–1997	M. Jamiolkowski	Italy
12	1997–2001	K. Ishihara	Japan
13	2001–2005	W. Van Impe	Belgium
14	2005–2009	P.S. Sêco e Pinto	Portugal
15	2009–2013	J.L. Briaud	USA
16	2013–2017	R. Frank	France
17	2017–2022	C.W.W. Ng	Hong Kong SAR, China

Table 2. ISSMGE Board (2017–2022).

Position	Name
President	Charles W.W. Ng (Hong Kong SAR, China)
Immediate Past President	Roger Frank (France)
Vice President for Africa	Marcelin Kana (Cameroon)
Vice President for Asia	Eun Chul Shin (South Korea)
Vice President for Australasia	Gavin Alexander* / Philip Robins (New Zealand)
Vice President for Europe	Mario Manassero (Italy)
Vice President for North America	Timothy Newson (Canada)
Vice President for South America	Alejo Sfriso (Argentina)
Appointed Board Member	Pedro Sêco E Pinto (Portugal)
Appointed Board Member	Kok Kwang Phoon (Singapore)
Appointed Board Member	Mounir Bouassida (Tunisia)
Secretary General	Neil Taylor (United Kingdom)

*deceased and subsequently replaced by Philip Robins

Table 3. Board Level Committee Chairs/Co-Chairs.

Board-Level Committee	Chair/Co-chair
Innovation and Development Committee	Dimitris Zekkos (USA)
Technical Oversight Committee	Pierre Delage (France)
Young Members' Presidential Group	Lucy Wu (USA)
Corporate Associates' Presidential Group	Sukumar Pathmanandavel (Australia) / Peter Day (South Africa)
Award Committee	Roberto Terzariol (Argentina)
Professional Image Committee	Ikuo Towhata (Japan)

2 EDUCATION AND INNOVATION

2.1 *ISSMGE Virtual University (VU)*

To enhance "Education" and to promote soil mechanics and geotechnical engineering freely, effectively and efficiently to every corner of the globe, I outlined my plan at the opening address of the presidential term starting from September 2017 to establish the Virtual University (VU) of ISSMGE (see Fig. 1), which consists of three major components all offered for free: (i) online postgraduate (PG) courses, (ii) online library and (iii) online publisher. The VU was officially launched in March 2020.

(i) *ISSMGE Post-graduate (PG) Courses*
With the team work by Professor Mounir Bouassida, and Professors Dimitris Zekkos and Kok Kwang Phoon, who are the chair and co-chair of the Innovation and Development Committee, my proposed ISSMGE Virtual University has been taking shape. As of today, a total of 13 courses have been developed as follows:

- Course 1: Risk Mitigation, Monitoring & Observational Methods
- Course 2: In Situ Testing
- Course 3: Earthquake Engineering
- Course 4: Foundations
- Course 5: Soil Characterization
- Course 6: Geo-Engineering Education
- Course 7: Unsaturated Soil Mechanics
- Course 8: Geosynthetics 1
- Course 9: Landslides and Mitigations
- Course 10: Soil Behaviour
- Course 11: Probability Analysis in Civil Engineering
- Course 12: Numerical Modelling in Geomechanics
- Course 13: Risk Analysis and Machine Learning

Each PG course is comprised of three or more webinars giving students and engineers a total of no less than five hours of viewing. I would like to thank all the lecturers, who delivered the courses, for their significant contributions to the VU and ISSMGE. More courses are being developed. Please visit http://virtualuniversity.issmge.org/ for more details.

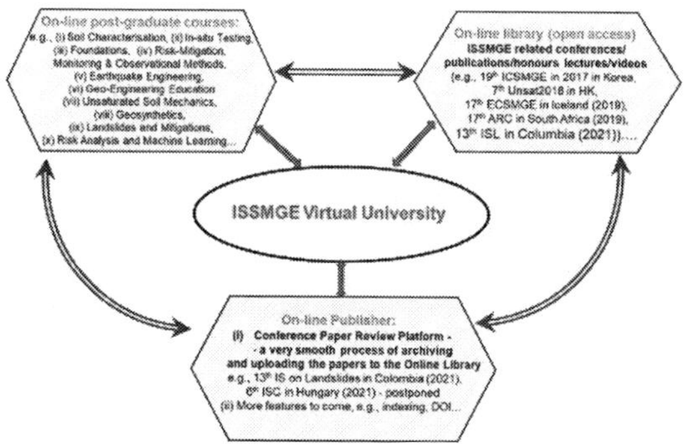

Fig. 1. ISSMGE Virtual University's three major components: (i) online postgraduate (PG) courses, (ii) online library and (iii) online publisher.

Due to the serious and continuous threat of COVID-19, many universities and training centres worldwide have remained closed, disrupting the necessary training and education of a large number of students and engineers. The VU offers a timely and free alternative for those staying at home or undergoing quarantine. Looking ahead, I would recommend that we should further explore and develop the VU to include some, if not all, of the following features to cater to the training needs under the new normal:

- Homework functionality
- Examination functionality
- Credit functionality
- Degree functionality

(ii) *ISSMGE Online Library*
The open access policy initiated by the immediate past President, Professor Roger Frank, has been continued and expanded rapidly in terms of contents and capabilities. Naturally, the open access materials form the ISSMGE online library of the VU. Various conference proceedings are now available in the online library, such as the 19th ICSMGE, the 7th International Conference on Unsaturated Soils and the 13th International Symposium on Landslides (JCT1). Moreover, many ISSMGE reports, honour lectures and videos have been uploaded to the online library. At the end of 2021, the VU held a collection of 143 videos with a total of 114,000+ views since the launch of the VU in March 2020.

About 80% of the papers in the online library has now been classified. A digital object identifier (DOI) can now be allocated to reports (e.g. from TCs, CAs, etc.) published in the online library, giving them permanent publication status. For more information about the online library, please visit https://www.issmge.org/publications/online-library.

(iii) ISSMGE Online Publisher

Under the leadership of Professors Dimitrios Zekkos and Kok Kwang Phoon, the Innovation and Development Committee (IDC) has been developing the VU's own publisher—the ISSMGE online publisher—with the intention of providing any organiser of a conference or symposium with a publishing platform. To date, the platform has been implemented and successfully adopted by three conferences and symposia such as the 13th International Symposium on Landslides held in 2021. This publishing platform enables accepted papers to be uploaded to our online library smoothly. Using the conference paper review platform and the online library, and by taking advantage of the ability to create DOIs, we are now in a position to push towards ISSMGE publications.

2.2 *ISSMGE Time Capsule Project (TCP)*

Under the initiative and outstanding leadership of Mr Sukumar Pathmanandavel, a co-chair of the Corporate Associates Presidential Group (CAPG) (https://www.issmge.org/corporate-associates/listing), the ISSMGE Time Capsule Project (TCP) is taking shape. The ISSMGE TCP has been developed with the central aim of creating and sustaining a strong level of conversation about the past, present and future of geotechnical engineering amongst, and for the benefit of, our 20,000 individual members. There is also the additional benefit of a legacy that could be added to and modified in the coming years and even decades. Some 90 Member Societies, 36 Technical Committees, the Young Members Presidential Group (YMPG) and ISSMGE Corporate Associates have been invited to contribute to the TCP. The leadership of the Member Societies and Technical Committees has been extensively briefed by the TCP Design Team between May and August 2021. A simple contents listing has been created with links to the following pages:

- Past Presidents' contributions
- Regional contributions—Africa, Asia, Australasia, Europe, North America & South America
- Technical Committees' contributions
- YMPG contributions
- Corporate Associates' contributions
- IDC contributions

A variety of new and existing materials from the contributors will be held and promoted on an online platform named ISSMGE Time Capsule 2022, which will be formally launched at the 20th ICSMGE. Information about the TCP and how to contact the TCP Design Team is available on the ISSMGE website: https://www.issmge.org/the-society/time-capsule.

2.3 *Technical Committees (TCs)*

I am grateful to the chair of the Technical Oversight Committee, Professor Pierre Delage, and his committee for reviewing new TCs and monitoring the performance of all existing TCs. Pierre has given regular reports to the Board for comments and approval. I am very pleased to inform you that five new and state-of-the-art TCs have been established since September 2017 as follows:

- TC219: System Performance of Geotechnical Structures
- TC220: Field Monitoring in Geomechanics
- TC221: Tailing and Mine Wastes
- TC222: Geotechnical BIM and Digital Twins
- TC309: Machine Learning and Big Data

On the contrary, TC207: Soil Structure Interaction and Retaining Structures was suspended due to its unsatisfactory performance. To learn more about all the TCs, please visit https://www.issmge.org/committees/technical-committees.

3 DIVERSITY (UNDER-REPRESENTED INCLUDING AGE, GEOGRAPHY AND ETHNICITY)

3.1 *Creation of The Bright Spark Lecture Series*

As our young members (less than 36 years of age) have been under-represented at major international keynote lectures for many years, I created the ISSMGE Bright Spark Lecture series to enable young members to take the main stage at plenary sessions at major conferences and symposia. I would like to thank the YMPG under the leadership of Lucy Wu for coming up with this wonderful title of the young members' lecture series. Guidelines on organising these lectures can be found at https://www.issmge.org/the-society/issmge-awards.

The first Bright Spark Lecture was delivered by Dr Lin Dong at the 7th Technical Conference in Eastern Asia on Geo-Natural Disasters held in Chengdu, China (see Plate 4). The conference was organised by Asian Technical Committee 3. I was very pleased to attend that lecture and to present a certificate to the lecturer. Since then, other Bright Spark Lectures have been organized and presented at various conferences and symposia (see also Plates 5 to 13 for examples).

Plate 4. Dr Lin Dong from the Lanzhou Institute of Seismology of China Earthquake Administration receives the first "ISSMGE Bright Spark Lecturer" certificate.

Plate 5. Dr Amin Askarinejad from TU Delft, the Netherlands, receives the certificate at the 7th International Conference on Unsaturated Soils held in 2018 in Hong Kong.

Plate 6. Dr Annan Zhou from RMIT University, Australia, receives the certificate at the 7th International Conference on Unsaturated Soils held in 2018 in Hong Kong.

Plate 7. Dr Chao Zhou from the Hong Kong University of Science and Technology, HKSAR, receives the certificate at the 7th International Conference on Unsaturated Soils held in 2018 in Hong Kong.

Plate 8. Dr Majid Ghayoomi from the University of New Hampshire, US, receives the certificate at the 7th International Conference on Unsaturated Soils held in 2018 in Hong Kong.

Plate 9. Dr Federico Pisano and Dr Mateo Ciantia from TU Delft and University of Dundee, UK, receive their certificates at the 17th ECSMGE held in September 2019 in Reykjavik, Iceland.

Plate 10. Dr Charles John MacRobert from the University of the Witwatersrand, South Africa, receives his certificate at the 17th African Conference on Soil Mechanics and Geotechnical Engineering held in October 2019 in Cape Town, South Africa.

Plate 11. Dr Tugce Baser, Canada/US, receives her certificate at the 16th Pan-American Conference on Soil Mechanics and Geotechnical Engineering held in November 2019 in Cancun.

Plate 12. Dr Carlos Omar Vargas Moreno, Mexico, receives his certificate at the 16th Pan-American Conference on Soil Mechanics and Geotechnical Engineering held in November 2019 in Cancun.

Plate 13. Dr Marlisio Oliveira Cecilio Junior, Brazil, receives her certificate at the 16th Pan-American Conference on Soil Mechanics and Geotechnical Engineering held in November 2019 in Cancun, Mexico.

To date, over 28 young members have been given a chance to deliver 20-minute keynote lectures each at plenary sessions at conferences and symposia. Dr Ashani Ranathunga from Sri Lanka and Dr Brendon Bradley from New Zealand will deliver their Bright Spark Lectures at the 20th ICSMGE, Sydney, to be held between 2nd and 5th May 2022. More details are available on the ISSMGE website: https://www.issmge.org/the-society/issmge-awards.

3.2 *Organization of workshops and forums to narrow the gap between academics and practitioners.*

Under the excellent leadership of Messrs Sukumar Pathmanandavel and Peter Day, co-chairs of the CAPG, many workshops and forums have been organized at major regional conferences. The goal is to narrow the gap between academics and practitioners so that they may find a common understanding and take collective action. The major regional conferences included the 17th African Regional Conference on Soil Mechanics and Geotechnical Engineering (SMGE) held in Cape Town, the 16th Pan-American Conference on SMGE held in

Cancun, the 16th Asian Regional Conference on SMGE held in Taipei and the 17th European Conference on SMGE held in Reykjavik. Another forum will be organized at the 20th ICSMGE to be held in Sydney between 2nd and 5th May 2022.

3.3 *Engagement with the Forgotten societies and Exploration of New Ones*

Promoting soil mechanics and geotechnical engineering to different parts of the world, engaging with our existing members and reaching out to new ones are central to my presidential plan. One way to achieve these goals is through ISSMGE international seminars. Under the outstanding leadership of Professor Pedro Seco e Pinto (Portugal), one ISSMGE International Seminar was held in Brest in Belarus, and another four in Asunción in Paraguay, Buenos Aires in Argentina, Mexico City in Mexico and León in Mexico in 2018. I would like to thank Pedro Seco e Pinto, Roger Frank (France), Liudas Furmanovich (Lithuania), Nikitenko Mikhail (Belarus), Sernov Vyacheslav (Belarus), Roberto Terzariol (Argentina), Alejo Sfrisio (Argentina), Tim Newson (Canada), Sukumar Pathmanandavel (Australia), Eun, C. Shin (South Korea) and Gabriel Auvinet (Mexico) for their outstanding contributions to these seminars. However, the international seminars were interrupted due to the outbreak of the pandemic since December 2019.

4 DUTY VISITS

During the first half of my presidential term (i.e., from September 2017 to February 2020) before the severe outbreak of the COVID-19 pandemic in December 2019, I travelled to over 40 cities on six continents (Africa, Asia, Australasia, Europe, North and South America) to attend and chair Board meetings and a Council meeting. I also made it a priority to visit Member Societies and attend their national, specialty TC conferences and regional conferences such as the 17th African Regional Conference on Soil Mechanics and Geotechnical Engineering (SMGE) held in Cape Town of South Africa, the 16th Pan-American Conference on SMGE held in Cancun of Mexico, the 16th Asian Regional Conference on SMGE held in Taipei and the 17th European Conference on SMGE held in Reykjavik, Iceland. Some examples of my activities are shown from Plates 14 to 35.

Plate 14. Attending the 1st International Conference on Press-in Engineering and meeting with key members of the Japanese Geotechnical Society on 19th September 2018 in Kochi, Japan.

Plate 15. Attending the Federation of the International Geo-Engineering Societies (FedIGS) meeting held on 24th September 2018 in San Francisco, US.

Plate 16: Delivering a keynote lecture at the 17th National Conference on Soil Mechanics and Geotechnical Engineering held on 26–28 September 2018 in Istanbul, Turkey.

Plate 17. Delivering a keynote lecture at GeoMEast International Congress and Exhibition held on 24–28 November 2018 in Cairo, Egypt.

Plate 18. Attending, on behalf of ISSMGE, the 1st joint conference organised by the Malaysian Geotechnical Society and the Geotechnical Society of Singapore held on 24–June 2019 in Kuala Lumper, Malaysia.

Plate 19. Delivering a keynote lecture at 1st joint conference organised by the Malaysian Geotechnical Society and the Geotechnical Society of Singapore held on 24–June 2019 in Kuala Lumpur, Malaysia.

Plate 20. Attending, on behalf of ISSMGE, the opening ceremony of the 13th National Conference held on 18–21 July 2019 in Tianjin, China.

Plate 21. Delivering a special invited lecture at the 13th National Conference held on 18 July 2019 in Tianjin, China.

Plate 22. Delivering a keynote lecture at the 4th Regional Symposium on Landslides in the Adriatic-Balkan Region held on 23–25 October 2019 in Sarajevo, Bosnia and Herzegovina.

Plate 23. Attending and chairing the Board meeting held on 5 October 2019 in Cape Town, South Africa.

Plate 24. Chairing the Council meeting held on 6 October 2019 in Cape Town, South Africa.

Plate 25. Delivering the President's report at the Council meeting held on 6 October 2019 in Cape Town, South Africa.

Plate 26. Delivering an opening address at the 17th African Regional Conference held on 7 October 2019 in Cape Town.

Plate 27. Visiting the exhibition booths during the 17th African Regional Conference held in Cape Town.

Plate 28. Delivering a keynote lecture at the 16th Asian Regional Conference (ARC) held on 14–18 October 2019 in Taipei.

Plate 29.. Meeting with some of the key leaders in the Far East during the 16th ARC held in Taipei.

Plate 30. Delivering an opening address at the 16th PanAmerican Conference held on 17–20 November 2019 in Cancun, Mexico.

Plate 31. Delivering the ISSMGE Presidential Lecture at the 16th PanAmerican Conference held in Cancun.

Plate 32. Presenting an ISSMGE tie to Professor William Van Impe (former President of ISSMGE) and Professor Gabriel Auvinet (former VP for North America) for their outstanding contributions to ISSMGE.

Since March 2020, I have not been able to travel due to pandemic control. I could only attend meetings and conferences online.

As of today, I have delivered over 30 keynote/special invited lectures and more than 40 opening addresses on behalf of the ISSMGE either face to face or online.

5 PUBLIC AND PROFESSIONAL IMAGE OF ISSMGE

5.1 *Congratulations to Professor Heinz Brandl*

I would like to congratulate Professor Heinz Brandl for receiving the Medal of Merit for Macedonia and for promoting the public and professional image of ISSMGE to the highest state level. The Board was invited by the then President of the former Republic of Macedonia, Dr Gjorge Ivanov, to join the award ceremony held at his Presidential Palace (see Plates 33 and 34). The Board was also invited by Professor Milorad Jovanovski, the President of Macedonian Association for Geotechnics, to attend the XVI Danube-European Conference on Geotechnical Engineering held between 6th and 8th June 2018 in Skopje and to convene our

Board meeting there. To promote our professional body, I presented an ISSMGE tie to Dr Gjorge Ivanov at the opening ceremony of the XVI Danube-European Conference (see Plate 35).

Plate 33. Professor Heinz Brandl receives the Medal of Merit for Macedonia at the Presidential Palace in Skopje.

Plate 34. A group photograph (from the left the President of ISSMGE, Professor Heinz Brandl (middle) and Dr Gjorge Ivanov (on the right)) taken at the Presidential Palace in Skopje.

Plate 35. The President of ISSMGE presents an ISSMGE tie to Dr Gjorge Ivanov at the opening ceremony of the XVI Danube-European Conference.

Plate 36. Dr Suzanne Lacasse is honoured with the Knight of the First Class, Order of the Falcon, Iceland, by Mr Ólafur Ragnar Grímsson.

5.2 *Congratulations to Dr Suzanne Lacasse*

I would like to congratulate Dr Suzanne Lacasse for receiving the Knight of the First Class, Order of the Falcon, Iceland, from Mr Ólafur Ragnar Grímsson, Iceland's President from 1996 to 2016, at the 17th European Conference on Soil Mechanics and Geotechnical Engineering held on 2nd September 2019 in Reykjavik, Iceland (see Plate 36). Previously on 27th December 2018, Dr Lacasse was awarded the Officer of the Order of Queen Elizabeth II. The award is Canada's highest civilian honour.

6 CONCLUDING REMARKS

It has been my tremendous honour and pleasure to have served you for more than four years. As of today, I am the 2nd longest serving president since our professional body was established in 1936. I believe I have delivered on my promise of improving education, innovation and diversity at ISSMGE. Obviously, I alone could not have achieved this. As Helen Keller, who was an American author, political activist and lecturer, once said: "Alone we can do so little; together we can do so much." I would like to thank all Board members, Board-Level Committees, Member Societies and member individuals for their immense contributions to ISSMGE.

Looking ahead, we should develop an ISSMGE app to facilitate more effective and reliable communication with individual members in different corners of the world. Also, we should continue to modernize our Society and start thinking about how to integrate ISSMGE into the Metaverse in the future.

Finally, I would like to end my report with a quote from Albert Einstein: "The value of a man should be seen in what he gives and not in what he is able to receive". I believe we should all give as much as we can both at work and in life.

Appendix 5: Report on the regional activities of ISSMGE in Africa

Etienne Marcelin Kana
ISSMGE Vice-President for Africa

SUMMARY: This report contains an overview of the state of the ISSMGE in Africa, and a summary of the highlights of its activities in this region, since the Council meeting held in Seoul, September 2017. It ends by general remarks and outlooks.

1 MEMBER SOCIETIES AND MEMBERSHIP

Currently, we have ten active national or transnational member societies of ISSMGE in the Africa region. They are namely: Association Algérienne de Géotechnique (ALGEOS), Association Tunisienne de Mécanique des Sols (ATMS), Comité Marocain de la Mécanique des sols et des roches (CMMSR), Comité Transnational des Géotechniciens d'Afrique (CTGA), Egyptian Geotechnical Society (EGS), Ghanaian Geotechnical Society (GGS), Sociedade Moçambicana de Geotecnica (SMG), Nigerian Geotechnical Association (NGA), South African Institute for Civil Engineer-Geotechnical Division (SAICE), and Sudanese Society for Soil Mechanics and Geotechnical Engineering (SSSMGE).

CTGA above mentioned, is a transnational geotechnical community gathering together at the moment, fifteen (15) African French speaking countries: Benin, Burundi, Cameroon, Congo, Djibouti, Comoro's Islands, Ivory Coast, Gabon, Mali, RD Congo, Chad, Centrafrique, Niger, Senegal and Mauritania.

Several attempts have been made, and are still going on, to form new national societies in Africa, some faded out, due to political unrest in these countries or other reasons (e.g., Angola, Libya, Tanzania and Uganda). Other attempts has been also made towards the revival of some dormant societies that existed previously, e.g., Zimbabwe and Kenya, Ethiopia.

Twenty seven (27) countries over forty (40), are therefore in this region, directly or indirectly affiliated to ISSMGE.

Communications with interesting geotechnical engineers in these countries, are established and plans are ongoing to help these societies to resolve their internal challenges.

The desired and expected level of activity would have been better achieved, if the current COVID-19's pandemic did not occurred. The implementation of the African member society's plan of actions, adopted in Seoul and reviewed during their meeting in Cape Town, was of course, brutally stopped by this pandemic. One way among others, to overcome this unexpected challenge is an increase of the use of news tolls offered by NTIC for virtual meeting. This is strongly encouraged.

Nevertheless, the total number of individual members of the African member societies has drastically pulled up. It is currently 1005, up from 791 members at the time of the Seoul Council Meeting on September 2017. That means **27% of growth**, which is non-negligible at all.

From the African perspective, there is a healthy membership outlook in terms of individual's members. The number of Corporate Associates is, however, still disappointing taking into consideration the low membership fee.

The categorization of the individual membership is giving in figure1 as bellow.

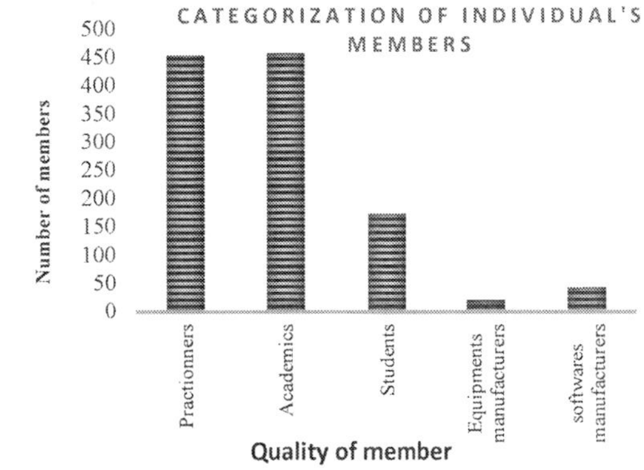

2 AFRICAN REGIONAL CONFERENCES

2.1 *African Young Geotechnical Engineer Conference (AYGEC)*

In order to increase the participation of younger geotechnical engineers in the ISSMGE's activities in Africa, a decision was taken at the Board Meeting of the African National Societies in Hammamet, April 2015. It stipulated to hold three African Young Geotechnical Engineers' Conferences (AYGEC) between successive Young International Geotechnical Engineers' Conferences (YIGEC). We got one in 2018, one in 2019, and the last one of this term was scheduled in 2020.

2.1 1 6TH *African Young Geotechnical Engineer Conference (6thAYGEC)*

The Sixth Young African Geotechnical Engineers Conference (6thAYGEC) was successfully organized , from 25th to 26th November 2018, in Khartoum (Sudan), by the Sudanese Society for Soil Mechanics and Geotechnical Engineering (SSSMGE), in collaboration with the Building and Road Research Institute (BRRI), the University of Khartoum and the ISSMGE's Board.

This conference was preceded by a workshop on "*Problematic Soils*" held at BRRI.

This event was honored by the attendance of many young researchers and engineers from various African countries, and the presence of distinguished geotechnical experts lead by the ISSMGE V.P. for Africa, and composed by Prof. Ikuo Towhata, Prof. Mounir Bouassida, and Prof. Fatma Baligh. The contribution of some universities staff members from Egypt and Sudan, and geotechnical engineers from the SSSMGE enriched the discussions. The number of overall participants tot up were seventy five (75).

2.1 2 7TH *African Young Geotechnical Engineer Conference (7[th]AYGEC)*

This event was brightly held on 5th October 2019, in Cape Town (RSA), at the Century City Conference Centre. It was, beyond the usual scheme of these conferences, an occasion to test the African Bright Spark Lecture awards nominees. This was done in relation with the YMPG who advertised and vetted applications.

The Bright Spark Lecture award has been initiated to foster geotechnical talent and accomplishment within the geotechnical community. The purpose of the Bright Spark Lecture award is to promote young and promising geotechnical engineers/academics by giving them a platform to deliver keynote and invited lectures at these conferences. It was done during the 17th ARC and the price was given to Dr Charles John MacRobert.

2.1 3 8TH *African Young Geotechnical Engineer Conference (8[th]AYGEC)*

Because of COVID-19 pandemic, this even, previously scheduled in 2020 was postponed. It needed to be rescheduled after the forthcoming ICSMGE in Sydney.

2.2 *African Regional Conference for Soil Mechanics and Geotechnical Engineering (ARCSMGE)*

It is the main regional event and a quadrennial conference, organized and hosted by one of the member societies of the region. The one of this term, the 17th ARCSMGE, was hosted by SAICE in RSA.

2.2 1 17TH *African Regional Conference on Soil Mechanics and Geotechnical Engineering (17[th]ARCSMGE)*

It was successfully held in Cape Town from 6th to 9th October 2019, under the main theme "*Innovation and sustainability in geotechnics for developing Africa*", at the Century City Conference Centre.

There were participants from 43 countries, including 23 countries from Africa: Algeria, Cameroon, Congo, Democratic Republic of Congo (DRC), Egypt, Ghana, Guinea, Kenya, Mali, Mauritania, Malawi, Morocco, Mozambique, Namibia, Niger, Nigeria, Rwanda, Senegal, South Africa, Sudan, Tanzania, Togo, and Uganda. All the other region were well represented also.

The 17th ARCSMGE was supported by more than 30 sponsors, and 24 stands in the exhibition. The conference program included:

- **2 Keynote lectures** on "*Recent advances in the application of quasi mechanistic approach for comprehensive design of geo-structures in Eastern Africa*"(Dr John Mukabi) and "*Advances in ground improvement and principles of Track Geomechanics for future railways* "(Prof Budhima Indraratna).
- **Jennings lecture** on "*Determination of unsaturated soil property functions for engineering practice*" (Prof Delwyn Fredlund)
- **Mercer lecture** on " *Geosynthetics for construction on soft foundation soils*' (Prof Kerry Rowe)
- **Bright Spark Lecture** on "Challenge to Judgment" (Dr Charles John MacRobert)
- **4 parallel sessions**:
  - Laboratory testing/soil characterization
  - Tailings dams and landfills
  - Lateral support, slopes, piled foundation and design
  - Mechanically stabilised ground

On Monday 7th October 2019, a special course was organized on the theme: *Unsaturated Soils mechanics"* delivered by Prof David Toll, Prof Antonio Gens, Prof Charles Ng, and Prof Samuel Ampadu.

The Vice President for Africa held a special meeting with the African Geotechnical members Societies, on 7th October 2019. 17 African countries, and the African corporate members were represented. The meeting was also attended by the ISSMGE President, Prof Charles Ng, Its Secretary General, Prof Neil Taylor, and the Co-Chair of the Board level committee in charge of ISSMGE's corporate members (CAPG). During this meeting, a review of ways to fight against the major challenges to the growth of the geotechnical engineering within our region were tackled, and a framework adopted.

Figure 2: Representatives of African member societies in Cape Town, Oct 2019

On Tuesday 8th October 2019, a **workshop dedicated to corporate members initiated by the CAPG,** took place under the theme: *Innovation in engineering practice*. 8 papers were proposed by corporate members for discussions.

The admirable organisation of Technical visits and other social events were very well appreciated.

The **18th ARC MSGE** will be held in Alger (Algeria) in 2024.

3 AFRICAN SUB-REGIONAL CONFERENCES

3.1 *African Days for Geotechnical Engineering (JAG)*

These very interesting conferences are, since 2010, organized every year by the CTGA, in collaboration with the African Association of civil engineering laboratories (ALBTP).

The **9th African Days** for Geotechnical Engineering (10ème édition des **JAG**), was held in Lomé (Togo), from 14th to 17th November, 2017, on the main theme: "*Relevance of innovative geotechnical engineering practices on the growth strategies of African countries*".

A workshop on the project of the revision of civil engineering codes and guidelines (P2RT project) was carried out and widely appreciated.

Under the patronage of the Togolese Minister of Public Works, this colloquium was attended by almost 225 delegates from 16 African countries. Three main lectures where given in plenary sessions. There are:

- *Growth of concrete pavement construction in the intertropical region of Africa*, by Engineer Jean PONDY
- *Prevention and attenuation of the risk of hydric erosion of soils occurrence for civil engineering constructions*, by Prof. Etienne-Marcelin KANA
- *Tropical soils specificities in road constructions in African intertropical regions*, by Prof. CAMAPUM DE CARVALHO and Prof. Etienne Marcelin KANA

During this colloquium, four workshops were also organized. And, in conformity with the ISSMGE's open access policy, the various presentation done can be downloaded on the website of CTGA: www.ctgaafrique.org,. A parallel session also took place, with some courses given under the theme: *Constructions on problematic soils in Africa.*

The **10th African Days** for Geotechnical Engineering (10ème édition des **JAG**), took place from 22nd to 25th October 2018 in Abidjan (Ivory Coast), under the main theme: "*Geotechnical engineering innovative practices in African tropical countries*".

About 250 delegates attended this colloquium from 18 countries namely: Benin, Burundi, Burkina Faso, Cameroon, Ivory Coast, Congo, Democratic Republic of Congo, Gabon, Niger, Mali, Senegal, Togo, Mauritania, Cape Verde, Comoro's islands, Madagascar, France and Germany.

Four technical workshop were organised on: Codes of Practice in African Countries; Soils in Africa: Lateritic Soils; Innovations in Geotechnical Engineering; Studies and Survey in Geotechnical Engineering. 8 Keynotes lectures were noted, and 56 papers accepted and published. 44 registration for the geotechnical engineering training workshop organized at once.

The **11th African Days** for Geotechnical Engineering (11ème édition des **JAG**), was held in the Conference Hall of Niamey (Niger), from 21st to 24th October 2019. The main theme was: "*Geotechnical Engineering and efficiency of sustainable development strategies in African tropical countries*".

About 300 delegates were count up, from 20 countries namely: Benin, Burundi, Burkina Faso, Cameroon, Ivory Coast, Congo, Democratic Republic of Congo, Gabon, Ghana, Niger, Mali, Senegal, Togo, Mauritania, Cape Verde, Comoro's islands, Madagascar, Egypt, France and Germany.

Four technical workshop were organised on: Codes of Practice in African Countries; Soils in Africa: Lateritic Soils; Innovations in Geotechnical Engineering; Studies and Survey in Geotechnical Engineering. 5 Keynotes lectures were noted, and 126 papers accepted and published.

To enable somehow, a fair participation of French or English speaking members, who were very excited to take part to the event, **the availability of translation services, English to French and vice versa, was assured**.

The **12th African Days** for Geotechnical Engineering (12ème édition des **JAG**), supposed to be hold in Dakar (Senegal) from 21st to 24th October 2020, has been postponed in 2023, due to the COVID-19 pandemic.

3.2 *International Conference on Advances in Structural and Geotechnical Engineering (ICASGE)*

The 4th International Conference on Advances in Structural and Geotechnical Engineering (ICASGE'19), organized by the *Egyptian Geotechnical Engineering Society* (EGS), in collaboration with the Structural Engineering Department of Tanta University, was held in Hurghada, Egypt from 25th to 27th March 2019, under the main theme : "*Together we can construct promising future*".

Figure 3: Participants to the ICASGE, HURGADA- MARCH 2019

There were 54 papers accepted by the Geotechnical Division. 10 Keynotes lectures noted, among them, 2 done by Prof Mohamed Elsharief (Sudan), and Prof E.M. KANA (Cameroon), 2 parallel sessions.

At the end of this event, a meeting was initiated and held between the EGS members and the ISSMGE VP for Africa, in order to search for solutions to the problems encountered by EGS. Among the main solutions taken was the ***enhancement of the communication*** between African individual members and the Board of ISSMGE, via at least, the regional VP or the SG of ISSMGE.

The **5th International Conference on Advances in Structural and Geotechnical Engineering (ICASGE'20)**, has been postponed to 2023, due to COVID-19 pandemic.

3.3 *2ndInternational Seminar on geotechnical engineering (II Seminario internacional de engenharia geotecnica)*

This geotechnical seminar, is organized by Mozambican geotechnical Society (SMG). It was held in Maputo (Mozambique), from 1st to 3rd October 2019,

The aim of this excellent seminar, was to bring together the "Community of geotechnical experts using the Portuguese Language ": (Mozambique, Angola, Equatorial Guinea, Guinea-Bissau and Cape Verde). Some core members of the ISSMGE's Board graced with their presence the seminar, and presented lectures or courses.

Were noted, 8 keynote lectures, 4 courses, and 10 papers. Around 54 persons attended the seminar that had been opened by the Mayor of Maputo.

3.4 *International Congress & exhibition (GeoMEast)*

The Egyptian Geotechnical Society (EGS), in collaboration with Egyptian government, the Housing and Building Research Center (HBRC), and the Soil-Structure Interaction Group in Egypt (SSIGE), organized annually, the International Congress & Exhibition (GeoMEast).

GeoMEast2018 was held, under the main theme: "*Sustainable Civil Infrastructures*", It took place from November 24th to 28th, 2018 in Giza, Egypt at Marriott Mena House hotel, in front of the Great Pyramids.

Over 900 experts from 71 countries representing the globe attended the conference. Thirty (30) companies from 20 countries took part to the exhibition. 570 full-papers were received, and 300 accepted and published in 16 Edited Volumes with Springer in SUCI book series.

During the GeoMEast2018, five official lectures was delivered by the current Presidents of ASCE, CSCE and ISSMGE, the former President of IGS, and the General Manager of GMA-IFAI. Twelve (12) Workshops/Courses and over 250 presentations were delivered.

GeoMEast2019 was held from 10th to 14th November 2019, under the leadership of Dr Eng HANY Farouk SHEHATA, CEO of SSIGE (Soil-Structure Interaction Group in Egypt). The main theme was : "*Sustainable civil infrastructures: Beyond the horizon*". 600 experts from 50 countries attended this event. It were noted: 2 Official lectures (ASCE lecture and IABSE official lecture), 6 Honors lectures, Courses and 3 workshops

(DFI Workshop, Introduction to tunneling engineering, Geosynthetics Engineering)

GeoMEast2020 was held from 8th to 12th November 2019.
It were noted: 2 Official lectures (ASCE lecture and IABSE official lecture), 4 Honors lectures, Courses and 3 workshops

4 ACTIVITIES OF SOME MEMBERS SOCIETIES

GGS (Ghana) :

An annual general meeting (AGM) involving both Ghana Geosynthetic Society and the Ghana Geotechnical Society (GGS) was organized on 24th February 2018. Ms. Samuella Denise Boadu was selected as the coordinator for the student chapters of the two societies. Due to a constitutional difficulty, the election of new officers could not come on as scheduled.

Being part of the Civil Engineering technical group of the Ghana Institution of Engineering (GhIE), the Ghana Geotechnical Society also sponsors activities under the GhIE Evening lecture series. Some of the events organized under the evening lecture series include

- Presentation on Geotechnical Design and Construction held on 25th October 2018 at the Engineers Center in Accra and it involved engineering experiences on geotechnical designs and construction and lessons learnt. It involved on how multidisciplinary engineers work together to deliver successful projects. The presentation was given by Monica Ansah-Osam, a senior geotechnical and tailings engineer with an Oil and Gas company.

- Presentations and discussions on recent earthquakes in Ghana was held on 25th April 2019 at the Kumasi National University of Technology (KNUST). There were four short presentations which were then followed by a discussion. The presentations and discussions were streamed online and were followed simultaneously by the northern group based in KNUST

Following the strategies at the AGM, GGS started a presentation series known as the Y-GeoPresentation which is given by the young geotechnical engineers. The first presentation was made on 27th April 2019 on the theme. "Expansive Soils of Ghana". Two presentations were made. The first presentation on "*stabilizing expansive soils using geogrid*" was presented by Ms Samuella Denise Boadu while Mrs Rosemary Kweisi made a presentation on the "*expansive soils of Pomadze*". The presentations were based on the MSc thesis research of the presenters. The presentations were followed by discussions of the topic. This was attended by 15 members and was chaired by Prof. S.I.K. Ampadu.

SAICE (South Africa):
The Geotechnical Division of the South Africa Institute of Civil Engineers (SAICE) is the largest member society in Africa. They held an AGM on 02 November 2017. In that AGM, discussions were held on the establishment of the South African Geotechnical Design code, Lateral Support Code, & Piling Code. It was agreed that the Geotechnical Society will adopt a Eurocode but have with South African annexures.

Table 1. Events organized by SAICE, Geotechnical Division in 2018.

Date	Nature of event & Comments
15 February 2018	Geotechnical Investigation course for no-geotechnical engineers at Central University of Technology presented by Peter Day
06, 12, & 14 March 2018	16th Memorial Jennings Lecture by Mike Jefferies in Pretoria, Durban and Cape Town
07, 08, & 09 March 2018	Critical Soil Mechanics workshop presented by Prof. Mike Jefferies and Dr Dawn Shuttle
16-18 July 2018	3 Days Plaxis Numerical Modeling course by Denis Waterman
28 August 2018	1 Day Finite Element Modelling Geosynthetics
28 February 2019	The Latest Development in Soil Improvement presented by Prof. Dr.-Ing. M.Sc, Wolfgang Jimmy Wehr

Table 2. Events organized by SAICE, Geotechnical Division in 2019.

Date	Nature of event & Comments
30 May 2019	Evening Lecture: Terzaghi Oration Lecture by Peter Day, Focus Rooms in Sunninghill
06 October 2019	African Young Geotechnical Engin eering Conference (AYGEC) at the University of Cape Town
07-09 October 2019	African Regional Conference (ARC) of the ISSMGE Conference in Cape Town
07/11/2019	Three awards will be awarded during our AGM 2019 • Barry van Wyk Award for the best final undergraduate thesis in Geotechnical engineering • Jennings Award- best technical paper published in Southern Africa during the 2018 year • Gold Medal his outstanding work in Geotechnical industry

The division has presented a course "Geotechnical Investigation course for non-geotechnical engineers" at the Central University of Technology in Free State aimed at educating students about the importance of conducting geotechnical investigation for all civil engineering projects

For the ***student outreach program***, the division presented two courses aimed at educating students about the importance of conducting geotechnical investigation for all civil engineering projects on the respectives themes

- "The importance of Geotechnics in Civil Engineering "at the University of the Witwatersrand;
- "Geotechnical Investigation course for non-geotechnical engineers" at the Central University of Technology in Free State

EGS (Egypt)

The EGS has particularly organized on 28th July 2017, a workshop entitled: A *Geotechnical Day,* hosted by the College of Engineering and Technology, Arab Academy for Science, Technology and Maritime Transport (AAST) in Abo Qir, Alexandria. A lecture on **Collapsible Soils: Great Challenge** was delivered by Prof. Dr. Tarek Mostafa, Egypt. During this event *Prof. Fathi AbdRabbo* of Alexandria University was honored for his contribution to Geotechnical Engineering in Egypt, and Eng. *Mohamed Salah Morsy* was nominated for the **Outstanding Young Geotechnical Engineer Award.**

ATMS (TUNISIA)

Under the auspices of ATMS, the Research Laboratory in Geotechnical Engineering of Tunisia organized in Hammamet, from 8th to 11th March 2020, the 4th International Conference on Geotechnical Engineering (4th ICGE'20). During this conference, chaired by Prof Mounir Bouassida, 8 keynote lectures were presented either in Visio conference or in live.

- Effects of horizontal and vertical stress relief on the capacity and deformation of frictionpiles (Prof Charles Ng)
- Data-driven decision making in geotechnics-Myth or relaty? (Prof Kok Kwang Phoon)
- Geometric aspects of Hazards monitoring (Dr Michel Kasser)
- Geotechnical hazards-lessons learned from case histories (Prof Pedro Sêco e Pinto)
- Geotechnical challenges and solutions to container port construction (Prof Fai Leung Chun)
- The thermo-hydro-mechanical behavior of clays and claystones (Prof Pierre Delage)
- Ultimate bearing capacity of soils, using pressure-deformation relationship of a plate test (Prof E.M. KANA)
- Integrated satellite, UAV and ground assessment of landslides in Lefkada, Greece, following the November 17, 2015 Mw 6.5 earthquake (Prof Dimitrios Zekkos)

05 parallel sessions were also organized on the themes:

- Geotechnical engineering: Foundations
- Geomatics, Georisk and environment
- Rock mechanics
- Geotehnical engineering: miscellaneous
- Geotechnical engineering: soils characterization

Due to the COVID-19 pandemic, the international attendance was very little. A Board meeting of ISSMGE organized also during this event was forced to be a virtual one.

5 TECHNICAL COMMITTEES (TC 107)

This TC107 was previously entitled "lateritic soils". It is since 2020 *known as TC on "Tropical Residual Soils*". It is the only international committee hosted by the African region. It is hosted, by the national ISSMGE member society of Morocco (CMMSG), and chair by Dr M. Ahmed Chraibi, with as the Secretary General, Prof Antony Leung. This Tc held a symposium on laterites and lateritic soils on 20th 21th September 2019- GOA, India.

6 GENERAL REMARKS AND OUTLOOKS

From the African perspective, there is a healthy membership outlook in terms of individuals. The number of Corporate Associates is, however, still disappointing taking into consideration the low membership fee.

But, the ISSMGE in Africa faced however, various defies which constituted stumbling blocks to efforts made to uphold the profession in that region. We have to go beyond for meaningful progress to be made. These challenges are:

✓ *Very low communication within the region*, hampered by lack of adequate roads and air travel links. This hikes ***the cost of transportation*** *to a level that, is almost unaffordable by private individuals. The use of new tools of NTIC could probably help to seek solutions for this matter*;

✓ ***Communication barrier*** imposed by linguistic dichotomy in the Region. The translation in each of the two official ISSMGE languages seems to be one of the best way to courageously solve the problem;

✓ The poor involvement of local corporate societies. ***The lack in general of corporate sponsorship*** for members in Africa Region, excepted for one or two national member societies;

✓ *Poor involvement of academics and local* ***experts*** in the design and the construction of major projects in the region

Finally, we totally take as ours in Africa, the view of Prof. Antonio Gens (ISSMGE VP for Europe 2013-2017) concerning the state of ISSMGE, indicated in his council report for Europe region in September 2017, I quote: "There are also general problems... that probably demand a global response from the International Society:

- the gap between academia and practice,
- the need to raise the public profile and general perception of geotechnical engineering and geotechnical engineers,
- and the importance of attracting and engaging younger professionals in our field."

Quite a large number of initiatives are already being implemented, addressing those issues but it is likely that a constant and permanent effort with always be required at all levels of our International Society.

10 ACKNOWLEDGEMENTS

I would like to thank the African Geotechnical Societies for their very much appreciated support and encouragement to me, my colleagues on the ISSMGE Board for their continuous help and support. For all them, I am forever grateful.

I extend my heartfelt wishes to the incoming ISSMGE Vice President for Africa, Professor MARAWAN SHAHIEN from Egypt. I can already anticipate that he will perform a great task in this role. I am sure that he will enjoy the support and encouragement from the entire African geotechnical community that I have personally felt throughout my period of office. Thank you all, so much.

Appendix 5: The activities of the ISSMGE for the Asian region (2017–2022)

Eun Chul Shin
Vice President of ISSMGE for Asia

SUMMARY: The Asian Region of ISSMGE has 26 Member societies which are situated in geotechnical conditions that consist of good ground condition or difficult ground conditions like soft ground, and earthquake prone area. The Asian Member societies represent the following nations and areas: Bangladesh, China, Chinese Taipei, Hong Kong, India, Indonesia, Iran, Iraq, Japan, Kazakhstan, South Korea, Kyrgyzstan, Lebanon, Malaysia, Mongolia, Nepal, Pakistan, Philippines, Singapore, Southeast Asia, Sri Lanka, Syria, Tajikistan, Thailand, Uzbekistan and Vietnam. The Philippines became a new member of ISSMGE in 2018. Mongolian Association of Soil Mechanics and Geotechnical Engineering was joined ISSMGE in 2020. This report describes about the geological and geotechnical problems of Asian continental, policy of low-cost conference, Asian Council Meeting, 19th ICSMGE, Asian Regional Conference on Soil Mechanics and Geotechnical Engineering, Asian Young Geotechnical Engineers Conference, Asian Regional Technical Committees, the types of selection for Time Capsule Project has been surveyed in Asian region and reported. The last part of this report is summarized the activities of vice president of ISSMGE for Asia.

1 INTRODUCTION

The political and mutual collaboration between societies has been described with the out-reach geotechnical engineering program. The geological situation and natural disasters around Asian region are briefly elaborated. The policy of Low Cost Conference (LCC) is emphasized to bring attention and thus applied for up-coming conference. In the past 50 years, many geotechnical related the social infrastructures such as highway, airport, port and harbor, subway, industrial complex have been constructed in the Asian region. The new technology and construction methodology are being developed for solving difficult problems in various ground conditions. Some of these findings are presented in the event of international conference and symposium as well as workshops.

1.1 *General scope*

The construction practice is related the political situation in the region. Instability of government in the western part of Asia is caused problem to the development of the country. Therefore, the countries in the region has constrained their ISSMGE activity.

The ISSMGE member societies in Asian region demonstrate a good collaboration model to the other part of member society throughout the joint conference and symposium as well as workshop together, because many societies are involved collaboration agreement between societies, particularly Southeast Asia, far eastern counties and central part of Asian countries, or neighboring countries. For example, the GeoSS and Malaysian Geotechnical Society could help to promote ISSMGE in neighboring countries in the region such as Cambodia, Laos, and Myanmar by organizing international seminars. In this respect, the ISSMGE could support the outreach program to establish their own member society.

1.2 *Climate change and natural disaster*

The two-thirds of entire population on the earth are lived in the Asian region. The large population can contribute to the emission of carbon dioxide due to the land cultivation and also the shortage of useful land in the urban area. Therefore, the living environmental of people are getting worse.

The great amount emission of carbon dioxide is contributed the temperature rise around the living space of earth. The coverage land of glacier in the cold region is dwindling and hence the sea level is rising. We should prepare the certain countermeasure against the inundation of low-land area. The irregular heavy rainfalls are caused of the landslides and debris flow in the mountain area.

The earthquake, heavy rainfall, and tsunami are the common natural disasters in the countries which situated along the circum-Pacific earthquake belt. Recently earthquakes were occurred in Palu and Sulawesi (Sept. 2018), Katabu and Rankasbitung (Jan., 2022) in Indonesia, Sichuan (May, 2008), Gansu (Jan., 2022) in China, Kobe (Jan., 1995), Fukushima (April, 2011), Kumamoto Huganada (Jan., 2022) in Japan. The liquefaction of ground, tsunami of sea-water as well as mud landslides are the deadly force to the human life and the property of government and private. Several earthquakes were occurred in India (April, 2019), Pakistan (Sept., 2013), Nepal (April, 2015) at the Himalayan Mountains range. The monitoring of volcanic eruption and natural disasters are another challenge to cope with it. The ISSMGE would support to dispatch the geotechnical expert to the place where needs a help and mitigation of problem.

1.3 *Low cost conference*

The registration fee of the international conference is relatively high these days. The people from low income countries have a problem to attend he international conference. The former Vice President for Asia, Ikuo Towhata initiated the Low Cost Conference (LCC). He actively involved the LCC policy to the Asian Regional Conference in Fukuoka, Japan in November, 2015 (USD 407). This LCC program is also adopted for the 16th ARC in Taiwan, Taipei, October 2019 (USD 450 for member of ISSMGE). This policy draws an attraction to bring as many participants as to the conference up to 1000 people in the 16th ARC. The members of Advisory Committee Meeting for the 16th ARC on 14th of October in Taipei suggested the LCC and the session topics related to sustainable development in geotechnical engineering. The registration fee for the Southeast Asian Geotechnical Conference which was held in Jakarta, November 2018 was about USD 400 for the ISSMGE member. I would like to emphasize the LCC policy for the forth coming international conference and symposium in Asian region. The epidemic of COVID-19 since March 2020 makes a difficult to travel, most of TC workshop and conference under the ISSMGE becomes webinar in the form of on-line and hence contributes as Low Cost Conference.

2 ASIAN COUNCIL MEETING

The Asian Council Meeting of ISSMGE was held two times, one was on September 20, 2017, during the 19th ICSMGE, Seoul, South Korea (participated 17 societies, non-participated 7 societies) and the other one was held during the 16th Asian Regional Conference in Taipei, Taiwan. Welcoming address made by new VP (EC Shin) and reported about Education, Innovation, Diversity from the ISSMGE Ad-hoc Board Meeting which was held on the 19th of September in Seoul, 2017. Each Asian Membership Society was introduced with new president and secretary of society. The presentation for the program of the 16th ARC in 2019 was made by Prof. Keh-Jian Shou. The matter of Eastern Asian and Western Asian Geotechnical Societies are discussed by the Immediate Past VP, Prof. Ikuo Towhata. He suggested that the 17th ARC could be held in the Western part of Asia. The Past VP for Asia, Prof. Askar Zhussupbekov discussed about the promotion activities for geotechnical societies in developing Asian countries including new membership as well as the venue of 9th AYGEC. Mr. Sohail Kibra, VP of Pakistan Geotechnical Engineering Society and Prof. S. Babu, President of Indian Geotechnical Society were reported on the activities of their societies, regional VP election, and venue of ARC.

The other Asian Council Meeting was held on October 15, 2019 during the 16th Asian Regional Conference in Taipei, Taiwan by the support of CTGS. Total 22 societies were attended and only 2 societies were absent. This meeting was opened by the introduction of each society and subsequently VP of ISSMGE for Asia, EC Shin reported about the recent activities of ISSMGE. Prof. Der-Wen Chang, Secretary General of 16th ARC presented about on-going program of 16th ARC. Dr. M.K. Chung from South Korea proposed new Asian Regional Technical Committee entitled "Smart Observation Method", with the Secretary of Dr. C.H. Choi and agreed with the title of AsRTC-14.

The venue of 17th Asian Regional Conference in the year of 2023 was decided by the ballot voting during the Asian Council Meeting. The Kazakhstan Geotechnical Society had a competition with the Thai Geotechnical Society (the Bangkok city). The result of ballot voting selected the Nur-Sultan city, the capital city of Kazakhstan becomes the venue of the 17th ARC on August 14-18, 2023. Prof. Keh-Jian Shou, National Chung-Hsing University in Taiwan is elected as the Vice President of ISSMGE for Asia in the period of 2022-2026.

3 19TH ICSMGE, 16TH ARC AND 8-9TH AYGEC

The 19th International Conference on Soil Mechanics and Geotechnical Engineering (Chairman, Prof. Dong-Soo Kim) was hosted by Korean Geotechnical Society at the Coex Convention Center in Seoul, Korea on September 17-22 in 2017. Total 1952 peoples were participated from 82 countries, and 65% of participants were from Asian region. The 19th ICSMGE were consisted of 17 Plenary sessions and 53 Discussion sessions, 32 Workshops and published 751 papers in the conference proceedings. The keynote lectures were recorded as ISSMGE Virtual University (Webinar).

The 16th Asian Regional Conference on Soil Mechanics and Geotechnical Engineering was successfully held in Taipei, Taiwan from October 14-18, 2019.There were 17 Plenary lectures including 4 keynote lectures,1 mentoring, 6 invited lectures, 4 theme lectures, and 2 special lectures. 7 Asian Regional Technical Committee sessions and 7 ISSMGE TC sessions, 2 CTGS sessions were also executed. Total 32 theme sessions and 2 Bright Spark lectures in YMGG/YGE sessions on October 13 were conducted. CAPG session; Are we over-designing? was offered along with 5 pre-conference courses. The quality of presented paper and execution of the conference were outstanding with participating 1000 people from 38 countries (Chinese Taipei 351, Japan 152, South Korea 83, China 58, India 40, Singapore 29, Vietnam 20, Kazakhstan 19, Malaysia, Philippines, Thailand, 18, Hongkong 16).

The Asian Young Geotechnical Engineers Conference (AYGEC) is holding a conference in every 4 years for geotechnical engineers and researchers below the age of 35 years from Asian countries. The AYGEC took place in Bangkok, 1991; Singapore, 1994 and 1997; Seoul, 2001; Taipei, 2004; Bangalore, 2008; Tokushima, 2012. The Kazakhstan Geotechnical Society hosted the 8th Asian Young Geotechnical Engineer Conference (8th AYGEC) in the capital city of Astana in August 5-7, 2016. Approximately 100 participants from 21 countries had an opportunity to exchange innovative ideas and development with 5 invited lectures and 59 paper presentations in the field of geotechnics during their time at the conference. The final day of the conference was dedicated to a technical trip to Baiterek monument and EXPO 2017 construction site. The international event was continued in Almaty as a satellite geotechnical workshop called "Joint Kazakhstan and Korea Geotechnical Workshop" at the Kazakhstan Highway Research Institute (President of KazDorNII, Dr. Bagdat Teltayev) on the 8th of August.

The 9th AYGEC was held at the city of Lahore on December 5-7, 2019 hosted by the Pakistan Geotechnical Engineering Society (established 1973) at the University of Engineering and Technology Lahore (UET Lahore). The 15th International Conference on Geotechnical Engineering was also taken place at the same period of time. Total number of participants was 500 people from 22 different countries and 6 keynote lectures as well as 84 papers in the field of advanced geotechnical testing, applied geomechanics, foundation & instrumented geotechnics, slope stability, dam engineering were presented. The 3rd day of conference made a field trip to the construction site of Lahore Orange Line Metro Train Project which has a 27 km long with 26 stations. Most of metro line is elevated above ground with the help of pile foundation which has a diameter of 1.2 m with the embedded depth of 12-25 m. Pakistan Lahore Ring Road is being operated now with the distance of 30.4 km and the technology of reinforced earth walls are also used in the overpass way along the highway. The site visits to Wall City and Fort of Lahore including Sheesh Mahal, Alamgiri Gate, Navlakha Pavilion, Moti Masjid are very much impressed.

4 ASIAN REGIONAL TECHNICAL COMMITTEES

(1) AsRTC-1: Geotechnical Mitigation and Adaption to Climate Change-induced Geo-disaster in Asia-Pacific Regions, Chair: Prof. Bergado, Dennes T. (Formerly AIT, Thailand) and Prof. Kazuya Yasuhara, Secretary Prof. Hazarika Hemanta (Japan). The 1st International Symposium on Construction Resources for Environmentally Sustainable Technologies (CREST 2020) was held on March 9-11, 2021. The CREST 2020 was executed by Zoom Links for Webinar, 2 Plenary & 11 Keynote lectures, 75 papers were presented (Organizing Chair, Hazarika Hemanta).

(2) AsRTC-3: Geotechnical for natural hazards Chair: Prof. Kazama Mokito (Japan). Secretary Prof. Hazarika, Hemanta (Japan).

(3) AsRTC-6: Urban Geoengineering, Chair Prof. OU Chang-Yu (Chinese Taipei), Secretary Bin-Chen Benson Hsiung (Chinese Taipei). 43 members from 13 member societies in Asia, 2nd ATC6 symposium in Changsha, Hunan Province in China, 24th and 25th of November, 2017 and 100 researchers participate, more than 50 papers. The 3rd ATC6 symposium was held in 14th to 18th of October, 2019 in Taipei, Taiwan, during the 16th Asian Regional Conference.

(4) AsRTC-7: Thick Deltaic Deposits, Chair Prof. Jiman Kim (South Korea), Secretary Prof. Jaehun Ahn (South Korea). Workshop on Geotechnical Issues and Practices on Deltaic Deposits and Highly Compressible Soils, in the 19th ICSMGE, Seoul, Korea, September 17-22, 2017. Soft Clays in South-West Area was held at Mokpo National Maritime University, on November 8-9, 2018. Several ATC7 members have been invited to deliver their lectures in 17 different academic seminars, conferences. The workshop was held on 17th of October, 2019 in Taipei, Taiwan, during the 16th Asian Regional Conference.

(5) AsRTC-8 on Groundwater Environmental and Quality Management is going to stop because of no activity.

(6) AsRTC-10: Urban Geo-informatics, Chair: Dr. Mimura, Mamoru (Japan), Secretary: Ms. Kitada, Naoko (Japan).

(7) AsRTC-18: Mega Foundation, Chair Prof.Sang-Seom Jeong (South Korea). ATC-18 workshop was held under Prof. Myoungmo Kim with Six presentation, 15th Asian Regional Conference in Fukuoka Japan. 2017 International Symposium on Design and Analysis of Piled Raft Foundations, 12-13 Sep. 2017 in Taipei, Taiwan. ATC-18 workshop was also held during the 16th ARC in Taipei, Taiwan.

(8) AsRTC-19: Geo-engineering for conservation of heritage monuments and historical sites Chair: Prof. Iwasaki, Yoshinori (Japan), Secretary: Prof. Mamoru Mimura (Japan). Special Workshop of ATC19 at Yunlin University, Taiwan, and also during the 16th ARC in Taipei, Taiwan in 2019.

(9) AsRTC-14: Smart Observation Method, Chair Dr. Moon-kyung Chung, Secretary Dr. Changho Choi (Korea), established in Oct. 2019 during the 16th ARC in Taipei, Taiwan.

5 TIME CAPSULE PROJECT

The types of selection for Time Capsule Project has been surveyed in Asian region. The Type A is involved with the significant content, interaction with individual members, and exploratory. The Type B is categorized with "off the shelf", in general reflecting past achievements. The Type C would be minimal, a couple of slides or link to existing website. The results of survey from Asian membership societies are summarized. Type A (9 societies; China, Chinese Taipei, Hong Kong, India, Indonesia, Kyrgyzstan, SEAGS, Sri Lanka, Uzbekistan), Type B (8 societies; Japan, Kazakhstan, Malaysia, Pakistan, Philippines, Singapore, South Korea, Tajikistan). Type C (2 societies; Iraq, Mongolia), and no response (7 societies; Bangladesh, Lebanon, Nepal, Syria, Vietnam, Thailand, Iran), respectively.
So far 3 societies, China, India, and Korea are submitted the TCP. The CISMGE submitted 16 reports, in which the first 13 reports focus on progresses in geotechnical profession in China from 1990s to 2022 (National Science and Technology, Natural Science Awards). The other 3 reports are related to the future prospects of geotechnical engineering. Indian Geotechnical Society is submitted the 6 reports regarding on the topic of tunnels in Himalaya and underground space for mining, three decades of geosynthetics in India, some challenging geotechnical projects such as steep slope in Lower Himalayas and coastal erosion, and construction of Tehri Hydropower Project in Himalaya region prone to seismic activity. The Korean Geotechnical Society are submitted 10 reports under the theme of "Geotechnical Engineering Milestone for Past Half Century" in which related to the projects such as Gyeongbu Expressway, Saemangeum reclamation project, Seoul Metropolitan Metro System, and Incheon International Airport. Other 6 reports are focused on the constructions of subsea tunnel, mega-foundation for high rise building, national high-speed railway system, long-span bridge, offshore wind power plants, and underground radioactive waste repository.

6 ACTIVITY OF VP OF ISSMGE FOR ASIA

The term of Vice President of ISSMGE for Asia was began since the September 2017, the 19th ICSMGE, Seoul, Korea and terminated by the Sydney 20th ICSMGE in May 2022. I have been traveled extensively for participating ISSMGE Board Meeting, ISSMGE TC, ATC, and national conference of membership societies which were held in Asian region until February 2020. The pandemic of COVID-19 makes a difficult to travel overseas. Therefore, the most of international activities including the ISSMGE Board Meeting as well as international conferences have been executed by on-line format or called Webinars.

The attending the international events and delivering lectures as the Vice President of ISSMGE for Asia during the period of September 2017 to May 2022 are total 50 times; 2017 (4), 2018(10), 2019(17), 2020(9), 2021(8), 2022(2), and summarized as follows.

2017:

Sept. 20 Organized ISSMGE Asian Council Meeting as the Vice President.

Oct. 31 Keynote Lecture, Indonesia Geotechnical Engineering Conference in Jakarta.

Nov. 20 Keynote lecture, Singapore Geotechnical Engineering Conference and attended GeoSS General Assembly Meeting, election new president of GeoSS.

Dec. 5 Delivered lecture and attended Indian Geotechnical Engineering National Conference at IIT Delhi and attended IGS General Assembly Meeting, elected new President of IGS (Prof. S. Babu, IIS).

2018:

Jan. 23 Zurich Institute of Switzerland Geotechnical Laboratory (Prof. Anastaspoulos Ioannis), Switzerland.

Jan. 26 Visiting Vienna Technical University Geotechnical Laboratory (Prof. H. Brandle), Austria.

Feb. 5 Special invitation lecture for Singapore Geotechnical Society at NUS. Attended GeoSS Board Meeting.

April 6 Invited speaker at the ISSMGE TC 202 Transportation Geotechnics at IIT Delhi, India.

April 4 Keynote Lecture at International Disaster Management Conference at Padang, Indonesia.

June 5 Attending ISSMGE Board Meeting at Skopje, Macedonia.

July 3 Keynote Lecture at International Disaster Reduction and Mitigation Conference at Sakhalin, Russia, organized by Russian Geotechnical Society, and Far Eastern Transportation University.

Nov. 5 Invited speaker, Southeast Geotechnical Engineering Conference at Jakarta, Indonesia.

Nov. 16 Attending ISSMGE Board Meeting and delivering Touring Lecture at Mexico City, Mexico.

Dec. 11 Attending ISSMGE TC 207, Soil-Structure Interaction and Retaining Walls and delivered invited lecture at Indian Geotechnical Engineering Conference at IIS, Bangalore, India. Attended IGS General Assembly Meeting and elected IGS president (Prof. S. Babu at IIS).

2019:

Jan. 28 Round Table Meeting with the Board Member of Vietnamese Geotechnical Engineering Society at Hanoi City. Discuss about Geotec Hanoi 2019 Conference in Nov. with Dr. Phung Duc Long (President of VSSMGE).

Feb. 11 Delivering Special Invitation Lecture at IGS-Chennai Chapter IIT Madras, India (Coordinated by Prof. Buminantan). Discussion about ISSMGE TC 301 Preservation of Historic Sites Conference in September, Chennai, India.

Feb. 15 Delivering Special Invitation Lecture at IGS-Cochin Chapter, Cochin, India (Coordinate by Dr. Anil Joseph)

Feb. 23 Keynote Lecture, ISSMGE TC 305 Geotechnical Infrastructures for Mega Cities and New Capitals. Symposium at Astana, Kazakhstan, Kazakhstan Geotechnical Society (Prof. Askar Zhussupbekov).

March 10 Attending ISSMGE Board Meeting and delivering Touring Lectures at Singapore. Round Table Board Meeting with GeoSS.

May 18 Memorial Lecture, Silver Jubilee Celebration Conference, 25th Anniversary of Nepal Geotechnical Society, Kathmandu, Nepal.

May 21 Invited Lecture, TRANSOILCOLD 2019 Conference, Transportation Soil Engineering in Cold Regions, Russian Geotechnical Society, St. Petersburg, Russia.

June 17 Keynote Speaker at ISCORD 2019, Cold Region Development Association Oulu, Finland.

June 26 Invited Lecture, ISSMGE TC 203 earthquake at Chania-Crete, Greece.

Aug. 29 Attend Unsaturated Soil Symposium at Okayama, Japan.

Sept. 17 Invited Lecture, ISSMGE TC 301 Preservation of Historic Sites, Chennai, India, IIT Madras.

Sept. 20 Invited Lecture, ISSMGE TC 107 Laterites and Lateritic Soils, Goa, India.

Oct. 5 Attending ISSMGE Council Meeting, Cape Town, South Africa.

Oct. 14 Invited Lecture and Asian Council Meeting, ISSMGE, 16th ARC, Taipei, Taiwan.

Nov. 28 Invited Lecture, 4th International Conference on Geotechnics, Geotec Hanoi, Vietnam.

Dec. 5 Keynote Lecture, 9th Asian Young Geotechnical Engineering Conference, Lahore, Pakistan.

Dec. 19 Invited Lecture, Indian Geotechnical Conference, SVNIT Surat, Gujarat, India.

2020:

Feb.6 Keynote Speaker, International Geotechnical Engineering Symposium at Manila Hotel, "Urban Geotechnics", Organized by Philippine Society for Soil Mechanics and Geotechnical Engineering (PSSMGE), 70 people attended.

March 8 ISSMGE Board Meeting by Zoom (Supposed to be in Tunis, Africa).

May 2 TC 305 Invited Speaker, International Webinar, Megacities and New Capital, Nur-Sultan City, hosted by Kazakhstan Geotechnical Society.

Aug. 23 Invited Speaker, TC305, Geotechnical Challenges of Mega Projects, Hosting by University of Kerbala and Iraqi Scientific Society of Soil Mechanics and Foundation Engineering, Bagdad, Iraq.

Sept. 7 ISSMGE Board Meeting, Participated by Zoom, London, England.

Oct. 15 International Webinar, TC 305, Geotechnical Engineering for Megacities, Russian Geotechnical Engineering Society, Moscow, Russia.

Dec. 12 Invited Speaker, International Webinar TC 302 Forensic Geotechnical Engineering, New Delhi, India (Prof. Sivakumar Babu).

Dec. 16 TC 213 Invited Speaker, International Workshop by Zoom, Scour and Erosion, Andhra University, Visakhpatnam, India construction of social infra-structures.

Dec. 16 Invited Speaker, Indian Geotechnical Society Annual Conference at Andhra University, Visakhpatnam, India.

2021:

Jan. 29 ISSMGE Board Meeting, participated by Zoom, London, England 20th ICSMGE is postponed from September, 2021 to May 1-6, 2022.

May 19 ISSMGE Board Meeting, participated by Zoom, London, England.

June 2 TC 216, Frost Geotechnics (Chair: Prof. Takashi Ono), TC305, Geotechnical Infrastructures for Megacities and New Capital, 2 Sessions, 12 Presentations, (Prof. Askar Zhussupbekov), Hosting Zoom by Kazakhstan Geotechnical Society, Nur-Sultan City, Kazakhstan.

June 22 2nd International Conference on Geotech. Engineering-Iraq-2021, Hosting Zoom By the Iraqi Scientific Society of Soil Mechanics and Foundation Engineering (ISSSMFE), University of Baghdad, Duhok Polytechnic University, University of Technology-Iraq (Akre-Duhok-Iraq).

Oct. 21 Invited Speaker, The 3rd International Scientific and Practical Conference "Russian Surveyor's Forum" was held in an off-line format in Moscow, Russia. EC Shin was awarded Gersevanov Medal by the Russian Society for Soil Mechanics, Geotechnics and Foundation Engineering (RSSMGFE) for his professional as well as educational contribution to the Russia and other parts of the world.

Oct. 26 Invited Speaker, The International Scientific and Technical Conference on Geotechnics was held in a mixed format, on-line (200 people) and off-line (200 people), in St. Petersburg State University of Architecture and Civil Engineering (SPBGASU). 150 papers were presented and speakers from 12 different countries were participated. EC Shin received the Honorary Doctorates Degree from SPBGASU.

Dec. 6 Attended with Congratulatory Message, the 3rd International Conference on Geotechnical Engineering (Previous conferences, Years 2007 and 2015), ICGE Colombo, 2020 organized by Sri Lankan Geotechnical Society. 5 keynote lectures, 2 Bright Spark Lectures, 100 papers were presented under the theme "Geotechnics in a challenging Environment".

Dec. 21 Attended ISSMGE Board Meeting by Zoom, London, England.

2022:

Jan. 24 Attended ISSMGE Board Meeting by Zoom, London, England.

Jan. 25 Invited Lecturer, The 1st International Conference on ACE (Architecture, Civil and Environmental) Forensic Engineering by Zoom, Korea University (Chairman, Prof. Jong-Sub Lee) in Seoul, under the auspicious of Korean Geotechnical Society (President; Dr. Moon-kyung Chung).

7 CONCLUDING REMARKS

The continental of Asian region covers the largest area among 6 different continentals with the 65% of world population. In the past 50 years, numerous social infra-structures such as highway, railway, airport, port and harbor, subway, and dams have been designed and constructed on land and also along the coastal area in Asian countries. The innovative geotechnical engineering technology is being applied in those projects. With conjunction of these developments, the international activities under the ISSMGE is pretty active through the ICSMGE, ARC, and AYGEC. The collaboration between geotechnical societies in Asian region is promising to contribute in the goal of ISSMGE. This report can see the glace of past 4 years and half activities under the auspicious of ISSMGE by the vice president of ISSMGE for Asia. I deeply thanks to ISSMGE and Korean Geotechnical Society for financial supporting to make me to participate and play a key role of the international activities in Asian region.

Appendix 5 : Report on Activities in Australasia

Rapport sur les activités en Australasie

Philip Robins
Vice President - Australasia, Technical Director, Beca, New Zealand philip.robins@beca.com

SUMMARY: This report contains an overview of the state of the ISSMGE in Australasia and summarises the activities that have taken place in the region since the Council meeting that was held in Cape Town in October 2019. The report also presents some general remarks on the state of the two member societies in the region and a view towards the future.

1 MEMBER SOCIETIES AND MEMBERSHIP

1.1 *Introduction*

The Australasian region of the ISSMGE, comprises two Member Societies, those of Australia and New Zealand. While it is the smallest ISSMGE region in terms of the number of Member Societies (2% of the total), its individual membership, at 1908, represents approximately 8% of the ISSMGE membership.

Australia and New Zealand are modestly populated counties and the current membership in the region represents 80 ISSMGE members per million inhabitants. This compares to an overall average of 3 ISSMGE members per million inhabitants for the Society as a whole. These statistics reflect both the high level of engagement with the ISSMGE by engineers in the region and the level of geotechnical engineering activity taking place in Australia and New Zealand.

1.2 *Australian Geotechnical Society*

The AGS (www.australiangeomechanics.org) is a joint technical society of the Institution of Engineers, Australia and the Australasian Institute of Mining and Metallurgy. The number of members has increased significantly over the last 4 years, and this is consistent with a general rebound in the infrastructure market in Australia during this period. At the time of writing this report the total AGS membership is 2,146. Of these, 1,200 (56%) are members of the ISSMGE.

1.3 *New Zealand Geotechnical Society*

The NZGS (www.nzgs.org) is a Collaborating Technical Society of Engineering New Zealand (previously the Institution of Professional Engineers NZ). It was recognised as the outstanding ISSMGE member society at the 2013 ICSMGE in Paris and remains extremely proud of this achievement.

At the time of writing this report, the NZGS has about 1,348 members. Of these, 824 (60%) are ISSMGE members. The NZGS works closely with the national structural engineering (SESOC) and earthquake engineering (NZSEE) societies to develop and deliver technical guidance and training. It also works closely with Engineering New Zealand and with Government in the development of policy and professional guidance.

1.4 Gavin Alexander

This report would be incomplete without my sharing that our former ISSMGE Vice President for Australasia, Gavin Alexander, passed away on 19th November 2020 after a battle with cancer, surrounded by his family.

Gavin was a Life Member of NZGS and served on the management committee for a decade taking on the roles of Treasurer and then Chair. He enjoyed his role as VP – Australasia and was really looking forward to the Conference in Sydney.

2 AUSTRALASIAN REGIONAL CONFERENCES

The conference series that serves the ISSMGE region is the Australia New Zealand (ANZ) Conference on Geomechanics, which is held once every 4 years and involves a rotation cycle, over a 12-year period, where two are held in Australia and one in New Zealand.

During the period 2017–2019, one ANZ Conference was held. The 13th Australia New Zealand Conference on Geomechanics, was held in Perth from 1 to 3 April 2019, and was hosted by the AGS. More than 450 registrants participated in the event. The conference was led by Michael Smith, Barry Lehane and Hugo Acosta and was very successful, both technically and socially. The next in the series of ANZ Conferences is to be held in 2023 in Cairns.

3 AUSTRALASIAN SUB-REGIONAL CONFERENCES

The NZGS holds a symposium once every two years. The 20th NZGS Symposium was convened in Napier, November 2107. No national NZ conference was held in 2019, due to the 13th ANZ Conference on Geomechanics in Perth.

The 21st NZGS Symposium was to be held at the Dunedin Conference Centre in March 2021 and was titled "Good Grounds for the Future". Total participation in the Symposium was 265 in person delegates and 46 virtual one-day passes.

The pre-Symposium event proved to be one of the most popular sessions and included a half-day workshop in Queenstown, followed by a one-day field study starting from Queenstown through Cromwell and Kawarau Gorges by bus, ending in Dunedin by train through the Taieri Gorge.

In the "Dunedin Centre" the 2-day Symposium included a packed technical programme; four international keynote presentations, special and plenary sessions. For the first time, and as part of the Symposium, the NZGS hosted a public talk "Natural Hazard Risk - Treatment Options for Resilience".

There was a breakfast with YGP followed by four presentations by the best paper winners of the NZGS regional YGP mini-Symposia and a wonderful Gala Dinner.

After a six-month delay caused by the pandemic it was wonderful to finally meet up with friends from around the country. Convener Eleni Gkeli and her organizing committee created an event that was a great success with quality technical presentations, fantastic keynotes beamed in from around the world, and many opportunities to socialize.

The AGS does not hold national conferences, instead, most of the AGS chapters hold an annual or biennial seminar on a wide variety of geotechnical engineering related themes. The AGS has also taken advantage of the COVID pandemic and created a series of webinars. The webinars are presented by leading practitioners and details of these are available from the AGS website.

Figure 1. Opening Ceremony at the 21st NZGS Symposium in Dunedin, March 2021.

4 20ICSMGE, SYDNEY, 2022

The Australian Geomechanics Society was delighted to have been unanimously awarded hosting rights for the 2021 ICSMGE. This conference is the highlight of the Presidential term, and Sydney is a truly stunning location for it.

The 20th ICSMGE will be held in Sydney from 1st to 5th of May 2022. The Local Organising Committee (LOC), led by Emeritus Professor John Carter with Graham Scholey (Conference Co-Chair) and Professor Mark Jaksa as Scientific Program Chair has put together an awesome programme of both technical sessions and other conference activities delivered on-site in Sydney and virtually in "hybrid mode".

The hybrid model ensures that all authors of accepted papers will have the opportunity to present, even if they are not able to travel to Sydney. Furthermore, for the very first time in the 85 years history of ISSMGE, all 21,000 individual members will have the opportunity to access the huge range of technical content presented on-line at ICSMGE 2022 - a remarkable opportunity for individuals to be part of this premier event!

As part of the Conference Advisory Committee (CAC), I had the chance to experience the hard work John's team has made towards ensuring the 20ICSMGE will be hugely successful. I have been amazed by the level of patience and tenacity the LOC has displayed while juggling everything the COVID pandemic has thrown at them. I am looking forward to attending the Conference in Sydney and encourage all those that can make it in-person, to join us there.

Darren Paul (YGP Convenor) and his team have also put together an amazing programme for the 7iYGEC. For the 7th time, young geotechnical practitioners will have the opportunity to participate in their own dedicated Conference on the days immediately prior to the 20ICSMGE in a relaxed, friendly and supportive environment to share their research, ideas and experience, build international networks and discuss the challenges facing young geotechnical engineers in a rapidly changing world.

The President, Secretary General and I had the opportunity to meet with the LOC in Sydney in February 2020. We were given a tour of the Sydney International Conference Centre (ICC). It is Australia's first fully integrated convention, events, exhibition, and entertainment centre. Featuring a striking contemporary design, leading technology and world-class meeting and exhibition spaces with wonderful views over Sydney's Darling Harbour.

5 YOUNG GEOTECHNICAL PROFESSIONALS

The NZGS and AGS also share responsibility for organizing and hosting the biennial Young Geotechnical Professionals (YGP) Conference series in the region. The conferences are intended for young professionals, 35 years old or younger, with a maximum of 10 years' experience. The number of attendees is generally capped at 50 delegates, to facilitate an intimate and nurturing conference experience.

Usually, 3 or 4 senior geotechnical engineers from industry and academia are actively involved in the conference and provide mentorship to the delegates. Unfortunately, the AGS organized the 13th YGP Conference planned for Cairns, Queensland, was cancelled due to travel restrictions imposed by the COVID-19 pandemic.

Figure 2. John, Neil and Philip at the Sydney ICC.

The level of over-subscription of these events has led to the introduction of local one-day events around New Zealand for young professionals who have not been able to attend the regional conferences. These local events provide similar opportunities to present to one's peers in a supportive environment, to develop professional networks, and to be mentored by more experienced practitioners.

6 TECHNICAL COMMITTEES AND BOARD LEVEL COMMITTEES

Both the AGS and NZGS have strong participation in the ISSMGE's technical committees. The Region's membership of the Technical Committees was comprehensively reviewed and refreshed through 2018 and 2019.

Sukumar Pathmanandavel of the AGS and Co-Chair of the Corporate Associates Presidential Group (CAPG), has been an extremely active and has worked tirelessly in this role to increase the CAPG. More recently, Sukumar has conceptualized and now now spear heads the Time Capsule Project (TCP).

A key focus of the TCP is the "process" by which there is connection and interaction with the individual members about the work of the contributors, an outcome of which is the "product" that is offered to the TCP for storage and display.

7 NZGS CLIMATE CHANGE SYMPOSIUM

At the beginning of November, NZGS delivered the 1st Geotechnical Symposium focused on Climate Change and Sustainable practices. This Symposium was organized by NZGS, in collaboration with MBIE and Engineering NZ, and attracted very interesting presentations showcasing that our industry is already preparing and adapting to the upcoming change. The discussions that followed the presentations highlighted the high potential and unique opportunities for our profession to thrive in this field, as there is a lot more yet to be done.

8 AUSTRALIAN GEOMECHANICS SPECIAL EDITION

Australian Geomechanics is the journal of the AGS. Published quarterly, it includes a selection of quality peer-reviewed papers with a particular focus on topics relevant to Australasia.

The AGS hosted several female speakers in 2020, including keynotes at local symposia, and Dr Jackie Skipper presenting her Glossop Lecture. To continue encouraging female professionals to contribute in this way and building on this increase in exposure for women in our industry, at end of 2021 the AGS produced a dedicated issue of the Australian Geomechanics. The special journal edition contained papers written by female geotechnical professionals (as the main author or a co-author).

9 INTERACTION WITH MEMBER SOCIETIES

The NZGS and AGS have a strong working relationship. They interact regularly, with the Chairs of both societies attending one another's committee meetings at least once each year. Moreover, the Vice Presidents of the 3 sister societies (ISSMGE, IAEG and ISRM) regularly attend both societies' committee meetings and work closely with local liaison officers from the other Society.

In addition, both societies collaborate on the development of joint codes of practice and guidelines, and the AGS and NZGS are in regular communication regarding visiting overseas speakers and visitors so that both societies can exploit opportunities for presentations in both countries.

An example of this is the Rankine Downunder series of presentations, where the Rankine Lecturer from the previous year is invited to tour both Australia and New Zealand in order to present their lecture to all chapters and regions of the AGS and NZGS.

10 YOUR NEW VP FOR AUSTRALASIA

For the next presidential term, and my successor, will be Graham Scholey from AGS, and he will represent the Australasian region on the ISSMGE Board.

Graham is a principal geotechnical engineer with over 30 years of experience providing geotechnical and applied geological consulting services, primarily for civil infrastructure projects such as ports, railways, roads, utilities and building developments.

I have known Graham personally for over 25 years and I am sure he will continue the energetic and engaged tradition of Vice Presidents from our region.

11 GENERAL REMARKS AND OUTLOOK

The Australasian region is vibrant, active and its members are engaged with the profession and the ISSMGE. It is exciting to see the leadership of our two societies lead by two women: Nina and Eleni. It has been awesome to see how well the Societies have adapted to the ever-changing COVID pandemic, always looking for different and creative ways to connect with their members, and other technical societies. I am confident that the The Australasian societies and the members they represent will continue to grow and keep engaged with the profession and the ISSMGE.

For the next ISSMGE Presidential term, I would like to see the AGS and NZGS continue to focus on a couple of things, namely climate change and diversity. The pacific island nations are likely to be affected most, by climate change and I believe that as geotechnical engineers we have the potential and unique opportunity to creative solutions to help those places affected most by climate change.

I think that diversity of thought leads to diversity of action, and I encourage both the AGS and NZGS to encourage all of it members, especially female geotechnical engineering to challenge the status quo and join management committees, conference organizing committees and the like.

12 CLOSURE

I am extremely grateful and humbled to have been given the opportunity to serve both the AGS and NZGS, as Vice President for Australasia and Treasurer on the ISSMGE Board over these last two years.

Although I reluctantly took on the role from Gavin in 2019 when he stood down due to his health issues, I have thoroughly enjoyed this experience. I wish the ISSMGE and all its member societies every success for the future.

Table 1. Technical Committee Representation (*corresponding member)

Technical Committee	AGS Representatives		NZGS Representatives
TC101 – Laboratory Testing	Mizanur Rahman*		Gabriele Chiaro
TC102 – In-Situ Testing	Richard Kelly	Adrian McCallum*	
TC103 – Numerical Methods	Ali Karrech Arman Khoshghalb	Ali Parsa Pajouh Ha Bui	Dr CY Chin Ioannis Antonopoulos
TC104 – Physical Modelling	Muhammad Shazzad Hossain	Christophe Gaudin Ali Karrech*	Philip Robins
TC105 – Geo-Mechanics from Micro to Macro	Itai Einav	J S Vinod	Rolando Orense
TC106 – Unsaturated Soils	Adrian Russell Hadi Khabbaz Arman Khoshghalb	Nasser Khalili Daichao Sheng	
TC202 – Transportation	Daichao Sheng* Shanyong Wang* Mahdi Miri Disfani	Mizanur Rahman Mohamed Shahin	
TC203 – Earthquake	Ahmad Dehghanpoor Sichani Behrooz Ghahremannejad	Ivan Gratchev Nihal Vitharana* Jay Lee Sanjay Nimbalkar*	**Misko Cubrinovski (Chair)** Ioannis Antonopoulos* Sjoerd Van Ballegooy Brendon Bradley
TC204 – Underground Construction	Jay Lee*		
TC205 – Safety and Serviceability	Nihal Vitharana		
TC206 – Interactive Design	David Zhang	Jinsong Huang	
TC207 – Soil-Structure	Ahmad Dehghanpoor Sichani* Chris Haberfield* Dino Sarac*	Chenhui Lee* Sanjay Nimbalkar* Ali Parsa Pajouh* Craig Butterworth	Ioannis Antonopoulos Jan Kupec
TC208 – Slope Stability	Ivan Gratchev* Allan Herse* Dino Sarac	Ali Karrech* Meysam Safavian	Ross Roberts
TC209 – Offshore	**Phil Watson (Chair)** Muhammad Shazzad Hossain*	Noel Boylan Christophe Gaudin James Doherty*	
TC210 – Embankment Dams	Behrooz Ghahremannejad	Meysam Safavian* Nihal Vitharana	James Burr
TC211 – Ground Improvement	Babak Hamidi Shanyong Wang* Bosco Poon Buddhima Indraratna	Ilhan Chang Hadi Khabbaz* Chenhui Lee	Martin Larisch
TC212 – Deep Foundations	Allan Herse Bosco Poon Hossein Ahmadi*	Barry Lehane* Dr. Bindumadhava*	Martin Larisch
TC213 – Scour and Erosion	Mahdi Miri Disfani* Ha Bui	Scott Draper	Bruce Melville
TC214 – Soft Soils	Shiao Huey Chow		
TC215 – Geo-Environmental	**Malek Bouazza (Chair)** Abbas El-Zein		
TC216 – Frost	Daichao Sheng	Adrian McCallum	
TC217 – Land Reclamation	David Zhang	Hadi Khabbaz	Anthony Fairclough
TC218 – Reinforced Fill Structures	Brett Gibbons Riccardo Musella*	Antonio Ramirez Martinez Gary Power*	
TC302 – Forensic	Dr. Bindumadhava Malek Bouazza	Sanjay Nimbalkar*	
TC303 – Floods	Buddhima Indraratna*	Hadi Khabbaz	
TC304 – Risk	Jinsong Huang Jianfeng Xue	Mark Jaksa	
TC305 – Megacities	Sanjay Nimbalkar		
TC306 – Geo Education	David Airey	Mark Jaksa	
TC307 – Sustainability	Mizanur Rahman	Abbas El-Zein	
TC308 – Energy Geotechnics	Malek Bouazza Asal Bidarmaghz*	Shanyong Wang Guillermo Narsilio	
TC309 – Machine Learning	Ali Karrech Jinsong Huang*	Sukumar Pathmanandavel Chongchong Qi*	Vic Kumaran

Appendix 5: Council report for Europe

Rapport du Conseil pour l'Europe

Mario Manassero
Department of Structural, Geotechnical and Building Engineering, Politecnico di Torino, Italy, mario.manassero@polito.it

SUMMARY: This report contains an overview of the state of ISSMGE in Europe and a summary of the main past and future activities in the region since the Council meeting held in Seoul in September 2017.

1. MEMBER SOCIETIES AND MEMBERSHIP

Presently, there are 38 European Member Societies (out of a total of 90), from 39 countries, since the Czech and Slovak Republics joined as a single society. Moreover, it is worth remembering that in the membership list submitted to the Seoul Council in September 2017 by the President, Roger Frank, and the Vice-President (VP) for Europe, Antonio Gens, Cyprus became a new ISSMGE Member Society with the name Cypriot Society for Soil Mechanics and Geotechnical Engineering (CSSMGE). The total number of individual members of the European Member Societies is currently 8751, which is a 7.5% higher number than the 8142 European societies at the time of the Seoul Council Meeting. They represent about 40% of the total ISSMGE membership.

Moreover, it is also worth mentioning that an active group of Moldavian geo-technicians, with the help and support of the Romanian Geotechnical Society, had made good progress in establishing the Moldavia society of soil mechanics and geotechnical engineering until the onset of the COVID-19 pandemic. During the pandemic, the institutional collaboration between the two countries, which had consisted of joint seminars, workshops and conferences, has significantly slowed down. Nevertheless, in spite of the many limitations that have arisen as a result of the COVID-19 pandemic, the willingness to apply for ISSMGE membership is still alive as well and, in the desired case of success, it will hopefully take place in the near future, thus Moldavia would become the 40th European Country Member of ISSMGE.

2. EUROPEAN REGIONAL CONFERENCES

The main event concerning geotechnical engineering in Europe was the XVII European Conference on Soil Mechanics and Geotechnical Engineering, which was held in Reykjavik from the 1st to the 6th September 2019. The Conference was organised by the Icelandic Geotechnical Society and it was a great success, thanks, in great part, to the hard work of an efficient and dedicated Conference Organising Committee (COC) chaired by Mr. Haraldur Sigursteinsson.

The venue was the Harpa Conference Centre, situated in the Reykjavik harbour. The general theme of the Conference was *Geotechnical Engineering, foundation of the future*, a theme that was able to accommodate a wide variety of topics grouped under 6 different headlines: A: Modelling and experimental assessment of geomaterials; B: Geotechnical construction and soil improvement; C: Geohazards, earthquakes and mitigation; D: Environment, water and energy; E: Historical heritage preservation and F: Special and specific Issues.

The Conference was structured over four days and the technical programme included: 5 Keynote lectures, 9 Invited lectures (including Bright Spark lectures), 7 Main Sessions and 42 Discussion Sessions, which were anticipated by 7 Pre-Conference Technical Committees Workshops on Sunday 1st September. In addition, a number of Meetings were organised by the Technical Committees (TC) and the European Regional Technical Committees (ERTC).

The five Keynote lecturers were Suzannne Lacasse, Sigurður Erlingsson, Antonio Gens, Jorgeb G. Zornberg and Lyesse Laloui, whereas the Invited lectures were delivered by Gioacchino Viggiani, Lidija Zdraykovic, Maurice Bottiau, Mike G. Winter, Sebastiano Foti, Ólafur G. Flóvenz and Sigurður Reynir Gíslason, and the two young geotechnical engineers, who were awarded with the Bright Spark Lecture, were Federico Pisanò and Matteo Ciantia.

No limits were set for the number of papers that each Member Society could submit. Out of the 1120 Abstracts initially proposed, 679 papers (from 34 European countries and from 22 countries outside the European Region) were finally accepted for publishing in the corresponding printed and digital Proceedings. As many as 410 papers were presented orally, and 192 were offered as poster presentations. It is worth noting that all the papers were available for downloading from the conference website at the very beginning of the conference in order to allow the audience to follow all the presentations on both the main screens and/or on their computers, tablets or smart phones. In addition, a series of Technical Tours and an extensive Social Programme completed the Conference activities.

A total of 852 participants from 60 countries attended the event and, moreover, 58 Technical Exhibition stands were occupied.

During the Conference, a meeting of the European Member Societies was held on the 3rd September. A number of issues of common interest were presented and discussed.

As had already occurred during the last edition of ECSMGE (Edinburgh 2015), an important topic was considered to deserve special attention during the meeting i.e. the National Registration of Ground Engineering Professionals viewed from an European perspective. Within his presentation, the coordinator of this initiative, Fitan Buggy, past chairman of the Geotechnical Society of Ireland, pointed out the very delicate and crucial theme of the level of educational qualification and the duration of post-academic experience, which should define a minimum professional competency for the Ground Engineering Professionals from European Countries. However, the convergence of views on this topic proved to be rather difficult, with a diverse range of opinions being expressed, and it was therefore preliminarily proposed to allow each National Standards Board to adopt its own requirements to reflect the definition of competency in their National Register. Obviously, the suitability of this provisional compromise will have to be monitored over time. At the end of the presentation and the related discussions, the fundamental role of ISSMGE in encouraging greater participation among European countries, as well as in integrating the European registration with world-wide systems, was stressed.

At the end of the meeting, the location of the subsequent European Conference was also decided on. There were three very strong and well-prepared bids from the Belgian, Portuguese and Turkish Member Societies. After the vote of the Member Society representatives, Lisbon (Portugal) was chosen to be the next venue of the XVIII European Conference on Soil Mechanics and Geotechnical Engineering, which will take place from the 25th to the 30th August 2024. The venue will be the Conference Centre situated on the right side of the Tagus River, which offers a wonderful view of the Vasco da Gama Bridge, among others. The theme of the conference is "*Challanges of Geotechncial Engineering to meet current and emergency society needs*", a theme that promotes a positive look towards the future in order to face it in a proactive way through Geotechnical Engineering contributions, among others, in terms of development, sustainability and capability to predict and avoid environmental disasters. More information is provided on the website: *http://www.spgeotecnia.pt.*

3. EUROPEAN SUB-REGIONAL CONFERENCES

In addition to the four-year ECSMGE series in the region, there are also a number of European Conferences involving sub-regional groups of Member Societies: the Danube - European Conference on Geotechnical Engineering, the Nordic Geotechnical Meeting and the Baltic Sea Geotechnical Conference.

The **16th Danube-European Conference on Geotechnical Engineering**, organised by the Macedonian Association for Geotechnics, was held at the Hotel Aleksandar Palace of Skopje on 7-9 June 2018. The Organising Committee was chaired by Professor Milorad Jovanovski and co-chaired by Professor Ljupčo Dimitrievski. The conference theme was: *Geotechnical Hazards and Risks: Experiences and Practices* (*Geotechnische Gefahren und Risiken: Erfahrungen und Praxis*). Participants from 46 countries took part, and there were 10 invited lecturers and more than 150 papers. During this event, the President of the Republic of Macedonia awarded the Medal for Merit to Prof. Heinz Brandl for his outstanding contribution to the establishment and to the general increase of the Macedonian geotechnical community.

The next **17th Danube - European Conference on Geotechnical Engineering,** originally scheduled for September 2022, was postponed because of the persistence of the COVID-19 pandemic and is now scheduled to take place in Bucharest (Romania) on 7 - 9 June 2023. The event will be organised by the Romanian Society for Geotechnical and Foundation Engineering.

Moreover, again due to the COVID-19 pandemic, both the **14th Baltic Sea Geotechnical Conference and the 18th Nordic Geotechnical Meeting** were run parallelly in full online mode on 18-19 January 2021. The events were organised by the Finnish Geotechnical Society under the chairmanship of Ville Raasakka.

4. EUROPEAN YOUNG GEOTECHNICAL ENGINEERS' CONFERENCES

One of the yearly highlights of the European meeting calendar is the European Young Geotechnical Engineers' Conference (EYGEC). The **26th EYGEC** was held at Hotel Klugbauer in Reinischkogel near Graz in Austria from the 11th to the 14th September 2018 and was organised by the Austrian Member Society under the chairmanship of Prof. Helmut Schweiger. It was attended by 45 delegates, nominated by 26 different countries. Four keynote lectures were delivered by Janko Logar (University of Ljubljana, Slovenia), Clemens Kummerer (Keller Kolding, Germany), Dietmar Adam (Vienna University of Technology, Austria) and Helmut F. Schweiger (Graz University of Technology, Austria). Overall, the presentations were of a high level and were well prepared by the young geotechnical engineers, and they covered a wide range of geotechnical problems. The participants represented a well-balanced group of academics and practitioners and, therefore, not only high-level research results were illustrated, but also interesting problems that arose from practice were discussed.

Just two weeks after the XVII ECSMGE, where great visibility had been given to young members through a joint main session, in cooperation with Corporate Members, which dealt with Bridging the Gap Between Academics and Practitioners, and through the Bright Spark Awards, the **27th EYGEC**, organised by the Turkish Geotechnical Society, under the chairmanship of Prof. Deniz Ülgen, took place in Bodrum, Muğla-Turkey, on the 26th and the 27th September 2019, in conjunction with another international event, planned in the same week and devoted to the young ISSMGE members, i.e. the **1st Mediterranean Young Geotechnical Engineers' Conference** organised in the same venue from the 23rd to the 24th September, 2019. In this way, a blended atmosphere, a bigger get together and a larger platform of collaboration were provided for the young attendees, together with "differently young" invited keynote speakers. Hopefully, this initiative should give rise, in the near future and possibly just after the pandemic limitation, to a new conference series that should be able to promote a lasting and profitable cooperation among all the participants for the common technical and scientific progress of the Mediterranean Regions within the Geotechnical Engineering discipline.

The EYGEC was attended by 36 young delegates, representing 23 Member Societies. The invited lectures were delivered by Pierre Delage, Andrea Dominijanni, İlknur Bozbey, Lyesse Laloui, George Gazetas and Wissem Frikha.

As a result of the limitations of the COVID-19 pandemic, the 28th EYGEC, which is going to be organised by the Russian Member Society and was originally planned to take place in Moscow, September 2020, has had to be rescheduled to take place in October 2023, also considering that, as is customary, there will be no EYGEC in 2022, due to the planned world-wide Young Geotechnical Engineers' Conferences which will be held in conjunction with the Sydney International Conference.

The 29th and 30th EYGEC have not been officialised yet, since the final choices and decisions will be in the hands of the next European VP, Lyesse Laloui. However, the Macedonian Association for Geotechnics and the Comité Français de la Mécanique des Sols et de Géotechnique have already expressed their willingness to organise the two aforementioned EYGEC editions, which at present are scheduled for the years 2024 and 2025, respectively.

5. OTHER INTERNATIONAL CONFERENCES IN EUROPE

In addition to the events mentioned in the previous sections, a large number of ISSMGE sponsored International Conferences have been held in Europe, often organised or supported by Technical Committees and/or by Member Societies:

- * 2nd International Symposium on Coupled Phenomena in Environmental Geotechnics (CPEG2). Leeds, the UK, 06-08 September 2017. (University of Leeds and endorsed by TC215).
- 3rd International Soil-Structure Interaction Symposium. Izmir, Turkey, 18-20 October 2017. (Turkish Society for Soil Mechanics and Geotechnical Engineering).
- Geo-Expo 2017 Scientific and Expert Conference. Sarajevo, Bosnia and Herzegovina, 26-27 October 2017. (Geotechnical Society of Bosnia and Herzegovina).
- 2nd International Conference "Challenges in Geotechnical Engineering". Kiev, Ukraine, 20-23

November 2017. (Kyiv National University of Building and Architecture, the University of Zielona Góra, Ukrainian Society for Soil Mechanics, Geotechnics and Foundation Engineering).

- Granular Matter Workshop - CEGD Workshop. Budapest, Hungary, 12-13 June 2018. (Óbuda University, Hungarian Academy of Sciences and BME Morphodynamics Research Group).
- 4th International Symposium on Cone Penetration Testing (CPT'18). Delft, The Netherlands, 21-22 June 2018. (Delft University of Technology and endorsed by TC102).
- 9th European Conference on Numerical Methods in Geotechnical Engineering. Porto, Portugal, 25-27 June 2018. (The University of Porto and endorsed by ERTC7).
- 9th International Conference on Physical Modelling in Geotechnics. London, The UK, 17-20 July 2018. (City University of London and endorsed by TC104).
- * China – Europe Conference on Geotechnical Engineering. Vienna, Austria, 13-16 August 2018. (Institute of Geotechnical Engineering, the University of Natural Resources and Life Sciences Vienna (BOKU) and University of Leeds).
- Urban Planning Below the Ground Level: Architecture and Geotechnics. Saint Petersburg, Russia, 19-21 September 2018. (St. Petersburg Council of Architects and endorsed by TC207).
- * International Symposium on Energy Geotechnics. Lausanne, Switzerland, 26-28 September 2018. (Swiss Federal Institute of Technology in Lausanne and endorsed by TC308).
- Geo-Expo 2018 Scientific and Expert Conference. Neum, Bosnia and Herzegovina, 18-19 October 2018. (Geotechnical Society of Bosnia and Herzegovina).
- International Scientific-Technical Conference "Geotechnics of Belarus: Science and Practice". Minsk, The Republic of Belarus, 23-26 October 2018. (Belarusian National Technical University and Belarusian Geotechnical Society).
- Geotechnical Challenges in Karst. Omiš, Croatia. 11-13 April 2019 (Croatian Geotechnical Society).
- Prague Geotechnical Days 2019. Prague, The Czech Republic. 13-14 May 2019 (Czech and Slovak Society for Soil Mechanics and Geotechnical Engineering, SG Geotechnika and Charles University).
- 4th International Conference "Transportation Soil Engineering in Cold Regions". St. Petersburg, Russia. 20-23 May 2019 (Emperor Alexander I Petersburg State Transport University).
- 14th International Conference on Underground Construction. Prague, The Czech Republic. 3-5 June 2019 (Czech Tunnelling Association and ITA-AITES).
- Construction in Historical Cities: Problems and Solutions. St. Petersburg, Russia. 6-7 June 2019 (TC 207).
- * 7th International Conference on Earthquake Geotechnical Engineering. Rome, Italy. 17-20 June 2019 (TC 203).
- 7th International Symposium on Deformation Characteristics of Geomaterials. Glasgow, The UK. 26-28 June 2019 (TC 101).
- 3rd International Conference "Challenges in Geotechnical Engineering". Zielona Gora, Poland. 10-13 September 2019 (the University of Zielona Gora and Kyiv National University of Construction and Architecture).
- * 1st Mediterranean Young Geotechnical Engineers Conference. Bodrum, Turkey. 23-24 September 2019 (Turkish Society for ISSMGE).
- 3rd International Conference on Information Technology in Geo-Engineering. Guimarães, Portugal. 29 September – 2 October 2019 (the University of Minho and Portuguese Geotechnical Society).
- Energy Geotechnics Mechanics of the Energy Transition. Delft, The Netherlands. 3 October 2019 (CRUX Engineering BV, Cohere Consult, Delft University of Technology, Eindhoven University of Technology, TC308).
- * International Conference on Geotechnical Engineering Education. Online event. 23-25 June 2020 (TC 306).
- 4th European Conference on Physical Modelling in Geotechnics. Luleå, Sweden (hybrid mode). 6-8 September 2020 (Luleå Technical University).
- 4th European Conference on Unsaturated Soils. Online event. 19-21 October 2020 (Instituto Superior Técnico, Delft University of Technology and Universitat Politècnica de Catalunya).
- * Mediterranean Symposium on Landslides. Online event. 7-9 June 2021 (Università di Napoli Federico II, Università di Milano Bicocca and Università della Campania L. Vanvitelli).
- 6th International Conference on Geotechnical and Geophysical Site Characterization. Budapest, Hungary. 26-29 September 2021 (Hungarian Geotechnical Society).
- Second Generation of Eurocode 7 - Improvements and Challenges. Online event. 28-29 September 2021 (ERTC10).

Besides the ISSMGE official International Conferences, several geo-engineering events have been held in Europe, under the endorsement of ISSMGE:

- Intensive Short-Course on "Energy Geostructures: Analysis and Design" - 2nd Edition. Lausanne, Switzerland. 6-8 March 2019 (Swiss Federal Institute of Technology in Lausanne).
- 6th International Course on Geotechnical and Structural Monitoring. Rome, Italy. 27-31 May 2019 (Natural HAZards Control and Assessment).
- 2nd International Conference on Natural Hazards and Infrastructure. Chania, Greece. 23-26 June 2019 (Innovation Center on Natural Hazards and Infrastructure).
- EUROCLAY 2019 - Geotechnical characterisation of clayey geomaterials from micro to macroscale: the role of microstructure and anisotropy. Paris, France. 1-5 July 2019 (French Clay Group).
- International Symposium on SPH and Other Particle-Based Continuum Methods and their Applications in Geomechanics. Vienna, Austria. 11-13 September 2019 (the University of Natural Resources and Life Sciences of Vienna).
- ReSyLAB & GEO-EXPO 2019. Sarajevo, Bosnia and Herzegovina. 23-25 October 2019 (Geotechnical Society of Bosnia and Herzegovina).
- 8th International Geotechnical Symposium. Istanbul, Turkey. 13-15 November 2019 (Turkish Chamber of Civil engineers and Turkish Society for ISSMGE).
- 2nd International Symposium on Seismic Performance and Design of Slopes. Edinburgh, The UK. 18-22 January 2020 (University of Edinburgh).
- DFI Deep Mixing 2021. Online event. 1-17 June 2021 (Deep Foundations Institute).

- 6th International Conference on Geotechnical Research and Engineering. Online event. 21-23 June 2021 (Hungarian Geotechnical Society).

The conferences with an asterisk were attended by the VP Europe.

The forthcoming Conferences in Europe in the period immediately following the 20th International Conference on Soil Mechanics and Geotechnical Engineering (Sydney, Australia) are planned as follows:

- 5th International Symposium on Cone Penetration Testing. Bologna, Italy. 8-10 June 2022 (Italian Geotechnical Society and University of Bologna).
- 3rd International Conference on Environmental Geotechnology, Recycled Waste Materials and Sustainable Engineering. Izmir, Turkey. 16-18 June 2022 (Dokuz Eylul University).
- 3rd International Symposium on Geotechnical Engineering for the Preservation of Monuments and Historic Sites. Naples, Italy. 22-24 June 2022 (TC 301 and Italian Geotechnical Society).
- 10th International Symposium on Geotechnical Aspects of Underground Construction in Soft Ground. Cambridge, The UK. 27-29 June 2022 (University of Cambridge).
- 5th International Conference on New Developments in Soil Mechanics and Geotechnical Engineering. Nicosia, Northern Cyprus. 30 June – 2 July 2022 (Turkish Society for ISSMGE and Near East University).
- 16th International Conference of the International Association for Computer Methods and Advances in Geomechanics. Turin, Italy. 30 August – 2 September 2022 (Politecnico di Torino).
- 11th International Symposium on Field Monitoring in Geomechanics. London, The UK. 4-7 September 2022 (TC220).
- 11th International Conference on Stress Wave Theory and Design and Testing Methods for Deep Foundations. Rotterdam, The Netherlands. 20-23 September 2022 (Royal Netherlands Society of Engineers).

6. TECHNICAL COMMITTEES

Fourteen (out of thirty-seven) ISSMGE Technical Committees (TCs) are currently hosted by European Member Societies:

- TC-101: Laboratory Stress Strain Strength Testing of Geomaterials. British Geotechnical Association, Chair: Matthew Richard Coop.
- TC-102: Ground Property Characterization from In-Situ Tests. Sociedade Portuguesa de Geotecnia, Chair: Antonio Viana da Fonseca.
- TC-106: Unsaturated Soils. British Geotechnical Association, Chair: David Toll.
- TC-201: Geotechnical Aspects of Dykes and Levees and Shore Protection. Netherlands Society for SMGE, Chair: Cor Zwanenburg.
- TC-204: Underground Construction in Soft Ground. Netherlands Society for SMGE, Chair: Adam Bezuijen.
- TC-206: Interactive Geotechnical Design. British Geotechnical Association, Chair: Duncan Nicholson.
- TC-211: Ground Improvement. Groupement Belge de la SIMSG, Chair: Noel Huybrechts.
- TC-212: Deep Foundations. Associazione Geotecnica Italiana, Chair: Alessandro Mandolini.
- TC-218: Reinforced Fill Structures. Associazione Geotecnica Italiana, Chair: Giulia Lugli.
- TC-220: Field Monitoring in Geomechanics. British Geotechnical Association, Chair: Andrew Ridley.
- TC-222: Geotechnical BIM and Digital Twins. Norwegian Geotechnical Society, Chair: Magnus Romoen.
- TC-301: Preservation of Historic Sites. Associazione Geotecnica Italiana, Chair: Renato Lancellotta.
- TC-306: Geo-engineering Education. Hellenic Society of SMGE, Chair: Marina Pantazidou.
- TC-309: Machine Learning and Big Data. Norwegian Geotechnical Society, Chair: Zhongqiang Liu.

The aforementioned TCs have undertaken a significant number of activities that have been reviewed in the Technical Oversight Committee (TOC) report.

In addition to the ISSMGE TCs, there are four European Regional Technical Committees (ERTC) focused on issues that are particularly relevant for European Geotechnics.

- *ERTC-3: Piles.* Chair: Maurice Bottiau (Belgium),

This is a very active Committee with a steadily expanding membership. The Committee has held a number of online and face-to-face meetings. In particular, a very successful meeting, hosted by the Belgian Member Society, was organised during the 19th ICSMGE in Seoul, on the 20th September 2017. Moreover, an information session about guidelines for a technical approval system of pile foundations was held in Brussels on the 22nd May 2018. One of the main aims of ERTC-3 is to promote an active interaction with the other two ERTCs, 10 and 12, within the pile design and construction standardisation framework established by the new generation of Eurocodes.

- *ERTC-7: Numerical Methods.* Chair: Helmut Schweiger (Austria)

The Committee organised a very successful 9th European Conference on Numerical Methods in Geotechnical Engineering (NUMGE) in Porto (Portugal) on 25-27 June 2018. The Committee has also organised a workshop in the XVII ECSMGE, Reykjavik, Iceland on Numerical Surveying, Compare and Discuss the Development and Applications of Numerical Methods in Geotechnical Engineering, on the 1st September 2019. The next event is planned for the 10th NUMGE, which will be held at Imperial College, London, June 26-28, 2023.

- *ERTC-10 Evaluation of Eurocode 7.* Chair: Trevor Orr (Ireland).

The main activity of ETC 10 was the planning and preparation of several workshops, together with TC205 and TC304, and face-to-face meetings. Following the resignation of Professor Trevor Orr, after a very long period of time and a really outstanding contributions to the implementation and application of Eurocode 7, the new chairman, Georgios Katsigiannis, was officially appointed at the XVII ECSMGE in Reykjavik, Iceland, in September 2019. The new Chair has already been able to organise an important and successful on-line event: 1st ERTC10 Seminar on "Second Generation of Eurocode 7 - Improvements and Challenges" on Tuesday 28th September 2021 with almost 600 registrations from all around the world. The success of this event highlighted the considerable international interest in the development of the next generation of European standards on geotechnical design.

- *ERTC-12: Evaluation of Eurocode 8.* Chair: Christos Vrettos (Germany)

The main purpose of ERTC-12 is to provide general, as well as specific comments on the EC8 code from the earthquake geotechnical engineering practice point of view, and in light of the latest research. Moreover, it intends to conduct comparisons with other national and/or international codes and to propose improvements and modifications to EC 8. The new chairman,

Professor Christos Vrettos, in the role of DIN delegate of Eurocode 7 and Eurocode 8 and as a member of several DIN Committees on calculation methods in geotechnics and the seismic design of buildings, is in the ideal position to re-assemble the committee and to start with a fruitful phase of activities. Presently, ETC12 concentrates on communicating these findings to the drafting teams working on the new version of EC8.

7. INTERACTION WITH MEMBER SOCIETIES

Most Member Societies organise National meetings (often with international participants), but they are too numerous to list here. Among those reported on the websites of the member societies, I personally attended the following ones:

- 16th National Congress of Geotechnics. Ponta Delgada, S. Miguel, Açores, 27th – 30th May 2018. *Geotecnia, Riscos Naturais e Geotécnicos e Sustentabilidade.* Sociedade Portuguesa de Geotecnia.
- 1st National Symposium on *Environmental Geotechnics*. Bucharest – Romania, 2nd July 2018. Romanian Society for SMGE.
- School of Architecture, Civil and Environmental Engineering. École Polytéchnique Fédéral de Lausanne. *Civil Engineering Seminar Series*, Switzerland, 12th October 2018. Swiss Geotechnical Society.
- XXV Geotechnical Conference of Turin, (Italy), Turin 8th – 9th November 2018. *Analysis and Design of Geotechnical Works in Seismic Areas*. Associazione Geotecnica Italiana.
- XXIV Széchy Károly memorial session, 9th February 2019, Budapest, Hungary. Hungarian Geotechnical Society.

I took advantage of those National Meetings and the International Conferences that I attended to interact with officers and members of most of the European Member Societies. The following issues have often been raised for discussion: application and use of Eurocode in the different countries, the availability of ISSMGE publications via open access, how to raise the profile and visibility of our profession, the national registration of geotechnical professionals and the improvement of the recognition, appreciation and remuneration of the capabilities, skills and contributions of Geotechnical Engineers. The latter three objectives are in some way closely linked to each other.

Unfortunately, due to the COVID-19 pandemic, my contacts, in presence, with the representatives and members of national societies have basically stopped since the end of 2019. Nevertheless, I hope to be able to get back in touch shortly with most of European colleagues and friends in the forthcoming ISSMGE events.

As far as the other complementary activities of the national member societies are concerned, it is possible to observe an increasing number of named lectures in the last years, organised by the member societies. This is an excellent way of recognising the merits of outstanding individuals and, at the same time, of paying tribute to legendary past figures in our discipline. Some of the lectures have a long tradition, but several others have been added in recent times:

- Austria: Terzaghi Lecture
- Croatia: Nonveiller Lecture
- Czech and Slovak Society: Prague Geotechnical Lecture.
- France: Coulomb Lecture.
- Hungary: Széchy Memorial Lecture
- Ireland: Hanrahan Lecture.
- Italy: Arrigo Croce Lecture
- Norway: Bjerrum Lecture
- Portugal: Manuel Rocha Lecture, Victor de Mello Lecture (together with the Brazilian Society)
- Slovenia: Šuklje Lecture
- Spain: Jiménez Salas Lecture
- The UK: Rankine Lecture, BGA Touring Lecture

Another significant feature of European Geotechnics is that several Member Societies publish refereed Technical Journals in order to enhance the diffusion of geotechnical research and practice. Here are some examples:

- Albania: Geotechnics
- Belgium: Geotechniek
- Bosnia &Herzegovina: Geotehnika
- Czech and Slovak Society: Geotechnika.
- Denmark: Bulletin of the Danish Geotechnical Society
- France: Revue Française de Géotechnique
- Germany: Geotechnik
- Italy: Rivista Italiana di Geotecnica
- Portugal: Geotecnia (with the Brazilian and Spanish Member Societies) and Soils and Rocks (with the Brazilian Member Society)
- Romania: Romanian Journal of Soil Mechanics and Foundations
- Slovenia: Acta Geotechnica Slovenica
- Spain: Boletín de la Sociedad Española de Mecánica del Suelo e Ingeniería Geotécnica

8. TIME CAPSULE PROJECT

An illustration of the proposals, planning and first outcomes of the activities related to the Time Capsule Project (TPC) initiative can be found on the ISSMGE website.

However, just to briefly sum up, I can here report that the basic aim of TPC is that of improving the profile and visibility of ISSMGE through the enhancement of its communication capability, and the effective use of modern media and the related tools that are today available. All these aspects should also be considered to ensure a wide and long-lasting conservation of the memory of the history of ISSMGE, coupled with a modern and proactive vision of the future of the new generations, in particular considering that it is almost 100 years since ISSMGE was first founded.

After the presentation of the first ideas concerning TCP to the ISSMGE Board by Sukumar Pathmanadavel, the Chair of the Corporate Associate Presidential Group (CAPG), and the subsequent approval and setting up of the general framework and the related activities, I have been deeply involved, with the initialisation and steering of the European Member Societies, in planning and implementing their TPC contribution to the XX ICSMGE.

The European Region has played a fundamental role in this initial phase of the project through the organisation of a number of conference calls devoted to the presentation of some showcase examples by the European Societies which, at that time, were already well underway concerning the definition and preparation of their TPC contribution to the XX ICSMGE.

More details about the present status and the future developments of the TCP are available in the specific report prepared by the TCP Presidential Group and included in the proceedings of the XX ICSMGE.

Considering its natural links with the TCP, it is worth mentioning a much appreciated initiative that was promoted and headed by the French Committee of Soil Mechanics and Geotechnical Engineering (CFMS) in cooperation with a number of supporters and sponsors: the Virtual Geotechnical Museum, which was completed and presented, in website form, at the beginning of 2019. This initiative is based on the "Les dessous des grands travaux" exhibition hosted by the Museum of Arts and Crafts in Paris, which was first inaugurated at the 18th ICSMGE held in 2013. A virtual visit is available, through the exhibition website, in French and English (http://expo.geotechnique.org). This installation is the first in France and abroad, and it illustrates the know-how of French engineering and construction companies in the geotechnical field.

This initiative, already in the present format, can be considered, to all intents and purposes, as a part of CFMS's contribution to TCP.

9. GENERAL REMARKS AND OUTLOOKS

The European Region has played, and should continue to play, a fundamental role within the ISSMGE since it is, by far, the most numerous ISSMGE Region, in terms of both individuals and member societies, although this is neither the only reason nor, probably, the most important one, as reported below in my closing remarks.

The general numbers and the related lists of events and activities shown in the previous sections appear to be more than satisfactory, but they are not always able to represent the whole geotechnical engineering status in Europe and, possibly also outside the European Region, which sometimes shows aspects that deserve further attention and possible corrective actions.

Let's try to ask ourselves what is really the most important problem at the worldwide scale today. In my opinion, the only possible answer is: the safeguarding and the preservation of the environment in order to keep our planet liveable and habitable for the next generations.

So, if we want to take care of our future, we must use the planet's resources in an extremely careful way in order to maintain the present quality of life and, at the same time, to preserve our natural resources, while taking into account the earth's fast growing population and the related needs.

In my opinion, the only way to overcome this possible contraposition is to further enhance research and activities in the science and technology fields.

Geotechnical engineering is one of the scientific and technical disciplines that can offer the most important contributions, for instance, to the adaptation and mitigation of climate change effects, in particular considering the European Region, where these kinds of problems first arose as early as two centuries ago due to its intensive industrial development and high population concentration. Since then, at least part of Europe has been characterised by very demanding life standards, in terms of generalised welfare and by the consequent need to intensely exploit environmental resources.

Therefore, as a direct consequence, it is now our duty to address our production and the related environmental activities in such a way that we achieve mitigation and sustainability, in order to cope with these problems, not just through a philosophical and purely theoretical approach, but rather in a very practical and realistic manner.

In this context, it is perhaps worth mentioning two of the reference terms that I proposed at the beginning of my mandate, in order to address the Geotechnical Engineering contribution of the European Region. Their main aims are addressed to shortening the present distance between professionals and researchers in the field of both geotechnical design and construction activities and to increasing and improving the relationships among the ISSMGE Member Societies within the Mediterranean area.

Moreover, referring to the aforementioned terms of reference, we should of course seek solutions, not only looking at the next few days, but rather at the next years and, from this perspective, I strongly believe in the potential of the new generations of geotechnical engineers.

The injection of young engineers with fresh minds and new ideas can effectively contribute, in my opinion, to bridging the gap between the ISSMGE Members from the academia and those from practice and corporate groups.

For these reasons, I have tried to promote the organisation of the main plenary sessions in many European conferences by joining the efforts of the Young Member and Corporate Associate Groups. These sessions have been devoted, among others, to the presentation of the European "Bright Spark" Lectures for the best contributions of the Young Members of ISSMGE and to promoting the interaction between the two groups of participants.

Last but not least, through this cooperation, other relevant advantages can also arise for both young and corporate members, including, occupation facilitations and opportunities for new experiences, for the benefit of the new generations, and, on the other hand, original contributions from fresh minds for the benefit of companies and professional firms in order to try to establish effective synergies for a long-lasting innovation trend.

The increase and improvement of the interaction and cooperation among the countries of the Mediterranean area is a huge problem, even at the international political level. Nevertheless, I think that any contribution in this direction, given by any international scientific and technical institution or association of these countries is both necessary and important, in particular when considering the long-term scenario.

As engineers, we have to try to transform difficult situations into new opportunities and possibly into future progress. For these reasons, it has always been my intention to promote any form of technical and scientific cooperation among the Mediterranean countries through events and activities that involve the participation of all the regions facing onto the Mediterranean Sea.

The first example of this kind of cooperation was that of the MYGEC event, already mentioned in section 4 of this report, which involved African, Asian and European Regions.

The conference, with the participation of a large number of young geotechnical engineers, mainly from Africa and Europe, was particularly successful and, hopefully, will eventually result in a new conference series that will be able to promote a lasting and profitable technical and scientific cooperation of the Mediterranean countries within the Geotechnical Engineering discipline to overcome, among others, the present problems related to the emigration/immigration phenomenon.

Apart from the aforementioned initiatives and the related organisation and management activities of the ISSMGE European Region, what is also important today, in my opinion, is to never forget the basic features of the Engineering professions and of Geotechnical Engineering in particular. These kinds of professions should operate on the basis of scientific and technical knowledge, which means being able to quantify, and not only to address in a qualitative way, the hydro-chemo-physical and mechanical phenomena, in order to frame them in an objective and rational manner, rather than through qualitative, biased and possibly misleading options, hidden behind empty words and/or just emotion-driven reactions, without knowing the most elementary scientific and technical concepts. Unfortunately, this communication trend is today widely spread by all kinds of media, thus allowing almost anyone to deal with any topic, without any specific competence or skill.

In order to maintain and protect our professional position, in my opinion, we need to strongly oppose this trend, in particular within our community, by maintaining and possibly improving the quality of education in our universities and professional companies.

For these reasons, it is important to not forget the importance of a broad-spectrum scientific and technical education for the young generations, founded on a solid theoretical basis, if we want to be excellent in our work and, then, to continue to substantially contribute to the sustainable development of the modern world.

In short, if we want to keep and eventually improve the authoritativeness and prestige of the Geotechnical Engineering community within modern society, this result mainly depends, in my opinion, on the real competences and skills that we are able to spend in our work.

On the other hand, as far as our visibility and attractiveness are concerned, they basically depend on our ability to communicate through the modern media. However, visibility and attractiveness should be considered as very useful features to strengthen our position, but they should always be regarded as accessory features, which, within our profession, do not have any meaning without the actual ability to guarantee excellent results in our work.

Quite a large number of initiatives are already in place in order to try to face these kinds of problems. Of course, it is a very difficult task to modify the current trends, but, with a constant and enduring effort at all levels, our International Society may effectively contribute towards changing this attitude, hopefully avoiding critical times in the future, when at that moment, even for "the man in the street" in a state of need, it would become very easy to recognise the relevance and the role of such scientific and technological professions as Geotechnical Engineering, if it is not already too late for environmental health and human welfare.

Finally, I would like to take this opportunity to thank my predecessor, Antonio Gens, for the precious and generous suggestions and indications he gave me at the handover of the vice-presidency and also for a long time afterwards. Moreover, in turn, I want to warmly welcome my successor, Lyesse Laloui from Switzerland. I could not have asked for a better replacement and I can already anticipate that he will perform a great task in this role. I am sure that he will enjoy the support and encouragement of the entire European geotechnical community that I have personally felt throughout my period in office. I will forever be grateful to ISSMGE for this support and encouragement and in particular to the members from the European Region.

APPENDIX
REPORTS SUBMITTED BY INDIVIDUAL MEMBER SOCIETIES IN EUROPE AND BY THE EUROPEAN REGIONAL TECHNCIAL COMMITTEES

Austrian Geotechnical Society

AUSTRIAN GEOTECHNICAL SOCIETY
(ÖIAV - ÖSTERREICHISCHER INGENIEUR- UND ARCHITEKTENVEREIN)
Eschenbachgasse 9
1010 Wien
Austria

The following activities have been organized or co-organized by the Austrian Geotechnical Society in the period from 2017 to spring 2022:

11th Austrian Geotechnical Conference (ÖGT) & VÖBU Fair, 02.-03.02.2017, Vienna, about 350 participants and almost 1000 visitors (fair), Vienna Terzaghi Lecture was delivered by Prof. Sarah Springman (ETH Zürich, Switzerland)

32nd Christian Veder Kolloquium (CVK), 19.-20.04.2017, Graz, 400+ participants

33rd Christian Veder Kolloquium (CVK), 05.-06.04.2018, Graz, 400+ participants

China – Europe Conference on Geotechnical Engineering, 13.-16.08.2018, Vienna (BOKU), about 400 participants

26th European Young Geotechnical Engineers Conference, 45 participants from 26 countries
Venue: Hotel Klugbauer, Austria, Reinischkogel, 11.09.2018-14.09.2018 (report see attachment)

40 Years of Roller Integrated Continuous Compaction Control (CCC) - Anniversary Symposium & Exhibition, 29.11.2018, Vienna, about 180 participants (organized by TU Wien in cooperation with KTH Stockholm)

12th Austrian Geotechnical Conference (ÖGT) & VÖBU Fair, 31.01.-01.02.2019, Vienna, about 800 participants including fair, Vienna Terzaghi Lecture was delivered by Prof. Jürgen Grabe (TU Hamburg, Germany)

34th Christian Veder Kolloquium (CVK), 25.-26.04.2019, Graz, 400+ participants

35th Christian Veder Kolloquium (CVK), due to the pandemic postponed from April 2020 to 23.06.-24.06.2021. Online event, Graz, about 230 participants

13th Austrian Geotechnical Conference (ÖGT) & VÖBU Fair, postponed from January 2021 to 19.-20.04.2022 due to the pandemic, Vienna, about 800 participants expected (including fair), Vienna Terzaghi Lecture will be delivered by Prof. Lyesse Laloui (EPFL Lausanne, Switzerland)

36th Christian Veder Kolloquium (CVK), 28.-29.06.2022, Graz, about 400 participants expected

Close cooperation and exchange with colleagues from neighboring societies, e.g. Prof. Luis Ibarra (The University of Utah, USA), visiting professor at TU Wien, 10/2017-06/2018

Summer schools, seminars and presentations given by colleagues from abroad in lecture series of TU Wien and Graz University of Technology.

German Report for the ISSMGE Council Meeting in Sydney, May 2022
Summary of activities in Europe

Report dated 20th January 2022

German Geotechnical Society (DGGT) (Deutsche Gesellschaft für Geotechnik e.V.)

The principal office bearers are:

Chairman:	Dr.-Ing. Wolfgang Sondermann
Vice Chairman:	Univ.-Prof. Dr.-Ing. habil. Christian Moormann
Managing Director/ Secretary General:	Ms. Dr. Kirsten Laackmann

The society counts with 2005 members. This number is made up by 1872 ordinary members, 117 corporate members (companies) and 16 promoting institutes. The national German group of the ISSMGE has 430 members.

National conferences (2017 - 2022):

Our largest and most well-known conference, the 35th Conference on Geotechnical Engineering (35. Baugrundtagung), was carried out in Stuttgart, September 26th to 29th, 2018).
The topics of the conference were:

- Research and Innovation
- Special underground construction
- Earthworks and foundation engineering
- Tunneling
- Infrastructure projects
- Contributions of geotechnical engineering to sustainable management
- Special applications in the observation method
- BIM in geotechnical engineering
- Standardization

An exhibition with 84 exhibitors, presenting their products and know-how, took place during the conference.
The visitors could enjoy an extremely interesting and varied lecture program with around 40 technical presentations from high-ranking speakers from science and practice. The event was held under the main theme "Geotechnics meets future. Promoting research - creating knowledge - securing the future", with the opening block on September 27th being dedicated to the infrastructure project "Stuttgart 21" and the topic "Building in the Stuttgart region".
One conference day was devoted to the Young Geotechnical Engineers from the German Geotechnical Society who reported about their research and practical engineering experiences in a special sessions.

The 36th Conference on Geotechnical Engineering (36. Baugrundtagung 2020), planned for September 16th to 19th, 2020 in the RheinMain CongressCenter Wiesbaden, had to be canceled due to the ongoing coronavirus pandemic.

A total of 21 lectures, which were planned for the 36th Conference on Geotechnical Engineering 2020 in Wiesbaden, were held in the form of two half-day online events as part of the DGGT Digital Lecture Events 2021.

The 37th Baugrundtagung will take place in Wiesbaden, Germany, October 05th to 08th, 2022 and the 38th Baugrundtagung is scheduled for September 25th to 28th, 2024 in Bremen, Germany

From October 29th to 30th, 2019 the German Geotechnical Society (DGGT) organized the 2nd Meeting of Expert Groups - Interdisciplinary Forum (Fachsektionstage Geotechnik – Interdisziplinäres Forum).
This event offered all participants the opportunity to participate in the following five Conferences / symposia at the same place:

- 21st Conference on Engineering Geology
- 23rd Symposium on Rock Mechanics and Tunnel Construction
- 2nd Ground Mechanics Conference
- 16th Information and Presentation Conference on Geosynthetics in Geotechnics FS-KGEO 2019
- 8th Symposium Environmental Geotechnics

The joined Conference was an opportunity for the interested participants to get informed about the latest experiences and latest findings and to get into contact and to change ideas with participants from other expert groups.

The 3rd Meeting of Expert Groups - Interdisciplinary Forum (Fachsektionstage Geotechnik - Interdisziplinäres Forum), which was planned to be held from September 23rd to 24th, 2021 in the Congress Centrum Würzburg, unfortunately could not take place as a face-to-face event due to the still ongoing coronavirus pandemic.

As an alternative, the DGGT held digital lecture events of its specialist sections in 2021 and the "Forum for Young Geotechnical Engineers 2020" was also included in these virtual offers.
In the months of March to December 2021, the DGGT held a total of ten free digital lecture events, which recorded a lively participation and extremely positive response.

Activities for young geotechnical engineers:

The working committee "Young DGGT", previously known as the working group "Young Members of DGGT", runs a blog on the Internet at www.junge-dggt.de and can be reached on Facebook at www.facebook.com/junge.dggt.

The working committee "Young DGGT" is a communication platform between young DGGT members, the chairs and institutes of geotechnics and engineering geology, and the corporate Members of the DGGT.

The networking enables the fast and efficient exchange of information about possible interesting excursion destinations (construction sites), final theses or job applications and the distribution to interested students and young DGGT members.

Every two years, the members of the "Young DGGT", with the support of the DGGT office, conduct a workshop for young scientists and participate in the Forum for Young Geotechnical Engineers and also the Forum for Young Engineering Geologists.

The 3rd workshop took place on March 28th, 2019 in the Cologne-Deutz youth hostel. The event was entitled "From the leap into the unknown - career prospects in geotechnical engineering".

Due to the pandemic, the fourth workshop on the topic of "career entry" was held as an online event for the first time on March 25, 2021.
219 students from Germany, Austria and Switzerland accepted the invitation. In addition to the presentations by the young engineers, the participants were able to get in touch with 17 companies in a virtual contact exchange.

Publications:

The periodical journal of the DGGT is the "*geotechnik*" with 4 issues each year.

Between 2017 and 2021 the German Geotechnical Society published 4 recommendations of its working committees as printed books, 4 pocket books on tunnelling (Taschenbuch für den Tunnelbau) and 4 conference proceedings:

Pocket books on Tunnelling:
- „Taschenbuch für den Tunnelbau", (pocket books on tunnelling), editions 2018, 2019, 2020 and 2021 (german editions)

Conference Proceedings:
- Abstractband und USB-Stick mit den schriftl. Beiträgen der Fachsektionstage Geotechnik 2017, Würzburg (Abstract book and USB stick with the written papers of the 1. Meeting of Expert Groups - Interdisciplinary Forum 2017, Würzburg – german edition)
- Tagungsband und CD mit den schriftl. Beiträgen der 35. Baugrundtagung, Stuttgart, 2018 (Conference proceedings and CD with the written papers of the 35th Baugrundtagung, Stuttgart, 2018 – german edition)
- Tagungsband und CD mit den schriftl. Beiträgen der Spezialsitzung „Forum für junge Geotechnik-Ingenieure" der 35. Baugrundtagung, Stuttgart, 2018 (Conference proceedings and CD with the written papers of the special session "Forum for young geotechnical engineers" of the 35th Conference on Geotechnical Engineering, Stuttgart, 2018 – german edition)
- USB-Stick mit den schriftlichen Vortragsbeiträgen der Fachsektionstage Geotechnik 2019, Würzburg (USB stick with the written papers of the 2. Meeting of Expert Groups - Interdisciplinary Forum 2019, Würzburg – german edition)

Recommendations:

- "Empfehlungen des Arbeitskreises Baugrunddynamik", 2. Auflage, 2019 ("Recommendations of the Working Group on Subsoil Dynamics", 2nd edition, 2019 - german edition)
- Geotechnisch-markscheiderische Untersuchung, Bewertung und Sanierung von altbergbaulichen Anlagen - Empfehlungen des Arbeitskreises 4.6 Altbergbau", 1. Auflage, 2020 ("Geotechnical-marking investigation, evaluation and rehabilitation of old mining facilities - recommendations of the working group 4.6 old mining", 1st edition, 2020, german edition)
- "Empfehlungen des Arbeitskreises Baugruben", 6. Auflage, 2021 ("Recommendations of the working group for construction pits", 6th edition, 2021 - german edition)
- "Kompendium zum Stand von Wissenschaft und Technik bei der Planung und Ausführung geotechnischer Barrieren für Untertagedeponien im Salinargebirge", 1. Auflage, 2021 ("Compendium on the state of the art in science and technology in the planning and execution of geotechnical barriers for underground storage sites in the Salinar Mountains", 1st edition, 2021 - german edition)

Activity report of the Hungarian Geotechnical Society

Period covered by this report: October 2017 – February 2022

The Hungarian Geotechnical Society has been an active period since the 19th International Conference On Soil Mechanics and Geotechnical Engineering, Seoul, Korea.

Széchy Károly Memorial Lectures

The annual Széchy Károly Memorial Lectures were held at the Hungarian Academic Sciences in Budapest. According to tradition, three speakers were invited: – a well-known International, a well-known Hungarian, and a young Hungarian engineer.. In February 2018 the **24th Széchy Károly Memorial Lecture** focused on the topic of numerical modelling. The agenda was as follows:

1) prof. Helmut Schweiger , Graz University, Austria - Advanced finite element modelling for practical geotechnical problems,

2) Manninger Marcel, Bohn Ltd., Hungary - Design of special deep foundation,

3) Szilvágyi Zsolt, Geoplan Ltd., Hungary as a young engineer - Dynamic parameters of Danube sands.

In February 2019 the **25th Széchy Károly Memorial Lecture** emphasized the topic of geotechnical testing including physical, laboratory and field tests. The international speaker for the Memorial Lecture was Charles W.W. Ng, the president of the ISSMGE. The agenda was as follows:

1) prof. Charles W.W. Ng, HKUST, Hong Kong - Understanding geotechnical phenomena to advance engineering design through centrifuge modelling

2) Móczár Balázs, BUTE, Hungary – Contributions to the interpretation of shear strength based on triaxial tests,

3) Nagy Péter, TU Wien, Austria - Experimental and theoretical investigation of vibroflotation.

In February 2020, the **26th Széchy Károly Memorial Lecture** was organized. Two international speakers and a Hungarian one presented. The agenda was as follows:

1) prof. Paul W. Mayne, Georgia Institute of Technology, USA - Use od in-situ geotechnical tests for foundation systems

2) Andrew J. Bond, Geocentrix Ltd., United Kingdom – 2020 vision: 2nd generation geotechnical design to Eurocode 7,

3) Károly Orosz, A-Híd Zrt., Hungary –Against each other (soil-structure interaction).

Organization of the **27th Széchy Károly Memorial Lecture** is in progress. The invited speakers will be António Joaquim Pereira Viana da Fonseca (Portugal), Edina Koch (Hungary) and Zoltán Bán (Hungary).

The Hungarian Geotechnical Society organized two scientific **symposiums in 2018.**

1) In June 2018. prof. Gyan N. Pande, Swansea University, United Kingdom, gave a short course on "Particle/pore size distribution & microstructure of saturation – key elements for rational description of mechanical behaviour of unsaturated soils".

2) In November 2018 prof. Ikuo Towhata, University of Tokyo, Japan, shared his presentation titled "Activities to solve the nuclear accident problems at Fukushima No.1

Power Station during the 2011 earthquake and tsunami disaster" with the members of the HNC.

Conference Organization

The **5th Kézdi Conference** was held on 27. May, 2019, organized by the HNC. The issue came up the 100th anniversary of Kézdi Árpád's birth. 26 papers were published in the proceeding and 12 selected presentations were organized on the one-day conference. The topics covered a wide range of geotechnical fields.

Under the umbrella of TC102, the International Conference on Geotechnical and Geophysical Site Characterization Conference (**ISC'6**) was held in Budapest, Hungary, 26-29 September, 2021. Due to the COVID pandemic situation, the originally announced date (September 2020) was postponed twice. Finally the conference was held in hybrid mode.

One of the aim of the conference was to facilitate more active cooperation between geotechnical engineers and geophysicists. 357 participants were registered from 52 countries and 174 delegates from 25 countries visited the conference in-person. The figure below shows the distribution of the 16 sessions.

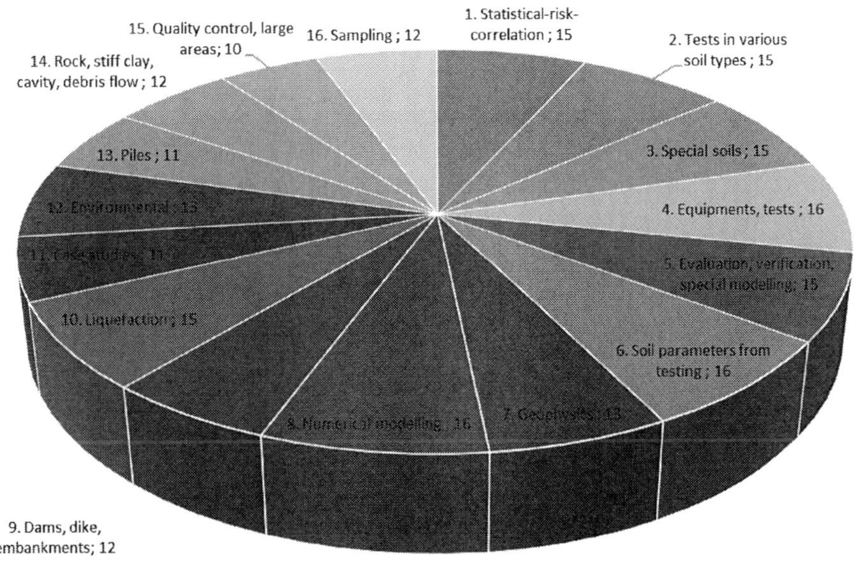

235 papers were submitted and will be published on ISSMGE Conference Review Platform.

The Mitchell Lecture was presented by Fernando Schnaid: On the geo-mechanics and geo-characterization of tailings.

There were 3 keynote lecturers: 1) Kenneth H. Stokoe: The Increasing Role of Seismic Measurements in Geotechnical Engineering 2) Jason Dejong: Integrated Site Characterization: Current Practices, Emerging Opportunities & Research Needs 3) Ray Wood: Integrating Recent Advances in Industry Site Characterization Capabilities to Reduce Unforseeability in Sub-Surface Conditions for Capital Works Projects.

The invited lecturers were 1) Helmut Schweiger - Marcos Arroyo: New Advances in Numerical Modelling of CPT 2) Catherine Jacquard: The pressuremeter : recent developments in testing and design methods.
4 parallel sessions were held in different rooms.
A new tradition-making event was introduced into the life of future ISC conferences, the Sylvano Marchetti Award. The winner of the I. Award was Dušan Berisavljević: Dilatometer and seismic dilatometer tests in different depositional environments.
18 exibitors raised the quality of the Conference.
The TC102 meeting was also organized during ISC'6 and the CPT'22 event was discussed.

TC activities

TC201 have been working on a report titled Failure paths for levees. The Hungarian members of TC201 contributed case studies to the report. The finalized document will be presented by TC201 on the 20th ICSMGE.

Plan for the future

The mandate of the current presidency of the Hungarian Geotechnical Society will expire this year. The future plan will be the main task of the new presidency.

ITALIAN GEOTECHNICAL SOCIETY (AGI) ACTIVITY FOR THE PERIOD 2017-2021

Technical Visits

- Technical Visit to Underground Line Metro C in Rome S. Giovanni – Roma, 22 June 2018

Young Geotechnica Engineers Annual Meetings

- Catania, 19-20 May 2017
- Udine, 18-19 May 2018
- Napoli, 10-11 May 2019
- Pisa, 3-4 September 2021

Workshops

- International Workshop on Complex Formations: Characterization and Case Studies – Turin 30 May 2019
- Webinar on Blast Test – 11 November 2021

Courses

- Indagini e Monitoraggi geotecnici per la definizione del modello del sottosuolo e per il controllo del comportamento delle opere - Milano, 11 March 2017 - Buildsmart
- Il ruolo delle indagini e dei controlli nella realizzazione e gestion delle discariche controllate – Ferrara, 20 September 2017- REemtech
- Linee guida rischio frana - Ferrara, 21 September 2017 – Remtech Esonda
- Recenti sviluppi nella progettazione di interventi di stabilizzazione di pendii e protezione da frane - Lecco 24 october 2017
- Barriere attive e passive: nuovi materiali e recenti sviluppi nell'ambito progettuale e tecnologico – Ferrara, 21 September 2018 - DI REMTECH
- Ponti e Viadotti – Bologna 17 October 2018 – SAIE
- Conferenza sul rischio idrogeologico ed ambientale e sulla gestione e prevenzione dei disastri naturali – 7 November 2018 – Ecomondo
- La geotecnica nelle stabilità delle discariche – Ferrara, 18 September 2019 – Remtech
- Aspetti geotecnici per la protezione dei rilevati arginali in occasione di grandi eventi meteorici - 5-6 November 2019 – Ecomondo
- La geotecnica nella stabilità delle discariche - 5-6 November 2019 – Ecomondo
- Webinar Monitoraggio geotecnico delle opere per la difesa del territorio e la tutela dell'ambiente - 3 November 2020
- Webinar Soluzioni geotecniche innovative connesse alla gestione dei sedimenti e dei fanghi - 4 November 2020
- Webinar Modello geotecnico degli ammassi rocciosi - 5 February 2021
- Webinar Relating physical and mechanical properties to temperature induced damage in carbonate rocks – Young members group - 1 April 2021
- Opere di mitigazione del rischio da frane rapide – 26 October 2021 - Ecomondo
- Il contributo dell'ingegneria geotecnica nella soluzione di problemi ambientali – 26 October 2021 - Ecomondo

Conferences

- MIR 2017 Innovazioni nella progettaxzione realizzazione e gesione delle opera in sotterraneo – Turin, 16-17 February 2017

- XXVI National Geotechnical Conference on "L'Ingegneria Geotecnica nella conservazione e tutela del patrimonio costruito" - Roma, 20-22 June 2017
- Sviluppi tecnologici ed applicativi nel monitoraggio geotecnico – Bolzano, 29 November 2017
- AGI-ISPRA Conference "La progettazione e la realizzazione di interventi di mitigazione del rischio da frana" – Bolzano, 21 November 2018
- VII International Conference on Earthquake Geotechnical Engineering – Roma, 17-20 June 2019
- Aspetti Geotecnici nella Progettazione e Manutenzione degli Argini Fluviali – Bolzano 27 November 2019
- Online conference Eurock 2021 – 21-24 September 2021

National IGS conferences

- XXIX National IGS Conference "Raccomandazioni AGI per l'impiego dei geosintetici nel rinofrzo del terreno in opere di sostegno e pendii ripidi" – Bologna 18 October 2018
- XXX National IGS Conference "Prestazioni dei materiali e delle opere" - Bologna, 15 October 2020
- XXXI National IGS Conference "Sostenibilità, aspetti ambientali e cambiamenti climatici" - Bari, 7 October 2021

In memory

- Study day in memory of Prof. Aldo Evangelista - Napoli, 19 October 2017
- Study day in memory of Prof. Pietro Colombo – Padova 29 March 2019

Arrigo Croce Lectures

- XVI Arrigo Croce Lecture: Prof. Ronald Kerry Rowe "Environmental geotechnics: looking back, looking forward" – Roma, 13 December 2017
- XVII Arrigo Croce Lecture: Prof. Augusto Desideri "Gallerie e movimenti di versante: una interazione complessa" – Roma 13 December 2018
- XVIII Arrigo Croce Lecture: Prof. Heinz Brandl "Structures in unstable slopes" – Roma 11 December 2019
- XIX Arrigo Croce Lecture: Prof. Paolo Simonini "La difesa della città di Venezia e dell'ecosistema lagunare: una sfida continua per l'ingegneria geotecnica" – Roma 16 December 2021

Next events

- XXVII National Geotechnical Conference "La Geotecnica per lo sviluppo sostenibile del territorio e per la tutela dell'ambiente" - Reggio Calabria, 13-15 July 2022
- 5th International Symposium on Cone Penetration Testing, CPT'22 – Bologna, 8-10 June 2022
- TC301-IS NAPOLI 2022 "Geotechnical Engineering for the Preservation of Monuments and Historic Sites" – Napoli 22-24 June 2022
- 12 International Conference on Geosynthetics, 12ICG "Geosynthetics: leading the way to a resilient planet" – Roma, 17-21 September 2023
- XX Arrigo Croce Lecture: Prof. Guy Houlsby "Recent developments in design for cyclic loading" – Roma December 2022

ДГМ
MAG
ДРУШТВО ЗА ГЕОТЕХНИКА НА МАКЕДОНИЈА
MACEDONIAN ASSOCIATION FOR GEOTECHNICS

To
Prof. Mario Manassero
Vice-President for Europe
ISSMGE

Subject: Report for activities of Macedonian Association for Geotechnics in the period September 2017 – February 2022

Dear prof. Manassero,

Regarding the preparation of report of societies' activities for the period September 2017 – February 2022, the template used for the Council Meeting in Edinburgh in 2015 was followed. Please find below the informal summary for Macedonian Association for Geotechnics (MAG):

[1-1] Name of your president, international secretary and their contact e-mail address:

- Jovan Br. Papić, papic@gf.ukim.edu.mk;
- Igor Peševski, pesevski@gf.ukim.edu.mk.

[1-2] Approximate number of your members (both domestic and international): 60

[1-3] How many student members approximately? 10

[2-1] Organizing International Conferences, Workshops, Seminars etc.:

- 16th DECGE (June 2018);
- International Tunneling and Underground Space Association (ITA) Seminar (June 2018);
- Meetings of Board Members of ISSMGE and Executive Board of ITA (June 2018).

[2-2] Organizing National Conferences, Workshops, Seminars etc.:

- Five annual meetings of MAG (with selected presentations given by academicians, experienced professionals and young researchers);
- Several invited lecturers (prof. Ahmet Saglamer – Turkey, prof. Ernest Olinic – Romania, prof. Dietmar Adam – Austria, prof. Boris Jeremić – USA, prof. Reşat Ulusay – Turkey);
- Workshops (realized by MOBA – Germany and TENSAR – UK);
- Conducted research project with Glasgow Caledonian University
- Round tables (on National Annexes to Eurocode 7, CEN/ISO technical standards, Pricelist of engineering services, Law for building etc.).

[2-3] Ongoing MOU with other member societies of ISSMGE:

- Mutual with Serbian, Bosnian, Croatian and Slovenian geotechnical societies.

[2-4] Current hosting of ISSMGE TC and ETC: Non.

[2-5] Periodical journals and magazines:

- special edition of Scientific Journal of Civil Engineering dedicated to 16th DECGE;

[2-6] Awards:

- Awarding of prof. Heinz Brandl by the President of the R. Macedonia (State Recognition: Medal of Merit) (June 2018)
- Prof. Milorad Jovanovski (immediate-past president of MAG) was awarded a Great Gold Medal for Outstanding Scientific and Practical Contribution to the Development of Engineering Geology and Geomechanics by the Austrian Association of Architects and Engineers (January 2019)
- MAG was awarded a Charter by the Serbian Society of Engineers and Technicians (SSET) (November 2019).
- Both prof. Vlatko Šešov (past president of MAG) and prof. Jovan Br. Papić (actual president of MAG) were awarded Charters by SSET (November 2019)
- Young members of MAG won the I prize at international student competition on engineering projects (Istanbul, 2021)
- MAG delivered plaques to prominent companies and professionals

[2-7] Other important achievements:

- Continual scholarship to students at first cycle of studies at geotechnical department;
- Memorandum for cooperation with local companies;
- Established prize for extraordinary dedication in developing geotechnics in the country;
- Established league of young geotechnical engineers;
- Membership in the Engineering Institution of Macedonia;
- Membership in the International Society for Rock Mechanics and Rock Engineering (ISRM);
- Membership in International Consortium of Landslides (ICL);
- Published dictionary of rock mechanics terms at ISRM.net;
- Honorary members: prof. Ahmet Saglamer, prof. Petr Koudelka, prof. Radomir Folić, prof. Ivan Vrkljan, prof. Reşat Ulusay;
- Distinguished academicians elected as visiting professors at Faculty of Civil Engineering in Skopje (prof. Dietmar Adam, prof. Reşat Ulusay);
- Participation at regional and international conferences and events, both for young and senior engineers, as well as visiting other international institutions;
- Composed database for landslides along roads;
- Development of Facebook profile GF.geotehnika;
- Maintaining of MAG's web-site mag.net.mk;
- Continual participation at technical committee for standardization;
- Reaccreditation of study program for geotechnics;
- Close cooperation with national sister-associations (covering dam engineering, roads, structural engineering etc.);
- After initiative by MAG, a street in Skopje is named in honor of renowned academic, civil engineer and designer of the first dam in Macedonia;

[3] Planned future activities (in 2022):

- Organization of 5th Symposium as 2nd Conference of regional geotechnical societies and as an ISRM Specialized Conference (June 2022);
- Planning of series of events "Geotechnical days" (twice per year);
- Organizing seminars for Eurocode 7 and BIM application;
- Applying to host the 12th International Symposium for Field Monitoring in Geomechanics in 2026;
- Preparation for hosting the EYGEC in 2023;
- Membership in the Balkan geophysical association;
- Organizing seminar on landslides' hazards and risks (in cooperation with national Directorate for protection and rescue);
- Completing the geotechnical part of national Pricelist for engineering services;
- Comments on preparation of new law for building from geotechnical point of view;
- Translation of geotechnical Lexicon;
- Support to organization of regional conference of students of geotechnology.

I hope that this review fits Your needs, but if not – I can fulfill the expectations with its extension or reduction.

While looking forward to hearing from You, I send warm regards from Skopje,

assoc. prof. Jovan Br. Papić,

President of MAG

NORWAY GEOTECHNICAL SOCIETY

REPORT FROM THE BOARD OF DIRECTORS 2017

(Annual report)

Meetings of the Board/Members
The Board held eleven meetings in 2017, and sent three newsletters to members. Information was also sent via e-post to about 95% of our members. We arranged a spring seminar with technical lectures etc., with approximately 50 participants, "Youth" day with approximately 40 participants, an Elders Forum with approximately 160 participants.

Information about the Society and our future meetings are published on the Society's web pages, www.ngf.no.

NGF's Economy
The accounts show a loss of NOK 161 678 compared to a budgeted loss of NOK 856 000. The difference is due to budget deviations both on the income and expenses side. Income from courses was less than budgeted, and there was less from the sale of guidelines because publication of the Excavation Guidelines and the revised Pile Guidelines will be in 2018.

On the expenses side, no scholarships were awarded in 2017 and several committees did not ask for as much support as they had indicated at the beginning of the year. Printing of the revised Pile Guidelines and Excavation Guidelines was postponed since they were not finished.

The Society's economy is satisfactory, with funds set aside to the committee fund among others to ensure continued support to the committees' contributions to the Society.

Technical and Administrative Committees
The Excavation Guidelines Committee, Pile Guidelines Committee, the Geosynthetics Committee, the Frost in Soils Committee, the Soil Investigation Committee, the Further Education Committee and the Geodynamics and Vibrations Committee were active in 2017. Details of committees work are given in the section "Virksomheten i NGFs komiteer og utvalg".

Many NGF members are also active in both European and international committees. Further information on these activities is given later. In December 2017, the board of ISSMGE approved a new Technical Committee, Machine Learning and Big Data, where NGF with Zhongqiang Liu is the Chair.

Members
The Society had 596 members at the end of 2017, including 80 pensioners. There are 21 more than 2016 (575 members). Work continues on removing those who no longer want to be a member from the membership list. There are still several unresolved memberships.

Support membership was established in 1987 for companies and institutions working within the geotechnical field. There are three subscription classes for support members. The membership entitles the members to a presentation of their logo on NGF's web pages http://ngf.no/?page_id=113 . Class 1 support members also have their logo rotating on the main page. NGF would like to thank all support members for their important contribution to NGF.

Companies registered as members as of 2018-01-01:

Class1

COWI AS (www.cowi.no)
Leca Norge AS (www.weber-norge.no)

Multiconsult ASA (www.multiconsult.no)
Norconsult AS (www.norconsult.no)
Norges Geotekniske Institutt (www.ngi.no)
Romerike Grunnboring AS (www.rgb.as)

Class 2

Entreprenørservice A/S (www.entreprenorservice.no)
Fundamentering AS (www.fas.no)
Glasopor (www.glasopor.no www.syklus.no)
Løvlien Georåd AS (www.georaad.no)
Rambøll Norge AS (www.ramboll.no)
SINTEF Byggforsk AS (www.sintef.no/Byggforsk)
SSAB Svenks Stål AS (www.ssab.no)
SWECO Norge AS (www.sweco.no)
Veidekke Entreprenør AS (www.veidekke.no)

Class 3

Cautus Geo AS (www.cautusgeo.com)
FinMeas Oy (www.finmeas.com)
Fugro Geotechnics AS (www.fugro.no)
Geomek Stockholms Geomekaniska AB (www.geomek.com)
Geonor AS (www.geonor.no)
GeoSafe as (www.geosafe.no)
GeoVita as (www.geovita.no)
Golder Associates AS Norge (www.golder.no)
Keller Grundläggning AB (www.kellergrundlaggning.se)
Trimble Solutions Gothenburg AB (www.constryction.trimble.com)

Courses, seminars and conferences

NGF supported IWLSC 2017, International Workshop on Landslides in Sensitive Clays, which was held 12th – 14th June 2017.

A course in "Practical Subsoil Drilling" was held 11th – 12th and 12th – 13th September 2017 and was a repetition of the course in 2016 due to a waiting list of participants.

A course "Geotechnical Design and Execution Standards" was held 17th – 18th January 2017 and a course "Permanent Supporting Structures" 9th – 10th November 2017.

The Geotechnics Day was again held in November and was as usual very successful. Our members held two presentation during the joint session and nine different presentations during the Geotechnics session. The day ended with a dinner, where approximately 130 members participated.

Frost Day was held 4th December 2017.

Scholarship

No scholarship was awarded in 2017.

Diploma prize

Two diploma prizes were awarded in 2017: Anders Lindgård and Christian Strømme Ofstad for thesis "Field and Laboratory Investigations to Evalutae the Coefficient of Earth Pressure at Rest, NGTS Flotten Quick Clay Test Site" and Johannes Osorio Mydland for his thesis "Numerical Framework for Modelling Cyclic Degradation of Soils".

Contact with the other Nordic Geotechnical Societies

The Society is in touch with the Danish (DGF), the Swedish (SGF), the Finnish (FGF) and the Icelandic (IGF) societies. No ordinary Nordic Board meeting was arranged in 2017, but a dinner meeting was arranged before the ISSMGE Council Meeting in Seoul 17th September 2017. Magnus Rømoen represented NGF.

REPORT FROM THE BOARD OF DIRECTORS 2018

(Annual report)

Meetings of the Board/Members

The Board held eleven meetings in 2018, and sent four newsletters to members. Information was also sent via e-post to about 95% of our members. We arranged a spring seminar with technical lectures etc., with approximately 50 participants, "Youth" day with approximately 40 participants, an Elders Forum with approximately 70 participants.

Information about the Society and our future meetings are published on the Society's web pages, www.ngf.no.

NGF's Economy

The accounts show a loss of NOK 291 216 compared to a budgeted loss of NOK 576 000. The difference is due to budget deviations both on the income and expenses side. Income from courses was less than budgeted, and there was less from the sale of guidelines because publication of the Excavation Guidelines and the revised Pile Guidelines will be in 2019. Membership fees are more than budgeted due to both more members, but also a tidy-up of the membership database and successful payment reminders.

On the expenses side, no scholarships were awarded in 2018 and several committees did not ask for as much support as they had indicated at the beginning of the year. Printing of the revised Pile Guidelines and Excavation Guidelines was postponed since they were not finished. Advance payment for the Excavation Guidelines in connection with the course held in November are given in the balance.

The Society's economy is satisfactory, with funds set aside to the committee fund among others to ensure continued support to the committees' contributions to the Society.

Technical and Administrative Committees

The Excavation Guidelines Committee, Pile Guidelines Committee, the Geosynthetics Committee, the Frost in Soils Committee, the Soil Investigation Committee, the Further Education Committee and the Geodynamics and Vibrations Committee were active in 2018. Details of committees work are given in the section "Virksomheten i NGFs komiteer og utvalg" and published on the web pages.

Many NGF members are also active in both European and international committees. Further information on these activities is given later.

Members

The Society had 624 members at the end of 2018, including 74 pensioners. There are 28 more members than 2017 (596 members). Work on removing those who no longer want to be a member from the membership list is completed.

Support membership was established in 1987 for companies and institutions working within the geotechnical field. There are three subscription classes for support members. The membership entitles the members to a presentation of their logo on NGF's web pages http://ngf.no/?page_id=113 . Class 1 support members also have their logo rotating on the main page. NGF would like to thank all support members for their important contribution to NGF.

Companies registered as members as of 2019-01-01:

Class1

COWI AS (www.cowi.no)

Leca Norge AS (www.weber-norge.no www.leca.no)
Multiconsult ASA (www.multiconsult.no)
Norconsult AS (www.norconsult.no)
Norges Geotekniske Institutt (www.ngi.no)
Romerike Grunnboring AS (www.rgb.as)
Mesta AS (www.mesta.no)
ÅF Engineering AS (www.afconsult.com)

Class 2

Entreprenørservice A/S (www.entreprenorservice.no)
Fundamentering AS (www.fas.no)
Glasopor (www.glasopor.no www.syklus.no)
Løvlien Georåd AS (www.georaad.no)
Maskinentreprenørenes Forbund, Fagavdeling Brønn- og spesialboring (www.mef.no)
Rambøll Norge AS (www.ramboll.no)
SINTEF Byggforsk AS (www.sintef.no/Byggforsk)
SSAB Svensk Stål AS (www.ssab.no)
SWECO Norge AS (www.sweco.no)
Veidekke Entreprenør AS (www.veidekke.no)

Class 3

Argeo AS (www.argeo.no)
Cautus Geo AS (www.cautusgeo.com)
ERA Geo AS (www.era-geo.no)
FinMeas Oy (www.finmeas.com)
Fugro Geotechnics AS (www.fugro.no)
Geomek Stockholms Geomekaniska AB (www.geomek.com)
Geonor AS (www.geonor.no)
GeoSafe as (www.geosafe.no)
GeoVita as (www.geovita.no)
Golder Associates AS Norge (www.golder.no)
Keller Grundläggning AB (www.kellergrundlagging.se)
PTO teknikk AS (www.pto.no)
Trimble Solutions Gothenburg AB (www.construction.trimble.com)

Courses, seminars and conferences
A course "Frost protection of soils" was held 7th – 8th February 2018.

The Pile Day was held 30th August 2018.

The Frost Day was held 10th September 2018.

A mini seminar on Geosynthetics was held 29th October 2018.

A course on the Excavation Guidelines was held 8th – 9th November 2018.

The Geotechnics Day was again held in November. Our members held two presentation during the joint session and nine different presentations during the Geotechnics session. The day ended with a dinner, where approximately 150 members participated.

Scholarship
At the end of the year a scholarship was awarded to Priscilla Paniagua for the subject *Characterization of Norwegian Peat*.

Diploma prize
No diploma prizes were awarded in 2018.

Contact with the other Nordic Geotechnical Societies

The Society is in touch with the Danish (DGF), the Swedish (SGF), the Finnish (FGF) and the Icelandic (IGF) societies. A Nordic Board meeting was arranged in Tampere, Finland 4th – 5th October 2018. Vegard Woldsengen and Geraldine Sørum represented NGF.

REPORT FROM THE BOARD OF DIRECTORS 2019

(Annual report)

Meetings of the Board/Members

The Board held eleven meetings in 2019 and sent four newsletters to members. Information was also sent via e-post to about 95% of our members. We arranged a spring seminar with technical lectures etc., with approximately 70 participants, "Youth" day with approximately 70 participants, an Elders Forum with approximately 140 participants.

Information about the Society and our future meetings are published on the Society's web pages, www.ngf.no.

NGF's Economy

The accounts show a loss of NOK 9 471 compared to a budgeted loss of NOK 253 000. The difference is due to budget deviations both on the income and expenses side. Income from courses was less than budgeted, and there was less from the sale of guidelines because publication of the Excavation Guidelines will be in 2020 and the revised Pile Guidelines was published at the end of 2019. Membership fees are less than budgeted, but payment reminders will be sent. Support membership has increased.

On the expenses side, no diploma prizes were awarded in 2019. Printing of the Excavation Guidelines was postponed since it was not finished. Advance payment for the Excavation Guidelines in connection with the course held in November 2018 are given in the balance. Due to a misunderstanding work by the Pile Committee in 2017 and 2018 was invoiced in 2019. The Soil Investigation Committee used more than budgeted due to planning of the Ground Drilling School.

Liabilities and equity from Laurits Bjerrum's Memorial Fund of 31 March 2019 are transferred to the balance.

The Society's economy is satisfactory, with funds set aside to the committee fund among others to ensure continued support to the committees' contributions to the Society.

Technical and Administrative Committees

The Excavation Guidelines Committee, Pile Guidelines Committee, the Geosynthetics Committee, the Frost in Soils Committee, the Soil Investigation Committee, the Further Education Committee and the Geodynamics and Vibrations Committee were active in 2019. Details of committees' work are given in the section "Virksomheten i NGFs komiteer og utvalg" and published on the web pages. New this year is that the Bjerrum Lecture committee replaces Laurits Bjerrum's Memorial Fund.

Many NGF members are also active in both European and international committees. Further information on these activities is given later.

Members

The Society had 647 members at the end of 2019, including 74 pensioners. There are 23 more members than 2018 (624 members).

Support membership was established in 1987 for companies and institutions working within the geotechnical field. There are three subscription classes for support members. The membership entitles the members to a presentation of their logo on NGF's web pages http://ngf.no/?page_id=113 . Class 1 support members also have their logo rotating on the main page. NGF would like to thank all support members for their important contribution to NGF.

Companies registered as members as of 2020-01-01:

Class1

COWI AS (www.cowi.no)
Leca Norge AS (www.weber-norge.no www.leca.no)
Multiconsult ASA (www.multiconsult.no)
Norconsult AS (www.norconsult.no)
Norges Geotekniske Institutt (www.ngi.no)
Romerike Grunnboring AS (www.rgb.as)
Mesta AS (www.mesta.no)
ÅF Engineering AS (www.afconsult.com)

Class 2

Akershus Grunnboring AS
Entreprenørservice A/S (www.entreprenorservice.no)
Franzefoss Minerals AS (https://kalk.no/)
Fundamentering AS (www.fas.no)
GeoVita as (www.geovita.no)
Glasopor (www.glasopor.no www.syklus.no)
Løvlien Georåd AS (www.georaad.no)
Maskinentreprenørenes Forbund, Fagavdeling Brønn- og spesialboring (www.mef.no)
Rambøll Norge AS (www.ramboll.no)
SINTEF Byggforsk AS (www.sintef.no/Byggforsk)
SSAB Svensk Stål AS (www.ssab.no)
SWECO Norge AS (www.sweco.no)
Veidekke Entreprenør AS (www.veidekke.no)

Class 3

Arya Consult (https://aryaconsult.no/)
Cautus Geo AS (www.cautusgeo.com)
ERA Geo AS (www.era-geo.no)
Fugro Geotechnics AS (www.fugro.no)
Geomek Stockholms Geomekaniska AB (www.geomek.com)
Geonor AS (www.geonor.no)
GeoSafe as (www.geosafe.no)
Golder Associates AS Norge (www.golder.no)
Keller Geoteknikk AS (www.keller-geoteknikk.no)
Keller Grundläggning AB (www.kellergrundlagging.se)
PTO teknikk AS (www.pto.no)
RIGeo AS (https://no.linkedin.com/company/rigeo-as)
Trimble Solutions Gothenburg AB (www.construction.trimble.com)
ViaCon AS (www.viacon.no)

Courses and conferences

The Ground Drilling School was held 4th – 7th March 2019.

A technical day on groundwater challenges was held 6th June 2019.

A course in reinforced soil was held 29th October 2019.

A pile course was held 6th – 7th November 2019.

The Frost Day was held 19th November 2019.

The Geotechnics Day was again held in November. Our members held two presentation during the joint session and nine different presentations during the Geotechnics session. The day ended with a dinner, where approximately 150 members participated.

Scholarship
At the end of the year a scholarship was awarded to Stefan Ritter for the subject *Biochar to stabilise Norwegian quick clays.*

Diploma prize
No diploma prizes were awarded in 2019.

Contact with the other Nordic Geotechnical Societies
The Society is in touch with the Danish (DGF), the Swedish (SGF), the Finnish (FGF) and the Icelandic (IGF) societies. A Nordic Board meeting was arranged in Reykjavik, Island 2nd September 2019. Carsten Hauser represented NGF.

REPORT FROM THE BOARD OF DIRECTORS 2020

(Annual report)

Meetings of the Board/Members

The Board held twelve meetings in 2020, and sent six newsletters to members. Information was also sent via e-post to about 95% of our members. We arranged an Elders Forum with approximately 70 participants and a jubilee webinar with approximately 230 participants.

Information about the Society and our future meetings are published on the Society's web pages, www.ngf.no.

NGF's Economy

The accounts show a profit of NOK 1 290 821 compared to a budgeted profit of NOK 92 000. The difference is due to budget deviations both on the income and expenses side. Income from courses was less than budgeted, while the sale of guidelines was more than budgeted.

On the expenses side, less was used on several items because of reduced activity due to the corona situation. No diploma prize or scholarships were awarded in 2020.

The Society's economy is satisfactory, with funds set aside to the committee fund among others to ensure continued support to the committees' contributions to the Society.

Technical and Administrative Committees

The Excavation Guidelines Committee, Pile Guidelines Committee, the Geosynthetics Committee, the Frost in Soils Committee, the Soil Investigation Committee, the Further Education Committee and the Geodynamics and Vibrations Committee were active in 2020. Details of committees work are given in the section "Virksomheten i NGFs komiteer og utvalg" and published on the web pages.

Many NGF members are also active in both European and international committees. Further information on these activities is given later.

Members

The Society had 695 members at the end of 2020, including 79 pensioners. There are 48 more members than 2019 (647 members).

Support membership was established in 1987 for companies and institutions working within the geotechnical field. There are three subscription classes for support members. The membership entitles the members to a presentation of their logo on NGF's web pages http://ngf.no/?page_id=113 . Class 1 support members also have their logo rotating on the main page. NGF would like to thank all support members for their important contribution to NGF.

Companies registered as members as of 2020-12-31:

Class1

Centrum Pile AB (https://www.centrumpile.se/)
COWI AS (www.cowi.no)
Leca Norge AS (www.weber-norge.no www.leca.no)
Mesta AS (www.mesta.no)
Multiconsult ASA (www.multiconsult.no)
Norconsult AS (www.norconsult.no)

Norges Geotekniske Institutt (www.ngi.no)
Romerike Grunnboring AS (www.rgb.as)
ÅF Engineering AS (www.afconsult.com)

Class 2

Akershus Grunnboring AS
Entreprenørservice A/S (www.entreprenorservice.no)
Franzefoss Minerals AS (https://kalk.no/)
Fundamentering AS (www.fas.no)
GeoVita as (www.geovita.no)
Glasopor (www.glasopor.no www.syklus.no)
Løvlien Georåd AS (www.georaad.no)
Maskinentreprenørenes Forbund, Fagavdeling Brønn- og spesialboring (www.mef.no)
Rambøll Norge AS (www.ramboll.no)
SINTEF Byggforsk AS (www.sintef.no/Byggforsk)
SSAB Svensk Stål AS (www.ssab.no)
SWECO Norge AS (www.sweco.no)
Veidekke Entreprenør AS (www.veidekke.no)

Class 3

Arya Consult (https://aryaconsult.no/)
Cautus Geo AS (www.cautusgeo.com)
ERA Geo AS (www.era-geo.no)
Fugro Geotechnics AS (www.fugro.no)
Geomek Stockholms Geomekaniska AB (www.geomek.com)
Geonor AS (www.geonor.no)
GeoSafe as (www.geosafe.no)
Golder Associates AS Norge (www.golder.no)
Keller Geoteknikk AS (www.keller-geoteknikk.no)
Keller Grundläggning AB (www.kellergrundlagging.se)
PTO teknikk AS (www.pto.no)
RIGeo AS (https://no.linkedin.com/company/rigeo-as)
Trimble Solutions Gothenburg AB (www.construction.trimble.com)
ViaCon AS (www.viacon.no)

Courses and conferences

No courses were held in 2020.

The Geotechnics Day did not take place in 2020 due to the corona situation. It was replaced by a webinar to mark the 70th jubilee of the society.

Scholarship

No scholarship was awarded in 2020.

Diploma prize

No diploma prizes were awarded in 2020.

DIBK

After several members took contact with the society, NGFs Board decided to send a letter to The Norwegian Building Authority (DIBK), 2019-11-27, about the challenges of central approval for geotechnical design. The members were informed of the letter in newsletter 1,2020. The letter resulted in a meeting with DIBK 2020-02-11. As a follow-up of this meeting, DIBK requested that NGF give a presentation on geotechnics for the Authority. The corona situation, also here, was a hinder and the presentation has been moved to 2021. In addition, the Board sent comments 2020-09-01 to the Ministry of Local Government and Modernisation in connection with Building Quality Committee's report "Acceptable Building Quality". The members were informed of this in newsletter 4,2020.

Contact with the other Nordic Geotechnical Societies

The Society is in touch with the Danish (DGF), the Swedish (SGF), the Finnish (FGF) and the Icelandic (IGF) societies. No Nordic Board meeting was arranged in 2020 since NGM was postponed due to the corona situation.

PORTUGUESE GEOTECHNICAL SOCIETY

Commemorating 50 years of activity

The Portuguese Geotechnical Society carried out several activities in the time span from 2017 to 2021. In the year of 2020, due to the pandemic, a gap in the regular events of the Society occurred but the Board rapidly introduced online activities, adapted to the sanitary restraints. In 2020 the new Board was elected, and Alexandre Pinto is the present President, succeeding to Manuel Matos Fernandes.

In the following pages a summary is made of the main events and activities.

1. National Geotechnical Conference (CNG)

This is the main National conference organized by SPG which occurs biennially. Due to the pandemic, the CNG planned for 2020 was postponed to 2021.

- The 16th CNG took place in May 2018, in Ponta Delgada, Azores.
- The 17th CNG occurred online in November 2021, in LNEC, Lisbon.

2018

2021

- The 18th CNG will take place in May 2023, in Évora and is under preparation.

2. Luso-Brazilian Geotechnical Congress (CLBG)

The CLBG is a result of the cooperation between the sister societies in Portugal (SPG) and Brazil (Brazilian Association for Soil Mechanics, ABMS), aiming to boost synergies between both countries geotechnical communities.

- The IX CLBG, focused on *Safety of Dams*, took place in 2018 in Salvador, Brazil.
- The X CLBG, under the subject *The Safety in Geotechnical Works* was held in Lisbon, in 2021, in association with the 17CNG.

3. Hispano-Portuguese Geotechnical Rounds (JHPG)

The JHPG is a result of the cooperation between the sister societies in Portugal (SPG) and Spain (Spanish Association for Soil Mechanics and Geotechnical Engineering, SEMSIG), aiming to boost synergies between both countries geotechnical communities.

- The VI JHPG, took place in 2018, in Ponta Delgada, Azores, in association with the 16CNG.
- The VII JHPG, under the subject *Challenges associated to the ground improvement* will took place in Mieres - Oviedo, in 2022, in association with the XI Spanish Symposium on Geotechnical Engineering.

4. Meeting of Young Geotechnicians (MYG)

The MYG is a one-day event in which young members of SPG present papers about the recent research developed. The José Folque / Young Geotechnical Awards aim rewarding works that constitute a contribution either to the progress of technical and scientific knowledge, or to the solution of national problems. The following MYG took place between 2017 and 2021:

- 8th Meeting of Young Geotechnicians, 2017, LNEC, Lisbon.
- 9th Meeting of Young Geotechnicians, 2018, Ponta Delgada, Azores.
- 10th Meeting of Young Geotechnicians, 2019, Academy of Sciences, Lisbon.
- 11th Meeting of Young Geotechnicians, 2021, LNEC, Lisbon.

5. Manuel Rocha Lecture (LMR)

The Manuel Rocha Lecture is a yearly event that occurs in the Autumn. The invited keynote speaker is in turn a Portuguese and an international specialist of recognized reputation. The following LMR lecture took place between 2017 and 2021:

Year	Number	Keynote speaker	Title
2017	XXXIV	Paul Mayne	Stress History of Soils from Cone Penetration Tests
2018	XXXV	Laura Caldeira	Internal erosion and the safety of dams
2019	XXXVI	Giovanni Barla	Performance assessment of underground caverns during excavation in view of completion
2020	XXXVII	José Mateus de Brito	The judgement in Geotechnical Engineering practice
2021	XXXVIII	Scott Barnes	Climatic Warming - using the past to prepare for the future of increased landslide and other geological disaster activity

2017

2018

2019

2020

2021

6. SPG Awards

There is several Prizes awarded regularly by SPG and GEOTECNIA journal, with variable periodicity, as presented below:

- **Ricardo Esquível Teixeira Duarte Award**, aiming at the recognition of the best geotechnical project, was granted for the biennia 2015-2016 and 2017-2018 during the opening ceremony of the 16CNG (2018) and 17CNG (2021).
- **Manuel Rocha Award**, for the best PhD in geotechnics for the quadrennium 2016-2019, was also delivered during the opening ceremony of the 17CNG (2021).
- **José Folque / Young Geotechnical Awards** in Portuguese and in English were granted at the end of the Meeting of Young Geotechnicians (MYG), that took place in 2017 (8YGM), in 2018 (9MYG), in 2019 (10MYG) and in 2021 (11YGM).
- **Úlpio Nascimento Award** was attributed for the biennia 2012-13, 2014-15, 2016-17, 2018-19 in the ceremony of the 50th anniversary of GEOTECNIA journal.

7. GEOTECNIA journal and SOILS AND ROCKS journal

GEOTECNIA is published every four months jointly by SPG, ABMS and SEMSIG. In 2020 the 50th anniversary was celebrated with a technical program supported by the special volume nº152 (2021). GEOTECNIA is available in open access since July 2021 (https://impactum-journals.uc.pt/geotecnia). The journal is now indexed in SHERPA/ROMEO.

SOILS AND ROCKS is published in English under the cooperation between ABMS and SPG. Four numbers are published every year and in 2021 a new platform for paper submission (SUBMIT) was adopted, aiming at facilitating and automatizing all the revision process, namely is double blind method. The journal is nowadays indexed in Scopus, Clarivate, CNki and Scielo (Soils and Rocks).

8. Newsletters and social network

The communication with SPG members, the technical community is made using different channels, by the quarterly Newsletter hosted at the SPG website (SPG (spgeotecnia.pt)). The Youtube channel (https://www.youtube.com/channel/UC2OKvQMwxac5VQAkcX-WjDQ), the Facebook page ((1) SPG - Sociedade Portuguesa de Geotecnia | Facebook) and the Twitter feed were all launched in 2021.

9. Time Capsule Project (TCP)

To better answer to this ambitious Project, SPG has created a task force, involving a significant number of colleagues from the universities, consulting companies and contractors. For the moment, the following two main themes are being developed:

1) Portuguese Geotechnical History

This theme contemplates the description of the general development of Portuguese geotechnics from 1945 to 2010, in the main areas of research and university education, geotechnical associations, geotechnical main national and international events that took place in Portugal, main Portuguese geotechnical publications, past and present main geotechnical consulting companies and contractors and past geotechnical personalities.

2) Emblematic Portuguese Geotechnical Works

This subject refers to the presentation of the evolution of the main types of geotechnical works: embankment dams, foundation of concrete dams, foundation of buildings and bridges, highways and railways, underground works, large urban excavations, slope stabilization, maritime and river works, mining facilities and rehabilitation; this will consist in the selection of some emblematic main geotechnical works in Portugal and abroad. At present, about 60 works were selected for the Portuguese Version and about 30 will be chosen for the English version.

10. XVIII European Conference on Soil Mechanics and Geotechnical Engineering

Following the decision, taken during the XVII ECSMGE in Reykjavik, regarding the appointment of Lisbon to hold the XVIII ECSMGE in 2023, the preparation of the Conference has already started.

Due to the pandemic consequences, the event was postponed to 2024, 25 to 29 August, at Altice Arena, in Lisbon.

Lisbon, February 2022

Alexandre Pinto

President of the Portuguese Geotechnical Society Board (2020-2024)

ROMANIAN SOCIETY FOR SOIL MECHANICS AND FOUNDATION ENGINEERING
SOCIETATEA ROMÂNĂ DE GEOTEHNICĂ ŞI FUNDAŢII

Bdul Lacul Tei 124, sector 2, Bucureşti
e-mail: srgf@utcb.ro, loretta.batali@utcb.ro, Tel.: 021/2421208/263

REPORT INTERNATIONAL ACTIVITIES
Sept. 2017 – February 2022

1. Scientific international events organised by SRGF Sept 2017 – Febr. 2022

(1) 2.07.2018 – **R.K. Rowe Lecture - Prof. Mario MANASSERO** (Politecnico di Torino, Vicepresident for Europe of ISSMGE) - *On the fabric, intrinsic and state parameters of active clays for contaminant control*"

(2) 9.11.2018 – **International** seminar **„New techniques and technologies for ground improvement"**.
- 120 de attendees
- 15 invites guests
- 2 keynote lectures - Prof. Jan Maertens (Belgium) and Dr. Serge Varaksin (France),
- 2 invited lectures – Assoc. Prof. Janos Szendefy (Hungary) and Dr. Serge Varaksin (France),
- 16 technical contributions
- Sponsors Carmeuse Holding, GT Ground Engineering & Construction Services, Iridex Group Plastic, Liebherr and CRH.

(3) 13.06.2019 – **International** workshop **on rigid and semi-rigid inclusions technique**
- 2 invited guests: Dr. Luc Thorel (IFSTTAR Nantes, France) and dr. Arash Khansari (BBG, Germany).
- 70 attendees

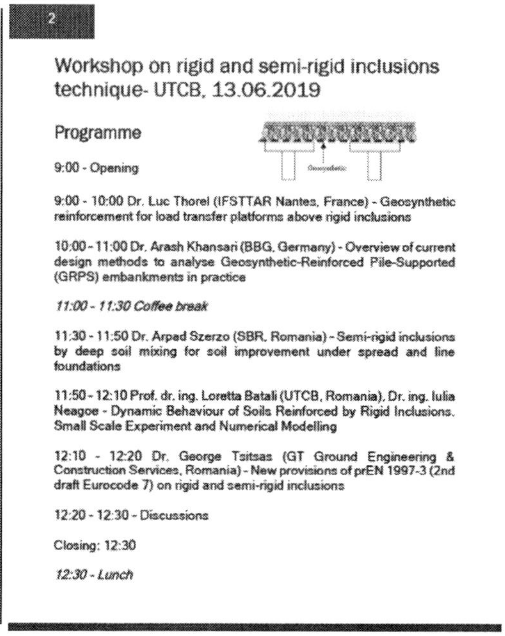

(4) 9.12.2019 – **Conferences**: Hydraulic fills with special focus on liquefaction and Deep excavations in urban areas – Prof. Antonio Gens (Technical University of Catalonia, Barcelona, Spain, former Vicepresident for Europe of ISSMGE), on the occasion of the award of the title Doctor Honoris Cause by TUCB

(5) 23.09.2021 – **Conference**: Pelicular Aspects of Soil Behaviour **Prof. Mario MANASSERO** (Politecnico di Torino, Vicepresident for Europe of ISSMGE) on the occasion of the award of the title Doctor Honoris Cause by TUCB

2. Other national events with international / European opening

(1) 9-10.05.2017: Seminar Ongoing Aspects in Geotechnical Engineering organized at the TUCN together with the Austrian Society of Geotechnics and TU Graz

(2) 24.11.2017: National symposium dedicated to Geotehnical design according to Eurocde 7
- 100+ attendees
- Organised in cooperation with Romanian mirror Group for Eurocode 7 in the standardization association (ASRO)
- Sponsors Octagon Contracting & Engineering S.A., Carmeuse and Popp & Asociații Inginerie Geotehnică.

Bdul Lacul Tei 124, sector 2, Bucureşti
e-mail: srgf@utcb.ro, loretta.batali@utcb.ro, Tel.: 021/2421208/263

(3) 2.07.2018 – **First National Symposium of Environmental Geotechnics and R.K. Rowe Lecture - Prof. Mario MANASSERO** (Politecnico di Torino, Vicepresident for Europe of ISSMGE) „*On the fabric, intrinsic and state parameters of active clays for contaminant control*".

- 90 attendees
- 17 technical contributions
- Sponsor: Naue Romania.

(4) 16.02.2021 Webinar „Evaluation of rock slope stability. related risks and seismic performance assessment of slopes in Romania"

(5) **2-3 June 2021 - CNGF2021 – XIVth National Conference of Soil Mechanics and Foundation Engineering – online**

XIV 2-3.06.2021 CNGF BUCUREŞTI Conferința Națională de Geotehnică și Fundații

- 2 invited keynote lectures: Dr. Andrew Bond (UK, former Chairman CEN/TC250/SC7) - Bridging the gap: bringing geotechnical and structural design together in the 2nd generation Eurocodes; Prof. Dietmar Adam (TU Vienna) - Retaining Structures in Challenging Ground Conditions - Case Studies from Europe
- Invited panelist: Dr. Sebastien Burlon (SETEC TERRASOL, France) - Seismic design : towards a displacement-based approach
- 2 plenary sessions, 4 discussion sessions, 2 workshops
- 45 technical presentations, 10 sponsors' presentations
- Technical exhibition: 21 virtual booths
- Proceeding: 93 papers, 700+ pages
- Virtual attendees: 275 (day 1); 216 (day 2)
- Virtual visitors technical exhibition: 189 (day 1), 132 (day 2)

ROMANIAN SOCIETY FOR SOIL MECHANICS AND FOUNDATION ENGINEERING
SOCIETATEA ROMÂNĂ DE GEOTEHNICĂ ŞI FUNDAŢII

Bdul Lacul Tei 124, sector 2, Bucureşti
e-mail: srgf@utcb.ro, loretta.batali@utcb.ro, Tel.: 021/2421208/263

3. Attendance to international events

- 19th ICSMGE2017- 17 – 22 Sept. 2017 Seoul, Coreea
- 6th iYGEC6 – 16-17 Sept. 2017 Seoul, Coreea
- XVIth DECGE, Skopje, June 2018
- 26th EYGEC – Graz, Sept. 2018
- International Symposium on Energy Geotechnics SEG2018 – Lausanne, sept. 2018
- 8th International Congress of Environmental Geotechnics, Hangzhou, Oct. 2018
- XVIIth ECSMGE Reykjavik, Sept. 2019.
- 27th EYGEC – Bodrum, Sept. 2019
- 40 Years of roller integrated continuous compaction – Vienna, 29.11.2018

4. Participation in the scientific committees of ISSMGE events
- Prof. Sanda Manea – ECSMGE 2019, XVI DECGE 2018, 20ICSMGE 2022

5. Other activities:
- Constantly informing our members about ISSSMGE (related) activities and events through newsletters, emails and Facebook posts
- Work for the Time Capsule Project and dissemination to the society members
- Organisation 17DECGE 2023, Bucharest

GEOTECHNIK SCHWEIZ
GÉOTECHNIQUE SUISSE
GEOTECNICA SVIZZERA

International Society for Soil Mechanics
and Geotechnical Engineering ISSMGE
Vice-President Europe
Prof. Mario Manassero
Politecnico di Torino
10129 Torino, Italy

Zurich, January 2022

Activities of the Member Society „Geotechnik Schweiz“; September 2017 to the Present.

Dear Prof. Manassero

Following your request of 15th January 2022 we present you our report.

We organized annual conferences in spring and autumn every year, which lasted one or two days. The conferences included presentations and discussions. The conference presentations are published in brochures and distributed to all our members and to the participants of the conference in question. The extra copies can be purchased from anyone who is interested in the subject and one year after the conference, they are placed on our website. The webinars and video-recordings of our conferences can also be found on our website.

Future events:

year	title of the conference	city	participants	publication «Mitteilungen der Geotechnik Schweiz» no.	available on our website
2022	Time Capsule Project for ISSMGE Council Meeting in Sydney We will contribute with articles about research, big projects of the past 60 years and lessons learnt				may 2022
2022	Geotechnische Herausforderungen bei weichen Böden/Défis géotechnique dans des sols mous	Muttenz		181	spring 2023
2022	Design Challenge and Meeting Geotech Abroad for Young Members	Lausanne		none	june 2022
2022	Exkursion 2. S-Bahn-Stammstrecke München	München		none	
2023	Wasserhaltung, Drainage, Abdichtung in der Geotechnik	Biel		182	fall 2023

Past events:

year	title of the conference of the past	city	partici-pants	publication «Mitteilungen der Geotech-nik Schweiz» no.	available on our website
2021	Exkursion Baustellenbesichtigung Mythenquai Swiss-Re, Zürich	Zürich	43	none	
2021	66th anniversary of the Swiss Geotechnical Society Lecture “A Tale of Two Towers – Pisa and Big Ben” by Prof. John Burland, Imperial College	Zurich	60	none	available
2021	Tiefbahnhof Bern / *Gare souterraine de Berne*	Webinar	128	180	spring 2022
2021	Vortragsreihe / *Cycle de conférences* *7 presentations to several topics (legal issues, projects, innovation,...)*	Webinar		none	
2020	BIM in der Geotechnik und im Strassen-/Infra-strukturbau / *BIM dans le domaine de la géotechnique, de la construction de routes et d'infrastructures* Conference organized together with VSS (Research and standardization in the field of road and transportation)	Live-Webinar	ca. 300	179	available
2019	Exkursion «Grand Paris»	Paris	51	none	
2019	Vereisungsmassnahmen in der Geotechnik *Congélation en géotechnique*	Bern	127	178	available
2018	Forschung und Innovation in der Geotechnik/ *Recherche et innovation en géotechnique*	Lausanne	98	177	available
2018	Beobachtungsmethode und Überwachung in der Geotechnik / *Méthode observationnelle et surveillance en géotechnique*	Yverdon	148	176	available
2017	Stützmauern / *Murs de soutènement* Conference organized together with VSS (Research and standardization in the field of road and transportation)	Bern	250	175	available

Yours faithfully

Geotechnik Schweiz

Vice-President's Report – North America Region

Timothy Newson
VP North America; Civil and Environmental Engineering, Western University, London, Ontario, Canada. tnewson@eng.uwo.ca

1 INTRODUCTION.

The international geotechnical activities in North America are organized by three professional groups. These are the Mexican Geotechnical Society (SMIG), the Canadian Geotechnical Society (CGS) and the Geo-Institute (G-I) of the American Society of Civil Engineers (ASCE). A number of other prominent organizations also host and sponsor technical and professional activities, including the Deep Foundations Institute (DFI), Association of Drilled Shaft Contractors (ADSC), Pile Drivers Contractors Association (PDCA), and Geosynthetics Institute (GSI). A brief summary of the activities by the SMIG, CGS, and G-I during my term (Sept 2017- May 2022) are described in the following sections.

2 THE NATIONAL SOCIETIES.

2.1 Mexican Society of Geotechnical Engineering (SMIG)

The membership for the society during this term has been approximately 500. Activities for geotechnical engineering are held every two years by the Mexican Society of Geotechnical Engineering and are posted at: www.smig.org.mx. Since the 19th ICSMGE in Seoul (Sept 2017), two national conferences have occurred:

- XXIX National Meeting of Geotechnical Engineering and XX National Meeting of Professors of Geotechnical Engineering was held from November 21st to 24th, 2018 in León, Guanajuato, Mexico.
- XXX National Meeting of Geotechnical Engineering and XXI National Meeting of Professors of Geotechnical Engineering (online) was held from 16th to 19th March 2021.

The SMIG has three invited lectures that are provided for the national events:

- The Nabor Carrillo Lecture;
- The Leonardo Zeevaert Lecture;
- The Raúl J. Marsal Lecture.

The SMIG has offered a number of technical courses, workshops and other events during the term. These include:

- 5th Colloquium of Young Geotechnicians and 2nd Meeting of Teachers: Challenges in the 60th Anniversary. October 27th - 28th, 2017.
- 6th Colloquium of Young Geotechnicians. October 25th – 26th, 2019.
- 7th Colloquium of Young Geotechnicians November 11th – 12th, 2021.
- 4th DFI-SMIG-GI-ISSMGE International Deep Foundations Symposium from November 15th-16th, 2017 in Mexico City.
- 5th International Symposium on Tunnels and Shafts in Soils and Rock. Held from August 16th-18th, 2018.
- 6th International Symposium on Tunnels and Shafts in Soils and Rocks, Mexico City form March 29th – 31st & April 5th – 7th, 2022.

An important event held in the North American region is our quadrennial regional conference that is alternately shared between North America and South America. The successful XVI Pan American Conference on Soil Mechanics & Geotechnical Engineering (PCSMGE) was held in Cancún from November 17th-20th 2019. There are a number of prestigious invited lectures associated with this conference; in particular the Casagrande Lecture was given by Professor Gabriel Auvinet and the Plenary Lecture was given by Professor Charles Ng. Seventeen keynote lectures were provided by global expert speakers. Three rising stars also gave the new Bright Spark Lectures. Seven hundred and fifty-five abstracts were submitted from forty-one countries.

As the pre-eminent conference in our discipline, these events provide a focus for the interests and activities of engineers, academics and contractors working in geotechnical engineering in the Americas. The next venue for the XVII PCSSMGE will be Concepción, Chile in October 2024.

2.2 *Canadian Geotechnical Society (CGS)*

The membership for the society during this term has been approximately 1200. National activities on geotechnical engineering are held every year by the Canadian Geotechnical Society. The national CGS conferences generally alternate between the eastern and western sides of the country, with details found at: www.cgs.ca. Over this term, the national conferences have included:

- GeoOttawa 2017: *70 Years of Geotechnics & GeoSciences.*
- GeoEdmonton 2018: *Transportation Geotechnique - Moving Forward.*
- GeoSt.John's 2019: *Under Land and Sea.*
- GeoVirtual 2020 (purely online): *Resiliance & Innovation.*
- GeoNiagara 2021: *Creating a Sustainable & Smart Future.*

The CGS has two invited lectures that are given at these national events:

- The R.M. Hardy Address;
- The CGS Colloquium.

Other major events and conferences were also held throughout this term.

The Cross Canada Lecture Tour (CCLT) of the Canadian Geotechnical Society has been delivered since 1965. The purpose of the lecture tour is to provide the CGS members & other geotechnical professionals an opportunity to attend high quality technical presentations by prominent Canadian and international geotechnical professionals. Typically, spring and fall tours are scheduled each year. The tour consists of presentations to 10 to 12 CGS local sections across Canada. Customarily one lecturer per year is a well-known Canadian CGS member and the other is a well-known international individual in the geotechnical field. The CCL

T has become nationally and internationally known and res pected. Although there was some disruption to the 2020 schedule, these lectures have continued during this term & the following individuals have been recently involved.

- 2019 Fall I Moore, Queen's University, Kingston, ON.
- 2019 Spring C Shackelford, Colorado State Univ., USA.
- 2018 Fall A Sy, Klohn Crippen Berger Ltd., Vancouver.
- 2018 Spring T O'Rourke, Cornell University, USA.
- 2017 Fall J-M Konrad, Université Laval, QC.

In lieu of the Cross-Canada Lecture Tour (CCLT), the Ca nadian Geotechnical Society (CGS) conducted the First CG S National Geotechnical Lecture featuring Dr. Doug Stead, Professor Emeritus of Simon Fraser University. It is hoped that these important events can be commenced in the near future.

There has also been tremendous effort directed towards the preparation/production of the next version of Canadian Ge otechnical Engineering Manual (CFEM 2021). The CFEM 2021 will include 25 chapters. Four new chapters have bee n added compared to the previous CFEM version. There a re over 75 CGS members actively contributing to the vari ous chapters, supported by a Senior Advisory Committee & external reviewers.

2.3 ASCE GeoInstitute

The membership for the society during this term has been approximately 3200. The ASCE Geo-Institute (www.geoinst itute.org) holds a national congress once a year, and on oc casion, additional specialty conferences are also held. Since the Seoul conference, the G-I congresses have included:

- IFCEE 2018, Orlando, Florida.
- GeoCongress 2019, Philadelphia, Pennsylvania.
- GeoCongress 2020, Minneapolis, Minnesota.
- IFCEE 2021, Dallas, Texas.
- GeoCongress 2022, Charlotte, North Carolina.

The ASCE G-I has a number of invited lectures that are given at these national events including:

- The Ralph B. Peck Medal Lecture;
- The H. Bolton Seed Medal Lecture;
- The Karl Terzaghi Lecture;
- The Mercer Award Lecture.

A number of other conferences and events have also occur red, including:

- PanAm UNSAT, November 12th-15th, 2017, Dallas, Texas.
- GEESD V, June 10th-13th, 2018, Austin, Texas.
- EMI/G-I joint conference, June 18th-21st, 2019, Pasadena, California.
- Strong participation has also occurred with the Transportation and Development Institute's Pavements Conferences:
  - August 27th-30th, 2017, Philadelphia, PA
  - July 21st-24th, 2019, Chicago, Illinois

The US has also hosted other international conferences, inc luding the 4th International Conference on Transportation G eotechnics (under the auspices of TC202) and the 10th Inte rnational Conference on Scour and Erosion (under the ausp ices of TC218). In addition, the US has published the pro ceedings of the 2nd joint technical workshops between the USA and Kazakhstan member societies.

One of the cycle of CAPG panel discussions was conducte d at 2021 IFCEE Dallas (*Knowledge gaps in geotechnical engineering design and construction and the role that the USA could play in influencing global practice*). This has been the only session of this kind to occur within a natio nal conference, rather than during a regional ISSMGE conf erence. Finally, the US has submitted a proposal to host the 21st ICSMGE in Washington, D.C. in 2026 (theme: Ge otech on Earth and Beyond!).

3 TECHNICAL COMMITTEES.

The North American region of the ISSMGE is currently hosting (or has previously hosted – shown in italic) a number of technical committees (TC) during this term. The committees and their associated chairs are shown below.

- TC 105 - Geomechanics from Micro to Macro: Kenichi Soga (United States);
- TC 202 - Transportation Geotechnics: Erol Tutumluer (United States);
- *TC 203 - Geotechnical Earthquake Engineering & Associated Problems: Ross Boulanger (United States);*
- TC 205 - Safety & Serviceability: Rodrigo Salgado (United States);
- TC 208 - Slope Stability in Engineering Practice: Andy Take (Canada);
- *TC 218 - Reinforced Fill Structures: John Sankey (United States)*
- TC 307 - Sustainability in Geotechnics: *Dipanjan Basu (Canada)*/ Anand J. Puppala (USA);
- *TC 308 - Energy Geotechnics: Marcelo Sanchez (United States).*

These TCs have been very successful and have undertaken a significant number of activities in the period that is reviewed in the final report of the Technical Oversight Committee (TOC). In addition, the Young Member Presidential Group (YMPG) has been expertly lead by the chair by Lucy Wu (United States).

4 GENERAL REMARKS.

Despite ongoing impacts from Covid-19, this term has been busy for the national societies. I have received ongoing responses from each member society and generally, they have all appeared to cope well with the situation, but have been required to reschedule or revise the modality of many events during this period. There have been some impacts of the pandemic on the regional economy, and therefore, the participation (assistance) of geotechnical engineers and students in the various activities has temporarily decreased. During the periods of the most significant lockdowns, face-to-face events were cancelled and this lead to the implementation of virtual events, using different digital platforms. In some circumstances there appeared to be advantages of these virtual modalities and one positive aspect of the pandemic may be the retention of the enhanced aspects of this activity.

It is gratifying to see that good cooperation and shared activities have still occurred in the region, including workshops, symposia, and conferences between the various North American professional organizations. I have taken advantage of the National Meetings and the International Conferences that I have attended to interact with officers and members of the North American member societies. The following items have been raised for discussion: membership, links with neighbouring and other countries, participation in International Conferences and Technical Committees, North American code provisions, how to raise the profile and visibility of our profession and the relationships between the national societies and the ISSMGE.

I have also made efforts to try to expose members of the national societies to the research and professional practice occurring in the other countries of the region through the technical publications of the societies (i.e. GeoStrata, Geotechnia and Geotechnical News). Special issues of these publications expounding geotechnical projects from the other regions have been published. New policies have also been put in place by the national societies to encourage the participation of the different leaderships to attend each other's national events. I sincerely hope this cross-border cooperation continues. It has been an honour to represent the North American region on the ISSMGE board and I thank the societies for their support and ongoing excellence to place our geotechnical engineering at the forefront of world practice.

ISSMGE activities in South America and the Caribbean 2017 – 2022

Alejo Oscar Sfriso
Vice President of ISSMGE for South America and the Caribbean

SUMMARY: The ISSMGE region of South America and the Caribbean has eleven member societies. This report contains a summary of the activities and interaction among the societies in the region and updates since the Council meeting held in Seoul in 2017. It closes out with an outlook of the future of ISSMGE in the region. Activities and interactions among the sister societies in the period 2017 – 2019 were omitted for brevity.

1 INTRODUCTION

The ISSMGE region of South America and the Caribbean has eleven member societies, namely Argentina, Bolivia, Brazil, Chile, Colombia, Costa Rica, Cuba, Guatemala, Paraguay, Perú and Venezuela.

This report summarizes the activities of these member societies, and of this VP, and presents some general remarks about the status of the region and it's outlook in the immediate future. Activities and interactions among the sister societies in the period 2017-2019 can be found in the 2019 Council Meeting Report and were omitted from this report for brevity.

The COVID-19 pandemic has disrupted the life and plans of everybody in the world, and of course also in the region. Suddenly, all of the in-person meetings that used to be the opportunity to keep people in close contact were postponed or cancelled in March 2020, and they remained so until the end of the term of this VP. The pandemic has also impacted the work and plans of this VP. It found me in the ISSMGE Board Meeting in Tunisia, and I was lucky to be able to rush home shortly before all airports were closed and flights cancelled.

As a result, the planned visits to the three member societies that were still outstanding by the time the pandemic started, Guatemala, Costa Rica and Venezuela, had to be postponed and will not happen before this VP hands office to the next appointed VP, Prof. André Pacheco de Assis. This VP sincerely apologizes for this situation. Also, the efforts to rescue the National Society of Ecuador and Dominican Republic, terminated in 2017, while advancing, have not yet come to a successful conclusion.

It was fortunate that the pandemic also found our member societies leaded by young and enthusiastic groups who were able to soon put in place alternate methods of communication and sharing of knowledge. As a result, we have observed a swift increment, rather than a reduction, in the technical activities in the region: the 51 events that took place in the region before the Council Meeting in Cape Town – and were reported there – were followed by +250 online events (+150 just in Brazil) in the period late 2019 to 2021. For instance, this VP was invited by National Societies and universities in the region to deliver eight courses and forty-one invited lectures during his term 2017-2022, twenty-five of them after the pandemic started. In those events, and as relevant in each case, the mission, activities and services provided by ISSMGE were shared with participants.

During the period 2017-2022 there was a close, intense and continuous interaction among the National Societies in the region. Twelve online meetings of the Presidents and Secretaries were organized and chaired by this VP, and two in-person meetings were held in Seoul and Brazil, the latter one during the Brazilian Conference in 2018 as informed in the report to 2019 Council Meeting. This VP is grateful to ABMS for hosting that event and inviting the officers of all sister societies to it, as it proved to be the last time we were all able to meet face-to-face. In these meetings, this VP reported back to officers the outcome and decisions of the various ISSMGE board meetings and the ISSMGE initiatives, including updates to the Virtual University, the Online Library, and the Time Capsule Project.
Into the future, more collaboration is required, including active support from our larger societies to bring our smaller groups up to speed in the organization of technical activities. Also, regional collaboration and effort is required to promote succession and growth in some of the societies where officers have been serving for several years already. This VP wishes Prof. Assis success with these initiatives and the many he will introduce during his term.

2 ACTIVITY OF THE MEMBER SOCIETIES

As mentioned, only events that happened since the Cape Town Council Meeting (2019) are listed below. Those already reported in 2019 are omitted for brevity.

2.1 Argentina

Sociedad Argentina de Ingeniería Geotécnica (SAIG) was founded in 1948 and has 127 members. The activities held in the period are:
- -2020: Eight lectures on "Geotechnical Applications in Power Generation"; Seminar on Latin American Geotechnical Challenges; Seminar on Geotechical Controls during Execution of Archaeological and Engineering Works.
- -2021: XXV Argentinean Conference on Soil Mechanics and Geotechnical Engineering; Workshop of Geotechnical Engineering Teachers of Argentina; Workshop CIRSOC 402: Argentinian Code of Foundations.

At the time of this report, the president is Diego Manzanal. Contact info: saig.org.ar.

2.2 Bolivia

Asociación Boliviana de Ingeniería Geotécnica (ABIG) was founded in 1975. After being terminated in 2017, it rejoined ISSMGE in 2018, and has 42 members. The activities held in the period are:
- -2020: Twelve Bolivian Geotechnical Lectures on various topics of geotechnical engineering, foundations, numerical modelling and site characterization; Colloquium "From CBR to Resilient Modulus"; 4 International Courses on In Situ Testing.
- -2021: Seven Advanced Courses on Design and Supervision in Pavement Geotechnics.

At the time of this report, the president is Juan Carlos Rojas Vidovic. Contact info: www.facebook.com/ABIGnacional/.

2.3 Brazil

Asociación Brasileira de Mecanica do Solos (ABMS) was founded in 1950 and has 1259 members. Being a very large society, ABMS has branches and regional committees that coordinate activities throughout the country. Their reaction to the pandemic was also swift and efficient, moving from in-person symposiums to webinars (146 in two years). The activities held in the period are:
- -2020: WebGeo, Digital Symposium on Geotechnics; ABMS 70th Anniversary; 1st Brazilian-Digital Congress on Geotechnics and Environment; sixty Webinars covering most relevant topics in Soil Mechanics and Geotechnical Engineering.

-2021: 5th Brazilian Tunnelling Congress; Indian-Brazilian Geotechnical Webinar; 10th Portugal/Brazil Geotechnical Congress; 3rd Panamerican Congress on Non-Saturated Soils; again, seventy six Webinars covering most relevant topics in Soil Mechanics and Geotechnical Engineering.

The list of webinars, too long to be included in this report, can be found at ABMS webpage and ABMS TV Channel, which also contains several interviews, educational material and recordings of many past technical events.

At the time of this report, the president is Fernando Schnaid. Contact info: www.abms.com.br.

2.4 Chile

Sociedad Chilena de Geotecnia (SOCHIGE) was founded in 1968 and has 100 members. The activities held in the period are:

-2019: Geotechnical Seminar in Curicó; Workshop on Tailings Dams (with AICE); Seminar on Design, Construction and Operation of Tailings Dams; International Course on Earthquake Geotechnical Engineering (with ACHISINA).

-2020: Seminars on Critical State Soil Mechanics; two Seminars on Ground Improvement; Seismic Characterization of Soils; Soil Liquefaction; Deep Excavations; Site Effects; sixteen seminars on Local Soils; Updates in National Codes for Soil Mechanics and Geotechnics; Case Histories of Failure of Tailings Deposits; Pressuremeter and its new Applications.

-2021: Downdrag of Piles; CSSM of Sands; Evolution of Mexico City Soils and its Impact in Earthquake Geotechnical Engineering; Impact of Geology in Engineering Projects; Low Enthalpy Geotechnics; First Meeting of Young Geotechnical Engineers; XI National Conference SOCHIGE.

At the time of this report, the president is Roberto Gesche Schüler. Contact info: sochige.cl.

2.5 Colombia

Sociedad Colombiana de Geotecnia (SCG) was founded in 1971 and has 141 members. The activities held in the period are:

-2019: III Course on Exploration and Ground Characterization; Course on Earthquake Geotechnical Engineering.

-2020: VI Regional Conference of Geotechnics; XIII International Symposium on Landslides.

-2021: IX International Symposium of Geomechanics (with ISRM); Webinar on Applied Geomechanics in Proactive Design; Seminar on Municipal Studies on Landslide Risk.

At the time of this report, the president is Ramiro Castellanos Jiménez. Contact info: www.scg.org.co.

2.6 Costa Rica

Asociación Costarricense de Geotecnia (ACG) was founded in 1979 and has 108 members. The activities held in the period are:

-2020: Seminar on Cross-hole Logging for Bored Piles; Liquefaction in subduction zones.

-2021: Ground Improvement and Foundation Design; Development of the Geotechnical Code of Costa Rica.

At the time of this report, the president is Adrián Fernández Castro. Contact info: geotecniacr.com.

2.7 Cuba

Comisión Técnica Nacional de Mecánica de Suelos e Ingeniería Geotécnica (CTNMSIG) was founded in 1983 and has 30 members. The activities held in the period are:

-2020: Invescons 2020, Conference on Geotechnical Engineering.

-2021: II International Convention of Engineering; Annual Meeting of the Technical Committee for Geotechnical Engineering.

At the time of this report, the president is Gilberto Quevedo Sotolongo. Contact info: quevedo@uclv.edu.cu.

2.8 Guatemala

Asociación Guatemalteca de Mecánica de Suelos e Ingeniería Geotécnica (AMSIG) was founded in 2010 and has 38 members. The activities held in the period are:

-2020: Lecture series on Geotechnical Engineering delivered to the Guatemala Chamber of Construction.

-2021: II Conference on Earthquake and Geotechnical Engineering.

At the time of this report, the president is Alberto Pérez Zarco. Contact info: www.amsig.org.gt.

2.9 Paraguay

Sociedad Paraguaya de Geotecnia (SPG) was founded in 1978 and has 30 members. The activities held in the period are:

-2020: First Conference on Dispersive and Sulphate Soils; Latest Advances in Technologies for Deep Foundation Construction.

-2021: Latin American Lecture Tour (with ISRM); Evaluation of Retention Structures using MEMs;

-2022: Latin American Rock Mechanics Symposium; Advances in the Monitoring of Geotechnical Projects; First Mine Expo.

At the time of this report, the president is Roberto Andrada. Contact info: www.spg.org.py.

2.10 Perú

Asociación Peruana de Ingeniería Geotécnica rejoined ISSMGE in 2017 and has 146 members. The activities held in the period are:

-2020: Lectures on Strength Parameters for Slope Stability Analyses; Construction Processes in Urban Excavations; Geotechnical Risks in Slope Stability; Microtunnels; Response Spectrum from Peruvian Standards, Comparison with Predictive Models; Settlement and Tilt of Structures in Liquefying Soils; Multi-directional Dynamic Response of Sands; Numerical Modelling of Seismic Settlement in Sands; Analysis and Design of Stone Columns; Excavation Support in Peru and New Technologies; Sustainability of Earth Reinforced Works employing Geosynthetics; Workshop on Field Testing and Instrumentation.

-2021: Workshops on Soil Properties; on Soil Exploration; on the Use of Geosynthetics in Transportation Problems; Current Practice in the Evaluation of Dam Break Risk; State of Practice of Earth and Rockfill Dam Construction; Geotechnical Reliability and Stabilization of Mining Slopes; Peruvian Geotechnical Engineering Conference (APGEO, COPEG, IGS).

-2022: Seismic Site Amplification in Lima.

At the time of this report, the president is Alvaro Javier Pérez Zúñiga. Contact info: apgeo.org.pe.

2.11 Venezuela

Sociedad Venezolana de Geotecnia (SVG) was founded in 1958 and has 32 members. The activities held in the period are:

-2021: First Venezuelan Conference of Geotechnical Engineering.

This VP congratulates SVG for organizing their first national conference, which was 100% online and a regional success. This VP also thanks SVG for inviting him to participate in a technical session in the conference.

At the time of this report, the president is Heriberto Echezuria. Contact info: civ.svdg@gmail.com.

3 OTHER REGIONAL ACTIVITIES

3.1 Pan American Conference on Soil Mechanics and Geotechnical Engineering

PCSMGE was held in Cancún, Mexico, in CC-CC, 2019. It was attended by +800 professionals from 32 countries. Out of the 335 papers, 300 of them coming from the Americas, and it was a big success in terms both of participation and technical quality of the sessions. This VP congratulates Sociedad Mexicana de Ingeniería Geotécnia (SMIG) for the organization of the event.

The next PCSMGE is scheduled to happen in late 2023 in Chile. The National Society organizing the Conference, SOCHIGE, appointed Daniela Pollak (dpollak@ gmail.com) and Gonzalo Montalva (gmontalva@gmail.com) as contact persons.

This VP invites all members of the Geotechnical Engineering Family to support this Conference, submitting papers and attending in person when practical, and notes that SOCHIGE is making a huge effort to produce a successful event while facing headwind due to the ongoing pandemic.

3.2 Symposium on Landslides

Sociedad Colombiana de Geotecnia organized the JTC1 XIII International Symposium on Landslides, an event that was schedule to happen in June 2020 in Cartagena de Indias, was forced into online mode due to the pandemic, and held in February 2021. The Symposium was a success, with 265 participants from 34 countries. It was also the first international use of the ISSMGE platform to handle papers and reviews. This VP congratulates SCG for the organization of the event.

3.3 PanAm UNSAT

The third Pan American Conference on Unsaturated Soils was organized by ABMS and held virtually in July 2021. 9 keynote and 5 state-of-the-art lectures were published in Soils and Rocks, the journal maintained by ABMS. All contributions are available open access in www.matec-conferences.org/articles/mateccconf/abs/2021/06/contents/contents.html.

3.4 International Seminars

An ISSMGE International Seminar Series – former Touring Lectures – was held in Paraguay, Argentina and Mexico in September 2018, and was well attended. The subject selected for this Seminar was "Codes for Foundations". Lectures were delivered by Roger Frank, Roberto Terzariol (also representing Pedro Seco e Pinto) and Alejo Sfriso.

3.5 Participation in Technical Committees

The region had a modest participation in the leadership of Technical Committees of ISSMGE, with only one vice-chair held by Bernardo Caicedo (Colombia) in TC2020 on Transportation Geotechnics.

Early in 2019, this VP proposed the creation of the new TC221 Committee on Tailings and Mine Waste which was incorporated by the ISSMGE Board in the meeting held in Tunisia in March 2020 (www.issmge.org/committees/technical-committees/applications/tailing-and-mine-wastes). The creation of this committee is a reaction by ISSMGE to the growing interest by the mining industry of a closer interaction between the industry-focused tailings groups and geo-professionals grouped in ISSMGE. TC221 is chaired by Ramón Verdugo, (Chile), and Fernando Schnaid (Brazil), thus increasing our presence significantly.

3.6 Young Geotechnical Engineers Conference

The Young Geotechnical Engineers Conference was planned to be held during COBRAMSEG 2020, the Brazilian National Conference. Unfortunately and due to the pandemic, COBRAMSEG 2020 and the regional YGEC were postponed and will be held in August 2022 in Campinas, Brazil.

3.7 Time Capsule Project

The Time Capsule Project (www.issmge.org/the-society/time-capsule/time-capsule) is an initiative of the ISSMGE Board that is fully explained elsewhere. TCP has been well received by the National Societies in the region, with 100% engagement and commitment to present content. Brazil, Colombia and Argentina had already created and shared fresh content in the form of interviews, articles and multimedia contributions. Other National Societies have already started their plans to contribute.

3.8 Corporate associates

Three CAPG members from South America – SRK Consulting, Expander Body Panama, and AOSA – were admitted into ISSMGE in the period.

4 OUTLOOK AND CLOSURE

One of the initiatives of this VP, announced in 2017, was a failure: the creation of a wiki-book, a source of knowledge and information in Spanish for the undergrad students of the universities in the region. The initiative was based on the fact that many universities still use written class-notes that have been around for many years and are therefore outdated in the light of the new technologies. After one year of planning and discussion, nothing real had happened, so this VP ditched the idea in 2018.

No failure comes without a lesson: we all learnt that contributors engage when they can share their personal experience and opinion, more than ghost-writing a reference that goes beyond their control – a feeling this VP also shares. Fortunately and during the period, ISSMGE Virtual University (virtualuniversity.issmge.org) expanded its contents to an extent that most of the modern knowledge is already there – now it's not creating content, it's just translation. This VP suggests that Spanish and Portuguese captions be added to Virtual University material and encourages the member Societies in the region to carry out this task, an initiative that is not difficult to accomplish with the aid of automated translation tools.

The future of ISSMGE and the National Societies in the region is bright. Practices that pandemic brought to stay have created a wealth of opportunities for our geo-professionals. More technical resources, made available by ISSMGE and other sources, have now been brought to attention to people that became suddenly isolated. "Working from home" also means "working worldwide" for all of us, and signs of South American geo-professionals taking advantage of this opportunity are ubiquitous. Finally, virtual technical events where worldwide experts share knowledge to local audiences have also created new avenues for recognition, collaboration and knowledge sharing. In this regard, the disruptive tragedy the pandemic was turned out to be also an opportunity, and our National Societies are working already into making this opportunity real for their individual members.

Being the Vice President for South America and the Caribbean of ISSMGE has been a fantastic experience, of which I will be always grateful to ISSMGE, to the National Societies that supported my nomination, and to the many officers that collaborated to bring this tenure to a successful completion. I was handed the torch of this Vice Presidency by Prof. Jarbas Milititsky and I hand it over to Prof. André Pacheco de Assis. André has long experience in leading large professional groups – he has been President of ABMS and President of the International Tunnel Association (ITA) – so I'm sure his contribution will bring our National Societies to a new, higher level of collaboration and growth. I wish him success and joy in the adventure he starts here.

Appendix 6: ISSMGE Virtual University program: report of activities 2017–2022

Programme de l'université virtuelle de la SIMSG : rapport des activités 2017l–2022.

Mounir Bouassida

University of Tunis El Manar, National Engineering School of Tunis], mounir.bouassida@enit.utm.tn

ABSTRACT: This report summarizes the activities developed during the program launched since 2017 in the framework of the virtual university (VU) of the ISSMGE. The report focuses on the adopted working plan to form on-line courses from existing webinars. Then, the first formed courses from existing webinars and their classification following specific soil mechanics themes. Further, courses delivered by eminent specialists and dedicated to specific themes in geotechnical engineering enriched the content of the VU. The final list of formed courses during the first contribution covering the 2017-2022 period is presented as well as recommendations for the next step of the VU of the ISSMGE.

RÉSUMÉ : Ce rapport résume les activités menées dans le cadre du programme de l'université virtuelle de la SIMSG nouvellement lancée depuis 2017. La démarche de travail adoptée pour la proposition de cours à distance à partir de webinars existants, la classification sous forme de cours spécifiques à des thèmes particuliers de la géotechnique constitue le point focal de cette nouvelle action. Les invitations faites à des spécialistes en géotechnique pour enrichir le contenu des thèmes choisis et de proposer des cours dédiés à une thématique choisie sont également commentés. La liste définitive des webinars proposés durant la première contribution de l'université virtuelle couvrant la période 2017-2022 est présentée, en outre quelques recommandations sont proposées pour le futur de ce nouveau programme.

KEY WORDS: webinar, course, credit, virtual university, on-line.

1 INTRODUCTION.

The President, Professor Charles Ng, announced his plan on Education, Innovation and Diversity in September 2017 at the ICSMGE in Seoul. The proposed "ISSMGE virtual university" has begun to take shape. It is the intention of the ISSMGE virtual university (VU) to provide free online structured postgraduate courses for anyone to use. There is neither assessment nor credit to deliver. The ISSMGE cannot be responsible for the use of any of the courses materials.

This report presents, first, the working plan. Second, it follows the contribution made that included the types of formed courses during the term 2017-2022, and recommendations for the next step of the VU program.

2 WORKING PLAN

As first stage, the existing ISSMGE webinars, before the launch of the VU program, have been reviewed and categorized into themed series of webinars or postgraduate courses. Each series comprises a minimum of three webinars giving students and engineers up to five hours of viewing. This viewing, by login to ISSMGE website, is equivalent to a fifteen hours course in a classroom.

The first five courses, approved by the ISSMGE board members, were identified from the delivered webinars before September 2017.

Themes of webinars and speakers are selected by the chairs of technical committees (TCs) and, at a final stage, approved by the ISSMGE board members.

3 CONTRIBUTION MADE

At mid-term, May 2019, the first six courses were publicly launched, i.e.:

1 Risk-Mitigation, Monitoring & Observational Methods
2 In situ Testing:
3 Earthquake Engineering
4 Foundations
5 Soil Characterization
6 Geo-Engineering Education.

The first course of the ISSMGE virtual university was experienced on February 06-09th 2018, at the National Engineering School of Tunis (University of Tunis El Manar), by 33 PhD candidates in civil engineering, mechanical engineering, electrical and industrial engineering. The viewed course was on "Risk Mitigation, Monitoring and Observational methods". Detailed report about this first experience including: organization of webinar sessions and evaluation of the synthesis of viewed webinars by the candidates was forwarded to the ISSMGE board members.

Then, followed courses of mixed content between available webinars and new ones recorded after September 2017. As examples, soil characterization, soil behavior, numerical modeling in geomechanics, and landslides-slope stability are among the formed courses of the ISSMGE virtual university program.

Full courses

This novel option provides more benefits for students and engineers to follow a complete course on a specific theme comprising five to six well linked lectures or webinars. Prof. Delwyn G. Fredlund (Canada) delivered the first full course on "Unsaturated soil Mechanics" that was been launched publicly on May 15th 2019. Seven PhD candidates, from the National Engineering School of Tunis, viewed the six lectures, at rate of two lectures per day, on May 15th, 16th and 17th 2019.

The second and third delivered full courses are "Probability analysis in civil engineering" and "Risk Analysis and Machine Learning in GE".

On 22nd March 2022, there were thirteen (13) available courses; appendix 1 of this report groups the titles of all formatted courses that can be viewed on: https://www.issmge.org/education/virtual-university

Appendix 2 of this report, details the content of each course as appearing on the above link. One can note that a supplement of material is available for some courses; i.e. short educational videos; etc. This summarizing table shows a description, in brief, for some courses. In turn, other courses need a description to highlight the viewers of the ISSMGE VU about the course objectives'.

4 CONCLUSION AND RECOMMENDATIONS

It was my great pleasure to contribute enthusiastically in the launch of the ISSMGE VU program and for being the first to experience its content.

For the next term (2022-2026), the development of the ISSMGE VU program requires the coverage of other I themes, like physical modelling, soil-structure interaction, etc. and other targeted full courses.

It is then, concluded that the new ISSMGE virtual university program is quite valuable in providing postgraduate courses of great help to PhD and MSc candidates to fulfill required credits by their institutions. Also,, it provides an enhanced learning tool to address quite challenging geotechnical engineering problems.

ACKNOWLEDGEMENTS

I am thankful to all board members of the ISSMGE, Dr K. Tsantilas (Elxis Group) and selected speakers of the delivered webinars-courses with whom I interacted to achieve the first step of the VU program of ISSMGE.

Appendix 1. Courses titles of the VU of ISSMGE (17th March 2022)

Course 1: Risk-Mitigation, Monitoring & Observational methods (4)
Course 2: In situ testing (4)
Course 3: Earthquake Engineering (3)
Course 4: Foundations (4)
Course 5: Soil Characterization (5)
Course 6: Geo-Engineering Education (5)
Course 7: Unsaturated soil mechanics (Full course, 6 chapters).
Course 8: Geosynthetics1 (5)
Course 9: Landslides and Mitigations (3), open
Course 10: Soil Behavior (3)
Course 11: Probability Analysis in Civil Engineering (Full Course, 4 chapters)
Course 12: Numerical Modelling in Geomechanics (4), open
Course 13: Risk Analysis and Machine Learning (Full course, 3 chapters), open.

Note: The number between brackets, after the course title, corresponds to the number of webinars (or chapters) included in the course. Further, the mention "open" means that other webinars will be added to the course in question.

Appendix 2: Detailed content of formed courses

Course 1: Risk-Mitigation, Monitoring & Observational Methods
Chap 1: Risk and Geotechnical Engineering (*Z. Medina-Cetina & M. Uzielli*)
Chap 2: Bridge Scour Depth Prediction and Levee Topping Erosion (*J-L. Briaud*)
Chap 3: Landslide Risk Mitigations with a Special Emphasis (*An-Bin Huangon*)
Chap 4: Challenges in Shoring Deep Excavations in Urban Areas (*M. Ballouz*)

Course 2: In-situ Testing
Chap 1: Geophysical Tests for Geotechnical Site Characterization (*S. Foti*)
Chap 2: In-Situ Testing in Geomechanics: Questioning Current Engineering Practice (*F. Schnaid*)
Chap 3: Introduction to Cone Penetration Testing (*P. Robertson*)
Chap 4: The Increasing Role of Seismic Measurements in Geotechnical Engineering (*K H. Stokoe*)

Course 3: Earthquake Engineering
Chap 1: Soil Dynamics and Seismic Geotechnical Engineering (*G. Gazetas*)
Chap 2: Earthquake Geotechnical Engineering-Mitigation of Earthquake Problems with Emphasis on Subsoil Liquefaction (*I. Towhata*)
Chap 3: Impacts of Liquefaction in the 2010-2011 Christchurch Earthquakes (*M. Cubrinovski*)

Course 4: Foundations
Chap 1: Pile Driving Analysis (*F. Rausche*)
Chap 2: Meeting the Challenges of Foundation Design for Tall Building (**H. Poulos**)
Chap 3: Multi-physical Processes and Design of Thermo-Active Foundations (*L. Laloui*)
Chap 4: Combined Loading of Offshore Foundations (*M. Cassidy*)

Course 5: Soil Characterization
Chap 1: Geotechnical Aspects of Peats (*C. Zwanenburg*)
Chap 2: Laterites and Lateritic Soil and the Potential Use of the Dynamic Cone Penetrometer (*S.I.K. Ampadu*)
Chap 3: Geotechnical Challenges in Mexico City Clay (*G. Auvinet*)
Chap 4: Compressibility and Consolidation of Clays: From Lab to Field Conditions (*S. Leroueil*)
Chap 5: Advanced Laboratory Testing to Characterize Stiff, Geologically Aged Clays (*R. Jardine*)

Course 6: Geo-Engineering Education
Chap 1: Eurocode-7 Past, Present, and Future (*A. Bond*)
Chap 2: Basic Geotechnical Engineering Skills - What Can Graduates Do? (*J. Atkinson*)
Chap 3: (What) to Teach or not to Teach - From Theory to Practice (*L.D. Wesley*)
Chap 4: Case Study "Porto Tolle test embankment - A full scale experiment (*C. Viggiani*)
Chap 5: What happens when soil compresses? (*M. Pantazidou*)

Course 7: Unsaturated Soil Mechanics (D.G. Fredlund)
Objective of the course is to introduce unsaturated soil mechanics within the framework of classical soil mechanics by incorporating soil suction.
Chap 1: Introduction to Unsaturated Soil Mechanics

Lecture 1 introduces unsaturated soils by illustrated practical problems. The link between field and laboratory conditions is established by the Soil-Water Characterization Curve (SWCC).
Chap 2: State variables and their Measurement
Lecture 2 presents the state variables by which the transition from the saturated soil condition to the unsaturated soil condition can be described, in particular the matric suction as the difference between pore-air and pore-water pressures.
Chap 3: The Soil-Water Characteristic Curve, SWCC.
Lecture 3 details the determination of SWCC in providing information about the amount of water and the energy state in water phase.
Chap 4: Seepage through Unsaturated Soils
Lecture 4 addresses the theory and solution dealing with water flow problems in unsaturated soil. Focus is made on the measurement of unsaturated coefficient of permeability function for design purposes.
Chap 5: Shear Strength of Unsaturated Soils.
Lecture 5 deals with the determination of shear strength of unsaturated soils. The latter is described by two independent state variables among which soil suction contributes to the cohesive component of soil.
Chap 6: Volume Change & Deformation of Unsaturated Soils
Lecture 6 investigates the theory with related solution of stress deformation problems. The main objective is to establish the relation between stress state and deformation strain variables for unsaturated soils. In this process, the use of SWCC enables the assessment of the behavior of unsaturated soil.

Course 8: Geosynthetics 1
Chap 1: An Overview of Geosynthetics and Their Major Applications (*R.M. Koerner*)
Chap 2: Geosynthetic Reinforced Walls: Overview, Failures and Items for Approval (*R.M. Koerner*)
Chap 3: Short and Long Term Leakage through Composite Liners. The 7th A. Casagrande Lecture (*K. Rowe*)
Chap 4: Geosynthetics-Reinforced Soil Structures - Developments from Walls to Bridges (*F. Tatsuoka*)
Chap 5: Geosynthetics in Roadway Applications (*J.G. Zornberg*)

Course 9: Landslides and Mitigations
Chap 1: Impact Mechanisms of Debris Flow on Barriers: Modelling, Analysis and Design (*C.W.W. Ng*)
Chap 2: The Classification of Landslides in Soils in a Mechanical Perspective (*L. Picarelli*)
Chap 3: Landslide Runout Analysis Current Practice and Challenges (*S. McDougall*)

Course 10: Soil Behavior
Chap 1: Unsaturated Soils: Basic Concepts and Applications: Parts 1 and 2 (*E. Alonso*)
Part 1 Hydromechanical behaviour of collapsible and expansive soils
The webinar provides a condensed and modern understanding of the hydro-mechanical behavior of unsaturated soils stressing the links among the relevant and unique features of the behavior of unsaturated soils: water retention, permeability, effect of suction on stiffness, volumetric response and strength.
Part 2: Computational tools and case histories. Embankment collapse, earth dam behaviour and foundation swelling
Geotechnical practice involving unsaturated soils, ranging from foundations on natural collapsible or expansive soils to compacted embankments and earth dams.
Chap 2: Thermal Issues in Clays and Claystones. Application to Radioactive Waste Disposal (*P. Delage*)
Chap 3: Stress-strain behavior of geomaterials (mainly granular materials) from elastic behaviour to shear banding. (*F. Tatsuoka*)
Chapter 3 introduces the major characteristic features of the stress-strain behaviour of geomaterials, mainly granular materials and also partially clays, cement-mixed soils and natural soft rocks.
Lecture 1 introduces the contents of the second to sixth lectures. It is explained that, in this course, it is attempted to bridge a gap in the stress-strain behaviour of geomaterials between the basic soil mechanics and the SOA geotechnical engineering practice.
Lecture 2 presents the stress-strain behaviour at very small strains. It is shown that the elastic properties from field shear wave velocities become the basis for non-linear FEM analysis in many geotechnical boundary value problems.
Lecture 3 presents the dilatancy characteristics in drained shear and their effects on undrained shear behaviour.
Lecture 4 presents the peak strength of granular materials
Lecture 5 addresses the shear banding, which starts immediately before the peak stress state and develops in the post-peak strain-softening regime.
Lecture 6, Two different time effects: i.e., rate-dependent stress-strain behavior and ageing effects, are explained based on results from series of stress-strain tests.

Course 11: Probability Analysis in Civil Engineering (J. Zhang)
Uncertainties are pervasive in geotechnical engineering. In the past decades, reliability methods have been developed as a powerful tool to model and quantify the effect of uncertainties in geotechnical engineering. The purpose of this course is to systematically introduce the reliability methods in the context of geotechnical engineering.
Chap 1: Analytical models for modeling uncertainties.
Chap 2: Reliability methods.
Chap 3: Reliability-based design.
Chap 4: Bayesian methods for uncertainty reduction.

Course 12: Numerical Modelling in Geomechanics
Chap 1: Introduction to Numerical modeling in GE (*H. Schweiger*):
Chap 2: Numerical Simulations on Energy Piles (*J.S. McCartney*):
This presentation focuses on numerical simulations of the thermal, thermo-mechanical, and thermo-hydro-mechanical response of energy piles and surrounding soils. Simulations include finite element analyses of heat transfer and coupled water flow in soils surrounding energy piles and load transfer analyses of thermo-mechanical soil-structure interaction in energy piles.
Chap 3: Discrete modeling of granular materials (*F. Nicot*)
Chap 4: Performance assessment of soils and structures by numerical analysis (*Toshihiro Noda*)

Course 13: Risk Analysis and Machine Learning
Chap 1: Machine Learning in Geotechnical Engineering (*P. Samui*)
Chap 2: Learning about a Site Using Sparse Site-Specific Data - Recent Advancements (*J. Ching & K.K. Phoon*)
Chap 3: Risk informed landslide Hazard mitigation in mountain highway design. (*LM Zhang*)

Appendix 7: 2017-2022 Report of the Innovation and Development Committee (IDC)

Dimitrios Zekkos
Chair of Innovation and Development Committee (IDC), Associate Professor, Department of Civil and Environmental Engineering, University of California at Berkeley, zekkos@berkeley.edu

SUMMARY: The Innovation and Development Committee (IDC) is a board level committee that was established in 2009 with the mission to develop innovations to better serve ISSMGE member societies, geoprofessionals and the profession as well as to increase the impact of ISSMGE. In this report the activities of the IDC during the 2017-2022 period are briefly presented. Through partnerships with several other board level committees, technical committees, and member societies, IDC has continued to push forward with efforts that can have a positive impact on the geotechnical engineering community at the global scale. These include: (a) the expansion in content and capabilities of the open-access online library that now has 16,000+ papers available; (b) the development of a conference paper review platform that is available at no cost and can support conference organizers in peer-review of conference proceedings; (c) the ability of ISSMGE to publish proceedings online, or even in print, and issue DOIs on conference papers or ISSMGE reports; (d) the development of ISSMGE's Virtual University that presently includes 149 recordings with original content; and (e) the continued growth of ISSMGE's online presence through its website, GeoWorld, its mobile application, and social media.

1 INTRODUCTION

The Innovation and Development Committee (IDC) of ISSMGE was established in 2009 as a Board Level Committee (BLC) by Past-President Jean-Louis Briaud. The first term of the committee was from 2009-2013. President Roger Frank renewed the IDC for 2013-2017 followed by a third renewal by President Ng for the 2017-2022 period. The following report outlines the activities of the IDC during the 2017-2022 period and provides some recommendations to ISSMGE and member societies for the future of IDC.

2 TERMS OF REFERENCE

The IDC charge was originally provided in 2009 by the ISSME President and Board and was summarized as "to think of ways to make ISSMGE progress in a manner which will increase its usefulness to the members and provide excitement for the future of geotechnical engineering in ISSMGE.". Th original charge was followed up by a number of tasks many of which have been completed. After eight years, with consultation with the President, the IDC revised its tasks, which were approved by the Board in 2017. The revised charges are:

- "Feel the pulse" of the geotechnical community through a survey that will help set new priorities.
- Expand the use of information technology (IT) tools to continue to enhance the value of the ISSMGE web site as a technical resource worldwide.
- Expand the open-access of ISSMGE publications database and explore innovative schemes to promote open access publications at a global scale.
- Implement educational initiatives under "ISSMGE University" (which was later termed Virtual University).
- Promote next-generation communication among Geo-Professionals through GeoWorld.

3 MEMBERSHIP

For the 2017-2022, the membership of the IDC was reduced compared to previous terms, to allow for a flexible and easy-to-meet and act committee that consisted of the following:
Chair: Dimitrios Zekkos (USA & Greece) – Associate Professor, Civil and Environmental Engineering, University of California at Berkeley; zekkos@berkeley.edu
Vice Chair: Kok-Kwang Phoon (Singapore) – Professor and Provost, Singapore University of Technology and Design
Jennifer Nicks (USA), Research Engineer, Federal Highway Administration
Roger Estephan (Lebanon), Director at Dar Al-Handasah (Shair and Partners).

In addition, to ensure a smooth transition of leadership, the IDC has been planning for Dr. Nicks to take over as Chair of IDC.

4 ACTIVITIES

4.1. Conduct IDC Survey

Since its establishment in 2009, IDC has been leading various initiatives. Major previous initiatives include the launch of a new ISSMGE website, the establishment of a webinar series, the launch of an open access library, the launch of GeoWorld, the development of an online Lexicon, as well as the development of a mobile app. It was, thus, deemed useful to "take stock" and assess past initiatives before setting new ones. Thus, the IDC conducted a survey with the objective to have the geo-community review previous IDC activities and their impact and solicit feedback on what should the priorities of the IDC be for the 2017-2021 period.

The survey was conducted from March 6th 2018 to May 1st 2018 with 701 participants and included 13 questions. The results of the survey have been analyzed and a public report on the survey can be found on the ISSMGE website.

Briefly, the survey indicated that all initiatives undertaken where generally valuable, but especially the ones that related to the ISSMGE website, open-access and educational resources (webinars, online lectures, etc.) were particularly appreciated. The survey respondents also provided feedback on proposed activities, and an outcome of this feedback are the six initiatives that IDC undertook in the 2017-2022 period and are listed in sections 4.2 to 4.7.

4.2. Expand Content and Capabilities of ISSMGE's Open Access Library

During the 2013-2017 tenure, the IDC established the Online Library that is available through ISSMGE's website (URL: https://www.issmge.org/publications/online-library). The online library provides open access to papers published in conferences organized under the Auspices of ISSMGE, but also papers published in other conferences.

4.2.1 Expansion of Online Library capabilities

In 2021, the IDC undertook a major upgrade of the Online Library. Specific new features of the library that have been incorporated are:

- Improved layout and indexing based on specific sub-disciplines/topics in geotechnical engineering. This allows for better indexing and collaboration between content available in the online library and virtual university.
- New conference proceedings added in the library have now a dedicated page for each conference. This supports ISSMGE's initiatives related to publishing.
- Each new publication/paper has its own URL allowing for the ability to issue DOI for new papers as well as reports published by ISSMGE.
- Each author is now an "entity" within the library that is also integrated with the individual's ISSMGE and/or GeoWorld profile (if desired). This allows for someone to search the library using the author name, and find an author's profile.
- Changes to the online library that support indexing of conference proceedings in databases such as Scopus.

4.2.1 Increase in Online Library's content

The IDC continued to work with many member societies, Technical Committees, and other groups to expand the papers included in the database of the online library. As of February 2022, the online library has 16,387 papers from 68 conferences. From November 2017, when the platform was launched, to December 2021 there were 1,683,000 downloads of the papers included in the Open Access Library, with about 350,000 paper downloads just in 2021.

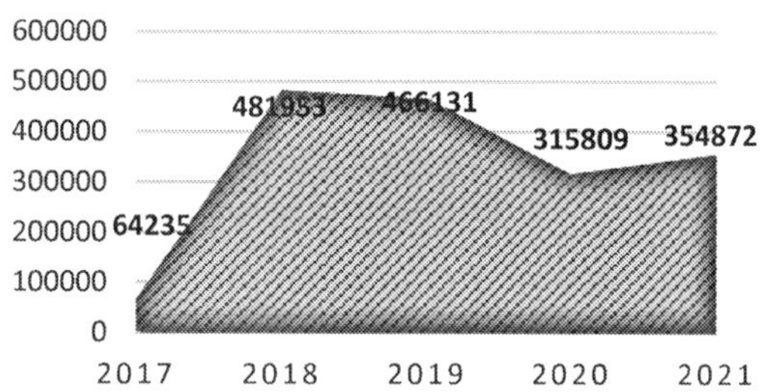

Figure 1: Paper downloads from the online library since its launch in 2017.

The following proceedings have been added to the online library since September 2017:

- 18th International Conference on Soil Mechanics and Geotechnical Engineering (Paris) (930 papers added)
- 19th International Conference on Soil Mechanics and Geotechnical Engineering (Seoul) (799 papers added)
- 5th International Conference on Forensic Geotechnical Engineering (42 papers added)
- 6th International Young Geotechnical Engineers' Conference (iYGEC6) (107 papers added)
- SFGE - 2016 (12 papers added)
- 14th Pan-American Conference on Soil Mechanics and Geotechnical Engineering 64th Canadian Geotechnical Conference, October 2-6, 2011, Toronto, Ontario, Canada (503 papers added)
- EUROFUGE 2016 (Nantes, France) (58 papers added)
- 3rd International Conference on Performance-based Design in Earthquake Geotechnical Engineering (Vancouver 2017) (238 papers added)
- 6th International Conference on Earthquake Geotechnical Engineering (Christchurch, New Zealand) (371 papers added)
- 7th International Conference on Unsaturated Soils 2018 (Hong Kong) (219 papers added)
- 17th Nordic Geotechnical Meeting (NGM-2016) (142 papers added)
- 5th International Conference on Earthquake Geotechnical Engineering (Santiago, Chile) (182 papers added)
- 26th European Young Geotechnical Engineers Conference (51 papers added)
- 13th Australia - New Zealand Conference on Geomechanics (Perth, 2019) (201 papers added)
- 1st Australia - New Zealand Conference on Geomechanics (Melbourne, 1971) (81 papers added)
- 17th European Conference on Soil Mechanics and Geotechnical Engineering (679 papers added)
- 2nd Australia - New Zealand Conference on Geomechanics (Brisbane, 1975) (62 papers added)
- 10th Australia - New Zealand Conference on Geomechanics (Brisbane, 2007) (203 papers added)
- 17th African Regional Conference on Soil Mechanics and Geotechnical Engineering (139 papers added)
- Field Measurements in Geomechanics 2018 (42 papers added)
- 11th Australia - New Zealand Conference on Geomechanics (Melbourne, 2012) (277 papers added)
- 7th Australia - New Zealand Conference on Geomechanics (Adelaide, 1996) (154 papers added)
- XVI Pan-American Conference on Soil Mechanics and Geotechnical Engineering (XVI PCSMGE) (356 papers added)
- IS SAO PAULO '17 (41 papers added)
- 5th Australia - New Zealand Conference on Geomechanics (Sydney, 1988) (114 papers added)
- 8th Australia - New Zealand Conference on Geomechanics (Hobart, 1999) (129 papers added)
- 4th Australia - New Zealand Conference on Geomechanics (Perth, 1984) (129 papers added)
- 25th European Young Geotechnical Engineers Conference (49 papers added)
- 7th International Symposium on Geotechnical Safety and Risk (ISGSR 2019) (145 papers added)
- 1st International Symposium on Geotechnical Safety and Risk (ISGSR 2007) (78 papers added)
- 6th Australia - New Zealand Conference on Geomechanics (Christchurch, 1992) (104 papers added)
- 12th Australia - New Zealand Conference on Geomechanics (Wellington, 2015) (159 papers added)
- 3rd Australia - New Zealand Conference on Geomechanics (Wellington, 1980) (110 papers added)
- International Conference On Geotechnical Engineering Education 2020 (GEE2020) (29 papers added)
- 9th Australia - New Zealand Conference on Geomechanics (Auckland, 2004) (128 papers added)
- 7th International Conference on Earthquake Geotechnical Engineering (Roma, Italy) (674 papers added)
- 13th International Symposium on Landslides (ISL2020) (125 papers added)
- 3rd International Symposium on Coupled Phenomena in Environmental Geotechnics (CPEG2020) (83 papers added)
- 10th International Conference on Scour and Erosion (ICSE-10) (142 papers added)

A complete listing of the conference proceedings available in the ISSMGE Online library is provided in Appendix A. The IDC remains committed to expand the content of the open access library and, as described subsequently, aims to support open-access publications through its Conference Paper Review Platform.

4.3: Support open-Access Publication: Development of a Conference Paper Review Platform

The IDC has also been actively working towards supporting the geo-community in expanding open-access publication of papers for future events. As part of that initiative, the IDC has developed, following authorization by President Ng and the Board, a Conference Paper Review Platform (URL: https://www.issmge.org/publications/review-platform/ongoing-conferences).

The ISSMGE Conference Review Platform is part of ISSMGE's cyber-infrastructure and aims to support the review of abstracts and papers for Conferences organized by ISSMGE Member Societies, Technical Committees and other groups affiliated with ISSMGE.

This platform is provided at no cost with the only requirement that published papers become available in open access through the ISSMGE Online Library. The platform supports the management of conference papers for ISSMGE and ISSMGE-affiliated events and allows also for the seamless integration between this paper review platform and the ISSMGE Online Library. The platform can be used by conference organizers of the ICSMGE, ISSMGE sponsored events, ISSMGE regional conferences, or conferences organized or sponsored by Technical Committees. Member Societies can also request from the Board to use the platform as long as they commit to make available the conference proceedings in open-access.

The platform is presently operational and has already been used successfully to support the following conferences:

- 6th International Conference on Geotechnical and Geophysical Site Characterization (ICS6), "Toward synergy at site characterisation", held on 26-29 September 2021 in Budapest, Hungary.
- XIII International Symposium of Landslides (XII ISL) held on 22-26 February 2021 virtually.
- 3rd International Symposium on Coupled Phenomena in Environmental Geotechnics (CPEG2020), held on 20-21 October 2021 virtually.

The proceedings published from these conferences are already available through the online library. IDC sees this platform as a major step towards a more integrated approach to open access publication (from submission to publication) in the field of geotechnics.

4.4. Towards ISSMGE Publishing

In the 2013-2017 period, IDC played an active role in setting the open access policy of ISSMGE. During this term, the IDC aimed to make open access in geotechnical engineering a reality. Through the conference paper review platform and the growth of the online library, ISSMGE provides the ability for papers to be submitted and published completely open access. Should ISSMGE make that decision, it can support not only the online, but also the printed publication of proceedings and other documents.

4.5. Establishment of ISSMGE Virtual University

A major activity of the IDC, following the request by ISSMGE President, Charles Ng, was the development of open-access, educational material to professionals globally. The activity, titled "Virtual University" is a step beyond the webinars that ISSMGE has launched in the last decade.

The purpose of the Virtual University platform is to educate geoprofessionals at a global scale, and support Universities that may not have the resources to provide detailed educational material on specific topics. The platform was publicly launched in 2020 and is available at http://virtualuniversity.issmge.org/ and through the main ISSMGE website. The platform has the capacity to deliver educational content such as educational videos, webinars, keynote lectures, or short courses freely to anyone. All content is archived based on sub-disciplines and topics that are also used in the online library. An ISSMGE account can be created to "save" content. The platform also bundles recordings in short courses and provide recommendations, based on previous videos watched.

The virtual university platform also has the following capabilities:

- Provides Video Analytics: Course instructors (who have an account as instructors) are able to review video analytics for each video uploaded within the platform. This allows them to better understand how users are interacting with their lectures. Statistics are provided for the number of enrolled users for each video, and what portion of the video recording is mostly watched.
- Exam Functionality: Virtual University can support simple or timed exams for each of the courses (or lectures). The exams are of "quiz" type and can be timed.
- Learner/Student Progress Functionality: This functionality allows the instructor to set homeworks for the learner that can also inserted at any point during the recording.
- Certificate Functionality provides the ability to issue certificates, should ISSMGE decide to do that in the future;
- Incorporates Collaborative Learning through Discussion Forums: Discussion forums are available for registered members to ask questions on specific topics.

The more advanced capabilities of the Virtual University listed above are not yet used extensively, but pave a direction for the future of geotechnical education. Out of the numerous recordings available, it would be important to highlight one course developed by Prof. Marina Pantazidou and TC306 titled "What happens when soil compresses", which uses some of these advanced capabilities. The video is segmented in short-duration subsections, each with a descriptive title, allowing for smaller subsections (recommended 7-15 min) that are likely to be well attended. Subsections are accompanied with quizzes of multiple-choice questions. The video is accompanied with the presentation slides and transcript for download. The transcript also appears as text next and below the presentation slides.

The IDC would like to acknowledge Board Member Prof. M. Bouassida who worked closely with the IDC and was responsible to reach out to individuals to develop educational content. Presently the Virtual University has 149 videos, and 11 short courses that aggregate videos of similar content. Average visits/pageviews for 2021 was 6,500 per month. In total, since its launch in 2020, ISSMGE videos have been played 114,000 times.

4.6. Upgrade ISSMGE website to increase its functionality, usage and impact

IDC has been continuously upgrading ISSMGE's online presence with the goal to provide relevant and valuable content to ISSMGE membership. The following major upgrades have been completed during the 2017-2022 period:

Visitors of the ISSMGE website have the capability to "sign-up" to receive updates from Technical Committees that fit their interests. Once content is posted by the TCs, it is automatically promoted through the website, but also, by integrating with GeoWorld, an e-mail is sent to the individuals who have opted-in to receive that content. This feature empowers technical committee to promote their work and allows ISSMGE to directly reach out to members and provide valuable content to those that need it.

Emphasis was also placed to further increase the use of ISSMGE website by TCs and easily assess level of activity. Towards this objective, the current website platform was expanded to allow TC Chairs to upload their progress reports online and make them accessible to TOC leadership.

4.7: Continue to support networking opportunities, and expand ISSMGE's reach through GeoWorld

The opportunities for international networking that ISSMGE provides are valued by ISSMGE members. Beyond the ISSMGE website, IDC has been trying to positively affect the careers of geoprofessionals by expanding ISSMGE's messaging and reach, through the following means:

GeoWorld: Since its launch in 2011, with the support of ISSMGE, GeoWorld has become the largest global network of geoprofessionals. GeoWorld has exceeded 25,000 members from 145 countries in January 2022 and is supporting many aspects of ISSMGE activities. Through GeoWorld, ISSMGE delivers content and information posted on the ISSMGE website. GeoWorld and ISSMGE continue also to publish annually the Geotechnical Business Directory, the largest geotechnical business directory in the field that includes an online format, an e-book format and a printed format and included all GeoWorld members who have opted-in. In June 2022 the 8th directory will be published.

Social Media Presence: ISSMGE's social media presence has also been increasing. Currently, there are 5566, 1527, and 11782 members on ISSMGE's Facebook, Twitter and LinkedIn account respectively.

Mobile Platform: ISSMGE's mobile platform, launched in September 2017, is presently used by 559 iOS and 790 Android users. Individuals who download the ISSMGE app, can receive push notifications of ISSMGE news.

5 RECOMMENDATIONS FOR THE FUTURE

For its first three terms (2009-2013, 2013-2017, and 2017-2022), the IDC has been able to undertake some major initiatives that have had a positive impact to the profession. As the first hair of the IDC all these years, I would like to thank the members of the IDC for their continued support and work, as well as the ISSMGE Boards, Presidents, and Chairs of Board-level Committees of ISSMGE that I worked with. As I step down from my role as Chair of IDC and I pass on the baton to the new Chair, Jennifer Nick, I would like to provide some recommendations for the future of IDC:

- IDC should be renewed and continue its work. I do not believe that a forward-looking IDC will ever fully complete its Mission. There are always many additional initiatives and activities that the IDC can pursue that can make an impact to the Profession, the member societies and geotechnical engineers.
- There is more work to be done for the IDC to further strengthen the content of the ISSMGE website with useful resources that the membership can use. This can be done with further engagement of Technical Committees, member societies, young members and corporate associates. In addition, the IDC should continue to support the open access vision of ISSMGE, whether it is through the Publishing initiative (conference paper review platform, online library) or the educational content (through the Virtual University), as well as additional initiatives.
- ISSMGE may also want to consider the establishment of a standing committee on Open Access Publishing to support the activities of ISSMGE. This may allow IDC to undertake additional new and exciting initiatives.

6 ACKNOWLEDGEMENTS

The IDC would like to gratefully acknowledge the support of the ISSMGE Board, the President and the Secretary General, and the Chairs of the other Board level Committees for their support in conducting this work.

7. APPENDIX

The following proceedings are currently available in open-access on the ISSMGE online library:

International Conference on Soil Mechanics and Geotechnical Engineering

- 19th International Conference on Soil Mechanics and Geotechnical Engineering (Seoul)
- 18th International Conference on Soil Mechanics and Geotechnical Engineering (Paris)
- 17th International Conference on Soil Mechanics and Geotechnical Engineering (Alexandria)
- 16th International Conference on Soil Mechanics and Geotechnical Engineering (Osaka)
- 15th International Conference on Soil Mechanics and Foundation Engineering (Istanbul)
- 14th International Conference on Soil Mechanics and Foundation Engineering (Hamburg)
- 13th International Conference on Soil Mechanics and Foundation Engineering (New Delhi)
- 12th International Conference on Soil Mechanics and Foundation Engineering (Rio De Janeiro)
- 11th International Conference on Soil Mechanics and Foundation Engineering (San Francisco)
- 10th International Conference on Soil Mechanics and Foundation Engineering (Stockholm)
- 9th International Conference on Soil Mechanics and Foundation Engineering (Tokyo)
- 8th International Conference on Soil Mechanics and Foundation Engineering (Moscow)
- 7th International Conference on Soil Mechanics and Foundation Engineering (Mexico)
- 6th International Conference on Soil Mechanics and Foundation Engineering (Montréal)
- 5th International Conference on Soil Mechanics and Foundation Engineering (Paris)
- 4th International Conference on Soil Mechanics and Foundation Engineering (London)

- 3rd International Conference on Soil Mechanics and Foundation Engineering (Switzerland)
- 2nd International Conference on Soil Mechanics and Foundation Engineering (Rotterdam)
- 1st International Conference on Soil Mechanics and Foundation Engineering

TC306 Conferences on Education
- ICETGES - 2008
- SFGE - 2012
- SFGE - 2016
- International Conference On Geotechnical Engineering Education 2020 (GEE2020)

TC204 Underground Construction Conferences
- IS NEW DELHI '94
- IS LONDON '96
- IS TOKYO '99
- IS TOULOUSE '02
- IS AMSTERDAM '05
- IS SHANGHAI '08
- IS ROME '11
- IS SEOUL '14
- IS SAO PAULO '17

International Conferences on Geotechnical Engineering (ICGE)
ICGE Colombo 2015

TC102 Conferences on In-situ testing
- Fifth International Conference on Geotechnical and Geophysical Site Characterization (ISC'5)

TC302 Conferences on Forensic Geotechnical Engineering
- 5th International Conference on Forensic Geotechnical Engineering

Young Geotechnical Engineers' Conferences
- 6th International Young Geotechnical Engineers' Conference (iYGEC6)
- 25th European Young Geotechnical Engineers Conference
- 26th European Young Geotechnical Engineers Conference

Regional Conferences on Soil Mechanics and Geotechnical Engineering
- 17th Nordic Geotechnical Meeting (NGM-2016)
- 17th European Conference on Soil Mechanics and Geotechnical Engineering
- 17th African Regional Conference on Soil Mechanics and Geotechnical Engineering

TC104 Conferences on Physical modelling
- EUROFUGE 2016 (Nantes, France)

TC203 Conferences in Earthquake Engineering
- 3rd International Conference on Performance-based Design in Earthquake Geotechnical Engineering (Vancouver 2017)
- 5th International Conference on Earthquake Geotechnical Engineering (Santiago, Chile)
- 6th International Conference on Earthquake Geotechnical Engineering (Christchurch, New Zealand)
- 7th International Conference on Earthquake Geotechnical Engineering (Roma, Italy)

TC106 Conferences in Unsaturated Soils
- 7th International Conference on Unsaturated Soils 2018 (Hong Kong)

TC220 Conferences in Field Monitoring in Geomechanics
Field Measurements in Geomechanics 2018

Pan-American Conferences on Soil Mechanics and Geotechnical Engineering (PCSMGE)
- 14th Pan-American Conference on Soil Mechanics and Geotechnical Engineering 64th Canadian Geotechnical Conference, October 2-6, 2011, Toronto, Ontario, Canada
- XVI Pan-American Conference on Soil Mechanics and Geotechnical Engineering (XVI PCSMGE)

Australia - New Zealand Conferences on Geomechanics
- 13th Australia - New Zealand Conference on Geomechanics (Perth, 2019)
- 12th Australia - New Zealand Conference on Geomechanics (Wellington, 2015)
- 11th Australia - New Zealand Conference on Geomechanics (Melbourne, 2012)
- 10th Australia - New Zealand Conference on Geomechanics (Brisbane, 2007)
- 9th Australia - New Zealand Conference on Geomechanics (Auckland, 2004)
- 8th Australia - New Zealand Conference on Geomechanics (Hobart, 1999)
- 7th Australia - New Zealand Conference on Geomechanics (Adelaide, 1996)
- 6th Australia - New Zealand Conference on Geomechanics (Christchurch, 1992)
- 5th Australia - New Zealand Conference on Geomechanics (Sydney, 1988)
- 4th Australia - New Zealand Conference on Geomechanics (Perth, 1984)
- 3rd Australia - New Zealand Conference on Geomechanics (Wellington, 1980)
- 2nd Australia - New Zealand Conference on Geomechanics (Brisbane, 1975)
- 1st Australia - New Zealand Conference on Geomechanics (Melbourne, 1971)

TC304 Conferences On Engineering Practice of Risk Assessment and Management
- 1st International Symposium on Geotechnical Safety and Risk (ISGSR 2007)
- 7th International Symposium on Geotechnical Safety and Risk (ISGSR 2019)

TC213 Conferences on Scour and Erosion
- 10th International Conference on Scour and Erosion (ICSE-10)

Other
- 13th International Symposium on Landslides (XIII ISL)
- 3rd International Symposium on Coupled Phenomena in Environmental Geotechnics (CPEG2020 - Kyoto, Japan)

Appendix 8: Report of the Technical Oversight Committee for the term 2017–2021

Pierre Delage
Chair, Technical Oversight Committee

This report of the Technical Oversight Committee (TOC) for the term 2017 – 2021 also contains some thoughts, conclusions and suggestions resulting from the two terms made by Pierre Delage as Chair of the TOC (2013 – 2017 under ISSMGE President Roger Frank and 2017 – 2021 under President Charles Ng).

1 INTRODUCTION.

The 37 Technical Committees (TCs) of the ISSMGE are the foundation on which the technical and scientifical activities of the ISSMGE are based. They are managed by a Chair and a Secretary and, if deemed necessary, a Vice-Chair. They are composed of four members appointed by the Chair (with voting right), of members (two members per Member Society,with voting right) and of corresponding members (with no voting right, no limitation in number). Members and corresponding members are nominated through their Member Societies, who are also in charge of putting their names in the TC website. The functioning of the TCs, based on a volunteer basis, is supervised by the Technical Oversight Committee and defined in the "Guidelines for ISSMGE Technical Committees and ISSMGE Honour Lectures" (revised in November 2019, called "Guidelines" in the following, see Section 5 below). The Guidelines state that TCs are "forum for discussing, developing and applying specialist geotechnical knowledge related to the behaviour of geo-materials, geotechnical engineering and engineering for society", with the following objectives:

- 1. Develop and disseminate knowledge and practice within the topic of the TC to the membership of the ISSMGE.
- 2. Establish guidelines and technical recommendations within the topic of the TC.
- 3. Assist with technical programs of international and regional conferences organised by the ISSMGE.
- 4. Interact with industry and overlapping groups working in areas related to the specialist area of the TC.

Briefly, it can be stated that, within active TCs during the past term, items 1 and 3 have been fairly well carried out, whereas item 2, which is much more time consuming, is not often done. In some applied TCs close to industry and practice, item 4 is correctly carried out. Item 1 can only be correctly performed when regular TC meetings are organised by TC officers. However, the question of the dissemination from the TC membership to ISSMGE membership arises. It should involve the contribution of Member Societies through their TC members, which is not always guaranteed and not easy to check.

Useful information aimed at introducing TC officers to the management of TCs is provided in the TOC website, together with some updates. TC officers are responsible for managing their TC website (including the TC membership), based on the User Guide for TC Officers prepared International Society for Soil Mechanics and Geotechnical Engineering Société Internationale de Mécanique des Sols et de la Géotechnique by the TC webmasters. In this regard, the TOC is very grateful for the two webmasters (Alexandros Tsavalas and Kostis Tsantilas) for their constant reactivity and efficiency.

2 COMPOSITION OF TECHNICAL OVERSIGHT COMMITTEE

The TOC is composed of the Vice Presidents of the ISSMGE Regions, who were the following during the last term:

- Africa: Etienne Marcelin Kana (CTGA - Comité Transnational des Géotechniciens d'Afrique)
- Asia: Eun-Chul Shin (Korea)
- Australasia: Philipp Robbins (New Zealand) – formerly Gavin Alexander, deceased in 2020
- Europe: Mario Manassero (Italy)
- North America: Timothy Newson (Canada)
- South America: Alejo Sfriso (Argentina)

The idea with this composition was to have the Vice Presidents following more closely the activities of the TC in their region. It wasn't finally not so successful, for some reasons

3 TC ACTIVITIES

An important concern of the TOC was to make sure that the activities conducted by the TCs under their officers' leadership was satisfactory and in line with the Guidelines, so as to be useful and beneficial for the ISSMGE and for geotechnical engineering practice. There is no use for ISSMGE in having inactive TCs. The functioning of TCs is based on a volunteer basis and it is not easy to impose anything. There is no problem with motivated officers, but some problems may occur in the case of officers (perhaps somewhat seduced by the prestige of the title) who might not have realised the expectations of the appointment, which is rather more than making use of their own technical and scientific expertise. TC inactivity can frustrate motivated TC members, though they often do not express this until asked directly. Besides inactivity, the lack of an international dimension (in cases where most officers coming from the same country, and/or managing all their events in their own country) is also poorly appreciated by TC members. All these issues are closely followed by the TOC and discussed when necessary during ISSMGE Board meetings. Note also that some TCs may have some dormant members, which is not useful and can now be corrected thanks to changes made in the Guidelines (see below).

The main conclusion drawn by the TOC is that the level and quality of TC activities strongly depend on the personal motivation of the Chairs. Given the way they are appointed (see below), most Chairs do have the relevant expertise. Managing of a TC is a volunteer activity, and some Chairs find it difficult to mobilise all the time and energy needed for developing relevant and satisfactory activity. The solution to this issue is not straightforward. It is indeed possible (but not easy) to change the TC Chair in some extreme cases, but not in "average" cases where it is obvious that the level of activity less than expected. In the case where the TC Chair has been changed (easier at the end of their term of appointment), it is not necessarily easy to find a replacement, even in the case of hot topics in which a TC is obviously needed.

During the two last terms (2013 – 2017 and 2017 – 2021), some changes have been successful, with good restarts of the TC, but some others not, depending on the personality of the new Chair. In some (rare) cases, a motivated Chair may also have been disappointed by the lack of reactivity of their members, who didn't follow the Chair proposals by lack of reactivity/motivation.

The monitoring of TC activity is made by using a Table in which various activities are recorded, as seen in Section 8. Points are given for various activities, finally resulting in an indicative but imprecise ranking. The Table is not used for TC awards (for which the TOC is, surprisingly, not consulted at all), but it provides a way to distinguish between very good TCs, good ones and problematic ones.

Three items are given more importance:

- The regular running of administrative meetings, which has been significantly favoured by the pandemic, with a generalised used of on-line meetings (3 points). Such meetings are considered essential to animate and promote the TC culture and the sense of community among TC members. Regular meetings are unfortunately still not managed by some TCs, though they have been strongly encouraged to do so in the future. They appear in the Table at low ranking levels;
- The running of a series of specific conferences regularly organised by the TC (3 points). These conferences are considered quite important since they regularly provide to TC members and to the profession an idea of the main progresses and achievements made in the field of the TC. They can be run by TCs well established in their field. Note however that some active ISSMGE TCs may not have their own series of conferences, in some domains like numerical methods, for instance. Publishing conferences proceedings in Open Access is strongly encouraged and most often followed by officers. Note that the ISSMGE IT Group led by Dimitrios Zekkos now provides through the ISSMGE website a useful system of paper management for ISSMGE Conferences;
- The maintaining of an updated website (1 to 3 points) with relevant information (either within the ISSMGE website or within specific TC website).

Other items include the managing of ISSMGE Honour Lectures and Bright Spark lectures. ISSMGE Honour Lectures are now only presented within the TC specific conferences, so TC with no specific conference cannot have an Honour Lecture. Participation to ISSMGE Conferences through specific workshops are also considered, together with contributions to other initiatives like the Virtual University, the Time Capsule Project, the Case History Journal, newsletters, surveys… Observation of the Table shows a clear distinction between active TCs, with scores above 10, and TCs with problems, with scores below 6 (note that some of them are starting ones, so the ranking does not apply). All good TCs hold regular meetings and many of them organise a specific conference series. The situation of low ranked TCs is discussed within the Board to end up with appropriate conclusions with respect to their status. Some exchanges are also made with the TOC to help clarify the reasons of poor ranking, in an attempt to improve things.

In the case of clearly identified problems, the TOC, following the Guidelines and after consultation and discussion during Board meetings, has two options:

- 1. Trying to restart the TC by changing the chairmanship;
- 2. Stopping the activity of the TC.

During the term, the Chair of TC206 (Interactive design) has been changed and the restart has been quite successful thanks to the motivation of the new chair. A change in Chair has also been made in TC214 (Soft soils) and, more recently in TC107 (with its name changed from "Lateritic soils" to "Tropical residual soils", with an on-line international seminar recently organised). Conversely, the change in chair of TC207 (Soil-Structure Interaction and Retaining Structures) has not been considered successful by the Board. Considering also a low level of past activity, this TC has been recently disbanded (note that any suggestion to re-activate it would be welcome).

4 ISSMGE HONOURS LECTURES

The procedure that TCs have to follow to propose a new ISSMGE Honour Lecture is described in detail in the Guidelines. The list of the ISSMGE Honour Lectures is presently as follows:

TC101: BISHOP Lecture – Laboratory testing (TC101 website)
TC102: MITCHELL Lecture – Site characterization (TC102 website)
TC104: SCHOFIELD Lecture – Physical modeling (TC104 website)
TC106: BLIGHT Lecture - Unsaturated Soils (TC106 website)
TC202: PROCTOR Lecture – Transportation Geotechnics (TC202 website)
TC203: ISHIHARA Lecture - Earthquake (TC203 website)
TC204: FUJITA Lecture – Underground construction (TC204 website)
TC209: McCLELLAND Lecture – Offshore geotechnics (TC209 website)
TC211: MENARD Lecture – Soil Improvement (TC211 website)
TC212: POULOS Lecture – Deep foundations (TC212 website)
TC215: ROWE Lecture – Environmental geotechnics (TC215 website)
TC220: DUNNICLIFF Lecture – Field Monitoring (TC220 website)
TC301: KERISEL Lecture – Monument preservation (TC301 website)
TC304: LACASSE Lecture - Risk (TC304 website)
TC306: BURLAND Lecture - Education (TC306 website)
TC308: Energy Geotechnics Honour Lecture (TC308 website)

5 CHANGES IN THE GUIDELINES (REV. 2019)

The changes made in November 2019 in the revised version of the Guidelines, with the substantial contribution of Prof. Neil Taylor, Secretary General, are summarised below. These changes, aimed at improving the functioning of the TCs, resulted from observations of how they work, from exchanges with TC officers during the TOC – TC meeting held during the Reykjavik European CSMGE in 2019 and from informal discussions and e-mails with TC officers and TC members.

The changes are as follows:

- The length of the term of appointment of TC members is 8 years. It can be renewed for 4 years upon proposal of the TC chair and approval by the TOC. Approval will be based on evidence of active and essential participation; - The chair may suggest to the ISSMGE President and the TOC to remove or replace inactive members;
- Normally, the venue for TC ISSMGE Honour Lectures shall be within a specific conference arranged by the TC;
- Newly created TCs are submitted to a two years probationary period prior to being formally confirmed by the Board;
- A new chair who has some responsibilities in other TCs (Vice-chair or Secretary) should resign from those positions;
- To reflect the international nature of TCs, TC chairs are strongly encouraged to select a Secretary and a Vice Chair from a different MS to that of the Chair.

6 CHANGES IN TC CHAIRS

The term of TC Chairs is of four years and can be renewed once, leading to a maximum Chairmanship of 8 years. Particular attention has been paid to ensure clear and sound changes in TC Chairs, with open calls among all the members (nominated, chair nominated and corresponding) followed by e-mail votes by TC members (nominated and chair nominated).

Although the participation of members was not necessarily significant in all cases (sometimes smaller than 50%, indicating low involvement of some members), the Chair was selected based on the largest number of votes received. In some cases, in which there was some risk of conflict of interest/loyalty with a Chair too much involved in favouring a candidate, a secret e-mail vote was organised by the TOC Chair in conjunction with the Secretary General. It seems that this nomination procedure worked well, providing the necessary legitimacy to the new Chair. Once selected among the TC, the personality has to get the support of his/her Member Society prior to be proposed by the TOC to the ISSMGE President for official appointment.

7 CREATION OF NEW TCS

As stated in the Guidelines, the creation on new TCs results either from personal initiatives of ISSMGE members (who propose a new topic that they find of interest and explain how they plan to deal with it within the new TC) or after some discussion within the Board about important topics to cover. If accepted by the Board (with possible modifications suggested), the TC is started. The new TCs started during the last term are the following:

- TC309 – Machine learning (2017), proposed and chaired by Dr. Zhongqiang Liu (NGI)
- TC219 - System performance (2019), proposed and chaired by Prof. Gang Zheng (Tianjin U., China)
- TC220 – Field Monitoring in Geomechanics (2019), proposed and chaired by Dr. Andrew Ridley (Geo-Observations) in the prolongation of a series of successful conferences already existing on this topic
- TC221 – Tailings and Mine Waste (2020), resulting from discussion within the Board and managed with the help of the Vice-President of South America, Prof. Alejo Sfriso. The Chair is Prof. Ramon Verdugo (Universidad de Chile)
- TC222 - Geotechnical BIM and Digital Twins (2021), chaired by Magnus Rømoen (NGI)

The start of these new TCs has been successful and they usefully complement the technical and scientific fields covered by the ISSMGE, extending the abilities of geotechnical engineers in topics of importance.

8 CONCLUDING REMARKS

The activity of TCs is of utmost importance for the technical and scientifical achievements of the ISSMGE. Significant attention has then been paid to facilitate and support the managing of TCs by officers, and to help ensuring satisfactory activities in TCs. Good activity is mainly based on the willingness of the TC chairs, who are nominated following a clear and open procedure.

The reactivity of TC members is another important component, and it is now possible, thanks to the revised version of the Guidelines, to refresh the TC membership when necessary. The close following of TC activities helps suggesting improvements, when needed, in conjunction with the Board, so as to ensure the significant technical and scientifical achievements of ISSMGE.

9 TABLE OF TC ACTIVITIES

Website	Honour lecture	Bright Spark lecture	Reg. admin. meeting telecons	Specific Conf. Series	Open access proceedings	Task force	Interaction TCs	Case History Journal	Newsletter, Journal, review papers, Youtube, on line workshop, survey	Wk or Session ERC Reykjavik	Session ARC Tai Pei	ISSMGE virtual U.	TCP - TC meeting	TCP - TC contribution	SoA, SoP, Guidelines	TC Workshop Sydney	Note
3	1	1	3	3	1	1	1	1		1	1	1	1		1	1	21
3	1	1	3	3	1	1	1	1		1	1	1	1	1	1		21
3	1	1	3	3	1		1	1	1	1		1	1		1		19
3	1		3	3	1		1		1	1		1	1		1	1	18
3	1		3	3	1	1		1	1	1			1	1	1		18
3	1		3	3	1		1	1	1	1		1	1	1			18
3	1	1	3	3	1	1			1	1	1		1				17
2		1	3	3		1	1	1	1	1			1	1		1	17
3	1		3	3	1		1			1			1		1	1	16
2	1		3	3		1	1	1		1	1		1	1			16
3		1	3	3	1	1	1	1		1			1				16
2		1	3	3	1	1	1	1					1		1	1	16
2	1	1	3	3					1	1			1	1	1		15
2	1	1	3	3	1			1		1		1	1				15
3			3		1		1	1	1	1			1		1	1	14
3		1	3	3	1		1	1					1				14
2	1	1	3	3	1					1		1	1				14
1	1	1	3	3	1		1			1		1	1				14
3			3				1			1	1	1	1	1	1		13
2	1	1	3	3	1					1			1				13
3	1		3	3			1		1				1				13
3			3				1	1		1		1	1	1			12
2			3			1	1		1	1			1	1		1	12
2			3	1					1	1		1	1	1			11
2			3			1	1	1		1			1	1			11
1		1	3			1	1		1				0				8
2			3				1		1				0				7
2	1			1				1		1			1				7
2	1						1	1		1			0				6
2			3										1				6
2							1	1	1				1				6
2					1			1			1		1				6
2										1	1		1				5
2							1			1		1	0				5
1										1			1				3
2							1						0				3
2																	2
1							1						0				2

Appendix 9: Report of the Young Members Presidential Group 2017–2021

Lucy Wu
ISSMGE Young Member Presidential Group (YMPG), Chair

SUMMARY: This report presents the work of the Young Members Presidential Group primarily from 2020 to 2022 with background provided on the work done in 2017 to 2019 as context. The report summarizes the major initiatives undertaken, the objective behind the initiatives, and the outcomes achieved. This report also describes how these initiatives supports the overall mission of the committee in its transition from an advisory committee to the President to becoming the nexus of younger member networks within the ISSMGE.

1 INTRODUCTION

The Young Members Presidential Group (YMPG) is a Board Level Committee within the ISSMGE working to engage the next generation of geotechnical engineers. This report primarily details the work of the YMPG from 2020 to 2022 in continuing the mission that was determined at the end of the 2017 to 2019 term. Where applicable, the work done during the 2017 to 2019 term was included to provide context and completeness.

2 MISSION STATEMENT

The Mission Statement that the YMPG adopted for the 2020 to 2022 term is as follows:

The mission of the YMPG is to engage the next generation of geotechnical engineers in the ISSMGE. This is accomplished through a wide network of liaisons and corresponding members around the world. The YMPG will act as the nexus of this network by:

1. Enabling the sharing of knowledge and ideas
2. Connecting groups with each other, and
3. Facilitating communication between younger members and ISSMGE leadership.

2 MEMBERSHIP

The committee is led by the Chair, appointed for a four-year term by the ISSMGE President. Membership is restricted to ISSMGE members 35 years of age or younger at the time of appointment. Members serve two-year terms, renewable for an additional two years depending on interest, activity level, and age eligibility. In previous terms, three members from each region were nominated by the Regional Vice Presidents, creating a committee of 19 members representing a broad global reach.

At the end of the previous term (2017-2019), a change to the composition of the membership was proposed to the Board to reduce the size of the Core Committee Team to less than ten members. The Core Team will still be very international in its composition but will not represent any region or Member Society. This Core Team would transition the YMPG from a group that creates initiatives to become the nexus of young member activities within the ISSMGE. This would be accomplished through the creation of official liaisons with all Member Societies and eventually with Technical Committees and other affiliated groups. For the 2020-2022 term, the Core Team comprised 7 people.

3 INITIATIVES AND ACHIEVEMENTS

This section describes the activities undertaken by the committee during this term.

3.1 *Bright Spark Lectures*

The Bright Spark Lectures award is an initiative started by the current ISSMGE President, Professor Charles Ng. The purpose of the Bright Spark Lectures award is to promote young and promising geotechnical engineers/academics by giving them a platform to deliver keynote and invited lectures at conferences.

During the previous term, the YMPG, with the support of the Vice Presidents and Conference Organizing Committees, advertised and vetted applications for four of the 2019 Regional Conferences: Europe, Africa, Pan-America, and Asia. The lectures at these conferences were well received. It was recognized that since the Regional Conferences are only held every four years, there is some inherent luck with applicants who are close to the age cutoff. These lectures were used as platforms to encourage Member Societies and Technical Committees to host Bright Spark Lectures at their own conferences. The award was promptly adopted by a number of conferences and its prestige has grown over this period.

Therefore, the selection of the Bright Spark Lectures for the Sydney ICSMGE became a center focus initiative for the YMPG in 2020 due to the conference's importance and prominence in ISSMGE activities. With the help of the Organizing Committee and Member Societies, the Committee aggressively advertised this award. A total of 45 nominees and 22 societies were received, far exceeding expectations. The nominations were very high quality and winners were announced in the ISSMGE Bulletin Volume 14, Issue 5, October 2020 and ISSMGE News.

3.2 *YMPG Alumni*

Since the YMPG has existed for more than 10 years, there is now a growing list of former YMPG committee members. This is a valuable group of people who are likely to be willing to serve on one-off task forces, provide mentorship, and willing to be panelists for younger member sessions at conferences. Maintaining this group as a community would allow the YMPG to continue to tap into the group as a resource.

A list of all past committee members has been created. No further action had been taken to reconnect with this group during this term. This committee believes that most YMPG alumni would highly appreciate targeted communications and also continue to contribute and interact with the Society.

3.3 *YMPG Liaisons*

One of the most important initiatives undertaken this term was the development of the network of YMPG Liaisons. The committee sought official nominations from each Member Society ensuring that the representative has support from his/her Member Society. This structure also better reflects the way the Society communicates through Member Societies rather than directly to individual members.

This network would be a venue to share ideas, allow groups to network with each other, and allow the Society to gain better feedback from its younger members. With full engagement from all Member Societies and Technical Committees, this would create 100-200 young member positions within the ISSMGE community, allowing more people to participate on the international level.

3.3.1 *Pilot Group*

An initial pilot group of seven liaisons were chosen from some of the most active societies. The nominated representatives took part in monthly meetings to explore different topics of interest, test methods of communication, and to get to know each other on an informal basis. One of the key findings was that even though the Liaisons were very active in their own Member Society, they did not seem to know about some key ISSMGE resources. The outcome was the creation of onboarding presentations about the ISSMGE and YMPG that were used to onboard subsequent YMPG Liaisons.

3.3.2 *General YMPG Liaison Meetings*

After the initial pilot group, nominations were accepted from all Member Societies. The group of YMPG Liaisons grew to about 30 individuals. Even though this does not come close to the total number of Member Societies in the ISSMGE (more than 90), this was deemed to be an adequate size group.

After initial onboarding meetings, meetings included guest speakers and specific topics. Guest speakers include Darren Paul (Chair of 7iYGEC), Sukumar Pathmanandavel (Co-Chair of CAPG), and KK Phoon (Vice Chair of IDC). Another particularly engaging meeting was a discussion about the structure of young member groups in each Member Society. The Liaisons described how their Member Society engages their younger members and opportunities ranged from having a position on the Board to dedicated student and young member groups. The global and diverse nature of the group created space for everyone to consider something different from their own norms. The group came to the conclusion that there is no single right way for professional society engagement but rather it is an ongoing effort to adapt to the needs of the members.

3.3.3 *Overdesign Survey/Challenge*

The confluence of this international collaboration and global events was the spark that influenced a series of "Are We Overdesigning" competitions that took place at the end of 2020.

The CAPG had been promoting the Overdesign Survey and it was recognized that the survey may get more responses if there is an actual event. The global pandemic normalized virtual events. The YMPG Core Team and Liaisons, both being "remote-first" groups where members meet virtually on a regular basis, were in a special position to take the technical work that the CAPG has started and disseminate it. After a few months of getting to know each other, the Liaisons were ready to take on a challenge. The following events were led and executed through the leadership of the YMPG Liaisons and other younger member representatives appointed by their Member Society (see Table 1).

Most notable was that the Latin American and Francophone events held entirely in Spanish and French, respectively. The leaders of these events took it upon themselves to translate all the technical material from English into French and Spanish. They considered this to be a critical part in making this event accessible to their members. Reception from the audience was overwhelmingly positive, many of whom remarked that the international nature of these events excited them to attend and participate.

Table 1. Overdesign Events/Competitions and Engagement

Region/Society	*Engagement*
Africa, led by South Africa	150 live attendees
Australia and New Zealand	480 registered, 300 live attendees
Southeast Asia (jointly hosted by Singapore, Indonesia, Malaysia, Hong Kong Societies)	500 registered, 300 live attendees, 600 views on YouTube
Latin America (jointly hosted by Paraguay, Chile, Argentina Societies)	3 separate events, 100 participants each
Francophone event (jointly hosted by France and Morocco Societies)	170 live attendees, 700 views on YouTube
Canada	300 live attendees

3.4 *Future of Geotechnics Event*

In 2021, the Committee centered its focus in the creation of the Future of Geotechnics, an online event that explored the future of the profession with an international audience. The event was inspired by the success of the Overdesign Challenges, the call-to-action for the Time Capsule Project, and the desire to create a capstone event to end the current YMPG term.

This event had four main objectives:

- To inspire younger geotechnical engineers to think beyond their own national borders
- To provide younger engineers the opportunity to interact with professors and researchers doing cutting edge research
- To demonstrate how geotechnics can contribute to big global problems
- To contribute to the Time Capsule Project by encapsulating what we think the future looks like

Three central themes, held as separate events, centered on the questions of "What will the world and the field of geotechnics look like in 40 years?" The themes are as follows:

- Mitigation of Climate Change: The current and future contributions of the geotechnics profession on the mitigation of global carbon emissions through supporting advances in renewable energy technologies, developing a circular carbon economy through research and development of energy geotechnics, and enabling carbon capture through sequestration.
- Adaptation to Climate Change: Increasing awareness on challenges that arise from climate change in the form of sea level rise, groundwater inundation, permafrost loss, and the increased frequency of natural disasters leading to an increased need for resiliency in our built environment.
- Technology, AI, Big Data: Realizing that the intersection of technology and geotechnical engineering will require geotechnical engineers to embrace and integrate advances in computing in their practice. This event was co-hosted with TC309, the Technical Committee on Machine Learning and Big Data.

The events were held as three separate, two-hour events approximately eight hours apart over the course of two days in October 2021. The timing was selected to allow an international audience to reasonably attend two out of the three events live.

The events featured participation from many groups within and affiliated with the ISSMGE. The leadership provided by TC309 in making the Technology event happen was invaluable.

The Adaptation event featured a special appearance from the International Geosynthetics Society (IGS), a sister society of the ISSMGE and a member of the Federation of International Geo-Engineering Societies (FedIGS). Several speakers were nominated by the Corporate Associates of the ISSMGE. The YMPG Liaisons promoted the event through social media and other channels in multiple languages.

Online registration for the event reached audiences in over 80 countries with more than 2,700 online tickets issued, far exceeding expectations. The live events were well attended and the recordings provided an additional way for members to engage with the content.

This event showcased a way to engage global members in a dialogue that is both meaningful and impactful. The YMPG envisages that this event is the start of a conversation where engineers consciously think about the future of the profession. As part of the Time Capsule Project, the YMPG also hopes that this event encapsulates the sentiments and concerns of the engineering community at this point in time.

4 CORRESPONDING MEMBERSHIP

The corresponding membership of the YMPG through Google Groups email list, GeoWorld Group, and LinkedIn Group have steadily increased over the past four years despite the lack of emphasis on increasing the membership numbers (see Figure 1). Previous terms have driven membership numbers and achieved a critical mass such that the groups continue to grow with little direct recruiting effort. The purpose of these communication channels was to enable direct communication to younger members worldwide as well as provide a platform for all members to share and disseminate resources and events happening in their own societies.

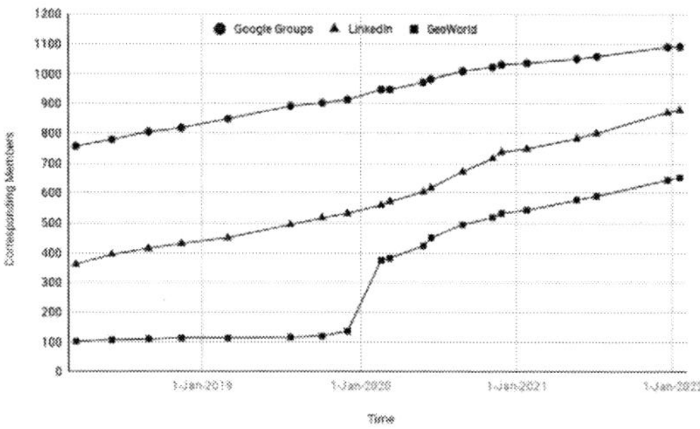

Figure 1. YMPG corresponding membership numbers over the past four years.

5 RECOMMENDATIONS

The Committee has started the transition of the committee's mission and composition to allow for wider participation of younger members in the Society. The following recommendations are provided for consideration for future YMPG committees:

- Continue to redefine the mission of the YMPG as a platform where initiatives are created, ideas area exchanged, and networks are formed.
- Build the YMPG Liaisons and Alumni networks to enable and elevate the participation of younger members on the national, regional, and international levels.
- Enhance the Society's impact on younger engineers through collaboration with other Board Level Committees and Technical Committees.
- Continue to promote the Bright Spark Lectures and the idea of allowing younger engineers to give prominent keynote lectures.

6 ACKNOWLEDGEMENTS

This committee would like to thank the efforts of other Board Level Committees, Board Members, and Technical Committees for supporting our initiatives. Special thanks go to the CAPG, IDC, and TC309 for their enthusiasm and unwavering support.

The group would also like to applaud the individual efforts of the YMPG members, both the Core Team and the Liaisons, who operated on a voluntary basis to advance the field of geotechnical engineering for younger members around the world. Their efforts and continued dedication have been remarkable especially given the challenging circumstances due to the global pandemic.

Appendix 10: Report of the Corporate Associates Presidential Group (CAPG) to the ISSMGE Council meeting, May 2022

Peter Day & Sukumar Pathmanandavel
Chair and Co-chair of CAPG

1 INTRODUCTION

The ISSMGE Corporate Associates Presidential Group provides a platform for leading corporations, consultancies, contractors, research institutions, equipment specialists, and product manufacturers in the geotechnical industry to participate directly in the activities of the ISSMGE. The CAPG now has 42 members https://www.issmge.org/corporate-associates/listing.

This report summarises the activities of the CAPG for the 2017 – 2022 Presidential term.

2 ISSMGE REGIONAL CONFERENCES

The CAPG hosted sessions at all five regional conferences in 2019.

- 13th Australasian Regional Conference, Perth, Australia – April 2019: Collaboration in geotechnical engineering – Impact on research and project delivery. http://issmge2014.ust.hk/jun2019/3b.Conference_report
- 17th European Regional Conference , Reykjavik, Iceland – September 2019: Bridging the gap between designers and constructors: how do we ensure effective 'transfer' of design into construction? http://issmge2014.ust.hk/oct2019/2.CAPG.pdf
- 17th African Regional Conference, Cape Town, South Africa – October 2019: Geotechnical innovation. 8 Papers on innovation available at https://www.issmge.org/publications/online-library Search under Regional Conferences tab, then the 17th African Regional conference, and then CAPG session.
- 16th Asian Regional Conference, Taipei, Taiwan – October 2019 Are we overdesigning? http://issmge2014.ust.hk/feb2020/2.Message.pdf
- 16th Pan-American Conference, Cancún, México – November 2019: Geo-education for the Future. http://issmge2014.ust.hk/apr2020/3.CAPGmessage.pdf.

3 OVERDESIGN SURVEY

During the 16th Danube Conference in Macedonia in 2018, members of the Asian Region raised the question "are we overdesigning?". In response, the CAPG launched a survey in which respondents provided solutions to 10 simple geotechnical problems on sand and clay sites – see https://www.issmge.org/news/are-we-overdesigning-a-survey-of-international-practice

The number of responses to the survey was significantly boosted by six online "geotechnical design challenge" events covering Africa, parts of SE Asia, Francophone countries, Canada, South America and Australia/New Zealand. These events were hosted mainly by young members and involved leading geotechnical experts in all regions. Just under 250 responses have been received. The results will be presented at the 20th ICSMGE in Sydney, Australia in May 2022. Thanks to the organisers of these events for their sterling efforts.

TC 304 made a major contribution by analysing all 10 survey questions from a reliability-based design perspective – see: https://issmge.org/files/reports/TC304-TC205-CAPG-YMPG-Report-on-overdesign-survey-problems.pdf.

3 FOUNDATIONS CONFERENCE AND EQUIPMENT EXPO (IFCEE)

The International Foundations Conference and Equipment Expo (IFCEE), an important event in the North American Geotechnical calendar, was held in May 2021 in Dallas, Texas, USA. With the assistance of the YMPG and local CAPG members, the CAPG hosted a speciality session entitled "Knowledge gaps in geotechnical engineering design and construction and the role that the USA could play in influencing global practice".

The event was hailed as a success and represents an important step towards greater involvement of the ISSMGE in Industry-sponsored speciality events in the USA.

4 TIME CAPSULE PROJECT (TCP)

The virtual time capsule project, led by CAPG Co-Chair Sukumar Pathmanandavel and the TCP Design Team, aims to record the past, understand the present and anticipate the future of geotechnical engineering and, in so doing, encourage dialogue at all levels of ISSMGE membership. Contributions are expected from past presidents, member societies, technical committees, the YMPG and the CAPG.

The time capsule will be officially launched at the 20th ICSMGE in Sydney, May 2022. Further details can be found at https://www.issmge.org/the-society/time-capsule

5 20TH ICSMGE, SYDNEY, AUSTRALIA – MAY 2022

With the kind permission of the Conference Organisers, the CAPG will be hosting a session on "Innovations and Achievements in Geotechnical Engineering" at which 13 papers by CAPG members will be presented. The CAPG will also use this opportunity to present the results of the "Overdesign survey".

Depending on the COVID-19 situation at the time, the CAPG intends hosting a social event for its members at the Sydney conference.

6 THANKS

The CAPG express its thanks to all our members, the CAPG core team who have met every 6 – 10 weeks for the past 5 years, the coordinators of these meetings, the organisers of the Overdesign online events, contributors to speciality and regional conference sessions and the YMPG for their assistance and valuable contributions.

Appendix 11: Report on the Activities of the ISSMGE Awards Committee (AWAC)

Committee period: 2017-2022

Roberto Terzariol
President of ISSMGE Awards Committee (AWAC)
Andre Archer
Secretary of ISSMGE Awards Committee (AWAC)

SUMMARY: This report outlines AWACs actives from August 2020 to January 2021 and acts as a guide for the awards currently on offer by the ISSMGE.

1 INTRODUCTION.

Professor Charles W.W. Ng, president effective from September 2017, appointed Roberto Terzariol as the new chairman of the Awards Committee (AWAC).

A new secretary, André Archer from South Africa, was nominated and his position as secretary was approved at the ISSMGE Board Meeting in Singapore on 10 March 2019. Other memberships to this committee were formed through consultation and nominations by regional vice-presidents.

The composition of this new AWAC consists of representatives from six regions.

2 MEMBERSHIP AND RESPONSIBILITIES OF AWARDS COMMITTEE

Committee members
Chairman: Roberto Terzariol (Argentina)
Secretary: André Archer (South Africa)

Regional representatives:

- Lidija Zdravkovic (Europe)
- Werner Bilfinger (South America)
- Hugo Acosta (Australasia)
- Teik Aun Ooi (Asia)
- Kerry Rowe (North America)
- Carlos Quadros (Africa)

The main responsibilities are to work with the ISSMGE Board to devise and improve awards guidelines and nomination procedures, to coordinate review processes, to conduct review of nominations and to make recommendations to the Board for consideration and approval.

3 TYPES OF ISSMGE AWARDS

The ISSMGE offers the following awards to recognize those individuals and bodies who have made important contributions to our profession, society and the world:

3.1. ISSMGE Outstanding Geotechnical Project Award
3.2. ISSMGE Outstanding Innovator Award (individual or team).
3.3. ISSMGE Outstanding Member Society Award
3.4. ISSMGE Outstanding Paper Published in the International Journal of Geo-Engineering Case Histories Award.
3.5. ISSMGE Outstanding Public Relations Award.
3.6. ISSMGE Outstanding Technical Committee Award.
3.7. ISSMGE Outstanding Young Geotechnical Engineer Awards (up to three awards).
3.8. ISSMGE Bright Spark Lecture award

Awards will be given once every four years, except for the Bright Spark lecture award, which is given yearly at various conferences.

Guidelines for the nomination of the various awards follow below.

All nominations must be made through the local ISSMGE Member Society (except the Outstanding Paper Award) and must reach the ISSMGE Secretariat by the deadline, which will be approximately one year before the International Conference on Soil Mechanics and Geotechnical Engineering (ICSMGE).

In this case the ICSMGE conference is to be held in Sydney, Australia in May 2022. The conference has been postponed due pandemic.

4 ASSESSMENT PROCEDURES

To avoid any potential conflict of interests, five sub-committees were formed to review and assess the suitability and quality of nominations and to make recommendations to the AWAC for considerations.

Additional committee members have also been chosen to assist.

The committee chairs and members will be asked on an ad-hoc basis to assist during the review period.

Nominations received will be distributed to selected committee chairs and members for an independent review.

Based on the outcome of the review process, the awardees will be communicated to the ISSMGE board for approval.

4.1. Sub Committees

The five sub-committees and committee members are listed as follows:

Sub-committee Chairs:

Dr Hugo Acosta (Australia)
Dr André Archer (South Africa)
Professor Kerry Rowe (Canada)
Professor Lidija Zdravkovic (UK)
Professor Dimitrios Zekkos (USA)

Committee members:

Dr Dennis Becker (Canada)
Dr Werner Bilfinger (Brazil)
Professor Jon Bray (USA)
Dr Johnny Cheuk (Hong Kong)
Dr Peter Day (South Africa)
Professor Russel Green (USA)
Professor Mark Jaksa (Australia)
Professor Mike Jamiolkowski (Italy)
Dr Zachieh Moh (Chinese Taipei)
Dr TA Ooi (Malaysia)
Professor K.K. Phoon (Singapore)
Professor Pedro Pinto (Portugal)
Professor Carlos Quadros (Mozambique)

The Bright Spark Lecture Award procedure is independent of AWAC with nominations sent directly to the Young Members Presidential Group (YMPG).

The YMPG will assess and short-list candidates, with the selected lecturers sent to the president for final approval.

5 NOMINATIONS RECEIVED

The deadline for award nomination was 1 July 2020. Due to the COVID-19 Pandemic the Awards Committee, in consultation with the President, decided to extend the submission deadline to 30 November 2020.

This was deemed a reasonable exception to allow for healthy competition and fair adjudication.

The following is a summer of the nomination received:

5.1. Outstanding Young Geotech. Engineer Award: 13

5.2. Outstanding Member Society Award: 3

5.3. Outstanding Geotechnical Project Award: 2

5.4. Outstanding Innovator Award: 3

5.5. Outstanding Tech. Committee Award: 3

5.6. Outstanding Public Relations Award: 1

5.7. Outstanding Paper Published in IJGCH: IJGCH have been contacted to nominate two papers for this reward.

5.8. Bright Spark Lecture Award: Nominations will go directly to YMPG for adjudication.

With the deadline extended, additional nominations were received, and this will allow for an equitable assessment and improved competition.

The adjudication process for the various awards have been completed.

To allow for a fair adjudication, and given the postponement of ICSMGE2021, it was decided the allow the adjudication process run until 31 March 2021.

The results will be conveyed to the ISSMGE Board for approval.

For the adjudication, the following sub-committee chairs have been nominated to coordinate a particular award, in order to avoid potential conflict of interest:

a. Outstanding Young Geotechnical Engineer Award: Prof. Lidija Zdravkovic (UK).

b. Outstanding Member Society Award: Prof. Dimitrios Zekkos (USA).

c. Outstanding Geotechnical Project Award: Dr André Archer (South Africa).

d. Outstanding Innovator Award: Dr Hugo Acosta (Australia).

e. Outstanding Technical Committee Award: Professor Pedro Pinto (Portugal).

f. Outstanding Public Relations Award: Dr André Archer (South Africa)

In general, AWAC believes the adjudication process for all the awards was fair and rigorous, and the variety of rewards meets the needs of the ISSMGE.

Appendix 12: Activity of the Professional Image Committee 2017–2022

Ikuo Towhata

PREFACE

The International Society for Soil Mechanics and Geotechnical Engineering (ISSMGE) has been seeking for a better public recognition of its profession for many years. In spite of many people who were engaged in this effort, the output was not satisfactory. This is in a clear contrast with the efforts of some other disciplines that have attained very good image among people. During the term of 2018-2022, its newly formed committee named Professional Image Committee, of which I was the Chair, attempted to propose a new approach towards the better recognition from a different perspective.

Because the international activity for this important theme was supposed to be difficult due to lack of frequent meeting and communication, a special domestic committee was established by the Japanese Geotechnical Society, of which I was the Chair as well, and this committee made efforts to propose what to do in order to improve the image of the geotechnical engineering profession among public.

This report is associated with a Power Point file that helps easy dissemination of the value of geotechnical engineering among public.

The outputs of this domestic committee, which consist of this report and the power point file, are distributed among and shared by the Member Societies of ISSMEGE so that they may use them for their domestic image. Certainly, the Member Societies can freely translate and/or edit the outputs into their local languages (except a few imageries with copyright notes).

The outputs and their edited/translated versions should be cited as "Ikuo Towhata (2022) For better professional image of geotechnical engineering, Output of the Professional Image Committee of ISSMGE Board in conjunction with the Japanese Geotechnical Society."

The major point of this report is that attentions are paid not only to the value of our profession, as has been stressed by the past efforts repeatedly without much success, but also to the weakness and shortcomings as well. The weakness stems from the nature of the ground that is heterogeneous and invisible, and makes engineering very difficult. The shortcomings are such as the low income of the geotechnical occupation and the insufficient public exposure. The income should be improved by stressing the importance of more subsurface investigation in order to mitigate the Georisk. The public exposure should be increased by saying "Do not be satisfied with customer satisfaction."

It is desired that the Member Societies of ISSMGE recognize the source of our weakness and use the power point file for better future of our YOUNGER colleagues.

Ikuo Towhata
Chair of the Professional Image Committee of ISSMGE and of the special committee established by the Japanese Geotechnical Society

1 INTRODUCTION

This report summarizes the achievements of the Professional Image Committee (PIC) from 2018 when the committee resumed its activities to the present time. The members of PIC other than myself are; Sam Mackenzie (Australia), Alessander Kormann (Brazil), Wenjie XU(China), François Depardon (France), Bruno Demay (France), Wolfgang Sondermann (Germany), Hemanta Hazarika (India), Tatsuya Nojima (Japan), Jin Man Kim (Korea), Stephen Crawford (New Zealand), Adam Latimer (UK), Donald A. Bruce (USA), Rick Deschamps (USA) and Trevor Green (South Africa). The major concern of PIC has been the improvement of the image of geotechnical engineering among people that is currently not very good in many countries. Ground or geotechnical sphere is invisible. Although it is essential to maintain the reliability and safety of human life and the civilization, people tend to overlook it or even unaware of it. From the technological viewpoint, ground is a product of the natural process in which no care is taken by nature of the homogeneity, engineering quality, quality certification, and material decay. As a consequence, it is very difficult to capture the three-dimensional underground structure, and empiricism as well as observational method is relied on. It is a pity that costsaving clients are reluctant to allocate enough budget to accurate site investigation. Accordingly, once an underground trouble occurs, they have to pay much bigger fee to solve the problem.

Due to shortage of subsurface information, construction projects often face 'georisk' troubles in which unexpectedly adverse soil/rock conditions cause additional cost and/or delay of the project. It is interesting that human is not aware of diseases that is going on in their organ and that medical doctors have to find it through health check so that suitable treatment may be made. This similarity (with many differences) between geotechnical practice and medical treatment will be touched upon in the later part of this report.

People suppose that geotechnical engineering is a 'dirty' engineering that is far behind the state-of-art modern technology such as information and biological technologies. In the worst case, people do not know anything about 'geotechnical' engineering.

Many geotechnical people made efforts to change this adverse situation. I have seen many advertisements by means of photos and motion pictures that demonstrate the great contribution of geotechnical engineering to the public welfare. Unfortunately, those efforts have not been very successful so far. People still do not care geotechnical engineering.

The unfavorable image of our profession has to be taken seriously because bright people of the next generation are not likely attracted to geotechnical engineering, and the shortage of talented people will make the scientific and technological developments difficult in the geotechnical profession.

In 2017, I decided to rethink the direction of such efforts and make different types of effort in order to improve the situation. It is expected today that the Member Societies pay attention to my output and start new types of effort in their countries.

It is encouraging that young citizens have fascinating image of geotechnical engineering. In 2009, the Japanese Geotechnical Society organized a picture contest for which many fantastic pictures were submitted by young fellows. Some of them are presented in this page but, for more, readers can visit the web site at: http://www.jgskantou.sakura.ne.jp/report/090819kaiga.htm.

Fig. 1.1. My dream (Sohta Tsuge) (property of JGS).

Fig. 1.2. Dinosaurs under my town (Ayane Ohno) (property of JGS).

Fig. 1.3. Earth is my life (Yuzu Inoue) (property of JGS).

Fig. 1.4. Hand contact with earth (Ayana Nitta) (property of JGS).

2 REVIEW OF THE PAST

It is obvious that the efforts have to be continued to disseminate the importance and the great contribution of geotechnical engineering to the public. The question is why the efforts in the past were not very successful in contrast to other fields of engineering and technology that did similar efforts and won good reputations among people. You can just imagine the good public image of computer & communication technology, space technology, biological technology, and many others. Probably, our efforts were not so efficient as theirs.

One possible reason for the difference is the 'dirty' appearance of geotechnical engineers who work in the field, touch 'dirt' and live in construction sites. However, the 'dirty' appearance is not the only reason of our unfavorable image because the rugby football players after match are 'dirtier' than us but are highly respected by people.

I make four points that have not been considered in our past efforts;

- Income of geotechnical engineers is not attractive to young and bright people.
- Geotechnical engineering is not visible and not known by people.
- Geotechnical engineering does not directly help people. It is working only with clients.
- Geotechnical engineering is not very scientific, relying on many empirical knowledge.

These points are discussed in more detail in the following sections.

3 INCOME AND GEORISK MITIGATION

As stated before, the relatively low income of geotechnical profession negatively affect the young and bright people and the shortage of good human resource makes development of geotechnical engineering more difficult.

Income is an important component of life because everybody has to maintain daily life of family, improve one's own professional skill and capability, and bring up the next generation (children). People sincerely with to do these duties. To do them, availability of money is important. Some people advocate by saying "You geotechnical engineers contribute a lot to the community. Your job has a supreme value. Do not insist on money!" But, they do not provide engineers with daily food and fees for children's education. Many family duties are supreme as well.

In general, people do not respect poor men. Note that it is not advisable just to claim more payment to the clients. It is necessary to show the clients that geotechnical engineering deserves more funding.

The following discussion concerns mainly the geotechnical consultants who are engaged in field investigation and data interpretation. This is because they are the major component of our Society and also belong to its Corporate Associates. However, geotechnical researchers are concerned as well. The aim of discussion is to show that more budget should be allocated to the subsurface investigation prior to construction so that unnecessary underground troubles may be avoided. This idea is named georisk management in this report. It is thereby expected that increased investigation business will bring more income to the geotechnical engineers and to make our profession more attractive and more respectable.

3.1 *What is georisk?*

In this report, georisk means the unfavorably increased cost of construction project and/or the elongation of the construction period as a consequence of underground troubles that are induced by insufficient subsurface information. It is herein expected that well-organized subsurface investigation can reduce, mitigate and manage those troubles. Several examples of georisk are presented in what follows. I believe that the readers can easily understand these example troubles and I do not provide their details.

Fig. 3.1. Leaning of a skyscraper in the center of a mega city.

Fig. 3.2. Minor tilting of a condominium building whose reconstruction cost US 400 million $ to the contractors and the developers.

Figs. 3.1 and 3.2 illustrate tilting of condominium buildings induced by heterogeneous subsurface conditions. Although the extent of buildings was minor in both cases, these troubles caused profound loss of the value of real estates. In this regard, the current level of geotechnical investigation is not high enough to predict a few-cm subsidence of huge buildings. Elaborately planned investigation program and sufficient funding are desired together with observational construction. In the case of Fig. 3.2, the site is close to a hill and the depth of the pile bearing layer was variable. The residents worried about insufficient pile length and the risk of total collapse during the big earthquake, which is coming soon, and the developers had to demolish the five buildings in the photograph, although only one was tilting. The financial loss of the developer and the contractors was disastrous.

(a) Surface depression

(b) Same site after restoration (the ongoing construction on the left-hand side of this street has nothing to do with the incidence)

Fig. 3.3. Collapse of shallow metro tunnel during construction (Fukuoka, Japan) (Photograph provided by the Asahi Shimbun Newspaper).

Fig. 3.3 shows an incidence in city center. A subway tunnel under construction suddenly collapsed and the street pavement depressed. Although soil condition had been investigated to a moderate level (not especially detailed), there was a localized geological weakness. When tunnel was excavated under the weak section, it collapsed under the effect of ground water pressure. The official investigation committee of this incidence (MLIT Investigation Committee on Depression of Subway Tunnel 2017) stated that

- more emphasis must be put on geotechnical investigation,
- ground water condition is important as well,
- deformation should be assessed by varying ground parameters in order to capture the worst possible situation (parametric study considering uncertainty and heterogeneity of ground parameters),
- information should be collected also from other projects in the vicinity (consideration of heterogeneous ground condition; importance of open-access geotechnical data base),
- monitoring is important during construction in order to detect any precursor of collapse, and
- stake holders should make good communication with one another, in order to avoid incidence of this type.

Fig. 3.4 schematically illustrates a complicated geological condition where lateral tectonic stress made the strata nearly vertical. Consequently, an unlithified/unconsolidated sandy layer became vertical and was filled with water. Because this sand chimney was thin, the preliminary subsurface investigation could not capture its existence. When the tunneling hit the bottom of this chimney, sand-water mixture flowed into the tunnel and flooded it (Fig. 3.5). Consequently, the tunneling project had to halt for more than one year.

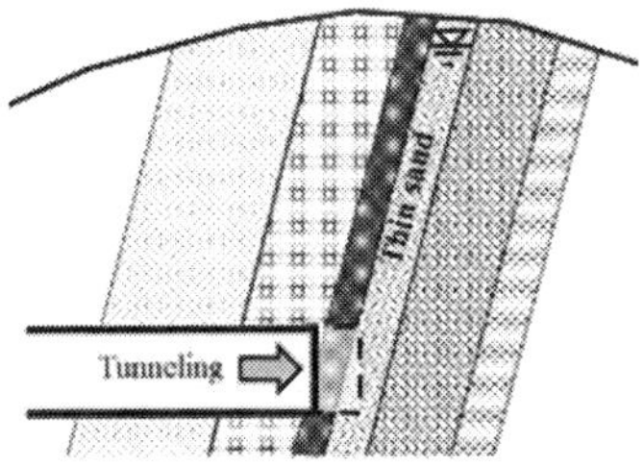

Fig. 3.4. Unlithified vertical sandy layer encountered by tunneling.

Fig. 3.5. Tunnel collapse after hitting the bottom of an unlithified sand pipe with water.

Obstacles hidden underground cause georisk troubles. The H-steel piles in Fig. 3.6 were installed fifty years ago as a foundation of temporary earth-retaining structures in deep excavation. After the excavation was completed, basement and building were constructed as usual and these temporary pile foundation was left under soil without recording its location. In contrast, the location of the pile foundation of the building was precisely recorded as usual.

After decades, the old building was demolished and new building construction started by firstly installing new pile foundations between old piles. Unfortunately, the old temporary H-steel piles were unexpectedly hit and the project halted for one year. A similar trouble was caused by plastic drains for acceleration of consolidation settlement in clayey subsoil. The problem is that

- no detailed record was made of the location of temporary structures,
- it is difficult to investigate the subsurface conditions under existing big buildings during new design prior to demolishing, and
- the same problem will be likely repeated from now on.

Fig. 3.6. H-steel pile foundation of temporary structures for basement excavation during the construction of previous building 50 years ago (their location was not documented for future construction).

Fig. 3.7. Plastic drain that was installed 50 years ago without documenting location and prevented jet grouting for recent construction.

3.2 *More subsoil investigation for georisk management*

The content of this section was obtained by a joint study that was carried out by the Japanese Geotechnical Society together with the Georisk Society and the Japan Geotechnical Consultants Association. For more details, refer to Towhata et al. (2021a and 2022).

The problem of insufficient subsurface investigation has been recognized by experts for a long time. Fig. 3.8 presents the results of pioneering works in this field and indicates clearly that the undesirable increase of construction cost can be avoided by allocating more budget on preliminary soil investigation. The readers should remember that '2%' of the construction cost is not a bad criterion of reasonable investigation expense.

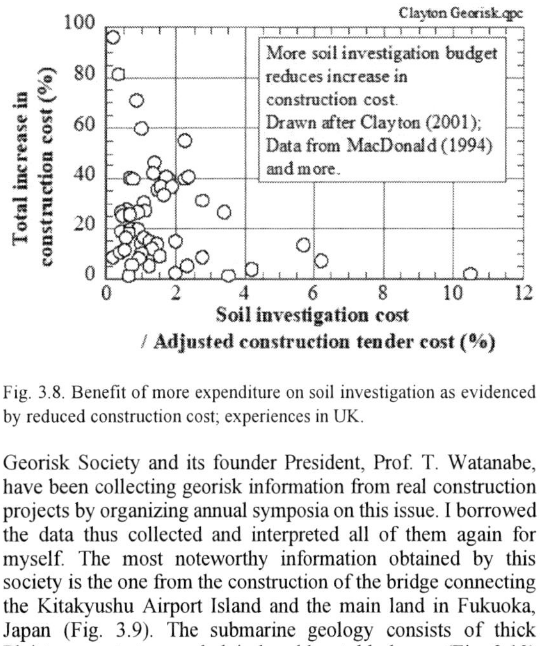

Fig. 3.8. Benefit of more expenditure on soil investigation as evidenced by reduced construction cost; experiences in UK.

Georisk Society and its founder President, Prof. T. Watanabe, have been collecting georisk information from real construction projects by organizing annual symposia on this issue. I borrowed the data thus collected and interpreted all of them again for myself. The most noteworthy information obtained by this society is the one from the construction of the bridge connecting the Kitakyushu Airport Island and the main land in Fukuoka, Japan (Fig. 3.9). The submarine geology consists of thick Pleistocene stratum underlain by older stable layers (Fig. 3.10) and the end-bearing piles designed for 28 piers on the basis of conventional SPT (Standard Penetration Test) N values were considered too costly (Tagami et al. 2010). The second choice was the use of shorter friction piles designed by SPT N again but it was considered costly again. Hence, the project decided to increase the subsurface investigation budget and to carry out more detailed field studies, including undisturbed soil sampling and laboratory tests. This effort made it possible to design shorter friction piles and reduce the total construction budget. As illustrated in Fig. 3.11, the increased investigation budget (US 2 million $) saved the total cost by 100 million $, for which the cost benefit ratio was 2:100.

Fig. 3.9. Bridge connecting the Kitakyushu Airport Island with the mainland.

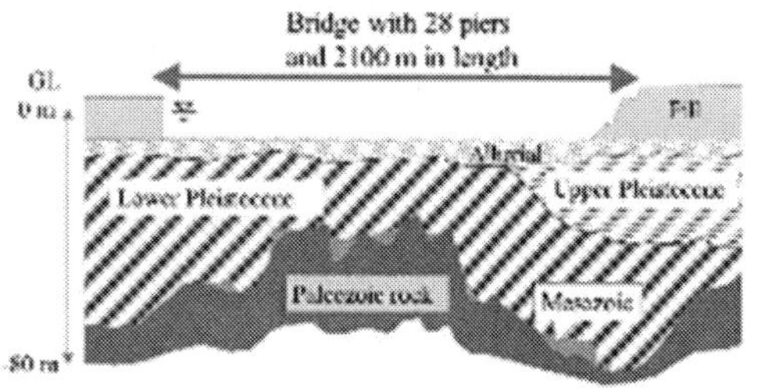

Fig. 3.10. Geological cross section along the Kitakyushu Airport Bridge (drawn after Tagami et al. 2010).

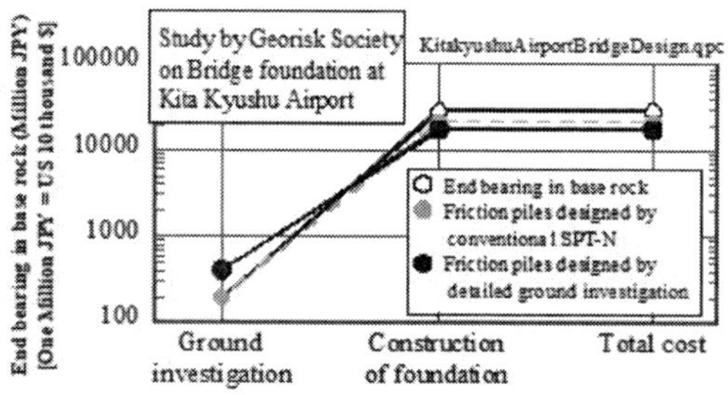

Fig. 3.11. Comparison of construction costs and expenditures with different extent of initial subsurface investigation (data by Tagami et al. 2010) (note that the original budget was decided by Japanese currency, while its conversion to US currency is subject to the varying currency exchange rate).

Georisk Society classified the studied cases into 4 groups;

- Group A: Original design was over-conservative and additional subsurface investigation helped reduce the cost, or georisk was anticipated after the initial design and additional ground investigation (georisk management) helped avoid troubles (59 cases),
- Group B: Risk (trouble) occurred during project and countermeasures increased the total cost (24 cases),
- Group C: Risk was anticipated during early stage of project and the additional investigation helped avoid the catastrophe (29 cases), and
- Group D: Detail is not clear (31 cases).

The following discussion addresses Groups A-C for which detailed information was available.

Fig. 3.12 shows the distribution of the types of structures for which georisk problem was reported. The majority is the slope instability and is followed by foundation. The detailed information shows that the unexpected slope instability occurred in a good number of cut slopes where instability was not anticipated prior to construction. Probably, weathering and hydration affected rock mass under the cut slopes to a substantial depth.

Georisk Society has been asking the site engineers about the original budget of construction, additional (increase) investigation budget, and two kinds of total construction cost that are the real expenditure and the hypothetical expenditure in case of troubles caused by georisk. Fig. 3.13 presents the summary for Group A for which georisk was successfully managed and georisk troubles were avoided. It is clear that the possible (hypothetical) high expenditure without georisk management was reduced to the real expenditure. The difference between these two expenditures is called 'Profit'. In one case, the real cost was reduced to as small as one million JPY because the anticipated ground instability was declined by detailed investigations and overdesign for safety was found unnecessary. Thus, safety confirmation is another type of benefit of detailed investigation.

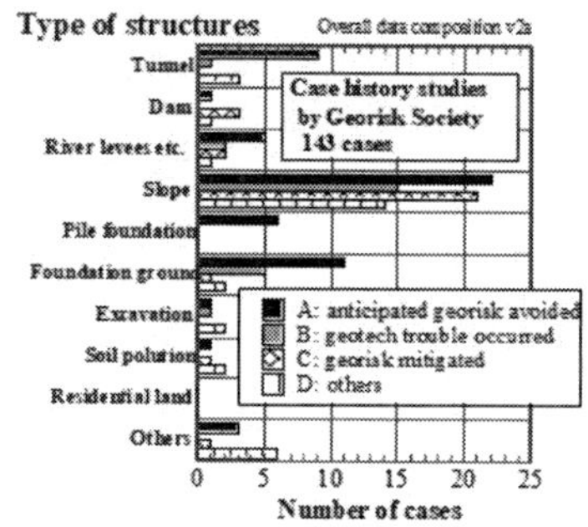

Fig. 3.12. Composition of Georisk Society's case history study

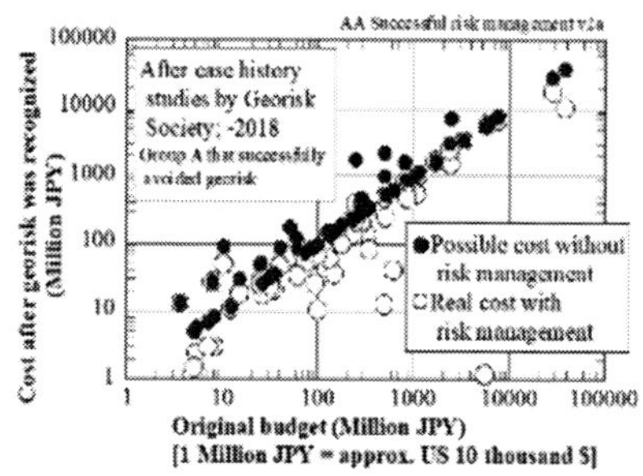

Fig. 3.13. Comparison of cost with and without successful risk management (Group A).

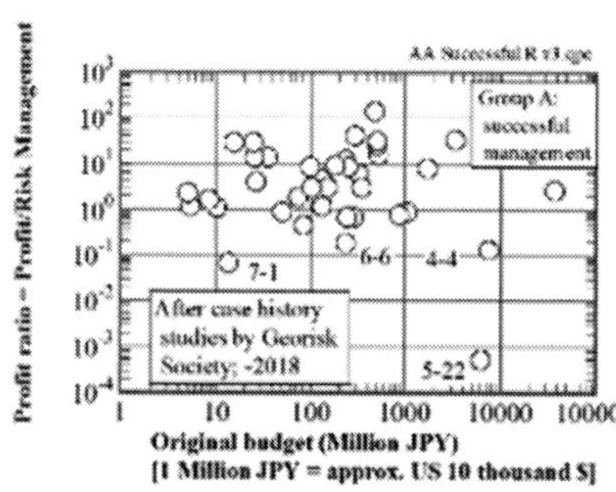

Fig. 3.14. Profit ratio versus original construction budget (Group A).

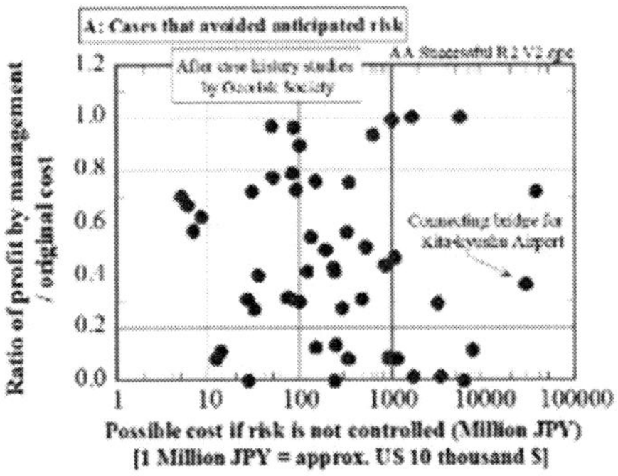

Fig. 3.15. Ratio of profit and original project budget plotted against total cost after possible risk manifestation (Group A).

The profit ratio for Group A is defined by 'Profit' / 'Additional expenditure on investigation' and is plotted in Fig. 3.14 against the original construction budget. In many cases, the profit ratio was greater than unity and the additional subsurface investigation was successful. Note that the ratio < 1 does not mean financial loss. 'Proft' is still positive and the total construction cost was saved. The very small profit ratio of the case of 5-22 in Fig. 3.14 is considered still successful by the concerned engineers because troubles other than money was avoided. Site engineers felt satisfied by avoiding headaches. Fig. 3.15 indicates that the ratio of profit against the original construction budget is independent of the worst-scenario cost in which georisk would have triggered troubles.

Fig. 3.16 shows the distribution of the types of additional investigation in response to the anticipation of georisk. Obviously, SPT (Standard Penetration Test) is the majority because of the tradition. However, eye inspection is considered important as well so that the overall situation may be captured.

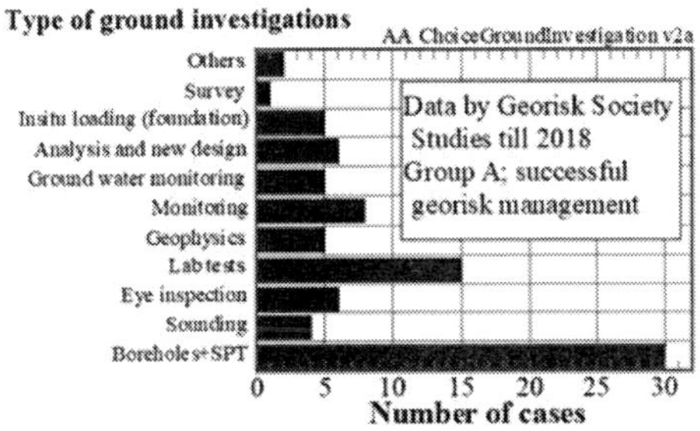

Fig. 3.16. Types of investigation employed for georisk management (Group A).

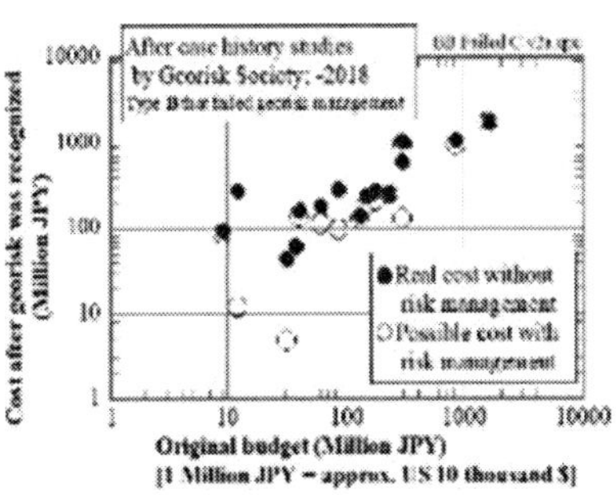

Fig. 3.17. Comparison of real cost increased by georisk and possible cost reduced hypothetically by missed risk management (Group B).

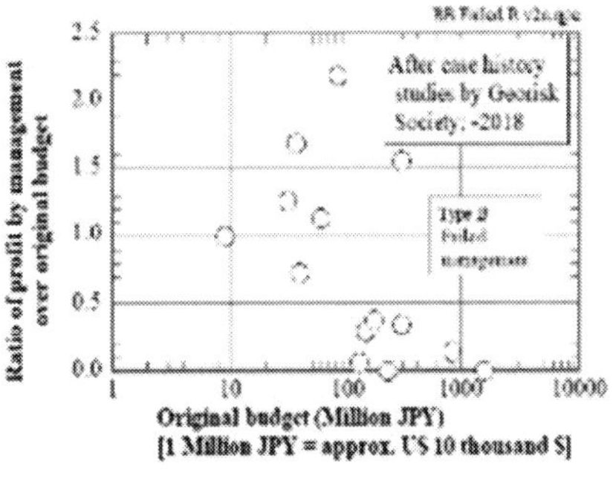

Fig. 3.18. Relative profit and size of project (Group B).

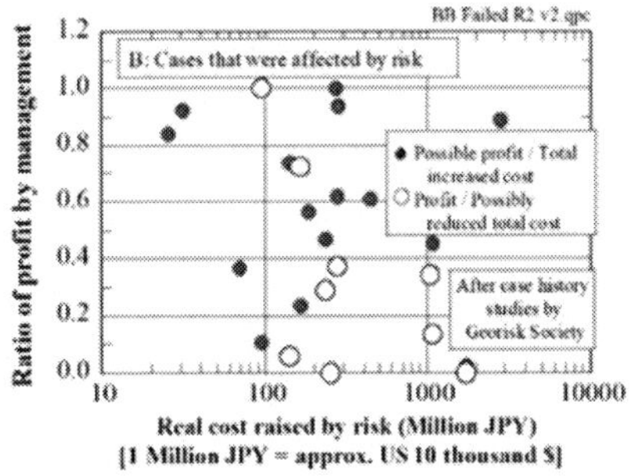

Fig. 3.19. Ratio of profit over two types of costs versus real cost increased by georisk (Group B).

Group B of georisk consists of the cases in which georisk was not properly managed and troubles occurred. The concerned engineers calculated two types of total construction costs that were the real cost as a consequence of the induced troubles and the hypothetical cost in case that georisk would have been managed in a reasonable way and troubles would have been avoided. They are plotted in Fig. 3.17 and their difference is called 'Profit' again. Then the ratio of this profit over the original construction budget is plotted against the project size (original budget) in Fig. 3.18. For smaller projects, this ratio can be greater than unity and the induced trouble was fatal to those projects. Even for bigger projects, 30 – 40 % of the original budget could have been saved if georisk had been recognized earlier and investigation had been carried out properly. Fig. 3.19 shows that the hypothetical ratio of profit construction cost is independent of the project size.

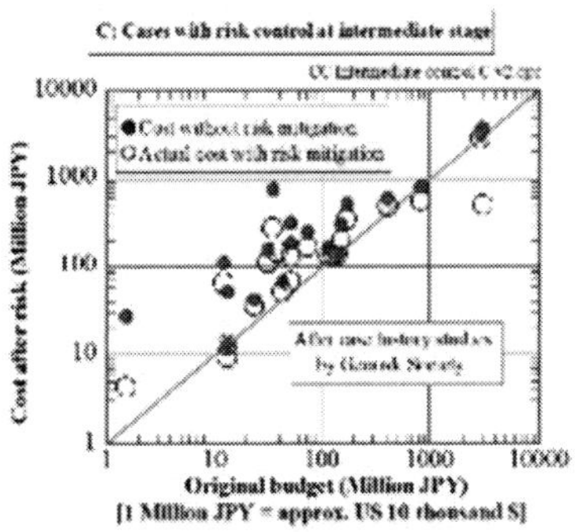

Fig. 3.20. Relationship between costs with and without georisk management and the original construction budget (Group C).

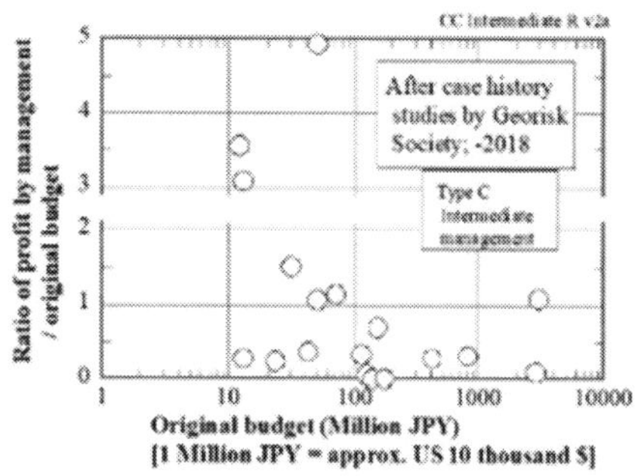

Fig. 3.21. Ratio of profit in Group C changing with the original budget.

Group C addresses the cases in which georisk was recognized at an intermediate stage of construction and subsurface investigation was conducted to avoid the worst situation. The meaning of 'intermediate stage' depends on individual judgement of site engineers but the timing of the georisk management was not so early as for Group A but not so late as for Group B.

In line with figures for other groups, Fig. 3.20 indicates that the total construction cost was reduced by the intermediate georisk management. The difference between the costs with and without georisk management (subsurface investigation) is called 'Profit' here. Fig. 3.21 illustrates the ratio of this profit over the original construction budget and, similar to Fig. 3.18 for Group B (failed georisk management), this ratio is substantially greater than unity for small projects. This means that the catastrophic situation was avoided in small projects by properly doing georisk management. The ratio of profit over total cost (worst-scenario cost without georisk management) is variable, being independent of the project size.

Fig. 3.23 plots 'Profit' of georisk management (subsurface investigation, changing design and construction methods) for three groups (A, B and C) against the additional investigation budget. The broken line in this diagram shows the 1:1 relationship and those data above this line indicates the cases in which the profit was greater than the additional investigation cost. In this regard, careful subsurface investigation deserves attention. Note that the data below this 1:1 line do not mean loss of money because their profit was positive and money was saved to a certain extent.

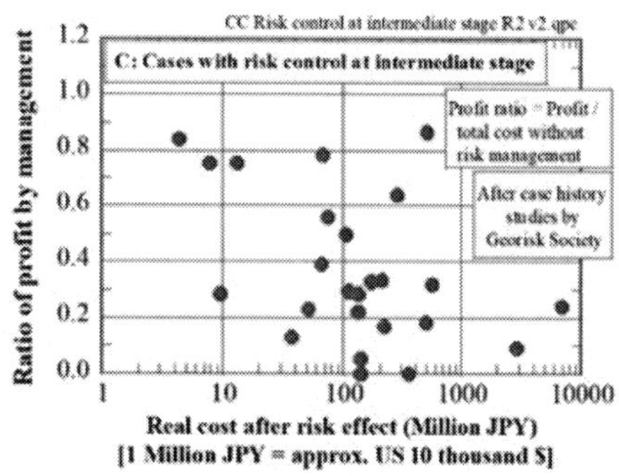

Fig. 3.22. Ratio management profit over total cost without management (Group C).

Fig. 3.24 demonstrates an important point. The ratio of expenditure on georisk management over the original construction budget is plotted here against the size of the project for the successful Group A cases. The concerned engineers feel that their georisk management was successful by spending this ratio of budget on additional subsurface investigation and related

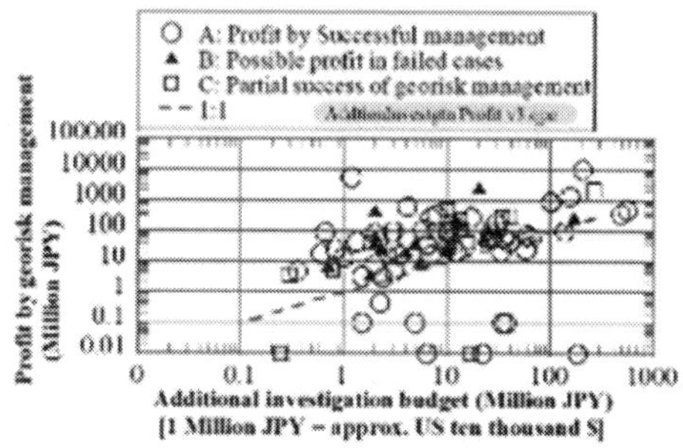

Fig. 3.23. Overall summary on profit of georisk management changing with cost for additional subsurface investigation (Groups A-C).

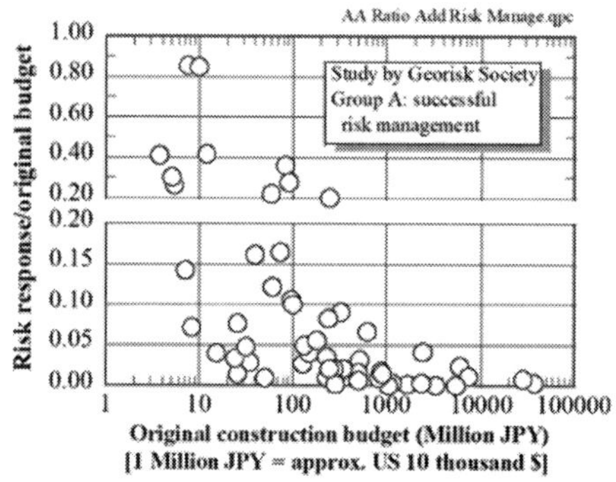

Fig. 3.24. Ratio of risk management cost and the original budget of the entire construction (Group A).

activities. By focusing on relatively big projects (>100 million JPY), it may be reasonable to state that 2-5 % expenditure on (additional) subsurface investigation helps avoid georisk troubles. By briefly inquiring several practitioners domestically and internationally, I found that, at the beginning before manifestation of georisk, most construction projects allocate on subsurface investigation 1% or less of the total construction budget. By adding this percentage to the additional 2-5% in Fig. 3.24 and then avoiding charging too much, I would state that 2% of the total construction budget is a reasonable amount to be allocated to subsurface investigation that helps avoid georisk troubles. Be recalled that 2% was a reasonable criterion as well in Fig. 3.8.

Delay of construction is another type of trouble in practice. Fig.3.25 plots the shortening of construction period (opposite to delay) by georisk management. For successful Group A, the shortening is mostly within one week only. This means that site engineers are not much interested in drastically shortening the construction period. It is good enough for them to complete the project on time. On the contrary, the unsuccessful Group B suffered from long delay such as several weeks. In this regard, sufficient amount of subsurface investigation helps avoid delay of construction.

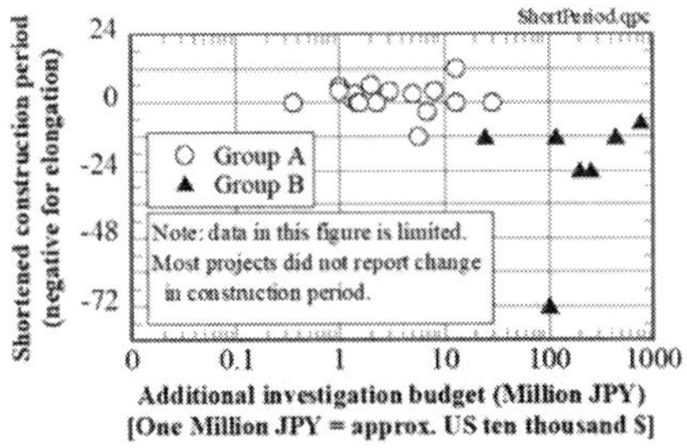

Fig. 3.25. Change in construction period after georisk management (Groups A and B).

3.3 *More thoughts on georisk management*

This report states that more budget should be allocated to subsurface investigation. The increased income will help improve the facilities of consulting farms as well as the professional image of geotechnical engineers. Thus, geotechnical engineering will be more scientific. However, geotechnical engineers have to make efforts that their job deserves more payment of the clients. Geotechnical engineers have to improve their professional skill and achieve good quality of job. One of the most serious problems in the current practice is the poor maintenance of equipment. In Fig. 3.27, the SPT device is not well maintained, and the edge of the sampling tube is rusted, making the penetration difficult and increasing the measured SPT-N values.

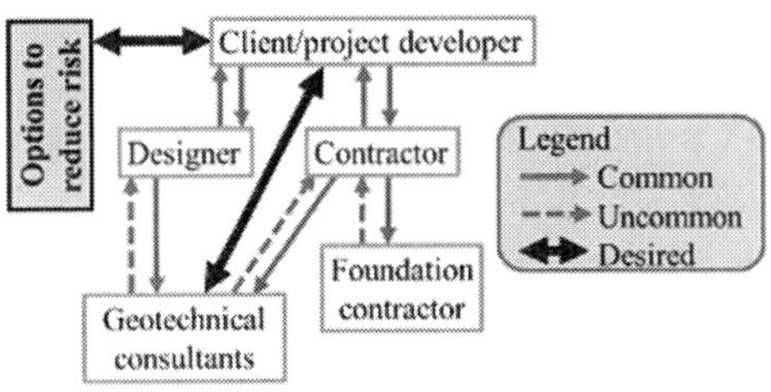

Fig. 3.26. Desirable human organization in construction projects.

Fig. 3.27. Example of poorly-maintained field investigation equipment.

Geotechnical consulting firms should establish a system of qualification and only good firms are allowed to enjoy the increased investigation budget. This qualification should be given not to individual engineers but to firms, similar to ISO (International Organization for Standardization) Certification because individuals cannot control the quality of company's practice.

Figs. 3.2 and 3.3 presented cases in which shortage of subsurface investigation led to troubles. While subsurface conditions have to be carefully investigated during the project, it is also important for the project managers and structural designers to understand the ground conditions and plan the project well in advance 'prior to bidding'. Then, the problem is the difficulty to run any subsurface investigation when the project has not yet started. Fig. 3.28 indicates the location of bore holes (investigation data available to the public) around the site of the condominium tilting in Fig. 3.2. Before construction of the condominium, there was not bore hole data in the open-access data base that was available to the project.

Actually, there was a factory at the site of this condominium and bore hole investigation was conducted before the factory construction many decades ago. The problem is that the current laws and regulations protect such data under the name of copy right of private sectors. The open-access data in Fig. 3.28 are entirely from public projects such as road and bridge construction. It is important now to change the copy right regulations on bore hole data because the earth is a common property of the society and property data should be shared for the safety of the public. Somebody worry that investigation data obtained by somebody else decades ago is not reliable and that the open-access data base is not necessary. That is wrong. Although the quality of investigation may be questionable, it is still possible to get a brief idea on soil type and the depth of pile-bearing layers from the 'unreliable' investigation records.

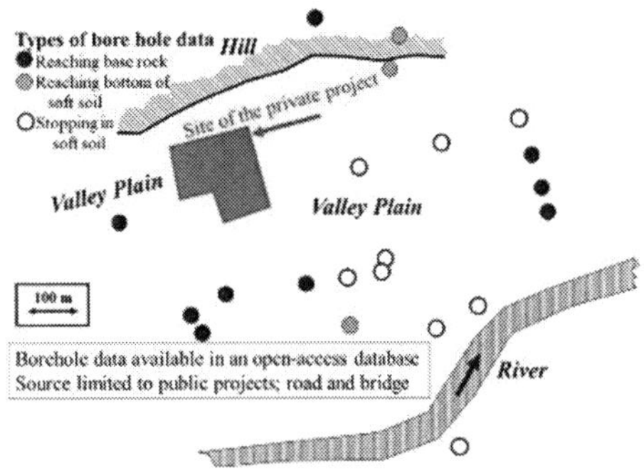

Fig. 3.28. Sites of bore hole data open to the public use.

The site of tunnel depression (Fig. 3.3) had locally weak geology and this problem had been recognized during one of the previous projects in its vicinity several decades before the incidence. One engineer in charge noticed this finding to his successor but, thereafter, this knowledge disappeared. This was a failure of knowledge transfer over generations. The issue is how to successfully transfer geotechnical knowledge over generations. As the case of Fig. 3.6 implies, the knowledge in the past may become essentially important after one life cycle of a structure that is typically 50 years or longer.

Many people think about the use of electronic media for conservation of knowledge and information. However, we have to be careful of the durability of electronic data for 50 years. The major electronic media 50 years ago (1970s) was a paper tape or punch cards and we do not have a device to read them anymore today. In 1980s, we had a floppy disk and do not have a floppy disk reader today. Although the value of electronic media is important from now on, we have to think about a supplementary medium for knowledge transfer.

We do not have to transfer much knowledge over 50 years. An exception was the case of Taga-Jo Castle near Sendai, Japan

(Fig. 3.3), where a remote branch of the central government was situated since AD 8th Century. An official chronicle clearly described a devastating tsunami disaster in AD 869 that water came to the human settlement in front of this site and destroyed everything. Unfortunately, the ancient stone statue in this site (Fig. 3.29b) did not refer to this disaster. Consequently, the modern people did not trust this knowledge written in the official chronicle, considering it as a typical exaggeration of ancient people in Kyoto who did not directly experience the tsunami. Finally, the disaster was repeated in 2011.

3.29(a) Entrance to the governmental section of Taga-Jo Castle in the eastern suburb of Sendai where the 2011 Tohoku earthquake caused tsunami disaster

3.29(b) Stone statue that describes the governmental branch here but does not refer to the tsunami in AD 869.

Fig. 3.29 (a & b) above. Archaeological site of the Taga-Jo Castle where a devastating tsunami disaster happened in Jogan Period (AD 869).

Clear writing of important knowledge on stone statues 'on site' is durable and more persuasive. Fig. 3.30 indicates a Buddhism statue that was built in memory of the victims of tsunami in AD 1677. Shown also in Fig. 3.31 is a historical stone statue that records the height of flood water in 1938. These two statues verify the long-term durability of records carved on stone. It is exciting that modern geotechnical engineering would record information in the same way as Ramesses II of Egypt recorded the Battle of Qadesh against Hittites in BC 1274; no progress for over 3,000 years?

Fig. 3.30. Stone statue in memory of tsunami victims who were killed by the disaster in 1677 (earthquake magnitude Mw = 8.5) (Kuju-Kuri in Chiba, east of Tokyo

Fig. 3.31. Stone statue that records the water level during the heavy rain and flood in 1938 (located to the north of Hakutsuru Museum, Sumiyoshi River, Kobe; pass by the museum, cross the Ochiai Bridge and take right. You will see this statue after 100 meters).

4 UNFORGETTABLE GEOTECHNICAL CONTRIBUTIONS TO HUMAN CIVILIZATION

Because of your geotechnical background, there is no need for me to write about the great history of our profession in details. Just for readers' information, I would like to present the photographs of the past great achievements so that you can show it to non-geotechnical people. Please feel free to add more photographs from your own album.

4.1 *Urban sewage*

Until the modern period, big cities in the world could not treat human wastes in a proper manner. Although draining the waste into rivers was practiced, the lack of proper treatment and purification resulted in the river pollution by pathogens, and smell (stink) was serious (Halliday 1999, Van Oosten 2016). This bad situation was remarkably improved by installation of modern sewer treatment systems (Fig. 4.1). Installation of underground pipeline network and treatment facilities were therein essential. Obviously, geotechnical engineering played one of the major roles therein and contributed to the improved hygiene of people. Probably, the geotechnical contribution to people's healthful life was as big as that of modern medicine. A similar drainage channel in Fig. 4.2 was successful as well in spite of the lack of modern waste treatment because the human waste was separated at the source (family) and purchased as organic fertilizer by farmers.

Fig. 4.1. The first modern sewage system in Paris (19th Century).

Fig. 4.2. Pre-modern sewage system in Osaka, which was originally constructed in the 16th Century as an open channel and was later covered.

4.2 *Irrigation and providing drinking water*

Cities have to provide sufficient drinking water to the citizens. When springs, wells, and rivers could not provide sufficient water, reservoirs and canals had to be constructed to acquire and transport water from far distance. Irrigation is another very important role of water supply. Construction of these facilities is a very important traditional mission of geotechnical engineering; see Figs. 4.3 to 4.7. Thus, we cannot stress too much the importance of the geotechnical contribution to water supply. In addition to these monumental structures, we should not overlook a small piece of the urban water system which is called lifeline and constructed/maintained by many people (Fig. 4.8).

Fig. 4.3. Tamagawa canal that provided water to present Tokyo, constructed in 17th Century

Fig. 4.4. All American Canal in southern California, USA , constructed in 1920s.

Fig. 4.5. Spillway at water intake of Dujiangyan, Sichuan Province, China (BC 3rd Century), from River Minjian

Fig. 4.6. Shushtar Irrigation Hydraulic System, Khuzestan, Iran (AD 3rd Century).

Fig. 4.7. Ganges Canal for irrigation in Utttarpradesh, northern India (construction in late 19th Century).

Fig. 4.8. One small piece of urban lifeline network undergoing maintenance (Zurich, Switzerland).

4.3 *Rockfill and earthfill dams*

Dams made of earth and rock fill are also the important monuments of geotechnical contribution to human welfare. Fig. 4.9 shows Wushantou Dam for irrigation system in southern Taiwan that drastically changed a dry plane to an agriculturally-rich field. Moreover, many dams contribute to hydraulic power generation (clean energy); see Fig. 4.10.

Fig. 4.9. Wushantou Dam for irrigation in Taiwan (completion in 1930).

Fig. 4.10. Mattmark Dam made of rock pieces in Switzerland (co mpletion in 1965).

Fig. 4.11. Aswan High Dam in Egypt made of rockfill (completion in 1970).

Fig. 4.12. Dike Kokaral in Aral Sea, Kazakhstan (completion in 2005).

In addition to irrigation and energy production, dams also contribute to flood and erosion control. It is well known that the Aswan High Dam in Egypt (Fig. 4.11) saved the Nile Delta from flood and consequent plague. The Kokaral Dike in Kazakhstan (Fig. 4.12) was constructed in the former Aral Sea in order to keep the fresh water from the Syr Darya River and create the Little Aral Sea. This was one of the practical solutions of the gigantic climatic disaster in which the former Aral Sea disappeared. It is interesting that this dam has a very gentle slope gradient which is as small as 2.5 %. Further from the disaster mitigation perspective, the Shiraiwa Dam in Japan (Fig. 4.13) was constructed in a volcanic valley in order to control the sediment disaster in the densely populated area in the downstream.

Fig. 4.13. Shiraiwa erosion control dam in Tateyama Caldera, Japan (completion in 1939).

4.4 *Transportation infrastructures*

Transportation by sea (vessels), by land (road and railway) and by air is essential in the modern world. Both passengers and cargos are transported for better life. All the transportation infrastructures rely on geotechnical engineering because they rest on land or installed underground.

Figs. 4.14 - 4.17 present canals. In the pre-modern period, ships and boats were the most powerful and efficient tools of mass transportation, and canals made inland transportation very easy. In particular, the Suez Canal in Fig. 4.17 changed the world, directly connecting the east and the west.

Fig. 4.14. Corinth Canal in Greece (completion in 1893).

Fig. 4.15. Jing Hang Great Canal in Hangzhou, China, connecting north and south of the country (history since Sui Dynasty in 6 – 7th Centuries).

Fig. 4.16. Caledonian Canal in Scotland (completion in 1822).

Fig. 4.17. Suez Canal in Egypt (completion in 1869).

Fig. 4.18. Entrance water path of the ancient military harbor of Carthage (Carthago) that was constructed by digging inland (in present Tunisia)

Fig. 4.19. Liquefaction-induced damage in Port Island of Kobe Harbor in 1995.

Harbor construction has a very long history which is more than 2000 years (Fig. 4.18) and are often constructed on soft marine soils. It is not easy to maintain the stability of harbor ground under gravity (consolidation settlement) and seismic force (mitigation of earthquake disaster). The pending of harbor operation exerts profound negative effects on the local and regional economy (Fig. 4.19).

There is not much need either to mention the important roles in the modern world played by land transport (Fig. 4.20). However, it should be stressed that roads and railways often have to cross mountains (Fig. 4.21), big rivers, and sea straits where difficult construction and maintenance are required. The famous aqueduct in Fig. 4.22 was constructed during the Roman Period. Although being installed underground and invisible, its foundation has been supporting the entire structure for nearly 2000 years. More recent bridges in Figs. 4.23 - 4.25 are supported by their invisible foundations for a long time as well. Without reliable bridges, land transport would be badly time wasting.

Fig. 4.20. Trans-Siberian Railway near Khabarovsk, Russia.

Fig. 4.21. Highway crossing Los Andes at Portillo in Chile.

Fig. 4.22. Masonry aqueduct crossing a valley in Segovia, Spain (early AD 2nd Century, Roman Period).

Fig. 4.23. Forth Bridge in Scotland (completion in 1890).

Fig. 4.24. Epoch-making Golden Gate Bridge with its central span = 1280 m (completion in 1937).

Fig. 4.25. Akashi Strait Bridge with its central span = 1991 m (completion in 1998).

The same is true of tunnels (Figs. 4.26 and 4.27). Nowadays, the technical development has made it possible to make the tunnel length longer than 50 km; Seikan submarine railway tunnel in Fig. 4.28 together with more recent Channel Tunnel connecting UK and France (50.45 km long and completion in 1994) and the Gotthard Base Tunnel under the Alps Mountains in Switzerland (57.10 km long and completion in 2016). Those tunnels drastically improved the efficiency of mass land transport by railway. Notwithstanding this, the invisible geological conditions trigger difficulties, troubles and even incidences during tunnel construction (Fig. 4.29). In landslide-prone mountainous regions, tunnels provide essential help for survival of local communities (Fig. 4.30).

Fig. 4.26. Construction of the Uodzu No. 2 Tunnel in Toyama, Japan, by means of the New Austrian Tunnelling Method.

Fig. 4.27. Shield tunnelling for urban motorway in Tokyo.

Fig. 4.28. Seikan submarine railway tunnel with total length = 53.85 km (completion in 1988).

Fig. 4.29. Tunnel collapse during construction due to unstable and heterogeneous geological condition.

Aviation is the latest version of transportation. As the airplanes become bigger and the number of tourists increases, the size of the airports becomes bigger as well. This means that many international airports are situated upon adverse ground conditions. Fig. 4.31 shows ongoing treatment of soft subsoil under a planned taxiway of the Soekarno-Hatta International Airport of Jakarta, Indonesia. The Kansai Airport of Osaka is situated on a huge manmade island that was constructed on thick consolidation-prone submarine strata (Nakase 1987). Although consolidation of the alluvial marine clay was well handled, the consolidation settlement in the deeper Pleistocene strata was difficult to treat. The subsidence of the island has been substantial, although the airport operation is maintained, and the extremely high sea wave during typhoon in September 2018 overtopped the sea walls and the airport was inundated by sea water (Figs. 4.32 and 4.33). After this problem, the sea walls and the runways were elevated.

Fig. 4.30. Protection of road traffic by tubular steel tunnel embedded in concrete (Sichuan Province, China)

Fig. 4.31. Ground improvement for expansion of the international airport in Jakarta, Indonesia.

Fig. 4.32. Kansai Airport of Osaka as seen from an airplane immediately before landing (September 21st, 2018).

Fig. 4.33. Dead reddish grass on the surface of the Kansai Airport after sea water inundation.

4.5 *Urban re-development*

Cities live long. Because the life style of people change with time, the plan and the infrastructures in the cities have to change and catch up with the new demands. Therefore, continuous efforts are needed to demolish or renovate the existing structures and to install new ones. This is the reason why the established city centers are continuously prone to big construction projects such as deep digging (excavation) for new underground shopping malls, railways and motorways (Figs. 4.34 – 4.36). Extreme care is necessary there not to trigger tilting or subsidence of adjacent buildings. If the local soil condition is too soft, special effort is made to improve the soil condition prior to construction (Fig. 4.37).

Fig. 4.34. Deep excavation for foundation of a new building (Yokohama, Japan).

Fig. 4.35. Big dig project in the center of Boston, USA, in 1990s.

Fig. 4.36. Construction of new underground tram (Köln, Germany).

Fig. 4.37. Ground improvement prior to excavation (Köln, Germany).

4.6 *Disaster mitigation*

Human history is full of natural disasters such as floods, earthquakes, landslides, high tides, volcanic eruptions, bush fires and droughts. It is important that geotechnical engineering has been, is and will be making remarkable contributions to prevent or mitigate many of them. This report has already touched upon the dams and irrigation channels to mitigate the effects of drought. Therefore, the following sections will introduce other geotechnical efforts concerning flood, earthquake, landslide and tsunami.

There are three types of infrastructures that can mitigate the effects of flood, which are dams in the upstream, the retention basins in the midstream, and levees in the mid- and downstream. Although some people believe in the value of emergency evacuation, it can help only human lives, while properties, workplaces, and community infrastructures cannot be saved by evacuation. In this regard, geotechnical engineering should emphasize the value of traditional measures that can save both human lives and infrastructures.

Fig. 4.38 demonstrates one of the most marvelous examples of levee (or causeway) that created a huge fresh-water Ijssel Lake which is isolated from the high tide in the North Sea. Many levees are protecting the human community from flood disasters (Fig. 4.39). To maintain its resistance against high water level during floods, the levees have to be continuously maintained by cutting grass (no animal living and no hole, burrow or den through which water may flow and trigger piping failure of the levee), filling holes, repairing subsidence, etc. Noteworthy is that the higher levee prevents flood but send more flood water to the downstream where the risk of overtopping increases. Therefore, the engineers always take care of the total safety of the communities along the river.

In this regard, the second mitigation measure is the retention basin in which the flood water is intentionally flooded and stored for some time in the midstream and, after the high water level ends, this stored water is sent back into the river (Figs. 4.40 and 4.41).

Fig. 4.38. Afsluitdijk in the Netherlands that produced a fresh-water lake disconnected from the North Sea (32 km long and completion in 1932).

Fig. 4.39. Levee along Danube River in Hungary.

Earthquakes triggers many kinds of disasters and, among them, geotechnical engineering concerns prevention or mitigation of slope failures (landslides) soil liquefaction and tsunami. Fig. 4.42 demonstrates the results of water drainage from an unstable slope where a previous earthquake in 1978 triggered a notable landslide disaster. In one part, no stabilization was made and the slope failed again (Fig. 4.42a), while in the other part, water had been drained out and no damage happened (Fig. 4.42b). Fig. 4.43 illustrates the slope reinforcement in a residential development fill that is intended to avoid landslide during future earthquakes.

Fig. 4.40. Intention of overtopping of flood water into a retention basin schematically illustrated by arrows (Midai River, Japan, 16th Century).

Fig. 4.41. Lower section of river levee where flood water overtops intentionally (Ohkubo Retention Basin, Japan).

Fig. 4.42 (a). Effects of drainage of ground water (removal of water) from an unstable slope during the 2011 Tohoku earthquake of Mw=9.0 in Shiroishi, Japan. Failure of slope without geotechnical mitigation

Fig. 4.42(b). Effects of drainage of ground water (removal of water) from an unstable slope during the 2011 Tohoku earthquake of Mw=9.0 in Shiroishi, Japan. Successful drainage and slope stabilization next to the damage site in (a)

Liquefaction is a phenomenon in which loosely-packed young sandy ground saturated with water undergoes strong seismic shaking and is converted into a very soft medium like liquid. Its consequence is intolerable extent of displacement and deformation such as subsidence of buildings, floating of underground structures, lateral displacement of sloping ground and consolidation settlement (Fig. 4.44). There are many mitigation technologies as practiced by geotechnical engineering nowadays, inclusive of ground water drainage as mentioned above, but this report cannot refer to all of them. For more information, refer to a summary paper (Towhata, 2021).

Fig. 4.43. Slope reinforcement to protect residential development from earthquake landslide (Takaradzuka, Japan).

Fig. 4.44. View after liquefaction disaster; see road undulation, ejected sand on pavement and tilting of structures (after 2011 Tohoku earthquake).

Fig. 4.45 (a). Contrast between densified and untreated parts of a sand compaction project (2011). Machines for densification of loose sandy subsoil

Fig. 4.45 (b) Contrast between densified and untreated parts of a sand compaction project (2011). Liquefaction and sand ejecta only at the surface of the untreated part (see arrows)

Fig. 4.46. Small jet grout machine for soil improvement in a narrow space.

Fig. 4.47. Ongoing jet grouting in a narrow space between houses (Urayasu, Japan).

Compaction/densification is a popular and reliable method to mitigate the liquefaction problem (Fig. 4.45a). Once densified, the treated ground will be resistant against liquefaction forever without maintenance. During the 2011 Tohoku earthquake in Japan, one project of ground densification was going on. Fig. 4.45b demonstrates a clear contrast between no liquefaction in the densified part and sand/water ejection (evidence of liquefaction) in the other to-be-densified part. Mixing sand with grout is another mitigation. Special liquid is ejected from such a device as in Fig. 4.46 (jet grouting) and mixed with liquefaction-prone sand. Because of its small size, this device can work in a very small space between houses (Fig. 4.47).

The lesson learnt from the project of Fig. 4.47 is that people desire more safety but are not willing to spend much money. Most people are affected by normalcy bias that disaster is somebody else's problem and does not occur to his/her own family. This attitude is difficult to change despite that media reports many earthquake and landslide disasters all the year around.

Fig. 4.48. Rock mass reinforcement prior to construction of Three Gorge Dam, China.

Fig. 4.49. Rainfall-induced landslide in Izu Oshima Island, Tokyo, Japan, triggered by 824 mm rainfall in one night (October 16th, 2013); arrows show the head scarp of many slope failures.

While coseismic landslide (triggered by earthquake shaking) is an important issue, more landslide problems are caused by gravity or heavy rainfall. To mitigate this hazard, water drainage is an efficient way (as shown in Fig. 4.42). However, there are situations in which water drainage is not sufficient and mechanical reinforcement is required therein (Fig. 4.48).

Heavy rainfall can trigger significant slope disasters and this problem is becoming more serious than before because of the (probably) ongoing climate change. Fig. 4.49 shows an example of such a disaster in Izu Oshima volcanic island in Tokyo (Towhata et al. 2021b).

Although many people are afraid of landslide behind their houses, they cannot get public fund to mitigate the problem because landslide is basically a personal problem for which public funding is not appropriate and also because there are too many hazardous slopes for the public sector to reinforce. In this situation, the people's choice is slope monitoring and early warning for evacuation. Fig. 4.50 indicates one of the sites of slope monitoring and, later, the landslide precursor was detected several hours before the final disaster (Uchimura et al. 2015, Towhata et al. 2015).

From the viewpoint of global warming, the risk of melting glacier ice is stressed. The melt water flows into slopes and increases the risk of slope failure. In the case of Fig. 4.51, the huge landslide blocked the river flow and formed a natural dam (Yin 2000; Zhou et al. 2016). After the dam breaching, the flood water flowed 500 km downstream along the Yarlung Tsangpo/Brahmaputra River Valley and caused flood in Assam, India (Delaney and Evans, 2015). It seems that melting of glacier will aggravate the risk of landslides at high altitudes.

Fig. 4.50. Slope monitoring with sensor atop the pole, aiming to foresee landslide and promote emergency evacuation (China).

Fig. 4.51. Yi Gong landslide site in Tibet, China, where landslide dam was formed and its breaching sent flood far down to Assam, India, in 2000.

Tsunami and high tide can destroy everything (Figs. 4.52 and 4.53). To mitigate this powerful and extreme disaster, the human choice is limited. One idea is construction of high sea walls along the vulnerable shore line (Fig. 4.54). This wall is made of earth and is a typical contribution of geotechnical engineering. It is aimed to fully protect the local community from tsunami of medium height; thus, infrastructures and community are not devastated by tsunami. However, there is always a possibility of extremely high tsunami that exceeds the wall height. Therefore, people should not feel safe with the wall and resume evacuation as quickly as possible when tsunami warning is issued.

Fig. 4.52. Banda Aceh, Indonesia, after tsunami devastation in 2004.

Fig. 4.53. Rikuzen Takata, Japan, after tsunami devastation in 2011.

Fig. 4.54. High sea wall in Rikuzen Takata constructed after the disaster in 2011 (2 km long and 12 m high).

Fig. 4.55. Earth fill as a tsunami shelter (Fukuroi, Japan).

Fig. 4.56. Traditional earth fill constructed as a shelter after typhoon and high tide in 1680.

Fig. 4.57. Embankment used for cyclone shelter in the coastal area of Myanmar

For evacuation, shelter has to be prepared in advance. Fig. 4.55 shows an earth fill where more than one hundred local people can stay during the next tsunami. Although this area is protected by high sea wall, the local community also prepare this shelter for the worst 'in case' situation. A gigantic earthquake of magnitude greater than 8 and tsunami are anticipated to occur before 2050 by seismologists. It should be noted that the local community here has a tradition to prepare such an earth shelter as shown in Fig. 4.56. The same idea is employed at other places in the world. Fig. 4.57 shows a cyclone shelter in the coastal region of Myanmar where there is no sea wall and the region is vulnerable to cyclone-induced high tides.

The advantage of such an earth mound is that construction is easy with local materials only and maintenance is inexpensive. In contrast, care must be taken of buildings as a tsunami shelter. Fig. 4.58 shows a building in Onagawa Harbor, Japan, where the tsunami in 2011 exerted high lateral force on this building and the foundation (pile foundation) could not resist the pull-out force. The consequence was toppling of the building. Because this site is a land reclamation, it is also possible that liquefaction occurred and skin friction along the pile shaft was drastically reduced.

Fig. 4.58. Building in Onagawa Harbor, Japan, toppled by tsunami in 2011; a steel pile is hanging next to the base slab (shown by arrow).

5 EXPOSURE OF GEOTECHNICAL ENGINEERING TO THE PUBLIC

Fig. 3.26 stressed the importance of communication between geotechnical engineers and the client who unlikely has engineering background. Nevertheless, geotechnical engineering does not have a good exposure to the public and suffers from many unfortunate misunderstanding by people. The problem lying behind this situation is the lack of communication tools, conversation channels and even common languages. In this regard, computer and information technologies are doing much better; see young children know about specifications of smart phones and even programming languages. Some geotechnical engineers may not want to communicate with people because they are not the customer of any project. It is true that construction projects come from public sectors and big industries. However, there is something more important.

People pay tax and the tax money is spent on construction projects. Public policies are governed by politicians who are elected by people through democratic processes. It is therefore important for geotechnical and construction engineers to be able to communicate with people. Nevertheless, most engineers have been and happy to satisfy the customers. In other words, they have been satisfied with customer satisfaction. They are not much interested in what people want and, therefore, cannot make a good proposal to the public sector in charge of construction policies.

On the other hand, geotechnical engineers do not have to entertain people. They are not comedians. It is important for them to make proposals whose significance can be understood by people. To do this, there are many major problems that the world is facing nowadays. Although all of them are not fully geotechnical, geotechnical engineering can collaborate with other disciplines and play important roles.

Fig. 5.1. Storage of used nuclear fuel in water pool inside a reactor building (Kashiwazaki-Kariwa Nuclear Power Plant).

Fig. 5.2. Fukushima No. 1 Nuclear Power Plant after disaster on March 11, 2011 (photograph taken on Nov. 30, 2016).

Fig. 5.3. Coolant pipelines for construction of frozen underground wall that prevents water from flowing into the radioactively contaminated basement of reactor buildings.

Fig. 5.4. Paved ground surface in order to prevent rain water from filtrating into soil and increasing polluted ground water.

Nuclear industries are seeking for safe and reliable repositories where the radioactive waste fuels and other wastes can be permanently dumped (Fig. 5.1). The tsunami disaster at the Fukushima No.1 Nuclear Power Plant is now undergoing many geotechnical issues (Figs. 5.2 – 5.4). The final dumping of radioactively-contaminated materials will be underground. The same is true of other kinds of waste (municipal waste after incineration in Figs. 5.5 and 5.6, industrial waste after necessary purification and even CO2 gas nowadays by means of Carbon dioxide Capture and Storage) that will be dumped underground as the ultimate treatment. Therefore, waste management is a chief job of geotechnical engineers who are familiar with mechanics, geohydrology, and environmental technologies and can make necessary safety measures against slope failure and leakage.

The Leuwigajah municipal waste landfill in Bandung, Indonesia, failed after rainy days (Fig. 5.7) (Koelsch et al. 2005, Lavigne et al. 2014). It seems to me that the fill was originally reinforced unintentionally to a certain extent by fibrous components in the waste (plastics and papers) but they were burnt down by fire. Afterwards, the slope stability was substantially reduced (Towhata 2007). Another fill made of construction wastes in Atami, Japan, failed during heavy rainfall in July, 2021. The number of victims was 26. After this disaster, it was found that the owner of this landfill did not take necessary care of safety as advised (not regulated) by related rules. The lesson was that we have to spend more money on safety of waste treatment and final dumping.

Fig. 5.5. Municipal waste landfill of mega city (Delhi, India).

Fig. 5.6. Collection of CH4 gas from municipal waste landfill in Delaware, USA.

Fig. 5.7. Municipal waste landfill of Leuwigajah in Bandung, Indonesia, that failed in 2005 during heavy rain and claimed 143 lives.

Fig. 5.8. Rainfall-induced collapse of fill of construction-waste earth in Atami, Japan that claimed 26 lives and one missing (July 2021).

One of the global problems that I have in my mind is the shortage of water for both drinking and irrigation. There are many countries where population is increasing rapidly and climate is arid; see India and Iran for example. Therefore, development of new water resource can help people who are prone to drought and shortage of food.

The present water management is very strict everywhere in the world. There is little chance to obtain surplus water from lakes and rivers in populated regions. One exception is the ground water that is flowing directly into the seabed; nobody claims to use it. Care must be taken, however, because uncontrolled pumping of ground water near the sea brings salty water into the underground aquifer and induces salt contamination of the subsurface environment.

Underground dam is the key and has a good tradition in the history (Åke 1988). As illustrated in Fig. 5.9, an underground dam is simply an impervious wall that is constructed by mixing cement with soil. The ground water level should be monitored for good water management (Fig. 5.10). Because both sides of the underground wall are supported by soil, mechanical strength of the wall is not important. Moreover, because water is stored underground, its evaporation loss is limited. The land use upon the surface is not affected by reservoir construction (Fig. 5.11). Good geotechnical expertise is important for underground dam construction with respect to site characterization, geohydrological study, construction quality, and operation. Possibly, the impervious wall does not have to reach the base rock as shown in Fig. 5.12. Fig. 5.13 indicates geological issues related with the underground dam in Miyako Jima Island.

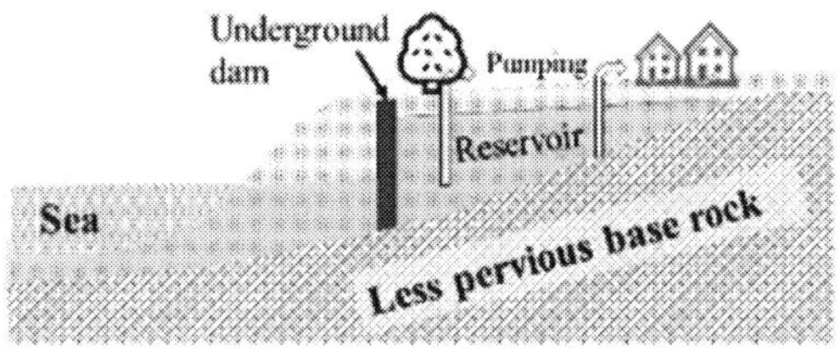

Fig. 5.9. Concept of the use of ground water near sea shore by installing underground dam.

Fig. 5.10. Monitoring of underground reservoir water in Miyako Jima Island, Japan.

Fig. 5.11. Site of underground dam that is not visible at the surface

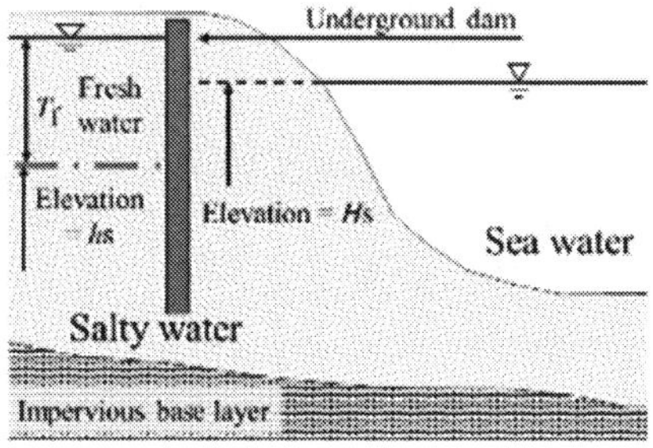

Fig. 5.12. Calculation of fresh water level achieved by underground partial impervious wall

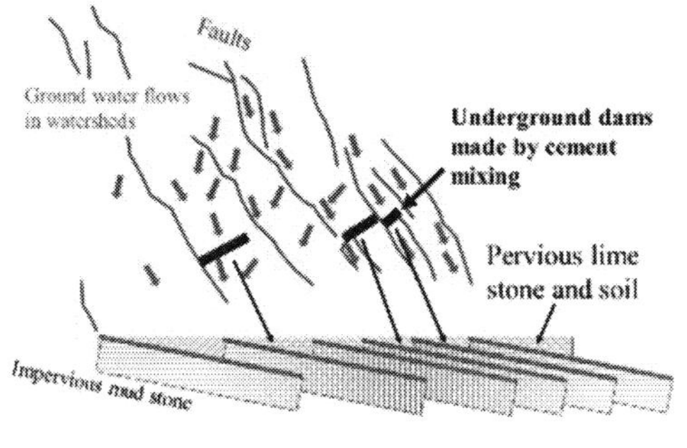

Fig. 5.13. Geo-hydraulic issues in Miyako Jima Island for underground dam planning (drawn after Tomita et al., 1987).

Fig. 5.14. Deteriorated steel bridge immediately before its scheduled demolition.

Fig. 5.15. Coseismic failure of road embankment made of crushed mud stone that absorbed water and got deteriorated (slaking) after many years (Niigata Chuetsu earthquake, 2004).

Deterioration of infrastructures is another urgent issue in many countries where construction was booming in 1950 – 1970 and the structures constructed in those days are now 50 – 70 years old and are prone to the deterioration problems (Fig. 5.14). Some people may imagine that the deterioration is a problem of steel and concrete structures but they are not very right. Soil and rock deteriorate as well through weathering and hydration (Fig. 5.15). The most vulnerable component of geotechnical structure is anchorage embedded in mudstone slopes where hydration and slaking mechanism may be affecting the safety factor. Asada (2005) studied the N values of Standard Penetration Tests at many residential development fills (Fig. 5.16) which were unfortunately constructed at saved cost. It was shown that the older fills have lower N values, implying that the fill is chiefly made of mud stone prone to slaking and becoming weaker with time. Fig. 5.17 illustrates another example of deteriorated infrastructure. The same problem is going on in many cities of the world.

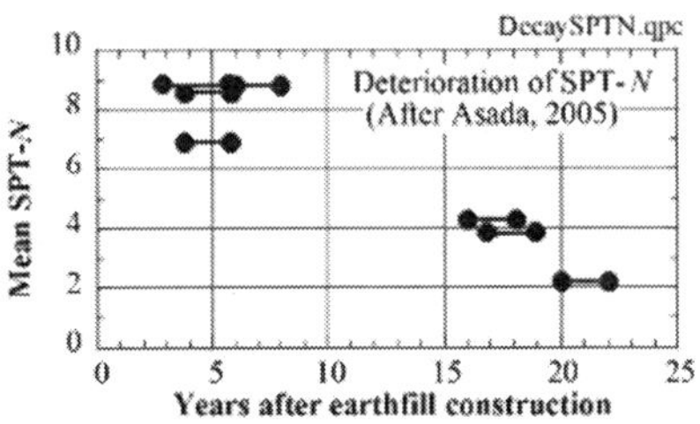

Fig. 5.16 Decreasing trend of SPT-N value with years after earth filling for residential development (Asada 2005).

Fig. 5.17. Collapse of sidewalk pavement caused by deterioration of underground water pipelines, leakage and consequent soil erosion.

Efforts to make maintenance periodically and to keep the quality of infrastructures above a certain level help elongate the life cycle of infrastructures and improve the cost-benefit ratio. In other words, those who are engaged in maintenance business should be respected.

As stated in Introduction, there are similarities and differences between geotechnical engineering practice and medical doctor's practice. This is presented in detail in Table 5.1. It is shown herein that the public image is substantially different between the two disciplines although the situation and problem are very similar. The important reason for this is that people can feel ill (pain, tiredness, discomfort etc.) once it starts, while people (public) do not see the geotechnical troubles. Even from the viewpoints of natural disasters, people are controlled by the normalcy bias and do not worry about the disasters, saying "it is somebody else's problem!" Many efforts have been made by disaster mitigation experts to change this bias by disseminating people about the consequence of natural disasters without much success.

Table 5.1. Similarity and difference in played roles between geotechnical engineers and medical doctors who carry out investigation/medical check prior to construction or treatment.

	Geotechnical	Medical
Limited available information	Local geology & geomorphology	Health/medical history
Target of work	Structures to be constructed	Current illness/surgery
First step of investigation	Monitoring from outside	Health monitoring from outside
Second step	Sounding & boring	Scanner, X-ray etc.
Third step	Undisturbed soil sampling and tests	Sample incubation and inspection
Action planning	Design	Treatment planning
Action	Construction	Treatment/surgery
Difficulty	High pressure to save cost	Low pressure; good result is respected
Bad scenario	Trouble and georisk	Unsatisfactory treatment
Self-discipline	Less	More
Independence	Controlled by company and client	High responsibility and authorization to make individual decision
People's idea	Ground is rigid without any disaster.	People do not mind spending money to recover health once problem is detected.
Income	Lower	Higher
Public image	Lower	Higher

6 ARE YOU ENJOYING YOUR JOB?

To make our occupation look attractive to others, we have to enjoy it. It is not enough to only talk about the social importance of your occupation. I arise a question "Are you enjoying your job?" It is very fortunate of me that I have found many interesting issues in geotechnical engineering. The following photographs show them. There is no need to talk about them in addition to showing the photographs. Earth and nature are amazing indeed.

Fig. 6.1. Obduction of the southern mountain mass over north; tectonic activity in Switzerland showing the dynamism of the earth.

Fig. 6.2. Rock powder mashed by collision between Indian subcontinent and Eurasia, resulting in many slope instabilities (Himalayan front in Nepal).

Fig. 6.3. Many piping holes from which ground water was ejected during heavy rainfall-induced landslide (Izu Oshima Island, Japan, in 2013), indicating complex geological-geotechnical-geo-hydrological system.

Fig. 6.4. Buddhism temple 'Mai-ji-shan shi-ku' carved by private funding in rock monolith (Period of Northern Wei Dynasty, 5-6 Centuries, Gansu Province, China).

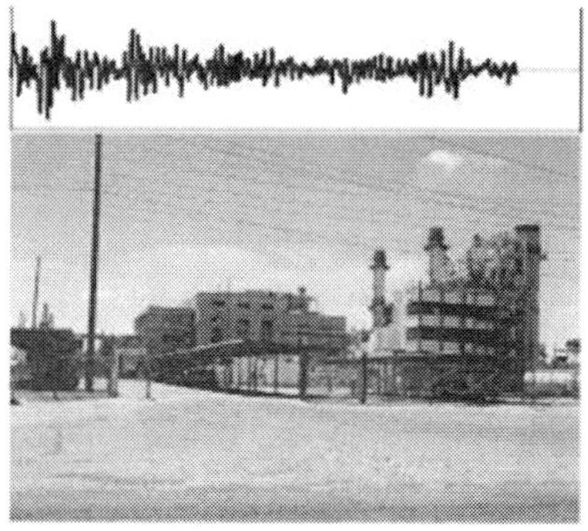

Fig. 6.5. Site of famous strong earthquake motion records in El Centro, USA

Fig. 6.6. Sandy ground in Amazon, Brazil; once vegetation is destroyed, recovery is not easy.

Fig. 6.7. Creeping rock slope in Bhutan.

Fig. 6.8. Creeping slope of Mt. Chogatake in Japan; the movement started more than 10,000 years ago and the rate of displacement is approximately 1 cm/year.

Fig. 6.9. Pyramid of Kush, Sudan, where the ancient local kings tried to do the same as the great Pharaohs of Egypt but in a much smaller scale.

Fig. 6.10. Thermopyles in Greece where the sea was close to the mountain in BC 480 but now is very far due to rapid deposition of eroded soil.

Fig. 6.11. One of many mud volcanos in Gobustan of Azerbaijan.

Fig. 6.12. Rock in Mahabalipuram, Tamil Nadu, India, presumably formed by weathering of a mother stone.

Fig. 6.14. Rift valley in Ethiopia where the African Continent is split into its eastern and western segments.

Fig. 6.15. Big pit of Chuquicamata Copper Mine in Chile (air is dusty).

Fig. 6.16. 'Sur le Pont d'Avignon' in France.

Fig. 6.17. Seidmareh gigantic prehistoric landslide in Iran with 22 billion cubic meters in volume.

7 RE-EVALUATION OF THE PIC ACHIEVEMENT EXAMINED AGAINST ITS ORIGINAL ACTION PLAN

The achievements of the committee are evaluated in Table 7.1 against the original plan that are described in its Terms of Reference.

Table 7.1. Self-assessment of achievements for respective aims proposed in Terms of Reference (dated September 4, 2018).

Item	Detail	Score+	Self-assessment
Objective 1: dissemination	Dissemination of contribution to the public welfare made by geotechnical engineering	4	See Chapter 3 of this report for contents. PPT file is under preparation and will be available for Member Societies to modify it freely. The score is not 5 because international dissemination was impossible due to pandemic.
Objective 2: Mitigation of Georisk	Analysis of case history studies by Georisk Society	5	Benefit by more detailed subsurface investigation was validated by interpreting case history data base and the desired budget for investigation was proposed.
Objective 3: Future possibility of geotechnical engineering	Proposal of new roles to be played by geotechnical engineering	4	Several ideas were proposed including disaster mitigation and water resource development. New Technical Committee for water technology was declined by the Board.
Objective 4: Production of materials for dissemination	PPT file for public presentation	4	Under preparation, paying attention to issues related with climate change, water and food shortage, population increase and decay of infrastructures. Most of its contents are included in this report.
Objective 5: Communication with public	Public dissemination	1	PPT file is not yet completed. Contact with public is very difficult due to pandemic. For domestic dissemination, initiative of Member Societies is desired.
Objective 6: Communication with clients	Dissemination	1	Difficult due to pandemic. Initiative of Member Societies is desired.
Objective 7: Awards	Promotion of enthusiastic efforts for better public image		Too early, not done.

+: 1 for lowest and 5 for highest.

8 EPILOGUE

This report reviews the activities of the Professional Image Committee since 2018 and suggests several points that should be appealed by Member Societies to the public. The major points of the suggestions are as what follows.

1) To improve our professional image of our discipline among public, it is essential to appeal our great achievements to the public. The problem is that this type of efforts in the past were not very successful. Something more should be done.
2) The public do not respect occupation with low income. It is important that geotechnical engineers should get reasonable income for what they do for the customers.
3) The public do not necessarily have to care what is going on in the invisible and heterogeneous underground space.
4) Subsurface investigation is an important procedure by which the hidden problems are detected in the invisible and heterogeneous underground space.
5) Georisk is a trouble caused by insufficient subsurface investigation and makes financial loss and/or delay in execution of projects. Customers should understand this risk and provide reasonable funds for investigation.
6) Georisk come from the fact that ground was made by nature without taking care of homogeneity, material quality and durability. These characteristics and invisibility makes subsurface investigation a difficult task.
7) Interpretation of georisk case histories suggests that about 2% of the construction cost should be allocated to the investigation.
8) Subsurface investigation helps avoid both financial loss and delay of construction projects.
9) On the other hand, geotechnical engineers have to do good jobs that deserve higher payment. Qualification should be granted on a company (consulting firm) basis, not on personal basis because individual engineers cannot control the company's policy.
10) Subsurface investigation should be supplemented by open-access bore-hole data base together with transfer of knowledge of land history (filling, digging, flood, landslide, liquefaction etc.) from generation to generation.
11) Do not be satisfied with customer satisfaction. Be exposed directly to the public and search what people want.
12) Geotechnical engineering has been doing many wonderful contributions to the human world. Because most of the achievements are hidden under ground, people are not aware of them unless special dissemination efforts are made of the geotechnical community.
13) To make our occupation attractive to the public, we must enjoy our occupation.

The figures and photographs were drawn or taken by myself except young fellows' pictures in Chapter 1, which belong to the Japanese Geotechnical Society, and Fig. 3.3(a) on depression of the subway construction site. I do not claim my copy right to them. Therefore, those who wish to use them and promote geotechnical engineering among public can freely use them only with a credit "Ikuo Towhata (2022) For better professional image of geotechnical engineering. Output of the Professional Image Committee of ISSMGE Board in conjunction with the Japanese Geotechnical Society".

9 ACKNOWLEDGMENT

The review of the georisk in real construction projects were made possible as a collaboration of the Japanese Geotechnical Society with the Georisk Society and Japan Geotechnical Consultants Association in Tokyo. The data on georisk has been collected by the long-term efforts of Georisk Society in which its founder president, Prof. T. Watanabe, has been playing the chief role. The value of these activities are deeply appreciated by the Author. My special thanks are extended to the members of the special committee that was established by the Japanese Geotechnical Society in order to promote the activities of PIC of ISSMGE. The names of the committee members are shown here: Messrs. S. Sawada, K. Azuno, T. Watanabe, T. Sumi, T. Shuku, H. Imanishi, S. Yamamoto, H. Hazarika, M. Ohishi, N. Hiraoka, M.

Tozawa, T. Yasuda, T. Nozawa, K. Kato, K. Watanabe and K. Nakayama.

10 REFERENCES

Åke N (1988) Ground Water Dams. Intermediate Technology Publications Ltd., ITDG Publishers, ISBN 1 85339 050 X.

Asada A (2005) Assessment of seismic risk in residential areas under future Miyagi-ken-Oki earthquake and its mitigation. Draft report, p. 112 (in Japanese).

Clayton CRI (2001) Managing geotechnical risk: time for change? Proceedings of the Institution of Civil Engineers-Geotechnical Engineering 149(1): 3-11.

Delaney KB, Evans SG (2015) The 2000 Yigong landslide (Tibetan Plateau), rockslide-dammed lake and outburst flood: review, remote sensing analysis, and process modelling. Geomorphology 246: 377-393.

Halliday S (1999) The Great Stink of London. Publ. Sutton, ISBN 0 7509 2580 9.

Koelsch F, Fricke K, Mahler C, Damanhuri E (2005) Stability of landfills - The Bandung dumpsite disaster. Proceedings of the 10th International Landfill Symposium, Cagliari (Italy).

Lavigne F, Wassmer P, Gomez C, Davies TA, Sri Hadmoko D, Iskandarsyah TYWM, Gaillard JC, Fort M, Texier P, Heng MB, Pratomo I (2014) The 21 February 2005, catastrophic waste avalanche at Leuwigajah dumpsite, Bandung, Indonesia. Geoenvironmental Disasters 1(10).

MacDonald MS (1994) Study of the Efficiency of Site Investigation Practices. Project Report 60, Transport Research Laboratory, ISSN 0968 4093.

MLIT Investigation Committee on Depression of Subway Tunnel (2017) Report on depression of metro construction site in Fukuoka. Public Works Research Institute (in Japanese).

Nakase A (1987) Kansai International Airport - Construction of man-made island. Proc. 8th Asian Regional Conference on Soil Mechanics and Foundation Engineering, Kyoto 2: 87-101.

Tagami H, Ochiai H, Yasuda S, Maeda Y, Yasufuku N, Migita T, Shirai Y (2010 Importance of subsurface investigation with consideration on local characteristics. Proc. 1st Conference on Georisk Management, Georisk Society, Tokyo, 29-34 (in Japanese).

Tomita T, Imamura M, Osada J (1987) Underground dam plan in Miyako Jima (part 1). Journal of Ground Water Technology, 29(8): 10-21 (in Japanese).

Towhata I (2007) On failure of municipal waste landfill. Progress in landslide science, Springer, Berlin, Heidelberg: 147-149.

Towhata I (2021) Liquefaction mitigation measures: a historical review, Keynote lecture. Proc. 7th Int. Conf. on Recent Advances in Geotechnical Earthquake Engineering (7ICRAGEE), Bangalore, Latest Developments in Geotechnical Earthquake Engineering and Soil Dynamics, publ. Springer, 41-86.

Towhata I, Uchimura T, Seko I, Wang L. (2015) Monitoring of unstable slopes by MEMS tilting sensors and its application to early warning. International Symposium on Geohazards and Geomechanics, Warwick, IOP Conference Series: Earth and Environmental Science.

Towhata, I., Watanabe, T., Sumi, T., Sawada, S. and Azuno, K. (2021a) Promotion of ground investigation for avoidance of geo-risk and better construction management. 6th International Conference on Geotechnical and Geophysical Site Characterization, Budapest, Paper 382.

Towhata I, Goto S, Goto S, Akima T, Tanaka J, Uchimura T, Wang G, Yamaguchi H, Aoyama S (2021b) Mechanism and future risk of slope instability induced by extreme rain-fall event in Izu Oshima Island, Japan. Natural Disasters 105(1): 501-530.

Towhata, I., Azuno, K., Imanishi, H., Sawada, S., and Nojima, T. (2022) Towards improved image of geotechnical profession. 20th International Conference on Soil Mechanics and Geotechnical Engineering, Sydney.

Uchimura T, Towhata I, Wang L, Nishie S, Yamaguchi H, Seko S, Qiao J-P (2015) Precaution and early warning of surface failure of slopes by using tilt sensors. Soils and Foundations, 55(5):1086-1099.

Van Oosten R (2016) The Dutch Great Stink: The end of the cesspit era in the pre-industrial towns of Leiden and Haarlem. European Journal of Archaeology 19(4): 704-727.

Yin YP (2000) Characteristics of Bomi–Yigong huge high speed landslide in Tibet and the research on disaster prevention. Hydrogeology and Engineering Geology 27(4): 8 – 11 (in Chinese).

Zhou JW, Cui P, Hao MH (2016) Comprehensive analyses of the initiation and entrainment processes of the 2000 Yigong catastrophic landslide in Tibet, China. Landslides 13(1): 39-54.

Appendix 13: 2022 Report of the International Journal of Geoengineering Case Histories

Dimitrios Zekkos

Editor-in-Chief & Associate Professor, Department of Civil and Environmental Engineering, University of California at Berkeley, zekkos@berkeley.edu

Jean-Louis Briaud

Co-Editor in Chief & Professor, Department of Civil and Environmental Engineering, Texas A&M University, briaud@tamu.edu

SUMMARY: The International Journal of Geoengineering Case Histories is an ISSMGE journal and the only diamond open access journal in the geotechnical engineering field. The journal focuses on geotechnical practice though the careful documentation of case histories. A review of the activities undertaken in the past four years, along with statistics for the year 2021 are provided in this report. In 2021, the 124 papers published in the journal were downloaded 84,000+ times. Efforts are underway to better integrate the journal with the International Society. To facilitate this integration, the role of Technical Committee Liaisons was established and Special Issues organized by Technical Committees were published.

1 INTRODUCTION

The International Journal of Geoengineering Case Histories is an official journal of the International Society for Soil Mechanics and Geotechnical Engineering (ISSMGE) that covers the broad area of practice in geotechnical engineering (soils and rocks).
It is a peer-review, diamond-level open access journal on the geotechnical engineering practice with a focus on careful documentation of case histories and emphasis on observations and data collected during and after project construction. Papers are reviewed by a distinguished international Editorial Board and additional reviewers.

All paper published in the journal are accompanied by electronic data for better documentation of each case history. The journal is the only diamond open-access journal in the geotechnical engineering field. Diamond open-access is the highest level of open-access where no publication costs are allocated to authors and paper accessibility is also provided at no-cost to the readers.

Each paper published in the journal is also accompanied by datafiles that include all the data presented in the figures of the paper. This is intended to increase the value of the published papers to the scientific community. In addition, each case history is positioned in an interactive map that allows visitors to find papers based on geographic terms. A view of the map showing the location of the case histories published in the journal to date is shown in Figure 1

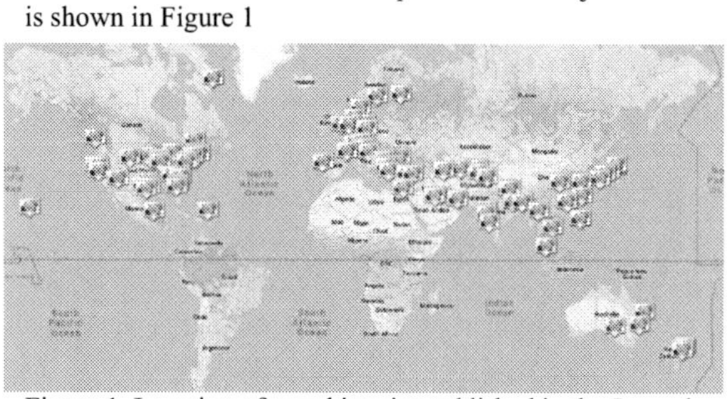

Figure 1. Location of case histories published in the Journal.

2 RELEVANT STATISTICS

The journal is available at: https://geocasehistoriesjournal.org/ To date, the journal has published a total of 124 papers. During 2021 these papers were downloaded 84,799 times, which is equivalent to an average download rate of each paper 684 times in one year. The most downloaded paper was downloaded 2719 times.

3 EDITORIAL BOARD TEAM

For the 2017-2022 period, the Editorial Board consistent of an international review board as follows:

Editor-in-Chief: Dimitrios Zekkos (USA & Greece) – Associate Professor, Civil and Environmental Engineering, University of California at Berkeley

Co-Editor-in-Chief: Jean-Louis Briaud, Professor, Texas A&M University, USA

Honorary Editor: Shamsher Prakash, Emeritus Professor, University of Missouri-Rolla, USA

Ombudsman: Michele Jamiolkowski, Emeritus Professor, Technical University of Torino, Italy

Editorial Board Members:
G L Sivakumar Babu
Professor
Indian Institute of Science, India
Also TC302 Liaison

Rudolph Bonaparte
President & CEO
Geosyntec Consultants, USA

Antonio Bobet
Professor
Purdue University, USA

Jonathan D. Bray
Professor
Univ. of California at Berkeley, USA

Ethan Cargill
Manager SE USA and Technical Director
ConeTec , USA

Peter Day
Geotechnical Consultant
Jones & Wagener, South Africa
Adjunct Professor
University of Stellenbosch, South Africa

Roger Estephan
Director
Dar Al-Handasah (Shair & Partners)

Russell Green
Professor
Virginia Tech, USA

Debra Laefer
Associate Professor
University College Dublin Ireland

Robert Mair
Professor
Cambridge University, U.K.

Priscilla Nelson
Department Head and Professor
Dept. of Mining Engineering, Colorado School of Mines, USA

Sissy Nikolaou
NIST, USA

Achilleas G. Papadimitriou
Associate Professor
National Technical University of Athens, Greece

Kok Kwang Phoon
Professor
National University of Singapore, Singapore

Salah Sadek
Professor
American University of Beirut, Lebanon

4 LIAISONS WITH TECHNICAL COMMITTEES

During the last four years, in an effort to better integrate the journal with ISSMGE, the journal has established liaisons with technical committees of the ISSMGE. The responsibilities of the Liaisons are:

- To review up to 5 papers per year;
- To publish one (or more) special issue under the auspices of the TC in the next four years;

In addition, a new role of a "Special Issues Coordinator" was established to work with TCs in coordinating the preparation of the special issues. The appointed Special Issues Coordinator is:
Shehab Agaiby
Cairo University and Dar Al Handasah

Current liaisons consist of:
Malek Bouazza
Professor
Monash University
TC215 Liaison

Ivan P. Damians
Dr
Universitat Politècnica de Catalunya (UPC-BarcelonaTech)

Noël Huybrechts
Professor
Belgian Building Research Institute
TC211 Liaison

Sang Seom Jeong
Professor
Department of Civil & Environmental Engineering, Yonsei University
TC212 Liaison

Apiniti Jotisankasa
Dr
Kasetsart University
TC106 Liaison

Andy Y.F. Leung
Dr
Hong Kong Polytechnic University
TC304 Liaison

Akira Murakami
Professor
Kyoto University
TC103 Liaison

Akira Murakami
Professor
Kyoto University
TC103 Liaison

Miguel A. Pando
Professor
Drexel University
TC210 Liaison

Dimitris Pitilakis
Professor
Aristotle University of Thessaloniki
TC203 Liaison

Helmut Schweiger
Professor
Graz University of Technology
TC103 Liaison

Erol Tutumluer
Professor
University of Illinois at Urbana-Champaign
TC202 Liaison

Shinji Sassa
Head of Soil Dynamics Group and Research Director
National Institute of Maritime, Port and Aviation Technology Japan
TC213 Liaison

4. PUBLICATION OF SPECIAL ISSUES

In 2021 the following special issues were published:

- TC103 on Numerical Methods in Geomechanics by Guest Editors Giovanna Biscontin, Ryosuke Uzuoka, Akira Murakami

- TC304 on Engineering Practice of Risk Assessment and Management by Guest Editor Andy Yat Fai Leung

4.1. JOURNAL FUNDING

As mentioned previously, the Case Histories journal is the only diamond open-access journal in the geotechnical field. Diamond open-access is the highest level of open-access where no publication costs are allocated to authors, and paper accessibility is also provided at no-cost for the readers. Funding for the journal is generated by a small group of forward-looking companies and organizations that support the Mission of the journal. In 2021, the funding for the journal was provided by the generous contribution of: Geosyntec Consultants, ENGEO, DAR Group (Al Handasah), and Conetec. Companies and organizations supporting the journal, are acknowledged in the back-page of all papers published by the Journal, as well as its website.

Appendix 14: ISSMGE Foundation

Michael C R Davies
Chair of ISSMGE Foundation Trustees

The attached document is a report on the activities of the ISSMGE Foundation that has been prepared for donors to the Foundation. As an independent charity, governance of the ISSMGE Foundation is conducted by a Board of Trustees and it is not overseen by Council. However, since the ISSMGE is a major donor to the Foundation and the beneficiaries are members of the ISSMGE Member Societies, it is highly appropriate that this report is presented to the ISSMGE Council for its information.

The ISSMGE Foundation was created to aid individual ISSMGE members throughout the world to enhance their geotechnical engineering knowledge and practice by providing financial support for participation in technical and professional activities. The report indicates that, since it was established, applications for an award have been received from individual members from all the ISSMGE regions and 43 of the 89 Member Societies. Whilst the global spread of awards is very good and increasing, the Trustees welcome applications for financial support for appropriate activities from members of all national geotechnical societies and request members of Council to help publicise the ISSMGE Foundation within their National Society.

As indicated in the report, as a consequence of the COVID-19 pandemic, fewer requests for funding were received over the last twenty months. However, the Trustees were very pleased to be able to make awards to applicants who requested funding to enable them to participate in online conferences and symposia. Recipients of these awards all indicated that they had gained significant educational and professional benefits from attending these events. A number of extracts from their reports are included in the attached document.

In addition to considering applications to the Foundation for awards, one of the major duties of the Trustees is to ensure the sustainability of the ISSMGE Foundation so it can continue to meet its objectives. To this end, the policy of the Trustees is to build an endowment so that awards can be made from the income that it earns. To achieve the situation where the Foundation does not have to rely on charitable donations to make awards but only on the income from the endowment, the Trustees believe that the endowment needs to be of the order of £1M. This figure has been set at as this target for fundraising for the ISSMGE Foundation. In 2019 the Trustees appointed Tilney Ltd (UK) to help establish an endowment fund and, as the attached accounts indicate, this now has a value of £574k. The Trustees are very grateful to all donors who have contributed to the Foundation since it was established by Professor Jean-Louis Briaud during his presidency of the ISSMGE. We are now making a new appeal to all members of the ISSMGE family to contribute to building an endowment that will secure the future of the Foundation. We seek the support of members of Council to help us achieve this.

The Trustees are also most grateful to the ISSMGE for its regular donations to the Foundation that have allowed the funds donated by corporate and private donors to grow to the current level whilst still allowing the Foundation to maintain a success rate of 89% for applications for awards.

The Trustees commend this report to Council.

REPORT TO DONORS

The ISSMGE Foundation

A charitable incorporated organisation (CIO) associated with the International Society for Soil Mechanics and Geotechnical Engineering

Charity Commission of England and Wales - No. 1153795

BACKGROUND

The concept of the ISSMGE Foundation dates back to September 2009, when Professor Jean-Louis Briaud, the newly-elected President of the International Society for Soil Mechanics and Geotechnical Engineering (ISSMGE) in his acceptance speech to the ISSMGE Council Meeting (Alexandria, Egypt), introduced his vision of a foundation that would enable geotechnical engineers from around the world to make better use of resources that may otherwise be inaccessible to them. These resources would include participation in conferences, attending knowledge development courses, etc., with a view to enhancing the awardees own education and professional formation and, as a result, their contribution to society at large. In short, and as stated in the Foundation constitution, its objective is "the advancement of education and educational training in the field of geotechnical engineering and its applications for the public benefit".

In September 2013, The ISSMGE Foundation achieved charitable status and was entered onto the Register of Charities of the Charity Commission of England and Wales, with the Registered Charity Number 1153795.

STRUCTURE AND MANAGEMENT

The ISSMGE Foundation currently has four trustees: Professor

Michael C.R. Davies (Chair), Professor R. Neil Taylor, Professor Mark Jaksa and Mme Valérie Bernhardt, all of whom contribute to decisions concerning the granting of awards. In the future and as the Foundation develops, the Trustees may find it necessary to name additional trustees. The new trustees may be recruited from the ISSMGE Membership, from its own (past) beneficiaries, or from other independent institutions, depending on the needs arising and skills and experience required to consolidate the activities of the Foundation. Professor Jean-Louis Briaud resigned as trustee in June 2019, at which point Mme Bernhardt was appointed.

Professor Taylor is also in charge of the day-to-day management; the ISSMGE Foundation does not have salaried staff members. Administrative support is provided by the ISSMGE.

ACTIVITIES

The purpose of the Foundation is to award grants to applicants who fulfil the requirements as specified below.

PROCEDURE FOR AWARDING GRANTS

The financial assistance is in the form of a bursary which the successful applicant can use to pay registration costs, travel and/or accommodation expenses. Although membership of the ISSMGE is not a prerequisite for receiving support from the Foundation, applicants are strongly encouraged to become members of their national geotechnical societies, and subsequently of the ISSMGE.

The following procedure has been developed to assess applications for ISSMGE Foundation bursaries:

a. Applications can be submitted four times yearly (at 3-month intervals) on a standard form, to the Secretary General of ISSMGE;
b. These applications are sent to the trustees (Professor Michael Davies, Professor Neil Taylor, Professor Mark Jaksa and Mme. Valérie Bernhardt);
c. The trustees assess the applications of the basis of the following criteria:

  i. Demonstration of need;
  ii. Quality of the activity;
  iii. Cost-sharing;
  iv. Potential impact of the activity on the career progress of the individual and of his or her Member Society.

d. Decisions are made and communicated normally within three weeks following the application round.
e. After the successful applicant has completed his/her nominated activity, they provide a written report to the Trustees on the benefits that they and their Society have gained from the award of the bursary.

The criteria and procedure described above were established in February 2014, though may be modified when considered necessary.

To date (Sept 2021), there have been a total of 219 applicants with 176 awards being approved at a success rate of 89%. Of these, 176 bursaries (in total £152,063) have been paid with a further 8 due when awardees have completed their activities. Some of this funding took place in the early days through the ISSMGE, prior to the Foundation achieving charitable status.

Not surprisingly and as a result of the pandemic, fewer requests for funding were received over the last twenty months. As activities are resumed both locally and internationally, it is expected that the number of applicants will increase. It is possible that the nature of the applications made and awards granted will evolve as more hybrid events (with possible online attendance) are rolled out.

The reports submitted by the awardees are available to read on the ISSMGE website: https://www.issmge.org/issmge-foundation/recipients-reports.

FINANCES AND FUND RAISING

Funding for the ISSMGE Foundation is generated from companies and individuals who concur with the objectives of the Foundation. The funding levels are as follows:

- Diamond: $50,000 and above
- Platinum: $25,000 to $49,999
- Gold: $10,000 to $24,999
- Silver: $1000 to $9,999
- Bronze: $0 to $999

The names of the donors are recognised with their corresponding levels on the ISSMGE web site (https://www.issmge.org/issmge-foundation/donors-list) and in the ISSMGE Bulletin on the page dedicated to the ISSMGE Foundation. They are also be recognised at International Conferences on Soil Mechanics and Geotechnical Engineering.

ISSMGE Foundation cash reserves at the end of 2020 were £576,273 (at the end of 2019 they were £554,771), as per audited accounts, extracts attached. Currently, the Board is considering awards to the value of approximately US $2000 (c. £1,630).

The ISSMGE Foundation trustees appointed Tilney Ltd (UK) in June 2019 to help establish an endowment fund which would enable the Foundation to make awards from the income earned, and not rely on charitable donations. The initial investment of £500,000 was valued at £574,584 on 31st December 2021.

IMPACT OF ISSMGE FOUNDATION ACTIVITIES

It is clear that the financial assistance provided by the ISSMGE Foundation has had a very positive impact on the individuals receiving the bursaries and consequently on their Member Societies.

As awareness of the awards becomes more widespread among our Member Societies, the number of applications has increased. We expect the number of applications to rise with the approach of the next International Conference, and the associated Young Engineers' Geotechnical Conference (both in Sydney, in May 2022).

Historically, the number applications received per year was increasing, but the pandemic has halted that trend. As mentioned earlier, we would expect the number of applications to eventually rise to their pre-pandemic level, but it is likely that different types of events will be organised (online, hybrid, in-person) and thus a greater range of values of the awards (as travel and accommodation may no longer be factors in some requests) will be considered by the trustees.

The ISSMGE Foundation is fulfilling its objective of providing financial help to geotechnical engineers throughout the world who wish to further their knowledge and practice. We are already seeing individuals who have benefitted from Foundation awards becoming valuable contributors to the future of ISSMGE via their national member societies, and through ISSMGE technical committees and presidential groups.

From our awardees:

The reports submitted by the awardees are available to read on the ISSMGE website: https://www.issmge.org/issmge-foundation/recipients-reports. The following are extracts from some of the reports.

"The time schedule was well defined and allow people from all over the world to attended it. The key lectures were very interesting and it was great to attend to many relevant researchers' lectures. The videos recorded for the themed sessions were available during the conference and two weeks after ..and that gave me the opportunity to watch again some of them and those that it wasn't possible to watch during the event because of schedule conflicts."

Malena D'Elia Otero

XIII International Symposium on Landslides (COLOMBIA), 2021

"Everything was preplanned. The organizers asked for the presentation recording so if the participants have any issue regarding network connectivity, they can simply play their presentation. There were also preview rooms for each and every parallel sessions, in which participants have to show their entire presentation so any issues can be avoided".

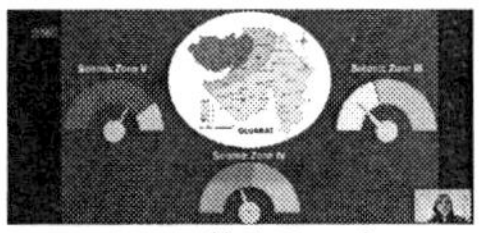

Manali S. Patel

CREST2020, 1st International Symposium on Construction Resources for Environmentally Sustainable Technologies Location of (JAPAN), 2021

"Although due to COVID Pandemic it was not able to an in person interaction but even in virtual mode I had able to interact with academicians and practicing engineers from various countries and leading researchers and stalwarts in the field of engineering and gain an insight into their on-going research activities."

Chaidul Haque Chaudhuri

17th World Conference on Earthquake Engineering (17WCEE) (JAPAN), 2021

"This was the first conference I have attended (although virtually), I was very delighted to learn about so many wide topics from researchers all over the world. I have read papers by the pioneers for my research, being able to listen to them live about their research interests was a unique experience for me."

Sanchari Mondal

6th International Conference on Geotechnical and Geophysical Site Characterisation (HUNGARY), 2021

"...the conference provided me with an ideal opportunity to interact with many young researchers and doctoral students which resulted in the exchange of ideas and feedback on our doctoral research work."

Mohammed Ashfaq (India) with Prof. Anand Puppala

GeoCongress 2020 (Minneapolis, USA)

"The knowledge gained in the conference and from the papers presented shall in no small measure guide me in my PhD research work. I am going to share the information gathered from the conference with other students for the furtherance of scientific research and development in the University of Port Harcourt."

Will Ibm Beresibo (Nigeria)

17th African Regional Conference on Soil Mechanics and Geotechnical Engineering, 2019 (Cape Town, South Africa)

"... this conference enriched me with guidance on improving my research skills ... (an) opportunity to meet and network with colleagues, senior professors, practising and professional geotechnical engineers, which may open the door for me for future collaboration and career development."

Imad Alainachi (Canada)

XVI Panamerican Conference on Soil Mechanics and Geotechnical Engineering 2019 (Cancun, Mexico)

"In the conference, I have made many friends with researchers from different countries which will help me significantly in the near future in exchange of valuable ideas and thereby leading to possible research collaborations."

Kaustav Chatterjee (India)

XVI Asian Regional Conference on Soil Mechanics and Geotechnical Engineering 2019 (Taipei, Taiwan)

"I was grateful for the career advice from senior professors and practising engineers."

Santhoshkumar Gunasekaran (India) on the left with Prof. Ikuo Towhata (far right)

XVI Asian Regional Conference on Soil Mechanics and Geotechnical Engineering 2019 (Taipei, Taiwan)

"... the conference allowed us to continue learning from recognized professionals worldwide and interact with them in order to share knowledge to put into practice in our daily work. The most important of these types of events is that it promotes the continuous education of all professionals in our area."

Jonathan Echevarria Vargas (Costa Rica)

ECSMGE 2019, "Geotechnical Engineering, foundation of the future" (Reykjavik, Iceland)

"I got an excellent opportunity to meet and talk to many professors, researchers and engineers from different countries and many other potential research collaborators as well."

Anna Kupreichyk (Ukraine)

27th European Young Geotechnical Engineers Conference 2019 (Bodrum, Turkey)

"... this conference gave me immense opportunity to network with my peers and receive comments on my research work. ... I also received opportunity to network with professors working in research areas that bear similarity to my areas of interest and other geotechnical fields. Also, ... I got an opportunity to learn about research being carried out in different universities that can also help me improve my work and get an exposure of developments in other fields of earthquake geotechnical engineering."

Kavita Tandon (India),

7th Internat Conf. Earthquake Geotechnical Engineering, 2019 (Rome, Italy)

"Now I know more people and I have invitations from other conferences, which is great opportunity for my future research."

Diana Egiazarova (Georgia)

China – Europe Conference on Geotechnical Engineering, 2018 (Vienna, Austria)

"CPT'18 allowed me to meet, network and learn from the world's expert professional academics, researchers, consultants, practitioners etc. I also believe that participating this conference broadened my horizons in the field of geotechnical engineering by sharing and following the contemporary/state of art research and practices."

Cagdas Bilici (USA)
4th Internat Symposium on Cone Penetration Testing, 2018 (Delft, Netherlands)

Number and value of awards (GBP k) per country

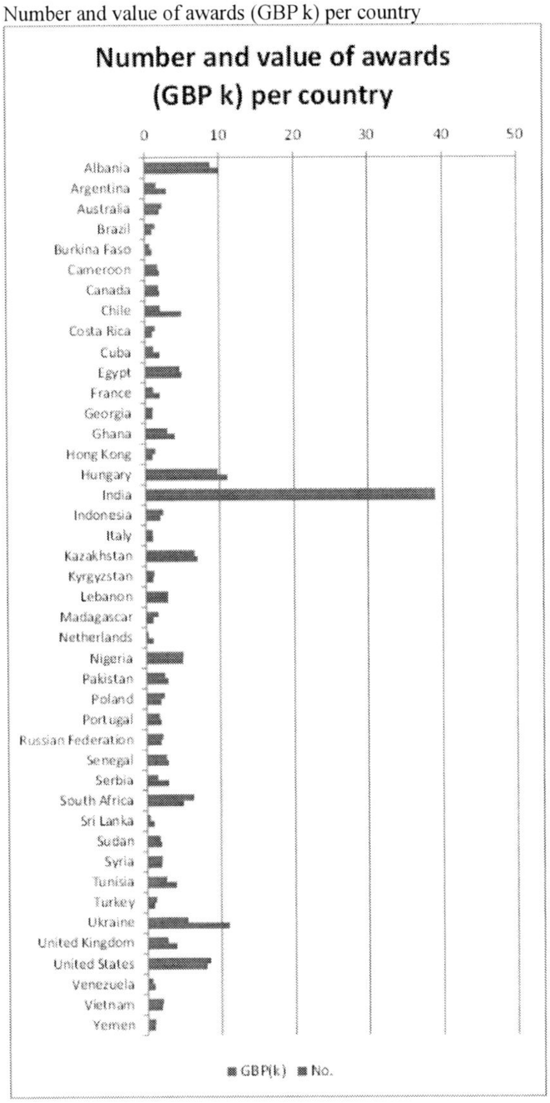

Proportional distribution of Value of Awards

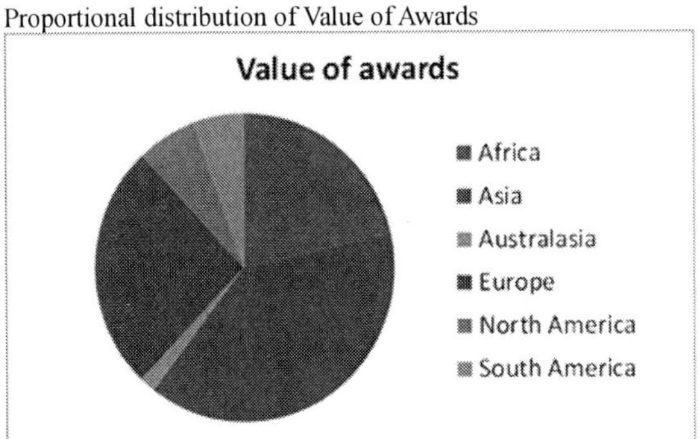

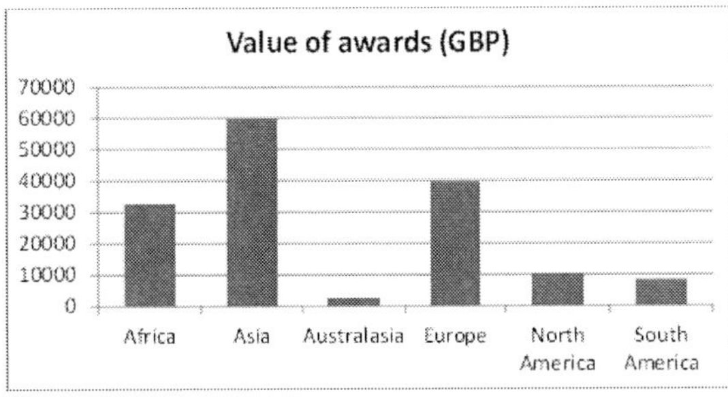

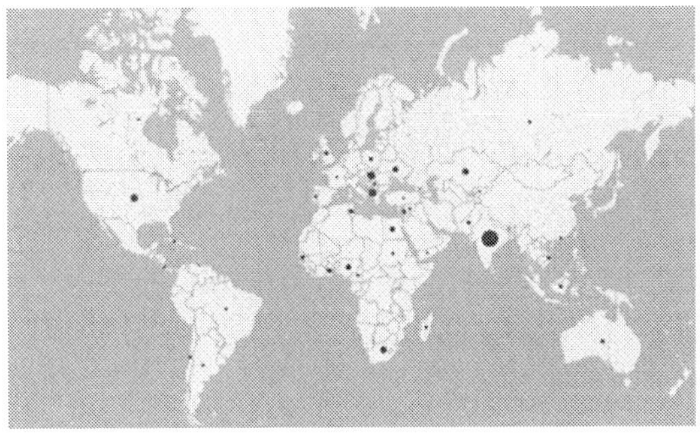

Proportional Distribution of Number of Awards

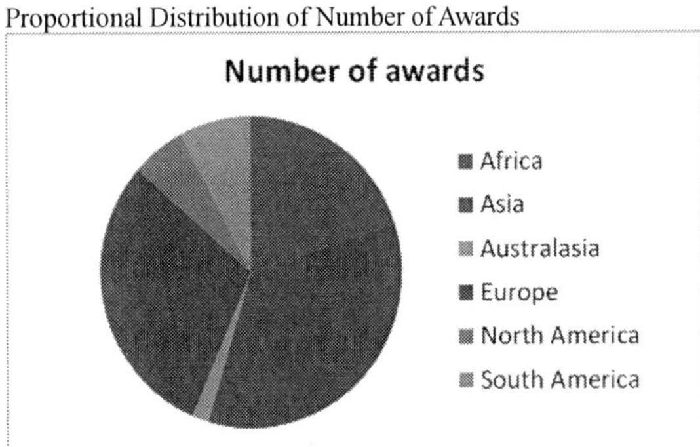

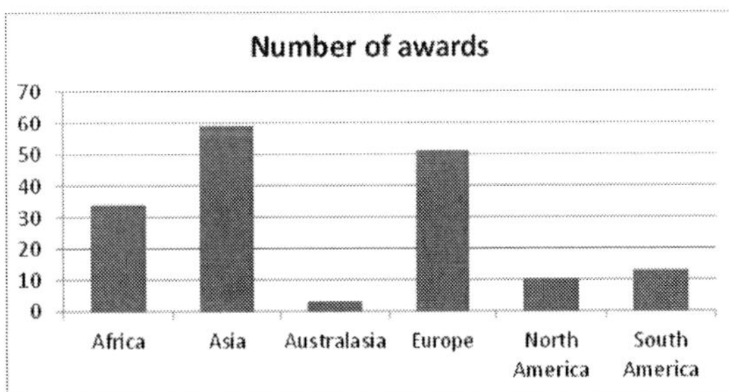

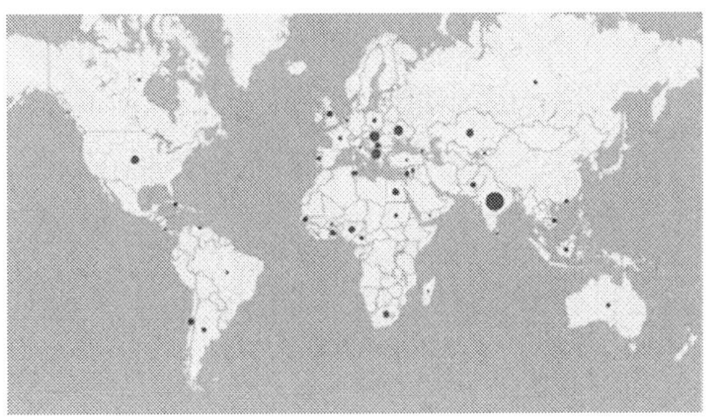

ISSMGE Foundation Accounts for the year ended 31 December 2019 - (As submitted to the Charity Commission)

INCOME	**Endowment Fund** £	**Unrestricted Funds** £	**Total Funds** £
Donations	-	41,000	41,000
Income from investment portfolio	2,200	-	2,200
Bank Interest		576	576
Total incoming resources	**2,200**	**41,576**	**43,776**
EXPENDITURE			
Costs of raising funds:			
Investment management fees	1,584	-	1,584
Expenditure on Charitable activities:			
Awards made	-	31,289	31,289
Bank Charges	-	664	664
Total expenditure	**1,584**	**31,953**	**33,537**
NET INCOME BEFORE RECOGNISED GAINS	**616**	**9,623**	**10,237**
Net gain/(losses) on investments	6,868	-	6,868
Transfer between funds	500,000	(500,000)	-
NET MOVEMENT IN FUNDS FOR THE YEAR	**507,486**	**(490,377)**	**17,107**
TOTAL FUNDS BROUGHT FORWARD	-	**537,664**	**537,664**
TOTAL FUNDS CARRIED FORWARD	**507,486**	**47,287**	**554,771**

ISSMGE Foundation Accounts for the year ended 31 December 2020 - (As submitted to the Charity Commission)

INCOME	**Endowment Funds** £	**Unrestricted Funds** £	**Total Funds** £
Donations	-	-	-
Income from investment portfolio	6,848	-	6,848
Bank Interest		-	-
Total incoming resources	**6,848**	-	**6,848**
EXPENDITURE			
Costs of raising funds:			
Investment management fees	4,597	-	4,597
Expenditure on Charitable activities:			
Awards made	-	4,384	4,484
Bank Charges	-	68	68
Total expenditure	**4,597**	**4,452**	**9,049**
NET INCOME BEFORE RECOGNISED GAINS	**2,251**	**(4,452)**	**(2,201)**
Net gain/(losses) on investments	23,703	-	23,703
Transfer between funds	-	-	-
NET MOVEMENT IN FUNDS FOR THE YEAR	**25,954**	**(4,452)**	**21,502**
TOTAL FUNDS BROUGHT FORWARD	**507,484**	**47,287**	**554,771**
TOTAL FUNDS CARRIED FORWARD	**533,438**	**42,835**	**576.273**

ISSMGE FOUNDATION DRAFT ACCOUNTS
For the year ended 31 December 2021

INCOME	Endowment Fund £	Unrestricted Funds £	Total Funds £
Donations		40,000	40,000
Income from investment portfolio	5,830		5,830
Bank interest			
Total incoming resources	**5,830**	**40,000**	**45,830**
EXPENDITURE			
Costs of raising funds			
Investment management fees	4,945		4,945
Expenditure on charitable activities			
Awards made		2,905	2,905
Bank charges		93	93
Total expenditure	**4,945**	**2,998**	**7,943**
NET INCOME BEFORE RECOGNISED GAINS	**885**	**37,002**	**37,887**
Net gain/(losses) on investments	40,273		40,261
Transfer between funds			
NET MOVEMENT IN FUNDS FOR THE YEAR	**41,158**	**37,002**	**78,160**
TOTAL FUNDS BROUGHT FORWARD	**533,426**	**42,835**	**576,273**
TOTAL FUNDS CARRIED FORWARD	**574,584**	**79,837**	**654,421**

Appendix 15: ISSMGE Time Capsule Project

Sukumar Pathmanandavel
Vice-chair CAPG

1 INTRODUCTION.

The Time Capsule Project (TCP) has the following aims:

a. To bring through some of the collective experience of the member societies and technical committees in an accessible way for the broader geotechnical community, and enable positive and uplifting conversations across the individual membership, in a challenging time for all of us
b. To assist the current and future generation of geotechnical engineers with a road map that helps them access the past, present and future in geotechnical engineering, globally, and stir their collective curiosity
c. To provide a platform for the up-and-coming cohort of geotechnical engineers to stand for, and to tell the story of their member society/ technical committee, and in this way take on the leadership of addressing the challenges of the communities of the future.

2 THE PROCESS

A key focus of the TCP is the "process" by which there is connection and interaction with the individual members about the work of the contributors, an outcome of which is the "product" that is offered to the TCP for storage and display.

Through the respective vice presidents, 71 member societies have confirmed that they will be submitting, with the following commitment percentages - Africa 50%; Asia 70%; Australasia 100%; Europe 80%; North America 100%; and South America 100%. Corporate Associates, Technical Committees, Young Member Presidential Group (YMPG), and past presidents of the ISSMGE and others are active in creating contributions.

It is ISSMGE's expectation that

- submissions offered for the time capsule have secured all appropriate authorisations within the submitting organisation, and are first displayed and promoted on the originators' websites, marked as submissions to the ISSMGE Time Capsule 2022
- secure and authorised links to the relevant web pages of the originator are to be notified to the ISSMGE IT department
- a backup electronic copy of each submission needs to be provided to the ISSMGE IT department for the purpose of archival with the ISSMGE.

The TCP Design Team is not involved in reviewing or authorising any submission for the TCP.

TCP material post conference (THE PRODUCT) will comprise those listed in the following three parts.

3 PART A - CONVERSATIONS BUILDING ON SUBMITTED MATERIAL

- A simple, contents listing, (with links to web pages), subdivided into contributions by past presidents, member societies, technical committees, YMPG, Corporate Associates and material offered by the Innovations and Development Committee (IDC). (WORD DOCUMENT)
- A trailer of the product, still rough, short (if possible) presentations/trailers of all the material. A trailer of the information. Something that is available long term. Trailer to point people to more interesting items. (PRE RECORDED VIDEO)

At the 20th ICSMGE (Sydney, 1- 5 May 2022)

- Discussions with Custodians - The ability to talk to someone about something in the time capsule. Custodians of items need to stand for and explain the role of the item. (LIVE RECORDING BY CONFERENCE ORGANISERS)
- Treasure hunt - Theme - on common lines/ divergences across the board; Quizzes and games - require recollection & finding certain pieces of information. (LIVE RECORDING BY CONFERENCE ORGANISERS)

4 PART B - ROAD MAP TO STIR CURIOSITY

- Past presidents' material - contributions of the 16 past presidents of the ISSMGE. (PRE RECORDED VIDEO/ PODCASTS/ WORD DOCUMENTS)
- Vice presidents (2017 - 2022) presentations on their region's process and product (PRE RECORDED VIDEO/ PODCASTS/ WORD DOCUMENTS)
- Future of Geotechnical Engineering, extracts from YMPG online 2021 event (PRE RECORDED VIDEO)
- "Resources and Activities of the ISSMGE and how to get involved", Webinar Series, 2022, proposed by IDC (PRE RECORDED WEBINARS)

5 PART C - PLATFORM FOR EMERGING AND INNOVATIVE LEADERS

- Discoverers - A team of 30 discoverers; Abstracters; To go through the material and highlight things they have found; Explaining how things link together; Need a cast of actors.
- Interactive displays - To help make sense of intense amount of information; Similar to interactive exhibits at a museum; Takes into account many people will not be interested with specifics.

6 ACKNOWLEDGEMENTS

Charles Ng, the President, ISSMGE Board and Board Level Chairs (BLCs) have played a leading part in the shaping and the execution of the project. Pedro Pinto and Roger Frank, both past presidents and Board members, have led the creation and collation of contributions of all past presidents of the ISSMGE.

A large number of individual members of the ISSMGE have worked on the contributions of the member societies and technical committees. Members of the YMPG Liaisons have been engaged in a number of activities related to the time capsule. The ISSMGE IT team has provided timely support, and will have a significant role to play in the launch of the time capsule.

In 2020, the Time Capsule Committee comprised the following: Sukumar Pathmanandavel (lead), Ikuo Towhata, Osamu Kusakabe, Ceres Chung, Philip Robins, Harry Poulos, Juan Ayala, Marc Ballouz, Menzer Pehlivan, Alejo Sfriso, Roberto Terzariol, Alessandro Flora, Fabio Tradigo, Laura Kerner, François Depardon, and Charles MacRobert. This committee played an important role in developing a common understanding of the concept of ISSMGE time capsule.

The TCP design team comprises Ceres Chung, Charles MacRobert, Emilio Bilotta and Sukumar Pathmanandavel (lead). In 2021, the TCP Design Team held an extensive number of online meetings to communicate the requirements of the time capsule to a variety of ISSMGE audiences. Now, the TCP Design Team is working on bringing together the product for launch at the Sydney 2022 conference.

The Time Capsule Project is still underway and updates can be accessed via the Time Capsule Project page on the ISSMGE website.

End of report, 2 February 2022

Appendix 16: Report on the ISSMGE Bulletin (May 2019 – Feb 2022)

Anthony Leung

EiC of the ISSMGE Bulletin Department of Civil and Environmental Engineering, Hong Kong University of Science and Technology, Hong Kong Special Administrative Region Department

Dr Anthony Leung (Hong Kong University of Science and Technology, Hong Kong SAR) was appointed as the Editor-in-Chief (EiC) of the ISSMGE Bulletin since Oct 2017, after he served the Editorial Board (Europe) for more than 3 years.

The bulletin is published bimonthly. During the report period, a total of 17 issues have been published. Each issue of the bulletin is open-accessed and can be downloaded freely from the following website: http://www.issmge.org/publications/issmge-bulletin. The publication of each issue was promoted via different paths of social media to maximise the publicity, The 17 issues have reported "Research Highlights" for the ISSMGE members to get to know more about the research activities from reputed geotechnical engineering research groups around the world, including University of Tunis El Manar (Tunisia; Vol(13), issue 3), Institute of Geotechnical Engineering, ETH (Switzerland, Vol(13), issue 5), Kyoto University (Japan, Vol(14), issue 1), The University of New South Wales (UNSW, Australia, Vol(14), issue 2), L. N. Gumolyov Eurasian National University (Kazakhstan, Vol(14), issue 4) and Saint Petersburg State University of Architecture and Civil Engineering (Russia, Vol(15), issue 1).

Thanks to the close liaison of the editorial board members with conference organisers, the Bulletin serves as an effective platform for sharing highlights of some major international conferences in "Conference reports". The content includes conference programme, keynote/invited speakers, conference tour and social events.

The Bulletin also has a section called "Young Member's Arena", for allowing young members to share their latest research activities such as the one-day technical event entitled "On the borderline between geotechnical and earthquake engineering" reported by a young member of the French Society for Soil Mechanics and Geotechnical Engineering (CFMS) at Vol(13), issue 4).

Activities organised by TCs of ISSMGE are reported under the section "TC Corner" in the bulletin. These include a new International Conference series on Embankment Dams organised by TC210 (Vol(13), issue 4), a joint TC304/309/210 workshop on "Machine Learning Dialogue for Geotechnics" (Vol(14), issue 1), an inauguration of the TC107 Symposium (Vol(14), issue 3), the first international administrative meeting of the TC305 (Vol(14), issue 3), the 5th ISSMGE Suzanne Lacasse Lecture organised by TC304 (Vol(14), issue 5), an International Online Symposium on Historical and modern applications in geotechnical engineering (Vol(14), issue 5), a report on an educational video "What happens when soil compresses" organised by TC306 (Vol(15), issue 1), a state-of-the-art review of inherent variability and uncertainty in geotechnical properties and models by TC304 (Vol(15), issue 2) and last but not least a joint TC304/309/210 workshop on "3 rd Machine Learning Dialogue" (Vol(16), issue 1).

The Bulletin also publishes "Board-Level Committee Activity". These include two reports written by the Corporative Associates Presidential Group (CAPG), one about an event held at the 17th ECSMGE, on an interesting session "'Bridging the gap between designers and constructors: how do we ensure effective 'transfer' of design into construction?" (Vol(13), issue 5) and the other one on an event held at the 16th ARC called "Are we overdesigning?" (Vol(14), issue 1). A regional session report was also presented by the CAPG in Vol(14), issue 2 to summarise the observations on the event called 'Geo-education for the Future' held at the XVI Pan-American Conference on Soil Mechanics and Geotechnical Engineering.

The Bulletin also received five "Reports from Member Society", including those from Hong Kong Geotechnical Society (Vol(13), issue 3 and Vol(15), issue 3), the Geotechnical Society of Bosnia and Herzegovina (Vol(13), issue 5), British Geotechnical Association (BGA; Vol(14), issue 4), Chinese Taipei (Taiwan) Geotechnical Society (Vol(15), issue 1) and Polish and Kazakhstan Geotechnical Society (Vol(15), issue 4).

With the new initiative of ISSMGE Time Capsule Project (TCP), the Bulletin offers a new platform for sharing articles of TCP Blog posts by five authors, Dr Charles MacRobert (Vol(15), issue 4), Dr Emilio Bilotta (Vol(15), issue 5), Ms Ceres Chung (Vol(15) issue 6), Dr Roger Estephan and Dr Grace Abou-Jaoude Estephan (Vol(16) issue 1) and Dr Marina Pantazidou (Vol(16) issue 2).

The Bulletin has a section called "Hot News", which allows members to share major awards and achievements such as the 2021 Terzaghi's Orator (Vol(14) issue 5), ISSMGE Spark Award (a new series created by the ISSMGE President for enabling young members to take the main stage at conference), call-for-papers, issues of new journals and books.

The Bulletin offers a platform for the awardees of ISSMGE foundation to share their experiences they have gained and learnt through the attendance of international conferences and workshops in a section called "ISSMGE Foundation report".

In every issue, an updated "Event Diary" is provided closer to the end of the bulletin to advertise all ISSMGE- and non-ISSMGE-sponsored events (mainly conferences and workshops) that are registered in the ISSMGE website. Updated lists of "Corporate Associates" and "Foundation donors" are also given in every issue to acknowledge their financial support to the ISSMGE.

On behalf of the Editorial Board, I cordially invite you to contact your Member Society and share any interesting events/activities related to soil mechanics and geotechnical engineering in the Bulletin. I am sure our members will be interested in hearing your exciting news over the world. Feel free to get in touch via my email: ceanthony@ust.hk

Anthony Leung EiC of the ISSMGE Bulletin Department of Civil and Environmental Engineering, Hong Kong University of Science and Technology, Hong Kong Special Administrative Region

Technical Paper

Proceedings of the 20th ICSMGE – Post-conference Volume – Rahman and Jaksa (Eds)
© 2023 Australian Geomechanics Society, Sydney, Australia, ISBN 978-0-9946261-7-2

Lessons learned from the Moroccan high-speed railway embankment constructed on soft soils

Leçons Tirées du Remblai de la Ligne à Grande Vitesse Marocain Construit sur Des Sols Mous

Ahmed Hamza Mridakh
Department of Mines, Rabat National School of Mines, Rabat, Morocco, ahmedhamzam@gmail.com

Fouad Lahlou
Department of physics, Ibn Tofail University, Kenitra, Morocco

Houssine Ejjaaouani & Hassan Labied
Laboratoire Public d'Essais et d'Etudes, Casablanca, Morocco

ABSTRACT: Challenges associated with excessive settlement are typically encountered in limited areas under the Moroccan high-speed railway embankment. A variety of mechanisms are controlling these soft soils depending on their deposition environments and their inherent characteristics. Therefore, each compressible area presents a unique behavior that requires a unique approach to be fixed during the deformation analysis. This paper presents a comparative study between two compressible areas, by highlighting the priorities to be followed during the settlement prediction process when using numerical simulations. The numerical modeling prediction results for both cases, for soft alluvial and lacustrine deposits, are compared with in-situ settlement measurements during and after the embankment construction. Based on the back-analysis, the time-dependent behavior in both cases was found necessary for capturing the in-situ behavior. However, soft lacustrine deposits are particularly complex in this case. For that, a systematic combination which consist of constructing dipped layers in the numerical model, selecting a constitutive model that include creep and adding artesian pressure effect can lead to accurate predictions during the numerical analysis of lacustrine soft soils.

RÉSUMÉ : Les défis associés au tassement excessif sont généralement rencontrés dans des zones limitées sous le remblai de la ligne à grande vitesse Marocain. Divers mécanismes contrôlent les sols mous en fonction de leurs environnements de dépôt et de leurs caractéristiques inhérentes. Par conséquent, chaque zone compressible présente un comportement unique qui nécessite une approche unique à fixer lors de l'analyse de la déformation. Cet article présente une étude comparative entre deux zones compressibles, en mettant en évidence les priorités à suivre lors du processus de prédiction du tassement par simulations numériques. Les résultats des prédictions de modélisation numérique pour les deux cas, pour les dépôts alluviaux et lacustres mous, sont comparés aux mesures de tassement in-situ pendant et après la construction du remblai. D'après la rétro analyse, le comportement dépendant du temps dans les deux cas s'est avéré nécessaire pour capturer le comportement in situ. Cependant, les dépôts lacustres mous sont particulièrement complexes dans ce cas. Pour cela, une combinaison systématique qui consiste à construire des couches inclinée dans le modèle numérique, sélectionner un modèle constitutif incluant le fluage et ajouter l'effet de la pression artésienne peut conduire à des prédictions précises lors de l'analyse numérique des sols mous lacustres.

KEYWORDS: Soft soil behavior, vertical drains, embankment, settlement, numerical modelling

1 INTRODUCTION

The high speed railway project involves the construction of rail lines linking between Tanger and Kenitra north of Morocco (Phase 1, section 1 &2), then to Marrakech and Agadir (Phase 2) in the south. Phase 1 of the current project has been completed during the year 2018 which covered a 200 km segment; this was divided into 2 sections. The first section is the one linking between Tanger and larache. The second section, which makes the object of this work, is the one linking between Larache and Kenitra. Phase 2 however, is currently under the early planning stages.

At numerous locations along the high-speed railway line, soft soil deposits are encountered mainly near river beds, lakes and small streams. For that, many solutions were chosen during the railway line construction. The later, depends on the type of the problematic behavior, its boundaries, depths and costs of the construction process.

In general, locations where soft soils are predominantly present near the major rivers in the northern section "section 1" (e.g. Loukkos river, El hachef river and el Mharhar river). Those sites were improved by deep piles under the concrete bridges Foundations. This improvement method is chosen for its efficiency to encounter the imminent challenges of bearing capacity failures and excessive settlements. The soft soils studied in this work are located in section 2 of the high-speed railway line. In this section soft soil deposits are improved by a combined effect of prefabricated vertical drains (PVDs) and preloading. For cost effective outcomes, this method is applied in both Sebou and Drader study areas selected for this work. This method main objective is to reduce the drainage length for rapid consolidation time (Walker et al. 2012, Indraratna et al. 2012). PVDs are used under surcharge "or preloading" to accelerate the dissipation of excess pore water pressure that accumulates during the embankment construction phases (Barron 1948), to ensure its safety and increase the subsoils shear strength by the end of the embankment construction.

To accurately capture the soft soils in both study areas 2D finite element analysis is used to predict the excessive settlements during the embankment construction for the PVD improved subsoils. Similar studies are routinely accomplished for PVD improved soft soil by many researchers (Rezania et al. 2017, Tschuchnigg and Schuweiger 2018), to examine some relevant approaches or advanced constitutive models in alluvial or marine deposits (Kim and Do 2010, Da Silva et al. 2017). However, very few publications have discussed the geological mechanism such as the deposition environment (Mridakh et al 2019), the artesian pressure (Kim et al. 2018), shallow tectonic structures effects on soft soil behavior and its influence on predicting the subsoil deformation, precisely excessive settlement.

Therefore, this paper aims at presenting a comparison between field measured settlement and predicted settlements of a well monitored embankment sections constructed on PVD-improved Sebou soft alluvial deposits and Drader lacustrine deposits.

2 GEOLOGY AND SOIL PROPERTIES

2.1 *Geological setting*

Both study areas are located in the Rharb basin, where Drader study area is at the northern basin limit and Sebou area is at the southern one (Figure 1). Drader area can be described as a subsiding sedimentary min-basin, that is controlled by neo-tectonic activities (Combe 1963, Le Coz 1964, Cirac 1985). The mini-basin filling process is marked by marine and continental sedimentary units. Those units are mainly plio-quaternary deposits (Combe 1963), where its continental origins are the cumulated sediments carried out from rif domain and Middle Atlas domain (Benmouhammadi 2007). In the Centrale part of drader mini-basin, the plio-quaternary deposits are mainly soft soils that are formed in lacustrine environment, where artesian pressures are documented in deep captive aquifers (Combe 1963).

Sebou area is located in the southern limit of the Rharb basin (Figure 1). Its geological configuration is marked by Neotectonic activity mainly due to the south-rifaine front structure. However, the deposition mechanism is controlled by both onshore and offshore environments (Le Roy et al 2014). Sedimentary systems geometries are mainly formed by the Plio-quaternary filling, which are controlled by sea level fluctuation and tectonic activities (Le Coz 1964, Cirac 1985, Flinch 1993). The sedimentary regime is mostly dominated by clays and sand Brought by the Sebou and Loukous rivers (Jaaidi 1993).

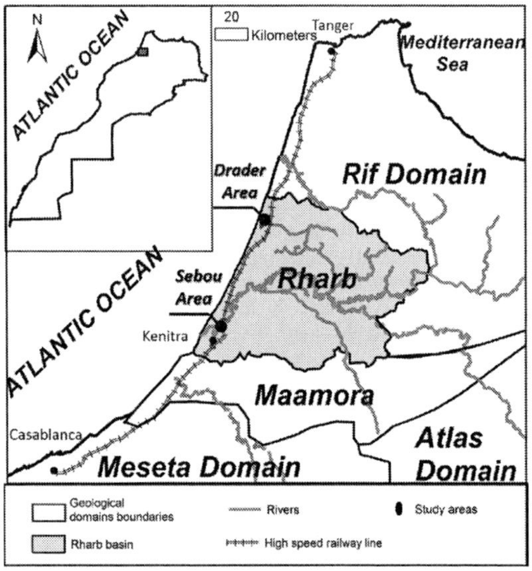

Figure 1. Localization of the study areas.

2.2 *Background information and data-base*

The selected high-speed railway line embankments were constructed both constructed in the right bank sides of Sebou and Drader rivers. Subsoils were characterized during the early geotechnical investigations that were followed by complementary fields tests (such as boreholes, cone penetration tests (CPTs), pressuremeters (PRs), vanes shear tests (VS)) and laboratory tests. Based on the analyzed database, stratigraphy models demonstrated rather undulated subsoil layers in the N-S orientation, and a rather uniform soil layers in the E-W orientation in Sebou area (Mridakh et al 2019). however, Drader area stratigraphy presented inclined layers in the W-E orientation following the deep syncline formation in Drader mini-basin (Combe 1975). The N-S orientation presents a depression shaped stratigraphy model that resulted from the flexure fault (Combe 1975).

The depositional history is described in case of Sebou compressible area as lightly over-consolidated for the shallower layers, and normally consolidated in deeper layers (Mridakh et al 2019). In Drader compressible deposits, subsoils are described as normally consolidated based on the interpreted odometers tests according to Casagrande method.

Ground water varies between –1 m during the humid period to -2m during the dry period in Sebou area. However, in Drader area ground water fluctuates between –0.5 m during the humid period and -1m during the dry period. In addition, Drader area is affected by a mean value artesian pressure of about 40 kPa that acts from a shallow aquifer underlying the compressible lacustrine soft soil deposits.

3 NUMERICAL MODELING

3.1 *Finite element models and approaches*

The numerical modeling is carried out based on the finite element (FE) software Plaxis 2D. For both cases, Class B prediction (predictions during the embankment construction) is applied at first by following the simplest modeling procedure possible. Then the modeling process develops by adding additional advanced soft soil behavior components during the back-analysis, by setting the priorities that should be added for each case study. This process is presented in a form of a chart as below:

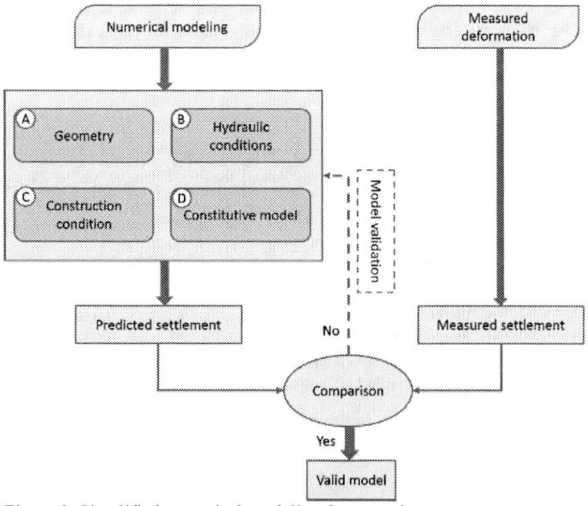

Figure 2. Simplified numerical modeling framework.

Only half of Sebou site embankment was modeled (20 m width) due to the subsoil layers symmetry, on 47 meters of different types of soft soils which are included in the FE model see Figure 3. Groundwater was assumed as hydrostatic at the ground water level (−1m). A 15-node triangular element were used following plane strain condition with very fine mesh option. The resulting elements number after meshing is 1682 and 13,681 nodes.

Mohr-Coulomb was used for both embankments (Sebou and Drader) granular and fill materials and the working platform layer.

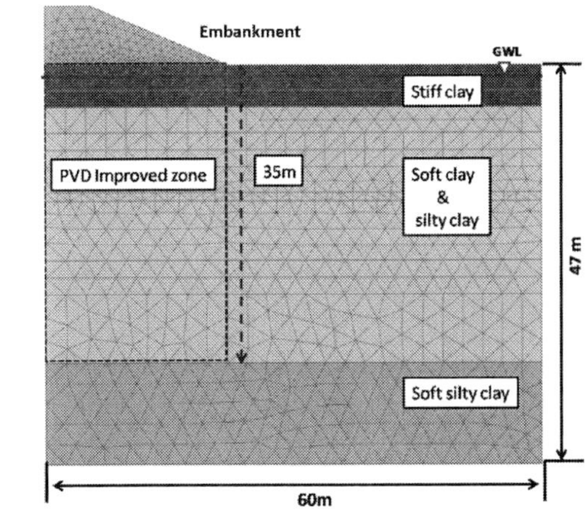

Figure 3. Sebou subsoil finite element mesh.

The full width of Drader site embankment was modeled (45 m width) due to the dipped subsoil layers. In this case 38 meters of multiple soft soil layers were modeled see Figure 4. Groundwater was assumed as hydrostatic at the ground water level (−0.5m). A 15-node triangular element were used following plane strain condition with very fine mesh option which resulted to 2319 elements and 18,807 nodes.

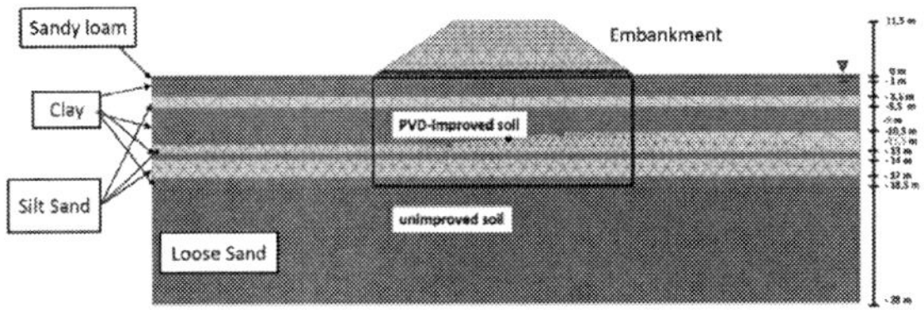

Figure 4. Drader subsoil finite element mesh.

3.2 *Matching method*

Including PVDs in the numerical modeling process is usually done as 3D problem under axisymmetric conditions. However, many researcher have resolve this problem to a 2D plain strain format to reduce time of computation (Hird et al. 1992, Kim and Lee 1997, Indraratna and Redana 2000, Chai and Miura 2001). In this study, Chai et al 2001 is used for both case studies. The later matching method proposed is based on calculating an equivalent vertical hydraulic conductivity k_{ve} as expressed in Eq. 1:

$$k_{ve} = \left(1 + \frac{2.5l^2 k_h}{\mu . D_e^2 . k_v}\right) . k_v \quad (1)$$

Where l= drainage of unit cell, D_e: diameter of unit cell. The value of μ can be expressed as:

$$\mu = \ln\left(\frac{n}{s}\right) + \frac{k_h}{k_s} ln(s) - \frac{3}{4} + \frac{2\pi . l^2 . k_h}{3 . q_w} \quad (2)$$

Where $n = D_e/d_w(d_w$ diameter of a PVD); $s = d_s/d_w(d_s$ is equivalent diameter of smear zone).

3.3 *Input parameters*

The models input parameters based on laboratory tests were directly used for the Class B analysis. However, a complementary study during the back-analysis is accomplished for the soft soil creep model parameters to match the field measured settlement for short- and long-term predictions.

For the drainage layer and the fill materials of the embankment the same parameters were fixed for both case studies as presented in Table 1.

Table 1. Embankment parameters values (For Sebou and Drader Embankments).

Material	$E'(MPa)$	v'	ϕ'	ψ'	$c'(kPa)$	$\gamma\ (kN/m^3)$
Fill	40	0,2	35	0	0.1	20
granular	100	0,2	40	0	0.1	20

Where E' is the young modulus, v' is the poison's ratio, ϕ' is the friction angle, ψ' is the dilatancy angle, c' is the cohesion and γ is the unit weight of the embankment materials.

The soil properties values for constitutive models (SSM and SSCM) used in Sebou and Drader soft subsoils are listed in Tables 2 and 3.

Table 2. Soil constant values for SSM and SSCM for Sebou subsoils.

Layer	Depth	K^*	λ^*	μ*	k_h	k_{ve}
	(m)				(m/day)	(m/day)
L1	0-3	0.02	0.05 (0.052)	4.0E-4 (5.0E-4)	9.0E-5	9.0E-5
L2	3-5	0.02	0.05	9.0E-4	3.4E-4	9.0E-5
L3	5-8	0.03	0.09	14E-4	4.0E-5	1.1E-5
L4	8-10	0.02	0.05	12E-4	5,0E-8	1.4E-8
L5	10-12	0.06	0.08	14E-4	1.4E-4	4.0E-5
L6	12-15	0.04	0.07	13E-4	1.6E-4	4.4E-5
L7	15-16	0.01	0.04	7.0E-4	2.4E-4	6.4E-5
L8	16-23	0.04	0.09	11E-4	1.3E-4	3.5E-5
L9	23-27	0.02	0.06	11E-4	4.0E-8	1.0E-8
L10	27-35	0.01	0.04	7.0E-4	2.4E-4	6.4E-5
L11	35-47	0.05	0.09	16E-4	1.0E-4	2.6E-5

(-) Back calculated parameters

Table 3. Soil constant values for SSM and SSCM for Drader subsoils.

Layer	Depth	K^*	λ^*	μ*	k_h	k_{ve}
	(m)				(m/day)	(m/day)
L1	0-1	-	-	-	2.5E-2	1.0E-1
L2	1-3.5	0.020	0.08	1.80E-3	2.6E-4	1.0E-2
L3	3.5-5.5	0.006	0.03	-	2.0E-3	6.4E-2
L4	5.5-10.5	0.016	0.07	1.9E-3	3,0E-4	3.5E-4
L5	10.5-13	0.04	0.02	-	5.5E-4	2.1E-2
L6	13-14	0.020	0.07	1.6E-3	1.2E-5	1.0E-5
L7	14-17	0.006	0.02	-	8.3E-3	1.5E-1
L8	17-18.5	0.04	0.09	1.2E-3	1.3E-4	1.1E-2
L9	18.5-38	-	-	-	2.5E-2	1.0E-1

Where λ^* is the modified compression index, K^* is the modified swelling index used by the soft soil model (SSM). The additional parameter used by the soft soil creep model μ* is the creep index, more details can be found in Neher et al (2001) describing the use of both constitutive models in soft soils.

4 RESULTS AND DISCUSSIONS

4.1 *Sebou site*

Class B prediction is based on using the soft soil model and the minoring data from the settlement cells for comparison. The SSM soil parameters are presented in Table 2. The settlement results included for the Sebou site used only half of the embankment, subsoil geometry as a result of the will stratified soil layers in the W-E direction.

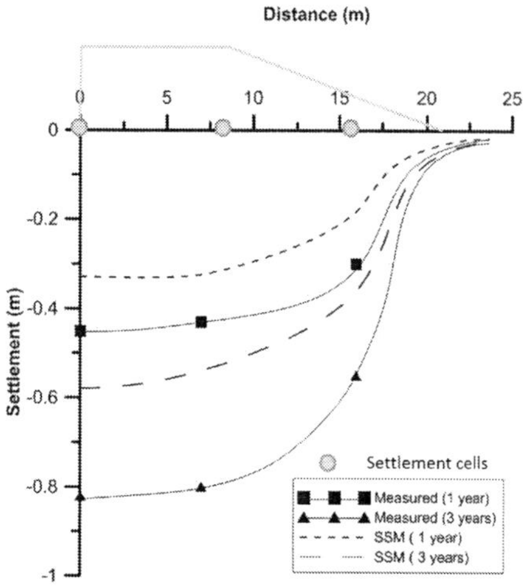

Figure 5. Predicted Settlements using the SSM compared with the measured settlements at the Sebou site.

The comparison between the predicted settlements and the monitoring data in Sebou site after 1 year and 3 years is presented in Figure 5. The measured maximum settlement at the center of the embankment is at 0.43 m after 1 year and 0.82 m after 3 years. The SS model predicted a maximum settlement of about 0.32 m after 1 year and 0.58 after 3 years. This represents an under estimation close to 25% at 1-year mark and 30 % at 3-year mark.

Therefore, the predictions based on soft soil model is getting less accurate with time, which indicate the necessity of including the creep effect into the analysis for Sebou soft soils.

When using the soft soil creep model (which include the creep effect based on the μ^* parameter), the maximum settlement at the center is about 0.36 m after 1 year and 0.76 after 3 years. Thus the underestimation settles at 17% after 1 year and 7% after 3 years. The results show that the prediction is at a good agreement especially in long term.

The remaining underestimation was analyzed in detail after a sensitivity analysis where it was found that layer L1 (see Table 2) parameters λ^* and μ^* were not following the general interval value variation of Sebou subsoils. In addition, the SSC model is not suitable for modeling the lightly over-consolidated state of L1 and L2 as discussed in (Mridakh et al. 2019). For that, we proceeded to increase the creep effect contribution in the discussed layers to obtain the near perfect results presented in Figure 6.

For that, prioritizing the use of constitutive models in Sebou area for prediction settlements is sufficient (Figure 2, D). more accuracy can be reach during the back-analysis (or the parameter sensitivity analysis) phase.

4.2 *Drader site*

Contrary to Sebou area, Drader site presented some high compressibility values for lacustrine Clay deposits and very low compressibility value for silty sand and sandy silt deposits. Thus, a different approach is fixed in this case by using both soft soil and soft soil creep models during the Class B prediction. The back-analysis phase is mainly reserved for the artesian pressure effect on settlement. In addition, the minoring data from the settlement cells is used for comparison.

The soil parameters for the SSM and SSCM are presented in Table 3. All values from the presented settlement cells measurement under the fall embankment is used in this case. This is due the complex subsoil geometry in the Drader subsoil which presented some dipped layers under the embankment in the W-E direction. In the analysis it was found that including the inclination in layer 4 is sufficient.

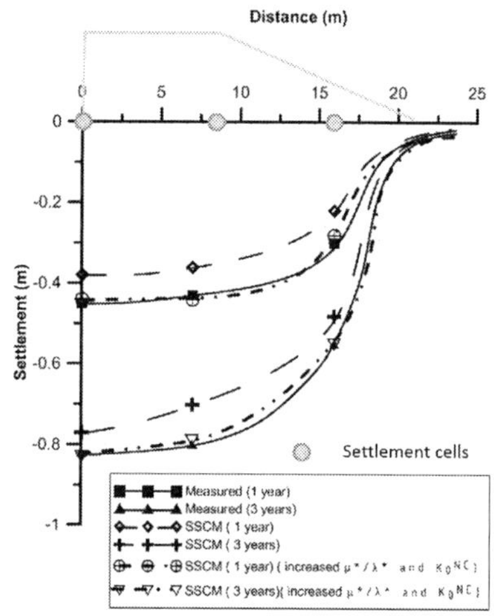

Figure 6. Predicted Settlements using the SSCM compared with the measured settlements at the Sebou site.

During Class B prediction (without including the artesian pressure (AP) effect), The maximum measured settlement at the center of the embankment is at 0.62 m after 1 year and 0.85 m after 2.5 years. The maximum predicted settlement without the AP at the center of the embankment at 0.51 after 1 year and 0.78 after 3 years. This represents an underestimation of the settlement of about 18% after 1 year and 8% after 2.5 years. Thus, is seems that the constitutive models used in combined manner produced acceptable predictions when compared to the measured settlement in short term, and in a good agreement in long term.

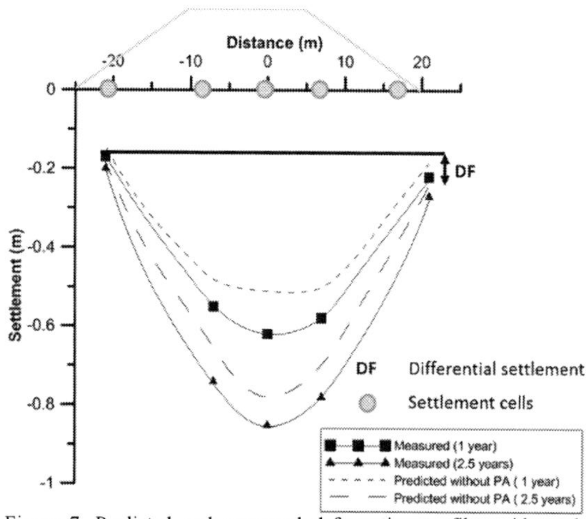

Figure 7. Predicted and measured deformation profiles without the artesian pressure effect at the Drader Site.

Based on some excess pore pressure measurement and field work, artesian pressure (AP) was estimated at a mean value of 40 kPa during the 2.5 years. Therefore, during the back analysis the authors included the AP effect in the numerical modeling to assess its effect on settlement. Figure 8 presents the results of adding the AP in the numerical modeling process, which results in a precise prediction in both short- and long-term settlements.

It can be added that the inclination used in layer 4 led to producing the exact differential settlement measured under the embankment. Therefore, multiple priorities are fixed to obtain such a good match between the predicted and the measured settlements. Those can be resumed as: having inclined layer or layers during the numerical model construction, including the

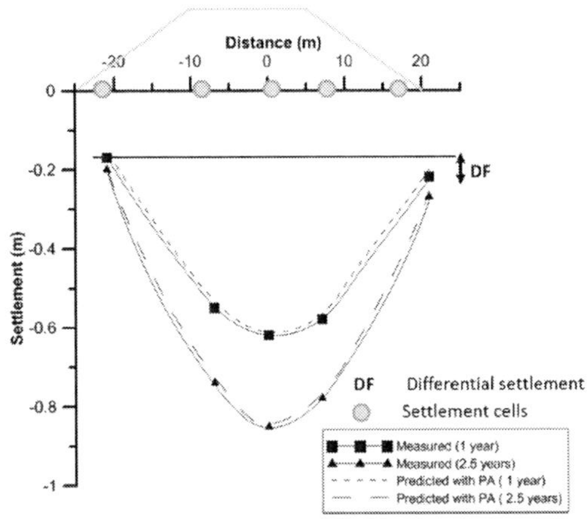

Figure 8. Predicted and measured deformation profiles with the artesian pressure effect at the Drader site.

artesian pressure effect and using two types of constitutive models is the approach that can lead to accurate predictions (Figure 2, A, B, D).

5 CONCLUSIONS

This study analyzed the issue of excessive settlement under two sections of the Moroccan high-speed railway embankment constructed on compressible subsoil improved with PVDs and preloading. Each compressible area (Sebou and Drader) had some specific priorities that should be fixed to accurately predict settlements for short and long term. For that, the following conclusions can be drawn:

- Each compressible area is unique by its geology, deposition environment and hydraulic conduction. Thus, each site should have its own numerical modeling approach for soil behavior prediction.
- Many aspects of soil behavior can affect the overall soft soil behavior in both sites. For that, priorities should be fixed for each site, to simplify the modeling process.
- The back-analysis showed that in Sebou compressible area, time-dependent behavior (i.e. creep) has an important effect on soil behavior prediction. Thus, choosing the constitutive model that includes creep effect such as soft soil creep is a priority in the Sebou compressible area.
- Drader site presented multiple priorities that should be fixed for accurate subsoil behavior prediction, due to its complex geology and hydraulic condition.
- Modeling the full embankment in the Drader area is a must, due to the dipped subsoil layers, however it was found that inclining the Clay layer between 5.5 m and 10.5 m depth, is sufficient to capture the differential settlement under the embankment in short and long term.
- Combining between the soft soil creep model for soft clays and soft soil model for low compressibility silty sands and sandy silts, can produce some good agreement with the measured settlements at the Drader site.
- During the back-analysis it was found that including an artesian pressure of about 40 kPa during the numerical modeling at the Drader site, can produce some near perfect results for settlement prediction in short and long term.

6 ACKNOWLEDGEMENTS

The work presented was supported by the LPEE (Laboratoire Public d'Essais et d'Etudes). The author would like to thank all personnel in the LPEE for providing all data necessary for the completion of this work.

7 REFERENCES

Barron. R.A. 1948. Consolidation of fine-grained soils by drain wells, *Am. Soc. Civil Eng.*, 113(1), 718-742.

Benmohammadi. A. Griboulard, R.,Zourarah, B. Carruesco. C. Mehdi. K. Mridekh. A. El Moussaoui. A. Mhamdi Alaoui, A. Carbonel. P. Londeix. L. 2007. Hyperactive neotectonic near the South Rifian front Lifted Late Quaternary lagunal deposits (Atlantic Morocco). *C. R. Geoscience*, Volume 339, pp. 831-839.

Combe. M. 1963. Etude hydrogéologique de la région de Lalla Mimouna, Kénitra: Office National des Irrigation.

Combe. M. 1975. Bassin du Gharb–Maâmora. In ressources en eau du Maroc Tome 2. Notes et 501 mémoire du Service Géologique (Maroc) 231: 93-128.

Chai, J.C. and Miura, N. 2001. Investigation of factors affecting vertical drain behaviour, *J. Geotech. Geoenviron. Eng.*, 125(3), 216-226.

Cirac. P. 1985. Le bassin sud-rifain occidental au Néogène supérieur. Evolution de ladynamique sédimentaire et de la paléogéographie au cours d'une phase decomblement. Thèse de doctorat d'état. Université Bordeaux I.

Da Silva, E.M., Juste J.L., Durant, P., Justo, E. and Vazquez Boza, M. 2017. The effect of geotextile reinforcement and prefabricated vertical drains on the stability and settlement of embankment. *Geotext. and Geomemb.* 45:447-441.

Flinch, J.F. 1993. Tectonic evolution of the Gibraltar arc. PhD Thesis Rice University. Houston Texas USA.

Hird, C.C., Pyrah, I.C. and Russell, D. 1992. Finite element modelling of vertical drains beneath embankments on soft ground, *Géotechnique* 42(3), 499-511.

Indraratna, B. and Redana, I.W. 1997. Plane strain modeling of smear effects associated with vertical drains, *J. Geotech. Eng.*, 123(5), 474-478.

Indraratna, B., Rujikiatkamjorn, C., Balasubramaniam. A.S. and McIntosh, G. 2012. Soft ground improvement via vertical drains and vacuum assisted preloading, *Geotext. and Geomemb.* 30, 16-23.

Jaaidi. E.B. 1993. La couverture sédimentaire post-glacial de la plateforme continentale atlantique ouest-rifaine (Maroc Nord-Occidental). Exemple d'une séquence transgressive (DSc thesis). Univ Mohammed V Faculté des sciences de Rabat.

Kim, Y.T. and Lee, S.R. 1997. An equivalent model and back-analysis technique for modelling in situ consolidation behavior of drainage-installed soft deposits", *Compos. Geotech.*, 20(2), 125-142.

Kim, Y. T. and T. H. Do. 2010. Effect of leaching on the compressibility of Busan clay. *Geotech. Eng. KSCE J. Civ. Eng.* 14 (3): 291–297.

Kim, Y.T., Nguyen, B.P. and Yun, D.H. 2018. Effect of artesian pressure on consolidation behavior of drainage-installed marine clay deposit. *ASCE J. Mater. Civ. Eng.* 30(8):04018156. 1-13.

Le Coz, J. 1964. Le Rharb. Fellahs et colons. Etude de géographie régionale. Inframar 1. Rabat 481pp.

Le Roy, P., Sahabi, M., Maad, N., Rabineau, M., Gutsher, M.A., Babonneau, N., VanVlietLanoe, B., AitBrahim, L., M'hammdi, N., Trentesaux, A., Dakki, M. and Hssain, M. 2014. 3D architecture of Quaternary sediment along the NW Atlantic Moroccan Rharb continental shelf: a stratal pattern under the dual control of tectonics and climatic variations. *Mar. Petroleum Geol.* 49:129-142

Mridakh, A.H., Lahlou, F., Ejjaaouani, H., Mridekh, A. and Labied, H. 2019. Soil improvement of alluvial deposits under high speed railway embankment: field case study, *Geotech. and Geol. Engrg.* 37(5), 3589-3603.

Neher, H.P., Wehnert, M. and Bonnier, P.G. 2001. An evaluation of soft soil models based on trial embankment. *Proc 10th Int. Conf. on Computer Methods and Advances in Geomech.*, Tucson 373-378.

Rezania, M., Bagheri, M., Nezhad, M.M. and Sivasithamparam, N. 2017. Creep analysis of an earth embankment on soft soil deposit with and without PVD improvement. *Geotext. and Geomemb.* 45:537-547.

Tschuchnigg, F. and Schweiger, H.F. 2018. Embankment prediction and back analysis by means of 2D and 3D finite element analyses. *Comput. Geotech.* 93:104-14

Walker, R., Indraratna, B. and Rujikiatkamjorn, C. 2012. Vertical drain consolidation with non-Darcian flow and void-ratio-dependent compressibility and permeability, *Géotechnique* 62(11), 985-997.